FLORA ZAMBESIACA

Flora terrarum Zambesii aquis conjunctarum

VOLUME TWO

FLORA ZAMBESIACA

MOZAMBIQUE

MALAWI, ZAMBIA, RHODESIA

BECHUANALAND PROTECTORATE

VOLUME TWO

Edited by

A. W. EXELL, A. FERNANDES and H. WILD

on behalf of the Editorial Board

Published on behalf of the Governments of

Portugal

Malawi, Zambia, Rhodesia

and the United Kingdom by the

Crown Agents for Oversea Governments and Administrations,

4, Millbank, London, S.W.1

1963–66

Volume Two: Part One: May 23, 1963
Volume Two: Part Two: March 18, 1966

Printed at the University Press Glasgow
by Robert MacLehose & Company Limited

FLORA ZAMBESIACA

Flora terrarum Zambesii aquis conjunctarum

VOLUME TWO: PART TWO

BERSAMA SWYNNERTONII

FLORA ZAMBESIACA

MOZAMBIQUE
MALAWI, ZAMBIA, RHODESIA
BECHUANALAND PROTECTORATE

VOLUME TWO: PART TWO

Edited by

A. W. EXELL, A. FERNANDES and H. WILD

on behalf of the Editorial Board:

Published on behalf of the Governments of
Portugal
Malawi, Zambia, Rhodesia
and the United Kingdom by the
Crown Agents for Oversea Governments and Administrations,
4, Millbank, London, S.W.1
March 18, 1966

Printed at the University Press Glasgow
by Robert MacLehose & Company Limited

CONTENTS

LIST OF FAMILIES INCLUDED IN VOL. II, PART 2

ANGIOSPERMAE

LIST OF NEW NAMES PUBLISHED IN THIS WORK

MALAWI, ZAMBIA AND RHODESIA

It was not possible for technical reasons to give the new names for these countries in the text of Volume 2, part 2.

51. AQUIFOLIACEAE

By E. J. Mendes

Trees or shrubs, mostly evergreen. Leaves alternate, simple; stipules absent or very small. Flowers frequently dioecious, actinomorphic, cymose, fasciculate, subumbellate or rarely solitary. Calyx-lobes imbricate. Petals free or connate at the base, imbricate or valvate. Stamens free, alternate with the corolla-lobes; anthers 2-thecous, opening lengthwise. Disk absent. Ovary superior, 3- or more-locular; style terminal or stigma sessile; ovules 1–2 in each loculus, pendulous from the apex. Fruit drupaceous. Seeds with copious fleshy endosperm and small straight embryo.

ILEX L.

Ilex L., Sp. Pl. **1**: 125 (1753); Gen. Pl. ed. 5: 60 (1754).

Evergreen trees or shrubs. Stipules very small. Inflorescences axillary. Petals shortly connate at the base, imbricate. Stamens as many as the petals. Style absent. Ovules 1 in each loculus. Fruit of 4–5(9) single-seeded pyrenes.

A genus of c. 400 species occurring mainly in tropical and subtropical regions of both hemispheres.

Ilex mitis (L.) Radlk. in Rep. Brit. Ass. **1885**: 1081 (1886).—Loesener in Engl., Pflanzenw. Ost-Afr. **C**: 246 (1895); in Engl. & Prantl, Nat. Pflanzenfam. ed. 2, **20b**: 65 (1942).—Bak. f. in Journ. Linn. Soc., Bot. **40**: 43 (1911).—Engl., Pflanzenw. Afr. **3,** 2: 218 (1921).—Burtt Davy, F.P.F.T. **2**: 445, fig. 69 (1932).—Burtt Davy & Hoyle, N.C.L.: 32 (1936).—Hutch., Botanist in S. Afr.: figs. pag. 232 and 233 (1946).—Brenan, T.T.C.L.: 58 (1949); in Mem. N.Y. Bot. Gard. **8,** 3: 235 (1953).—Exell & Mendonça, C.F.A. **1,** 2: 348 (1951).—Wild in Proc. & Trans. Rhod. Sci. Ass. **43**: 56 (1951); S. Rhod. Bot. Dict.: 95 (1953).—Pardy in Rhod. Agric. Journ. **53**: 960 cum tt. (1956).—Keay, F.W.T.A. ed. 2, **1,** 2: 623 (1958).—Robyns f., F.C.B. **9**: 110, t. 11 (1960).—Boughey in Proc. & Trans. Rhod. Sci. Ass. **49**: 61, 70 and 78 (1961).—Goodier & Phipps in Kirkia, **1**: 58 (1961).—White, F.F.N.R.: 215 (1962). TAB. **75.** Type probably from S. Africa.

Sideroxylon mite L., Syst. Nat. ed. 12, **2**: 178 (1767). Type as above.

Ilex capensis Sond. in Harv. & Sond., F.C., **1**: 473 (1860).—Burkill, List Pl. Brit. Central Afr.: 239 (1897).—Monro in Proc. & Trans. Rhod. Sci. Ass. **8**: 147 (1908).—Sim, For. Fl. Port. E. Afr.: 22 (1909).—Eyles in Trans. Roy. Soc. S. Afr. **5**: 403 (1916). Syntypes from S. Africa.

Tree up to 20–30 m. high or shrub, sometimes straggling, with dark shining green foliage; bark light grey-brown with prominent whitish lenticels; dioecious. Leaf-lamina up to 13 × 4·5 cm., narrowly elliptic to lanceolate, apex acute, subrounded or emarginate and apiculate, obtuse to subacute at the base, entire or remotely denticulate (particularly towards the apex), glabrous, with numerous spreading looped lateral nerves; petiole c. 1 cm. long; stipules narrowly triangular, deciduous. Inflorescences axillary, solitary or congested. Flowers pedicellate, unisexual by abortion; male flowers (1)3–5(7), subumbellate with peduncles up to 1 cm. long; female flowers 1–3(5), subumbellate on shorter peduncles or subfasciculate. Sepals minutely ciliolate. Petals whitish. Stigma thick and sessile. Ovary (4)5–6(7)-locular. Fruit of (4)5–6(7) pyrenes, c. 7 mm. in diam., subglobose, red, partially enclosed at the base by the persistent sepals.

N. Rhodesia. N: Shiwa Ngandu, fr. 5.ii.1955, *Fanshawe* 1987 (K); W: Mwinilunga Distr., Kalene Hill Mission, Zambezi R., fl. 20.ix.1952, *White* 3303 (BM; BR; COI; EA; K). C: Mkushi, fr. 23.i.1955, *Fanshawe* 1835 (BR; K; SRGH). **S. Rhodesia.** N: Mazoe Distr., Tatagura R., fl. ix.1906, *Eyles* 436 (BM; BOL; SRGH). C: Salisbury, fl. 25.iv.1836, *Eyles* 8814 (BR; K; SRGH). E: Inyanga, Pungwe Falls, 2075 m., fl. 23.x.1955, *Chase* 5840 (BM; COI; K; LISC; SRGH). S: Zimbabwe, fr. 22.v.1951, *Mullin* 84/51 (SRGH). **Nyasaland.** N: Nyika Plateau, Lake Kaulime, 2200 m., fl. 23.x.1958, *Robson* 277 (BM; K; LISC; PRE; SRGH). C: Dedza, fl. 20.x.1956, *Jackson* 2075 (BM; BR; K; SRGH). S: Mt. Mlanje, fl. 11.x.1957, *Chapman*

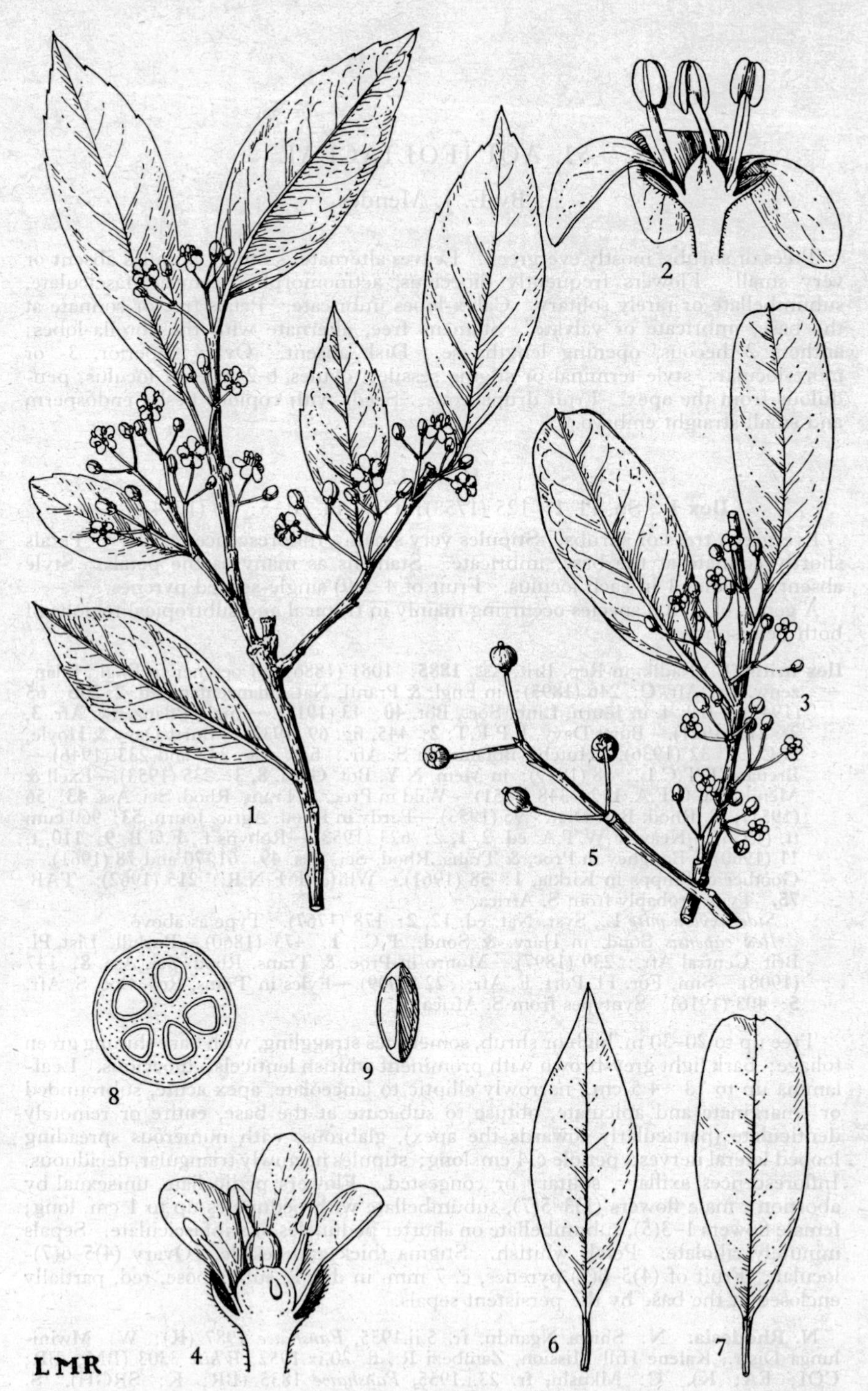

Tab. 75. ILEX MITIS. 1, branch with male flowers (×⅔) *Chase* 5840; 2, vertical section of male flower (×6) *Chapman* 464; 3, branch with female flowers (×⅔) *Eyles* 8814; 4, vertical section of female flower (×6) *Robinson* 4016; 5, branchlet with fruits (×⅔) *Fanshawe* 1835; 6, leaf (×⅔) *Chapman* 464; 7, leaf (×⅔) *Eyles* 8814; 8, transverse section of fruit (×4) *Fanshawe* 1835; 9, pyrene (×5) *Fanshawe* 1835.

464 (BM; BR; K). **Mozambique.** Z: Gúruè, Pico Namuli, fr. 9.iv.1943, *Torre* 5166 (BR; K; LISC; LM; SRGH). MS: Manica, Tsetsera, fl. 6.xi.1946, *Simão* 1153 (EA; LMJ).

From the mountains of Ethiopia to the Cape and from Guinée to the Congo and Angola; also in Madagascar. Evergreen mountain forest and in moist dense vegetation and deep gully and riverine formations at lower altitudes.

The female flowers sometimes have quite large anthers, but they are always barren as far as I have seen.

52. CELASTRACEAE

(incl. *HIPPOCRATEACEAE*)

By N. K. B. Robson

Trees, shrubs, shrublets or woody climbers, without tendrils, glabrous or with simple hairs, unarmed or with axillary shoots terminating in a spine, sometimes with rubber-like latex (gutta) in various parts appearing as elastic threads when a leaf is broken. Leaves simple, alternate or spiral to subopposite or opposite, sometimes fasciculate on short shoots, entire or with crenate or denticulate to spinose margins, penninerved; stipules small, simple or laciniate, sometimes united by a transverse ridge, usually deciduous, or absent. Flowers bisexual or polygamous or unisexual, monoecious or dioecious, actinomorphic, often fragrant, in axillary and/or terminal dichasial or monochasial cymes or panicles or thyrses, sometimes with accessory* branches, or fasciculate or solitary, usually bracteate; pedicels often articulated. Sepals (3)4–5(6), imbricate or rarely valvate in bud, free or united at the base, persistent. Petals (3)4–5(6), free or rarely united at the base, imbricate or rarely valvate in bud, usually persistent, sometimes with ventral grooves or hollows or appendages. Stamens (2)3–5(6–10), antisepalous, free or more rarely with filaments partly united to form a tube, inserted outside or on or inside the disk; anthers usually short, (1)2-thecous, extrorse or introrse, basifixed or dorsifixed or versatile, sometimes deciduous, dehiscing by longitudinal or oblique or horizontal slits; pollen simple or more rarely in tetrads or polyads. Disk nectariferous, annular, entire or angular or crenulate or lobed or covered with fleshy processes, concave to convex, rarely wholly or partly forming an androgynophore or discontinuous pockets, fleshy or membranous, very rarely absent. Ovary free or partly or wholly immersed in the disk, sessile or on a short androgynophore, syncarpous, completely or very rarely incompletely 2–5-locular, or rarely 1-locular by abortion, with 1–∞ erect or rarely pendulous ovules in 2 rows or rarely superimposed in each loculus; styles as many as the loculi, free or $\pm$ united, or absent; stigmas various, free or $\pm$ united. Fruit capsular, loculicidal, or of divergent $\pm$ dorsiventrally flattened dehiscent mericarps or baccate or drupaceous or dry, indehiscent and sometimes winged. Seeds with a fleshy or submembranous brightly coloured aril, or winged with the funicle free from the wing (Tab. 80 figs 6–7) or united to its base (Tab. 85 fig. A8) or neither arillate nor winged (usually in indehiscent fruits), with or without endosperm; embryo erect, with cotyledons flat or fleshy, rarely united.

A family of about 60–70 genera occurring in all tropical and warmer temperate regions. *Hippocratea*, *Salacia* and some related genera have frequently been placed in a separate family, *Hippocrateaceae*; but none of the alleged diagnostic characters of this family (e.g. 3 stamens inserted inside the disk) has proved to be constant. Indeed it seems possible that the *Hippocratea* group (with dehiscent mericarps and winged seeds) and the *Salacia* group (with indehiscent drupaceous fruits) may have been derived from different parts of the *Celastraceae*.

Leaves all alternate or spiral; stamens outside the disk, isomerous; seeds of dehiscent fruits arillate; spines and fasciculate shoots sometimes present:
  Fruit dehiscent, capsular; anthers persistent or, if deciduous, then leaves usually acuminate and showing latex threads when broken:
    Capsules without emergences, usually dry; leaves rarely entire; spines and fasciculate shoots often present:

* Branches additional to the normal one in the bract axil.

Ovules 6 per loculus (in our species) - - - - **1. Putterlickia**
Ovules 2 per loculus - - - - - - - **2. Maytenus**
Capsules with horn- or wing-like emergences, sometimes somewhat succulent; leaves entire; spines and fasciculate shoots absent - **3. Pterocelastrus**
Fruit indehiscent, drupaceous; anthers eventually deciduous; leaves obtuse to rounded, without latex - - - - - - - - **4. Mystroxylon**
Leaves opposite, at least on flowering shoots, or very rarely all alternate but then stamens inside the disk and fewer than the petals; seeds of dehiscent fruits winged; spines and (usually) fasciculate shoots absent:
Fruit dehiscent, capsular; inflorescence pedunculate, dichasial, sometimes with accessory branches:
Capsule elongated, narrow; funicle not attached to the seed wing; stamens 5, outside the disk - - - - - - - - - - **6. Catha**
Capsule expanded laterally, forming 3 dehiscent mericarps; funicle at base of the seed wing; stamens (2)3–5, inside the disk, or disk absent:
Stamens (2)3(4), inside the disk; latex absent (in our species) **11. Hippocratea**
Stamens 5; disk absent; latex threads present in broken leaf **12. Campylostemon**
Fruit indehiscent, various; inflorescence pedunculate or sessile, variously branched, without accessory branches:
Fruit a small nut, with a persistent lateral style; stigma peltate; ovary with 1 fertile loculus - - - - - - - - **7. Pleurostylia**
Fruit drupaceous or baccate, with a terminal stylar scar; stigma not peltate; ovary with 2–4 fertile loculi:
Fruit baccate; plant (when mature) with small alternate leaves on climbing branches and larger opposite leaves on deflexed flowering branches; flowers pseudotubular - - - - - - - **5. Allocassine**
Fruit drupaceous; plant not climbing or, if so, then leaves not heteromorphic; petals ± spreading:
Stamens 4–5, outside the disk; leaves homomorphic (in our species); sepals and petals 4–5 - - - - - - - **8. Elaeodendron**
Stamens (2)3–4, inside the disk or almost so; leaf-form various; sepals and petals 3–6:
Sepals and petals 3–4; stamens isomerous; seeds with endosperm; trees or shrubs - - - - - - - - **9. Crocoxylon**
Sepals and petals (4)5(6); stamens meiomerous; seeds without endosperm; shrubs, climbers or shrublets - - - - - **10. Salacia**

1. PUTTERLICKIA Endl.

Putterlickia Endl., Gen.: 1086 (1840).

Shrubs, spreading or straggling, without latex, glabrous; branches lined or angular, sometimes ending in a spine, frequently condensed to form short shoots. Leaves spiral or fasciculate (on short shoots), entire or with margin denticulate, petiolate; stipules free, small, caducous. Inflorescence pedunculate, dichasial or sometimes becoming monochasial, many- to few-flowered, simple, in axils of foliage leaves or in clusters on short shoots; bracts persistent. Flowers bisexual, pedicellate, with pedicels articulated. Sepals (4)5, subequal, imbricate, free or united at the base, with margin finely ciliolate. Petals (4)5, white, imbricate in bud, spreading or suberect. Disk intrastaminal, single, convex, shallowly (4)5–10-lobed. Stamens (4)5, with filaments thin, free, united with base of disk; anthers versatile, introrse, with separate thecae dehiscing longitudinally. Ovary superior, sessile, $\frac{1}{4}$–$\frac{1}{2}$-immersed in the disk, 3(4–5?)-locular, with (2?)3–6 erect ovules in 2 rows in each loculus; style ± elongated, cylindric, simple; stigma 3(4–5?)-branched. Fruit capsular, coriaceous to woody, without emergences, dehiscing loculicidally to the base. Seeds 6–18, glossy, red-brown, with a laciniate fleshy aril forming a complete covering, with fleshy endosperm.

A genus of 2 species in southern Africa. In one species (*P. verrucosa* (E. Mey. ex Sond.) Szyszyl.) the number of ovules in each loculus appears to be constantly 6, but in the other (*P. pyracantha* (L.) Endl.) it varies from 3 to 6 and may sometimes be reduced to 2 (cf. Davison's descriptions of *Gymnosporia saxatilis* Davison and *G. integrifolia* (L.f.) Glover). Apart from the number of ovules *Putterlickia* cannot be distinguished from *Maytenus*.

Putterlickia verrucosa (E. Mey ex Sond.) Szyszyl., Polypet. Disc. Rehm. **2**: 34 (1888).— Davison in Bothalia, **2**: 338, t. 13 fig. 1 (1927). TAB **76** fig. C. Syntypes from Natal and Cape Prov.

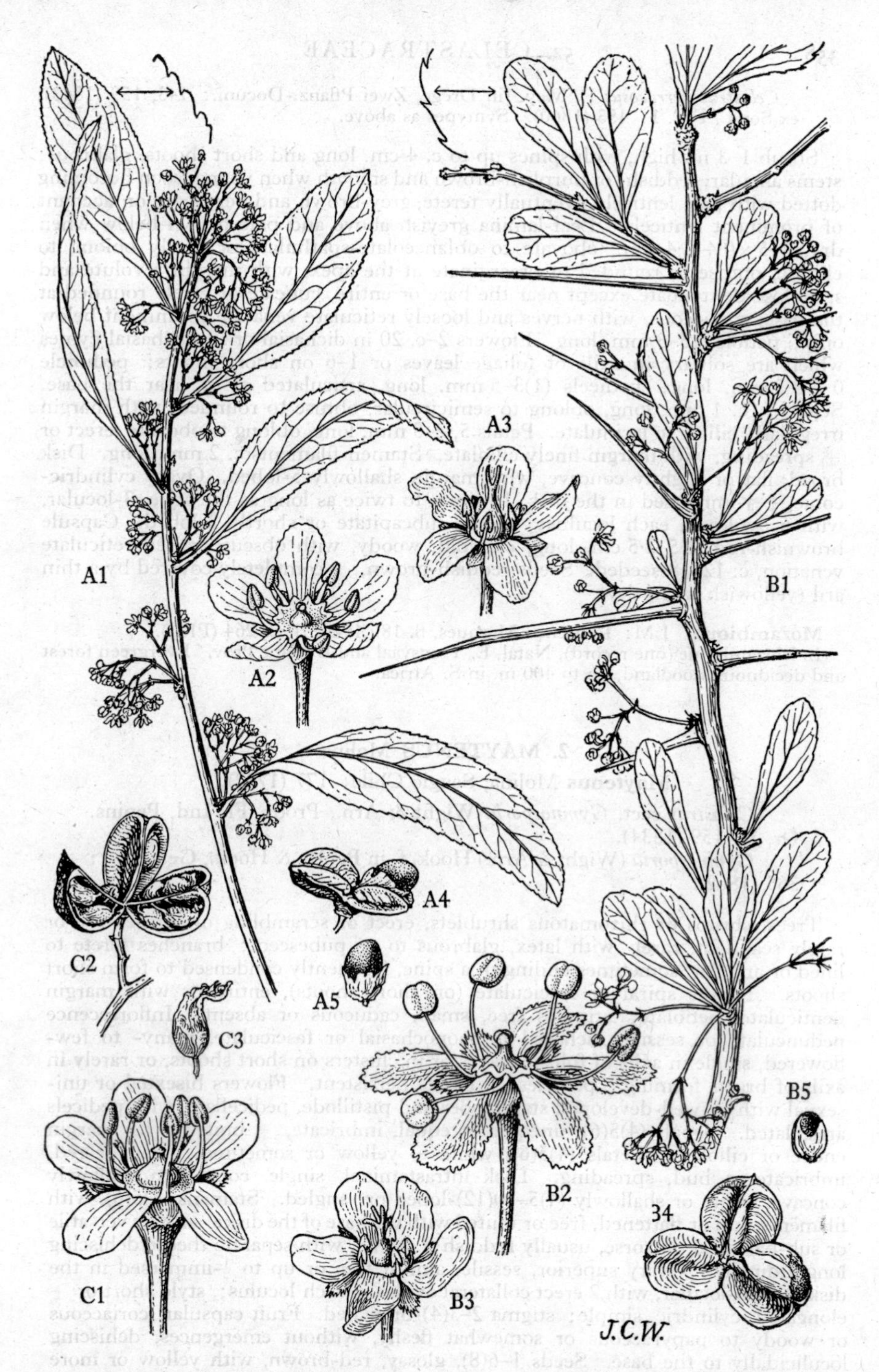

Tab. 76. A.—MAYTENUS SENEGALENSIS. A1, shoot with ♂ flowers ($\times\frac{2}{3}$); A2, ♂ flower ($\times 6$), both from *Faulkner* 200; A3, ♀ flower ($\times 6$), *Rand* 1413; A4, fruit ($\times 2$); A5, seed ($\times 2$), both from *Kafuli* 36. B.—MAYTENUS HETEROPHYLLA. B1, shoot with ♂ flowers ($\times\frac{2}{3}$); B2, ♂ flower ($\times 6$), both from *Rand* 138; B3, ♀ flower ($\times 6$), *E.M. & W.* 484; B4, fruit ($\times 2$); B5, seed ($\times 1$), both from *Hutchinson & Gillett* 3490. C.—PUTTERLICKIA VERRUCOSA. C1, flower ($\times 4$), *Rudatis* 1167; C2, fruit ($\times\frac{2}{3}$); C3, seed ($\times 1$), both from *Pegler* 602.

Celastrus verrucosus E. Mey. [in Drège, Zwei Pflanz.-Docum.: 143, 159 (1843)] ex Sond., F.C. **1**: 453 (1860). Syntypes as above.

Shrub 1–3 m. high, with spines up to c. 4 cm. long and short shoots, glabrous; stems angular, reddish- or purplish-brown and smooth when young, soon becoming dotted with pale lenticels, eventually terete, grey-brown and verrucose on account of prominent lenticels. Leaf-lamina greyish above and pale brown below when dry, 1–5 × 0·4–2·4 cm., obovate to oblanceolate-spathulate or rarely oblong to elliptic, obtuse to rounded or emarginate at the apex, with margin revolute and spinulose-denticulate except near the base or entire, cuneate or rarely rounded at the base, coriaceous, with nerves and loosely reticulate venation prominent below only; petiole 0·5–3 mm. long. Flowers 2–c. 20 in dichasial or subdichasial cymes which are solitary in axils of foliage leaves or 1–6 on short shoots; peduncle 0·7–3·5 mm. long; pedicels (1)3–5 mm. long, articulated at or near the base. Sepals 5, c. 1 mm. long, oblong to semicircular, obtuse to rounded, with margin irregularly ciliate to lacinulate. Petals 5, 2–3 mm. long, oblong to obovate, erect or ± spreading, with margin finely ciliolate. Stamen-filaments c. 2 mm. long. Disk broad, flat or slightly concave, with margin shallowly 5-lobed. Ovary cylindric-conic, $\frac{1}{4}$–$\frac{1}{3}$-immersed in the disk, 1$\frac{1}{2}$ times to twice as long as the style, 3-locular, with 6 ovules in each loculus; stigma subcapitate or shortly 3-lobed. Capsule brownish-red, 1·5–2·5 cm. long, obovoid, woody, with obscure raised reticulate venation, c. 12–18-seeded. Seeds reddish-brown, ± completely covered by a thin aril (yellowish when dry).

Mozambique. LM: Lourenço Marques, fl. 18.i.1920, *Borle* 264 (PRE).

In Mozambique (one record), Natal, E. Transvaal and E. Cape Prov. Evergreen forest and deciduous woodland, up to 400 m. in S. Africa.

2. MAYTENUS Molina

Maytenus Molina, Saggio Chile: 177 (1782).

Celastrus sect. *Gymnosporia* Wight & Arn., Prodr. Fl. Ind. Penins. Or. **1**: 159 (1834).

Gymnosporia (Wight & Arn.) Hook. f. in Benth. & Hook., Gen. Pl. **1**: 365 (1862).

Trees, shrubs or rhizomatous shrublets, erect or scrambling or sarmentose or rarely scandent, rarely with latex, glabrous to ± pubescent; branches terete to lined or angular, sometimes ending in a spine, frequently condensed to form short shoots. Leaves spiral or fasciculate (on short shoots), entire or with margin denticulate, petiolate; stipules free, small, caducous or absent. Inflorescence pedunculate or sessile, dichasial to monochasial or fasciculate, many- to few-flowered, single in axils of foliage leaves or in clusters on short shoots, or rarely in axils of bracts forming " panicles "; bracts persistent. Flowers bisexual or unisexual with ± well-developed staminodes and pistillode, pedicellate, with pedicels articulated. Sepals (4)5(6), unequal or equal, imbricate, ± united, with margin entire or ciliolate. Petals (4)5(6), white to yellow or sometimes green or red, imbricate in bud, spreading. Disk intrastaminal, single, convex to shallowly concave, entire or shallowly (4)5–10(12)-lobed or -angled. Stamens (4)5(6), with filaments thin or flattened, free or united with the base of the disk; anthers versatile or sub-basifixed, introrse, usually reddish if sterile, with separate thecae dehiscing longitudinally. Ovary superior, sessile, almost free or up to $\frac{1}{2}$-immersed in the disk, 2–3(4)-locular, with 2 erect collateral ovules in each loculus; style short or ± elongated, cylindric, simple; stigma 2–3(4)-branched. Fruit capsular, coriaceous or woody to papyraceous or somewhat fleshy, without emergences, dehiscing loculicidally to the base. Seeds 1–6(8), glossy, red-brown, with yellow or more rarely white or purple aril forming a fleshy basal cupule or a partial or complete thin covering, with or without fleshy endosperm.

A genus of c. 200 species in the tropics and subtropics of both hemispheres. It is very closely allied to *Celastrus* L., in which the ovary is always free from the disk, the central axis of the capsule is persistent, the habit always scandent and the inflorescence frequently racemose or paniculate. Indeed *Celastrus* subgen.

2-locular), truncate, chartaceous, smooth. Seeds 2–3, purplish- or reddish-brown, rugose, with a yellow aril at the base.

Bechuanaland Prot. SE: Kgatla Distr., 19 km. N. of Sikwane, 1050 m., fl. & fr. 28.xi.1955, *Reyneke* 449 (BM; K; PRE).

Also in the Transvaal. Woodland, thickets and scrub on sandy or stony soil, 900–1260 m. (to 1500 m. in the Transvaal).

M. tenuispina appears to be a derivative of *M. oxycarpa*, differing from it in size of leaves, flowers, etc., in the fruit, in having dioecious flowers, and sometimes in the presence of an indumentum.

3. **Maytenus putterlickioides** (Loes.) Exell & Mendonça in Mem. N.Y. Bot. Gard. **8,** 3: 238 (1953); C.F.A. **2,** 1: 4, t. 28 (1956).—Wilczek, F.C.B. **9**: 120 (1960). Type from Tanganyika (Kiwanda).

Gymnosporia putterlickioides Loes. in Engl., Bot. Jahrb. **17**: 544 (1893).—Engl. & Loes. in Engl., Pflanzenw. Afr. **3,** 2: 228 (1921).—Burtt Davy & Hoyle, N.C.L.: 38 (1936).—Brenan, T.T.C.L.: 128 (1949). Type as above.

Gymnosporia fischeri var. *magniflora* Loes., tom. cit.: 543 (1893).—Brenan, T.T.C.L.: 127 (1949). Type from Tanganyika (Irangi).

Gymnosporia fischeri var. *parviflora* Loes., loc. cit.—Brenan, T.T.C.L.: 128 (1949). Type from East Africa.

Gymnosporia borumensis Loes. in Bull. Herb. Boiss., **4**: 430 (1896).—Schinz, Pl. Menyharth.: 60 (1905).—Gomes e Sousa, Pl. Menyhart.: 78 (1936). Type: Mozambique, Boroma, *Menyhart* 1156 (Z).

Maytenus welwitschiana sensu Brenan in Mem. N.Y. Bot. Gard. **8,** 3: 237 (1953).

Shrub or small tree 1–3(6) m. high, often bushy or straggling, with spines up to 1·8 cm. long, terminal or axillary on short branches, without latex; branches flattened, angular, reddish-brown, reddish-brown- or fawn-pubescent or -puberulous when young, becoming terete, often longitudinally striate, with numerous pale lenticels and eventually greyish. Leaves frequently fasciculate, petiolate; lamina 1·8–7·5(9) × 1–4 cm., elliptic or oblong-elliptic to obovate or oblanceolate, obtuse to rounded or emarginate at the apex, with margin subentire to shallowly crenulate or remotely glandular-denticulate, cuneate to angustate at the base, membranous to chartaceous, fawn-pubescent or -puberulous on both sides, with lateral veins and dense reticulate venation slightly prominent on both sides; petiole 1–5 mm. long, pubescent or puberulous. Cymes regularly dichasial or submonochasial, solitary and axillary or 2–5 in fascicles on short shoots, with peduncle (3)5–18(36) mm. long, pubescent; pedicels c. 4–10 mm. long, articulated near the base, pubescent; flowers c. 7–20(40) in each cyme, (6)8–10 mm. in diam., scented. Sepals 5, subequal, c. 1 mm. long, broadly ovate to subcircular, rounded, pubescent outside. Petals 5, white or cream, 4–5(6) mm. long, obovate to oblong, sessile, with margin irregularly ciliolate-fimbriate. Stamens 5, with filaments 1·5–2 mm. long, slender, arising below the disk. Disk red, broad, convex, ± shallowly 5-lobed. Ovary 3-locular, subglobose, puberulous, c. ½-immersed in disk; style c. twice as long as ovary, 3-fid at the apex; stigmas small. Capsule orange, 4–6 mm. long, flattened-subglobose or obconic-3-gonous, pubescent. Seeds (2)3, reddish-brown, completely enclosed by an aril.

N. Rhodesia. N: Abercorn to Tunduma, km. 37, 1500 m., fl. 26.viii.1960, *Richards* 13156 (K; SRGH) E: Katele Distr., near Kabuula village, fl. 22.viii.1957, *Grout* 182 (FHO). S: Namwala, near Banga pontoon, fl. 26.ix.1963, *van Rensburg* 2486 KBS (K). **S. Rhodesia.** N: Mazoe Distr., Chipoli, 840 m., fl. 17.x.1958, *Mowbray* 40 (K; SRGH). W: Shangani Distr., Gwampa Forest Reserve, c. 900 m., fl. xi.1957, *Goldsmith* 50/58 (K: SRGH). E: Chipinga Distr., Sabi valley, Chibunji, fl. & fr. 14.x.1958, *Phelps* 275 (K; SRGH). S: Ndanga Distr., Chiduma Clinic, 300 m., st. i.1959, *Farrell* 15 (SRGH). **Nyasaland.** C: Chenga Hill, 1600 m., fl. 9.ix.1946, *Brass* 17601 (K; SRGH). S: Chikwawa Distr., Chikwawa, 200 m., fl. 2.x.1946, *Brass* 17907 (K; PRE; SRGH). **Mozambique.** N: Montepuez-Balama road, fl. 28.viii.1948, *Andrada* 1304 (COI; LISC). T: Boroma, R. Zambeze, fr. v.1892, *Menyhart* 1156 (Z).

Also in Ethiopia, Kenya, Tanganyika, Congo (Katanga), Angola (Benguela, Huila) and the Transvaal (Kruger National Park). Dry deciduous woodland, thickets and termite mounds, 200–1600 m. in our area.

M. putterlickioides is easily distinguishable from all other species of *Maytenus* in our area except *M. pubescens* by its pubescent leaves and differs from the latter in having large flowers with a prominent disk. It is very closely allied to the Angolan *M. welwitschiana*

Exell & Mendonça, which has ovate to elliptic leaves with pubescence usually confined to the midrib and smaller flowers.

4. **Maytenus mossambicensis** (Klotzsch) Blakelock in Kew Bull. **12**: 37 (1957).—Marais in Bothalia, **7**: 385 (1960). Type: Mozambique, Inhambane and Lourenço Marques (Delagoa Bay), 23°–26° S., *Peters* (B†).
Celastrus mossambicensis Klotzsch in Peters, Reise Mossamb., Bot. **1**: 112 (1861).—Oliv., F.T.A. **1**: 362 (1868).—Bak. f. in Journ. Linn. Soc., Bot. **40**: 44 (1911).—Eyles in Trans. Roy. Soc. S. Afr. **5**: 404 (1916). Syntypes as above.
Gymnosporia mossambicensis (Klotzsch) Loes. in Engl., Bot. Jahrb. **17**: 547 (1893).—Hutch., Botanist in S. Afr.: 464 (1946). Syntypes as above.
Gymnosporia harveyana Loes. in Bull. Herb. Boiss. **4**: 430 (1896).—Davison in Bothalia, **2**: 313 (1927).—Burtt Davy, F.P.F.T. **2**: 448 (1932).—Hutch., op. cit.: 668 (1946). Type from Natal.
Celastrus concinnus N.E. Br. in Kew Bull. **1906**: 16 (1906).—Bak. f., loc. cit. Type from Natal.
Celastrus huillensis? sensu Steedman, Trees, etc. S. Rhod.: 41, t. 40 (1933).

Shrublet or shrub or small tree (0·3)1–8 m. high, sometimes sarmentose, with spines up to 8 cm. long, axillary or terminating short axillary branches, without latex; branches 4-lined, reddish-purple to reddish-brown with numerous pale somewhat prominent lenticels and sometimes puberulous when young, becoming eventually terete, pale grey or cream, glabrous, slender. Leaves fasciculate or not, petiolate; lamina bright green, slightly paler below, 1–6·6(9·7) × 0·6–3·5(4·3) cm., ovate or lanceolate to elliptic or subcircular, acute or rarely shortly and obtusely acuminate to obtuse or more rarely rounded at the apex, with margin shallowly and ± irregularly rounded-serrulate to acutely incurved-denticulate, rounded or rarely subcordate to cuneate or angustate at the base, submembranous to chartaceous, with lateral nerves and ± lax reticulate venation more prominent below than above; petiole 1·5–6(9) mm. long. Cymes dichasial at first, becoming monochasial, solitary, axillary, or 1–4 on short axillary shoots, ± slender, with peduncle (5)7–23 mm. long, glabrous or rarely puberulous; pedicels 2–5·5 mm. long, glabrous or rarely puberulous, articulated usually in the lower ½; flowers (1)3–c. 55 in each cyme, 2·5–4(5) mm. in diam., bisexual. Sepals 5, unequal, 0·4–1 mm. long, triangular to semicircular, obtuse to rounded, glabrous or rarely puberulous outside, with margin ciliolate to subentire. Petals 5, white (in our area), (1)1·5–2·7 mm. long, oblong to oblong-elliptic, with margin finely ciliolate to entire. Stamens 5, with filaments 0·7–1 mm. long, slender, arising below the disk. Disk yellow, small, convex, 5-lobed. Ovary 3-locular, subglobose, c. ¼-immersed in the disk; style very short, with 3 large divergent stigmas. Capsule white when immature, pink to red when ripe, 7–13 mm. long, pyriform to obconic-3-gonous or subglobose, rounded, subcoriaceous to somewhat fleshy, smooth. Seeds 3, reddish-brown, rugulose, completely enclosed by a bright orange aril.

From SE. Kenya, E. Tanganyika and Zanzibar to the Transvaal, Natal and E. Cape Prov. In understorey of evergreen forest, forest margins, fringing forest and sometimes deciduous woodland, 150–1950 m. in our area.

M. mossambicensis is a variable species which has given rise to four distinct pubescent varieties. Only one of these occurs in our area.

Var. **mossambicensis**

Stem, leaves and inflorescence axis glabrous. Plant a tree or shrub, rarely a shrublet.

S. Rhodesia. E: Vumba Mts., 1590 m., fl. 28.x.1956, *Chase* 6279 (K; LISC; SRGH). S: Lundi R., fr. 30.vi.1930, *Hutchinson & Gillett* 3248 (BM; K). **Mozambique.** N: R. Rovuma, 45 km. from coast, fl. 15.iii.1861, *Meller* (K). Z: Mocuba to Quelimane, km. 95·2, fl. 28.v.1949, *Barbosa & Carvalho* in *Barbosa* 2913 (K; LMJ). MS: Manica, Mavita to Vila Pery, km. 15, fl. & fr. 18.vi.1942, *Torre* 4339 (LISC). SS: Inhambane, Quissico, 150 m., fr. 1908, *Sim* 21154 (PRE).

Throughout the range of the species except Kenya and NE. Tanganyika.

In our area var. *mossambicensis* shows a cline from lowland plants with leaves single, ovate to lanceolate, rounded at the base, with margins shallowly rounded-serrulate, and fruits pyriform, subcoriaceous, to montane plants with leaves fasciculate, elliptic to subcircular, angustate at the base, with margins acutely incurved-denticulate, and fruits subglobose, ± fleshy.

Var. **gurueensis** N. Robson in Bol. Soc. Brot., Sér. 2, **39**: 10 (1965). Type: Mozambique, between Gúruè and Ile, *Torre* 5074 (LISC).

Stem pubescent. Leaves ± sparsely puberulous on both sides. Inflorescence puberulous. Plant a shrublet c. 35–50 cm. high, sometimes rhizomatous.

Mozambique. Z: serra do Gúruè, between Marrequilo and R. Nuire, fr. 22.ix.1944, *Torre* 2218 (LISC).
Known as yet from the Gúruè district only.

Other shrublet forms in deciduous woodland occur in SE. Tanganyika, but these are always glabrous.

5. **Maytenus buchananii** (Loes.) Wilczek, F.C.B. **9**: 125 (1960). Type: Nyasaland, *Buchanan* in Herb. Wood. 6990 (B†, holotype; E; K; PRE).
Gymnosporia buchananii Loes. in Engl., Bot. Jahrb. **28**: 153 (1900).—R.E. Fr., Schwed. Rhod.-Kongo-Exped. **1**: 128 (1914).—Burtt Davy & Hoyle, N.C.L.: 37 (1936). Type as above.
Celastrus ndelleensis A. Chev., Études Fl. Afr. Centr. Franç. **1**: 57 (1913) *nom. nud.*
Celastrus littoralis A. Chev., Expl. Bot. **1**: 129 (1920) *nom. nud.*
Maytenus edgari Exell & Mendonça in Bol. Soc. Brot., Sér. 2, **26**: 223 (1952); C.F.A. **2,** 1: 6 (1956). Type: N. Rhodesia, Mwinilunga Distr., Matonchi R., *Milne-Redhead* 3124 (BM; K, holotype; PRE).
Maytenus ovata var. *ovata* forma *pubescens* Blakelock in Kew Bull. **11**, 2: 240 (1956) pro parte quoad specim. ex N. Rhod.—Keay & Blakelock in Keay, F.W.T.A. ed. 2, **1**, 2: 625 (1958).—White, F.F.N.R.: 218 (1962).

Shrub or small tree 2·1–8 m. high, sometimes sarmentose or scandent, with spines up to 2 cm. long, terminating short axillary branches usually on older parts, without latex; branches ± flattened, ± angular, purplish-brown and chocolate- to fawn-puberulous when young, becoming terete, striate, with numerous pale lenticels. Leaves not fasciculate, petiolate, glabrous; lamina pale to yellowish- or mid-green, concolorous or paler below, (3)3·6–11·2(17) × (1·4)1·8–5·5(8) cm., elliptic or elliptic-oblong to ovate or oblanceolate, obtuse (or rarely shortly acuminate on juvenile leaves) to rounded at the apex, with margin shallowly crenulate-serrulate, angustate to cuneate or rarely rounded at the base, coriaceous, with lateral nerves and densely reticulate venation more prominent below than above; petiole 2–9 mm. long. Cymes dichasial or ± monochasial, solitary, axillary, with peduncle absent or up to 3·5(5) mm. long, whitish-puberulous; pedicels 2–5 mm. long, whitish-puberulous, articulated in the lower ⅓; flowers 5–25 in each cyme, c. 2·5 mm. in diam., polygamous. Sepals 5, equal, 0·5–1 mm. long, lanceolate to triangular, acute to subacute, puberulous outside, with margin ciliolate or fimbriolate. Petals 5, white to cream, 1·5–2·5 mm. long, oblong, with margin ciliolate to subentire. Stamens 5, exceeding or equalling or slightly shorter than petals, with filaments 1–3 mm. long, slender, arising below the disk. Disk narrow, convex, unlobed. Ovary 3-locular, ovoid, c. ¼-immersed in the disk; style up to twice as long as the ovary, 3-fid at the apex, with small stigmas, or shorter and sometimes capitate, undivided. Capsule pale pink to bright red, 4–10 mm. long, obconic-3-gonous, truncate or with lobes slightly ascending, coriaceous, smooth. Seeds (2)3, red turning brown, glossy, with a fleshy smooth white to pale yellow aril at the base.

N. Rhodesia. N: Samfya, near shore of Lake Bangweulu, fr. 21.viii.1952, *Angus* 266 (BM; COI; FHO; K; PRE). W: Mwinilunga Distr., Muzera R., 16 km. W. of Kakoma, fr. 30.x.1952, *White* 3434 (BM; FHO; K). C: Broken Hill, fr. vi.1920, *Rogers* 26165 (K; PRE). **Nyasaland.** N: Masuku Plateau, 1950–2100 m., fl. vii.1896, *Whyte* (K). S: Mlanje Mt., Palombe R., c. 900 m., fl. 21.v.1958, *Chapman* 577 (FHO; K; SRGH). **Mozambique.** T: Zóbuè, Angónia road, R. Mueredzi, fr. 20.vii.1949, *Andrada* 1790 (COI; LISC).
Also in Angola (Lunda); Ivory Coast, Nigeria and N. Cameroun; and Central African Republic, the Congo, Rwanda-Burundi, Sudan Republic, Uganda, Kenya and Tanganyika. Fringing forest, 900–1200 m. in our area.

M. buchananii is distinguished from the other species of the *M. ovata* group by a combination of characters including its puberulous young shoots and inflorescence axes, smooth white to pale yellow aril and narrow disk. The indumentum appears to be non-glandular (*pace* Wilczek, loc. cit.). There is an apparently continuous variation in floral structure from ♂ flowers with long stamens c. 1½ times as long as the petals and sterile (?)

ovary with very short capitate styles to ⚥ flowers with stamens shorter than the petals and slightly exceeded by the elongated 3-fid style. Both types of flower may occur on the same shoot.

The specimens from W. Africa and Ubangi-Chari usually have petals reflexed at the apex and may be subspecifically distinct.

6. **Maytenus heterophylla** (Eckl. & Zeyh.) N. Robson in Bol. Soc. Brot., Sér. 2, **39**: 17 (1965). TAB. **76** fig. B. Type from S. Africa (Cape Prov.).

Celastrus buxifolius L., Sp. Pl., **1**: 197 (1753) pro parte excl. tab. Plukenet.—Sond. in Harv. & Sond., F.C. **1**: 459 (1860) non *Maytenus buxifolia* Griseb. (1866). Type from S. Africa.

Celastrus multiflorus Lam., Encycl. Méth. Bot. **1**: 661 (1785).—Oliv., F.T.A. **1**: 364 (1868) non *Maytenus multiflora* Reiss. (1861) nec Loes. (1905). Type from Africa, cult. in Paris.

Celastrus ellipticus Thunb. in Hoffm., Phytogr. Blaett. **1**: 22 (1803).—Sond., tom. cit.: 458 (1860) non *Maytenus elliptica* Krug & Urb. ex Duss (1897). Type from S. Africa.

Celastrus cymosus Soland. in Sims, Curt. Bot. Mag. **46**: t. 2070 (1819) non *Maytenus cymosa* Krug & Urb. (1904).

Catha buxifolia (L.) G. Don, Gen. Syst. **2**: 10 (1832). Type as for *Celastrus buxifolius*.

Celastrus heterophyllus Eckl. & Zeyh., Enum. Pl. Afr. Austr. Extratrop. **1**: 120 (1834–35?).—Sond., tom. cit.: 458 (1860). Type from S. Africa (Cape Prov.).

Celastrus patens Eckl. & Zeyh., loc. cit. Type from S. Africa (Cape Prov.).

Celastrus goniecaulis Eckl. & Zeyh., loc. cit. Type from S. Africa (Cape Prov.).

Celastrus humilis Eckl. & Zeyh., loc. cit. Type from S. Africa (Cape Prov.).

Celastrus spathephyllus Eckl. & Zeyh., tom. cit.: 121 (1834–35?). Type from S. Africa (Cape Prov.).

Celastrus venenatus Eckl. & Zeyh., loc. cit. Type from S. Africa (Cape Prov.).

Celastrus empleurifolius Eckl. & Zeyh., loc. cit. Type from S. Africa (Cape Prov.).

Celastrus polyanthemos Eckl. & Zeyh., loc. cit. Type from S. Africa (Cape Prov.).

Celastrus rhombifolius Eckl. & Zeyh., loc. cit.—Sond., tom. cit.: 460 (1860). Type from S. Africa (Cape Prov.).

Celastrus parvifolius Eckl. & Zeyh., loc. cit. Type from S. Africa (Cape Prov.).

Catha heterophylla (Eckl. & Zeyh.) Presl, Bot. Bemerk.: 33 (1844). Type as for *Celastrus heterophyllus*.

Catha cymosa (Soland.) C. Presl, loc. cit. Type as for *Celastrus cymosus*.

Celastrus angularis Sond., tom. cit.: 460 (1860). Syntypes from Cape Prov. and the Transvaal.

Celastrus andongensis Oliv., tom. cit.: 361 (1868). Type from Angola.

Gymnosporia buxifolia (L.) Szyszyl., Polypet. Disc. Rehm.: 34 (1888).—Bak. f. in Journ. Linn. Soc., Bot. **40**: 44 (1911).—Eyles in Trans. Roy. Soc. S. Afr. **5**: 404 (1916).—Davison in Bothalia, **2**: 317 (1927).—Burtt Davy, F.P.F.T. **2**: 448 (1932).—Steedman, Trees etc. S. Rhod.: 42 (1933).—Loes. in Engl. & Prantl, Nat. Pflanzenfam., ed. 2, **20b**: 152 (1942).—Hutch., Botanist in S. Afr.: 89, 225, 228, 361, 506 (1946).—O.B. Mill., B.C.L.: 35 (1948); in Journ. S. Afr. Bot. **18**: 48 (1952).—Brenan, T.T.C.L.: 127 (1949). Type as for *Celastrus buxifolius*.

Gymnosporia woodii Szyszyl., tom. cit.: 35 (1888).—Loes. in Engl. & Prantl, Nat. Pflanzenfam., ed. 2, **20b**: 152 (1942). Type from Natal.

Elaeodendron glaucum sensu Szyszyl., tom. cit.: 36 (1888).

Cassine szyszylowiczii Kuntze, Rev. Gen. Pl. **1**: 114 (1891). Type from S. Africa (Cape Prov.).

Gymnosporia heterophylla (Eckl. & Zeyh.) Loes. in Engl. & Prantl, Nat. Pflanzenfam. **3**, 5: 207 (1892); op. cit., ed. 2, **20b**: 152 (1942).—Davison, tom. cit.: 317 (1927). Type as for *Celastrus heterophyllus*.

Gymnosporia brevipetala Loes. in Engl., tom. cit.: 546 (1893); in Engl. & Prantl, Nat. Pflanzenfam., ed. 2, **20b**: 149 (1942). Type from E. Africa.

Gymnosporia senegalensis var. *maranguensis* Loes. in Engl., Bot. Jahrb. **19**: 231 (1894). Type from Tanganyika.

Gymnosporia glauca Loes. in Engl., Bot. Jahrb. **28**: 154 (1900) *nom. illegit.* Type from S. Africa (Cape Prov.).

Gymnosporia buxifolioides Loes., op. cit. **30**: 344 (1901).—Brenan, loc. cit. Type from Tanganyika.

Gymnosporia capitata var. *tenuifolia* Loes. in Bull. Herb. Boiss., Sér. 2, **3**: 823 (1903). Type from SW. Africa.

Gymnosporia rhombifolia (Eckl. & Zeyh.) Bolus & Wolley-Dod in Trans. S. Afr. Phil. Soc. **14**: 247 (1903). Type as for *Celastrus rhombifolius*.

Gymnosporia condensata Sprague in Kew Bull. **1906**: 246 (1906).—Davison, tom. cit.: 305 (1927). Type from the Transvaal.

Gymnosporia angularis (Sond.) Sim, For. & For. Fl. Col. Cap. Good Hope.: 186 (1907). Syntypes as for *Celastrus angularis*.

Gymnosporia maranguensis (Loes.) Loes. in Engl., Bot. Jahrb. **41**: 303 (1908); in Engl. & Prantl, Nat. Pflanzenfam., ed. 2, **20b**: 149 (1942).—Brenan, tom. cit.: 126 (1949). Type as for *G. senegalensis* var. *maranguensis*.

Gymnosporia acanthophora Loes., tom. cit.: 299 (1908). Type from SW. Africa.

Celastrus polyacanthus sensu Eyles, tom. cit.: 404 (1916).

Gymnosporia elliptica (Thunb.) Schonl. in Bot. Surv. S. Afr., Mem. **1**: 73 (1919). —Davison, tom. cit.: 306, t. 10 fig. 2 (1927). Type as for *Celastrus ellipticus*.

Gymnosporia uniflora Davison, tom. cit.: 294 (1927). Type from Natal.

Gymnosporia crataegiflora Davison, tom. cit.: 314, t. 18 (1927). Syntypes from Natal.

Gymnosporia trigyna sensu Perrier, Fl. Madag., Celastrac.: 21 (1946).

Maytenus cymosa (Soland.) Exell in Bol. Soc. Brot., Sér. 2, **26**: 222 (1952).—Exell & Mendonça, C.F.A. **2**, 1: 3 (1954).—Marais in Bothalia, **7**: 385 (1960).—Wilczek, F.C.B. **9**: 122 (1960). Type as for *Celastrus cymosus*.

Maytenus angolensis Exell & Mendonça in Bol. Soc. Brot., Sér. 2, **26**: 224 (1952); C.F.A. **2,** 1: 4, t. 1 fig. B (1954). Type from Angola.

Maytenus brevipetala (Loes.) Wilczek, F.C.B. **9**: 118 (1960). Type as for *Gymnosporia brevipetala*.

Maytenus senegalensis sensu White, F.F.N.R.: 218 (1962) pro parte quoad syn. *M. cymosus*, et auct. plur.

Shrub or tree, often spreading or straggling, or more rarely a shrublet, (0·3)1–7(9) m. high, unarmed or with green to brown spines up to 24 cm. long, axillary or terminating short axillary branches, glabrous or very rarely with young shoots and leaves puberulous (var. *puberula*), without latex; branches lined or angular or rarely subterete, pale green or rarely reddish-purple or glaucous at first, becoming terete (or rarely remaining angular), grey-brown or dark grey to purplish-brown or whitish, usually without visible lenticels. Leaves fasciculate or not, petiolate; lamina pale to deep green or rarely ± glaucous above, concolorous or often grey-green below, often with pale midrib, rarely mottled when dry, 1–9·5 × 0·4–5 cm., ovate or oblong-ovate or elliptic or circular (on long shoots) to obovate or oblanceolate or linear-oblanceolate or often spathulate (on fasciculate shoots), subacute or obtuse to rounded or emarginate and often shortly apiculate at the apex, with margin acutely or rarely obtusely shallowly and often irregularly serrulate (especially towards the apex) to entire, cuneate or rarely rounded to angustate at the base, membranous to coriaceous, with lateral nerves and reticulation varying in prominence and density; petiole 1–10 mm. long. Cymes dichasial, solitary and axillary or 1–7 on short axillary shoots, or rarely in axillary panicles, with peduncle 1–30 mm. long or occasionally absent; pedicels 1–7 mm. long, articulated at or near the base; flowers 2–24 or more in each cyme or very rarely solitary, 2–5 mm. in diam., always (?) dioecious, unscented or sometimes apparently malodorous. Sepals 5(6), equal, 0·3–1(1·5) mm., circular to triangular-lanceolate, rounded to acute, with margin ciliolate. Petals 5(6), white, 1–3·5 mm. long, elliptic-oblong to oblong-spathulate, with margin finely ciliolate to entire. ♂ flowers with stamens 5(6), shorter or longer than petals, with filaments 1–2·5 mm. long, slender, arising below disk; disk relatively narrow, ± concave, 5(6)-lobed; pistillode small, globose, with short style without spreading stigmas. ♀ flowers smaller than the ♂ flowers, with staminodes 5 shorter than ovary; disk as in the ♂ flowers; ovary 2–3(4)-locular, 0·3–0·5 mm. long, globose, not or scarcely immersed in the disk; style 0·2–0·5 mm. long, shorter than or equalling the ovary, shorter than or exceeding the petals, with 2–3(4) spreading stigmas. Capsule yellow or yellow-tinged-red or red, 3–10 mm. long, obovoid or subpyriform to 3-gonous or subglobose, thinly to thickly coriaceous or slightly succulent, smooth or rugulose. Seeds 1–3(4), reddish-brown, glossy, with a thin yellowish aril obliquely covering the lower $\frac{1}{2}$–$\frac{2}{3}$.

From Ethiopia, the Sudan and E. Congo southward to the Cape and westward to Angola and SW. Africa; also in Madagascar and St. Helena (? adventive). In forest, fringing forest margins, thickets and woodland or on termite mounds or sand dunes, 0–2100 m. in our area.

A very variable aggregate species which has frequently been confused with *M. senegalensis*, from which it differs by several characters e.g. usually angular or lined and greenish young shoots, green (rarely glaucous) leaves with usually acutely denticulate margin, the frequently 3-locular ovary, and yellow to red capsules. Its nearest relatives are *M. nemorosa* (Eckl. & Zeyh.) Marais, a S. African species with polished brown spines, dull green foliage and usually bisexual flowers, and *M. pyria* (Willemet) N. Robson from Mauritius, which is also bisexual and forms a link with *M. senegalensis* (Lam.) Exell. The variation in *M. heterophylla* is difficult to analyse; but the populations in our area may be divided into 4 subspecies.

Branches green or brown when young, becoming dark grey to purplish-brown with smooth or shallowly striate bark; leaves on fasciculate shoots varying in shape, rarely entire, not usually glaucous, with nervation ± prominent, often discolorous:
- Stamens and style shorter than the petals - - - - subsp. *heterophylla*
- Stamens in ♂ flowers and style in ♀ flowers exceeding the petals:
  - Leaves and young shoots glabrous - - - - - - subsp. *arenaria*
  - Leaves (beneath) and young shoots puberulous - - - subsp. *puberula*

Branches reddish-purple when young, becoming pale grey and eventually cream-white with rough bark; leaves on fasciculate shoots spathulate, entire to subentire, ± densely glaucous, usually without prominent nervation, concolorous subsp. *glauca*

Subsp. **heterophylla.**

Tree or shrub, glabrous; branches green to brown when young, becoming dark grey to purplish-brown with smooth or shallowly striate bark. Leaves variable in shape and texture, concolorous or discolorous, with ± prominent nervation. Stamens and style shorter than the petals. Ovary 2–3-locular. Capsule coriaceous.

Bechuanaland Prot. N?: without precise locality, fl., *Rogers* 6209 (K). **N. Rhodesia.** W: Ndola, West Forest Reserve, Dola Hill, fr. 13.viii.1952, *Angus* 194 (BM; FHO; K; PRE). E: Lundazi Distr., Nyika Plateau, Kangampande Mt., fl. 2.v.1952, *White* 2560 (FHO; K). S: Livingstone road, within 5 km. of Victoria Falls, fl. 5.v.1948, *Rodin* 4497 (K; PRE; SRGH). **S. Rhodesia.** N: Urungwe Reserve, Zwirani area, st. 5.i.1948, *Goodier* 524 (K; SRGH). W: S. of Victoria Falls, fr. 9.vii.1930, *Hutchinson & Gillett* 3490 (BM; K). C: Hartley, Poole Farm, Umfuli R., fl. 25.iv.1948, *Hornby* 2956 (K; SRGH). E: Umtali, Stapleford, 1770 m., fl. 26.xii.1953, *Chase* 5173 (BM; K; LISC; PRE; SRGH). S: Victoria Distr., fr. i.1931, *Eyles* 6639 (K; SRGH). **Nyasaland.** N: Nyika Plateau, near Nganda Hill, 2250 m., fl. 7.ix.1962, *Tyrer* 850 (BM). C: Dedza Distr., Mua-Livulezi Forestry Reserve, Namkokwe, R., 640 m., fr. 19.iii.1955, *E. M. & W.* 1054 (BM; LISC; SRGH). S: Ncheu, Chirobwe Hill, fl. 22.v.1961, *Chapman* 1331 (K; SRGH). **Mozambique.** N: Nampula, fl. 22.v.1937, *Torre* 1459 (COI; LISC). Z: Milange, fr. 24.ii.1943, *Torre* 4832 (LISC). MS: Manica, Mavita, frontier at Rotanda, fr. 28.iv.1948, *Barbosa* 1622 (LISC). LM: Namaacha, near the waterfall, fl. & fr. 11.iii.1958, *Barbosa & Lemos* in *Barbosa* 8255 (COI; K; LISC; LMJ; SRGH).

From central Tanganyika to the Cape; absent from coastal Mozambique and Natal, and from south-western S. Rhodesia. Forest edges, woodland or scrub.

Subsp. **arenaria** N. Robson in Bol. Soc. Brot., Sér. 2, **39**: 21 (1965). Type: Mozambique, Inhambane, Mangorra, Malamba, *Barbosa & Lemos* in *Barbosa* 8514 (COI; K; LISC, holotype; LMJ; PRE; SRGH).

Gymnosporia senegalensis var. *stuhlmanniana* Loes. in Engl., Bot. Jahrb. **17**: 542 (1893). Type: Mozambique, Quelimane, *Stuhlmann*, ser. 1, 314 (B†; BM (sketch)).

Shrub or rhizomatous shrublet; branches as in subsp. *heterophylla*, often with very long spines, glabrous. Leaves often ± rhombic, sometimes with indurated margin, never entire. Stamens (in ♂ flowers) and style (in ♀ flowers) longer than the petals. Ovary 2–3-locular. Capsule somewhat fleshy (in some 3-locular forms) or coriaceous.

Mozambique: N: between Mocímboa da Praia and Maate, fr. 9.x.1948, *Barbosa* (LISC). Z: environs of Mocuba, fl. 11.xii.1942, *Torre* 4792 (LISC). SS: Inhambane, Inharrime, Ponta Zavora, fl. 4.iv.1959, *Barbosa & Lemos* in *Barbosa* 8503 (COI; K; LISC; LMJ). LM: Maputo, Ponta do Ouro, fl. 18.xi.1944, *Mendonça* 2946 (LISC).

Also in Natal. In dune scrub and in open woodland on sandy soils; lowland.

Subsp. **puberula** N. Robson in Bol. Soc. Brot., Sér. 2, **39**: 23 (1965). Type: S. Rhodesia, between World's View and Silozwe, *E. M. & W.* 1507 (BM, holotype; LISC; SRGH).

Shrub; branches red-brown-puberulous when young. Leaves narrowly oblanceolate to spathulate (on fasciculate shoots), with margin entire, glabrous above, whitish-puberulous beneath. Cymes few-flowered; peduncles short or sometimes absent. Stamens (in ♂ flowers) and style (in ♀ flowers) exceeding the petals. Ovary 3-locular. Capsule as yet unknown.

S. Rhodesia. W: Matobo Distr., Besna Kobila, 1410 m., fl. iii.1953, *Miller* 1682 (K; SRGH).

Apparently confined to the Matobo and Bulalima-Mangwe Districts of S. Rhodesia. In fringing forest.

Subsp. **glauca** N. Robson in Bol. Soc. Brot., Sér. 2, **39**: 23 (1965). Type: Mozambique, Lourenço Marques, environs of Namaacha, *Barbosa & Lemos* in *Barbosa* 7592 (COI; LISC, holotype; LMJ; SRGH).

Gymnosporia angularis var. *orbiculata* Davison in Bothalia, **2**: 316 (1927) pro parte quoad specim. *Gerrard* 1377.

Shrub or small tree, glabrous; branches reddish-purple when young, becoming metallic grey and eventually cream-white with rough bark. Leaves on fasciculate shoots spathulate, entire to subentire, ± densely glaucous, concolorous, usually without prominent nervation. Stamens and style shorter than petals. Ovary (2)3-locular. Capsule coriaceous.

Mozambique. LM: between Moamba and Boane, fl. 4.xii.1940, *Torre* 2240 (LISC).

Also in Swaziland, Natal and eastern Transvaal. In open woodland and stony places.

7. **Maytenus pubescens** N. Robson in Bol. Soc. Brot., Sér. 2, **39**: 24 (1965). Type: S. Rhodesia, Beitbridge Distr., Chikwarakwara, *Wild* 5347 (K, holotype; PRE; SRGH).

Shrub 0·6–2·4 m. high, bushy, with spines up to 6 cm. long, terminal or terminating short axillary branches, without latex; branches terete or subangular, reddish-brown and shortly whitish-pubescent at first, becoming terete, grey-brown or dark grey, eventually glabrous. Leaves fasciculate or not, petiolate; lamina pale green, 0·5–3 × 0·5–1·5 cm., obovate to spathulate-oblanceolate, rounded to emarginate at the apex, with margin shallowly denticulate to entire, cuneate to angustate at the base, softly chartaceous, ± densely shortly pubescent on both sides, with lateral nerves and relatively lax reticulate venation slightly prominent on both surfaces; petiole 0·5–2 mm. long. Cymes dichasial, solitary and axillary or 1–5 on short axillary shoots, with peduncle 2–8 mm. long, shortly pubescent; pedicels 1–2 mm. long, pubescent or puberulous, articulated at or near the base; flowers solitary or 2–7 in each cyme, 3–4 mm. in diam., dioecious. Sepals 5, subequal, c. 1 mm. long, broadly ovate to subcircular, rounded, dorsally puberulous, with margin ciliolate. Petals 5, white to cream, c. 1·5 mm. long, oblong to spathulate, with margin finely ciliolate. ♂ flowers with stamens 5, shorter than the petals, with filaments c. 0·7 mm. long, slender, arising below disk; disk narrow, convex, 5-lobed; pistillode small, ovoid-subglobose. ♀ flowers with staminodes 5 about as long as the ovary; disk as in ♂ flowers; ovary 3(4)-locular, ovoid, scarcely immersed in disk; style almost as long as ovary, 3(4)-fid at the apex, with small stigmas. Capsule yellow, 6–7 mm. long, globose to obovoid, ± woody, smooth, puberulous. Seeds 3–4, reddish-brown, glossy, with a thin yellowish aril obliquely covering lower $\frac{2}{3}$–$\frac{3}{4}$.

S. Rhodesia. E: Chipinga, Rupisi, st. 17.ii.1960, *Farrell* 149 (SRGH). S: Nuanetsi Distr., near Malipate, fr. 23.iv.1961, *Simon* 2 (BM; SRGH).

Also in the Transvaal (Soutpansberg). In *Colophospermum mopane* scrub and among rocks, 405–480 m.

M. pubescens appears to be derived from *M. heterophylla*, from which it differs principally in indumentum characters and habitat.

8. **Maytenus senegalensis** (Lam.) Exell in Bol. Soc. Brot., Sér. 2, **26**: 223 (1952).—Brenan in Mem. N.Y. Bot. Gard. **8,** 3: 238 (1953).—Exell & Mendonça, C.F.A. **2,** 1: 8 (1954).—Keay & Blakelock in Keay, F.W.T.A. ed. 2, **1,** 2: 624 (1957).—Wilczek, F.C.B. **9**: 121 (1960).—Marais in Bothalia, **7**: 384 (1960).—White, F.F.N.R.: 218 (1962) pro parte excl. syn. *M. cymosa*. TAB. **76** fig. A. Type from Senegal, cultivated in Paris.

Celastrus senegalensis Lam., Encycl. Méth. Bot. **1**: 661 (1785).—Oliv., F.T.A. **1**: 361 (1868). Type as above.

Celastrus montanus Roth apud Roem. & Schult. in L., Syst. Veg. ed. nov. **5**: 427 (1819). Type from India.

Celastrus coriaceus Guill. & Perr. in Guill., Perr. & Rich., Fl. Senegamb. Tent. **1**: 142, t. 36 (1831). Type from Senegal.

Catha senegalensis (Lam.) G. Don, Gen. Syst. **2**: 10 (1832). Type as for *Maytenus senegalensis*.

Catha montana (Roth) G. Don, loc. cit. Type as for *Celastrus montanus*.

Celastrus europaeus Boiss., Elench.: 29 (1838). Type from Spain.

Celastrus senegalensis var. *inermis* A. Rich., Tent. Fl. Abyss. **1**: 133 (1847). Syntypes from Ethiopia.

Catha grossulariae Tul. in Ann. Sci. Nat., Sér. 4, **8**: 99 (1857). Syntypes from Madagascar.

Gymnosporia montana (Roth) Benth., Fl. Austral. **1**: 400 (1863). Type as for *Celastrus montanus*.

Gymnosporia crenulata Engl., Bot. Jahrb. **10**: 38 (1888).—Brenan, T.T.C.L.: 126 (1949). Type from S.W. Africa.

Gymnosporia senegalensis (Lam.) Loes. in Engl. & Prantl, Nat. Pflanzenfam. **3**, 5: 207 (1892); op. cit. ed. 2, **20b**: 149, t. 39 (1942).—Bak. f. in Journ. Linn. Soc., Bot. **40**: 44 (1911).—Eyles in Trans. Roy. Soc. S. Afr. **5**: 404 (1916).—Davison in Bothalia, **2**: 320 (1927).—Burtt Davy, F.P.F.T. **2**: 449 (1932).—Steedman, Trees etc. S. Rhod.: 41, t. 41 (1933).—Burtt Davy & Hoyle, N.C.L.: 38 (1936).—Perrier, Fl. Madag., Celastrac.: 28 (1946).—O. B. Mill., B.C.L.: 35 (1948); in Journ. S. Afr. Bot. **18**: 48 (1952).—Brenan, T.T.C.L.: 126 (1949).—Suesseng. in Proc. & Trans. Rhod. Sci. Ass. **43**: 110 (1951).—Pardy in Rhod. Agr. Journ. **50**, 4: 324, cum tab. (1953).—Williamson, Useful Pl. Nyasal.: 63 (1955). Type as for *Maytenus senegalensis*.

Gymnosporia senegalensis var. *inermis* (A. Rich.) Loes. in Engl., Bot. Jahrb. **17**: 541 (1893).—Eyles, loc. cit.—Burtt Davy & Hoyle, loc. cit.—Brenan, T.T.C.L.: 126 (1949).—Suesseng., loc. cit. Type as for *Celastrus senegalensis* var. *inermis*.

Gymnosporia senegalensis var. *inermis* forma *coriacea* (Guill. & Perr.) Loes., tom. cit.: 541 (1893).—Schinz, Pl. Menyharth.: 60 (1905).—Gomes e Sousa, Pl. Menyhart.: 78 (1926).—Brenan, T.T.C.L.: 126 (1949). Type as for *Celastrus coriaceus*.

Gymnosporia senegalensis var. *inermis* forma *chartacea* Loes., tom. cit.: 542 (1893).—R.E. Fr., Schwed. Rhod.-Kongo-Exped. **1**: 128 (1914). Syntypes from Ethiopia and Sudan Republic.

Gymnosporia senegalensis var. *inermis* forma *macrocarpa* Loes., loc. cit.—R.E. Fr., loc. cit. Syntypes from Ethiopia.

Gymnosporia senegalensis var. *spinosa* Engl. ex Loes., loc. cit.—Eyles, loc. cit.—Brenan, T.T.C.L.: 126 (1949). Syntypes from Senegambia, Ethiopia and Sudan Republic.

Gymnosporia grossulariae (Tul.) Loes., tom. cit.: 543 (1893). Type as for *Catha grossulariae*.

Celastrus saharae Batt. in Bull. Soc. Bot. Fr. **47**: 251 (1900). Type from Algeria.

Gymnosporia baumii Loes. in Warb., Kunene-Samb.-Exped. Baum: 291 (1903). Type from Angola.

Gymnosporia dinteri Loes. in Bull. Herb. Boiss., Sér. 2, **3**: 823 (1903). Type from SW. Africa.

Gymnosporia eremoecusa Loes. in Engl., Bot. Jahrb. **41**: 299 (1908); in Engl. & Prantl, Nat. Pflanzenfam., ed. 2, **20b**: 149 (1942). Type from SW. Africa.

Maytenus baumii (Loes.) Exell & Mendonça, tom. cit.: 6 (1954). Type as for *Gymnosporia baumii*.

Shrub or tree or rarely a shrublet, (0·15)1–9(15) m. high, unarmed or with spines up to 4 cm. long, axillary or terminating short axillary branches, glabrous, without latex; branches unlined, ± flattened, reddish-purple to reddish-brown, often densely glaucous, and sometimes with numerous pale indistinct lenticels at first, becoming terete and eventually grey-brown. Leaves fasciculate or not, petiolate; lamina pale green, usually glaucous, sometimes with reddish midrib, often mottled (at least when dry), 2–11·6 × 0·4–5·7(6·3) cm., oblong or rarely ovate to obovate or oblanceolate or oblong-elliptic or rarely subcircular, rounded or retuse to obtuse or apiculate at the apex, with margin ± densely regularly obtusely serrulate or serrulate-crenulate to subentire or very rarely entire, cuneate to angustate at the base, coriaceous or subcoriaceous, with lateral nerves and relatively lax reticulate venation slightly prominent or not prominent; petiole 3–13(20) mm. long. Cymes dichasial at first, becoming monochasial, solitary and axillary or 1–6 on short axillary shoots or occasionally in axillary panicles, with peduncle 1–16 mm. long; pedicels 0·7–4(6) mm. long, articulated in the lower ⅓ or rarely ½; flowers 3–60 or more in each cyme, 2–6 mm. in diam., dioecious or rarely monoecious, scented. Sepals 5, subequal, 0·3–1·2 mm. long, oblong-lanceolate to ovate-triangular or subcircular, obtuse to rounded with margin ciliolate. Petals 5, white or greenish-white to pale yellow, sometimes tinged pink, 1–3·5 mm. long, elliptic to oblong-elliptic or oblanceolate, with margin finely ciliolate. ♂ flowers with stamens 5, shorter than the petals, with filaments 0·5–1 mm. long, slender, arising below disk; disk relatively broad, flat or slightly concave, shallowly 5–10-lobed; pistillode small, globose. ♀ flowers with staminodes 5, shorter than ovary; disk as in ♂ flowers; ovary 2(3)-locular, globose, scarcely immersed in the disk; style about as long as ovary, 2(3)-fid at the apex, with small stigmas. Capsule pink to deep red, 2–6 mm. long, globose to pyriform, coriaceous, smooth. Seeds 1–2, dark reddish-brown, glossy, with a fleshy smooth rose-pink aril obliquely covering the lower $\frac{1}{3}$–$\frac{2}{3}$.

Bechuanaland Prot. N: Serondela, near Chobe R., fl. 25.vii.1950, *Robertson & Elffers* 45 (K; PRE). SW: Gemsbok, fl. 7.v.1930, *van Son* in Herb. Transv. Mus. 28808 (BM; PRE). SE: Mochudi, fl. ii.1914, *Rogers* 6439 (BM). **N. Rhodesia.** B: Balovale, Chavuma, fl. & fr. 4.viii.1952, *Gilges* 180 (K; PRE; SRGH). N: 11 km. NW. of Abercorn, 1410 m., fl. 19.vii.1930, *Hutchinson & Gillett* 3911 (BM; K; LISC; SRGH). W: Ndola, fl. 9.i.1955, *Fanshawe* 1787 (K; SRGH). C: Chilanga Distr., Quien Sabe, fl. & fr. viii–ix.1929, *Sandwith* 9 (K; SRGH). E: Jumbe, c. 1000 m., fr. 12.x.1958, *Robson* 59 (BM; K; LISC; PRE; SRGH). S: Livingstone, 900 m., fr. 28.viii.1911, *Rogers* 7420 (BM; K; SRGH). **S. Rhodesia.** N: Urungwe, Kansate R., fl. 31.x.1956, *Phipps* 160 (K; SRGH). W: Plumtree, Nata, 1320 m., fl. iv.1953, *Davies* 526 (K; LISC; SRGH). C: Hartley, Poole Farm, Umfuli R., fr. 11.viii.1946, *Hornby* 2957 (K; PRE; SRGH). E: Inyanga, Van Niekerk Ruins, 1500 m., fl. 4.viii.1950, *Chase* 2853 (BM; SRGH). S: Gwanda, Beitbridge, near Masera Camp, fl. 31.viii.1958, *West* 3713 (K; LISC; SRGH). **Nyasaland.** N: Vipya Mts. between Champoyo and Lwafwa, fl. 24.vi.1960, *Chapman* 786 (BM; K; SRGH). C: Dedza Distr., Chongoni Forestry School, fl. 13.v.1960, *Chapman* 699 (BM; SRGH). S: Chikwawa Distr., Lower Mwanza R., 180 m., fr. 6.x.1946, *Brass* 18007 (K; PRE; SRGH). **Mozambique.** N: between Namapa and Posto do Lúrio, fl. 11.x.1948, *Barbosa* 2375 (LISC). Z: Mocuba, Namagoa, 320 km. inland from Quelimane, 60–120 m., fl. & fr. v–viii.1945, *Faulkner* 200 (BM; K; PRE; SRGH). T: Tete, Chicoa road, fl. 25.vi.1949, *Andrada* 1637 (COI; LISC). MS: Chimoio, between Zembe and R. Revuè, fl. 30.iv.1948, *Andrada* 1216 (LISC). SS: Bilene, environs of Maniqueque, fr. 10.x.1957, *Barbosa & Lemos* in *Barbosa* 7982 (COI; K; LISC; LMJ). LM: Marracuene, Costa do Sol, fl. 10.viii.1959, *Barbosa & Lemos* in *Barbosa* 8662 (COI; K; LISC; LMJ; PRE; SRGH).

Africa south of the Sahara from Senegal to Eritrea and southward to northern SW. Africa, Bechuanaland Prot., Transvaal and Natal; also in Aldabra I. and N. Madagascar; and in S. Spain, Morocco, Algeria, Egypt, Arabia, Afghanistan, W. Pakistan and India. In deciduous woodland, thickets, scrub and wooded grassland, and also on river-banks and swamp margins; 0–1800 m. in our area (to 2400 m. in the Sudan Republic).

M. senegalensis is a very variable species which has frequently been confused with *M. heterophylla*, from which it can be distinguished *inter alia* by its unlined reddish or glaucous young shoots, rarely elliptic leaves with densely crenulate or obtusely serrulate margins, and almost constantly 2-locular ovary. (A very few specimens have 1- or 3-locular ovaries among the 2-locular ones.) The variation appears to be continuous throughout the range of the species, although some extreme forms seem distinct at a first glance e.g. the shrublets in Angola ("*G. senegalensis* var. *pumila*") and the damp habitat form occurring from Abercorn and Nyasaland to Angola ("*M. baumii*").

9. **Maytenus acuminata** (L.f.) Loes. in Engl. & Prantl, Nat. Pflanzenfam. ed. 2, **20b**: 138 (1942).—Brenan in Mem. N.Y. Bot. Gard. **8**, 3: 236 (1953).—Wilczek, F.C.B. **9**: 116, t. 12 (1960).—White, F.F.N.R: 218 (1962). Type from S. Africa (Cape Prov.).

Celastrus acuminatus L.f., Suppl. Pl.: 154 (1781).—Sond. in Harv. & Sond., F.C. **1**: 454 (1860). Type as above.

Celastrus populifolius Lam., Encycl. Méth., Ill. **2**: 94 (1797). Type from S. Africa (Cape Prov.).

Celastrus ? plectronia DC., Prod. **2**: 9 (1825). Type from S. Africa (Cape Prov.).

Celastrus rupestris Eckl. & Zeyh., Enum. Pl. Afr. Austr. Extratrop. **1**: 119 (1834–35?). Type from S. Africa (Cape Prov.).

Celastrus mucronatus Eckl. & Zeyh., loc. cit. Type from S. Africa (Cape Prov.).

Catha acuminata (L.f.) C. Presl, Bot. Bemerk.: 33 (1844). Type as for *Celastrus acuminatus*.

Catha rupestris (Eckl. & Zeyh.) C. Presl, loc. cit. Type as for *Celastrus rupestris*.
Celastrus acuminatus var. *microphyllus* Sond., loc. cit. Syntypes from S. Africa (Cape Prov.).
Gymnosporia acuminata (L.f.) Szyszyl., Polypet. Disc. Rehm.: 33 (1888) non Laws. (1875).—Burtt Davy, F.P.F.T. **2**: 448 (1932).—Burtt Davy & Hoyle, N.C.L.: 37 (1936).—Hutch., Botanist in S. Afr.: 225, 230, 668 (1946).—Brenan, T.T.C.L.: 124 (1949). Type as for *Maytenus acuminata*.
Gymnosporia lepidota Loes. in Engl., Bot. Jahrb. **17**: 549 (1893). Type from Uganda (Ruwenzori).
Gymnosporia bukobina Loes., op. cit. **41**: 305 (1908).—Brenan, T.T.C.L.: 124 (1949). Type from Tanganyika.
Gymnosporia amaniensis Loes., loc. cit.—Brenan, loc. cit. Type from Tanganyika.
Gymnosporia acuminata var. *lepidota* (Loes.) Loes., tom. cit.: 307 (1908).—Brenan, loc. cit. Type as for *Gymnosporia lepidota*.
Gymnosporia populifolia (Lam.) Dummer in Gard. Chron., Ser. 3, **54**: 248 (1913). Type as for *Celastrus populifolius*.
Gymnosporia lepidota var. *ruwenzorica* Loes. in Notizbl. Bot. Gart. Berl. **9**: 489 (1926). Syntypes from Uganda.
Gymnosporia lepidota var. *kilimandscharica* Loes., loc. cit., pro parte excl. specim. Natal. Syntypes from Kenya and Tanganyika.
Gymnosporia acuminata var. *microphylla* (Sond.) Davison in Bothalia, **2**: 313 (1927). Type as for *Celastrus acuminatus* var. *microphyllus*.
Maytenus bukobina (Loes.) Loes. in Engl. & Prantl, loc. cit. Type as for *Gymnosporia bukobina*.
Maytenus amaniensis (Loes.) Loes. in Engl. & Prantl, loc. cit. Type as for *Gymnosporia amaniensis*.
Maytenus lepidota (Loes.) Robyns & Lawalrée in Robyns, Fl. Parc Nat. Alb. **1**: 496 (1948). Type as for *Gymnosporia lepidota*.
Maytenus lepidota var. *kilimandscharica* (Loes.) Robyns & Lawalrée, loc. cit. Syntypes as for *Gymnosporia lepidota* var. *kilimandscharica*.
Maytenus rhodesica Exell in Kew Bull. **8**: 103 (1953). Type: N. Rhodesia, Mwinilunga, Matonchi R., *Milne-Redhead* 3129 (K, holotype; PRE).

Shrub or tree (0·3)1–15 m. high, unarmed, glabrous, with latex threads in branches, leaves, flowers and fruits; branches ± markedly 4-lined when young, eventually terete and longitudinally striate. Leaves not fasciculate, petiolate; lamina dark green and ± glossy above, paler below, (1)1·7–12·1(13·5) × (0·5)0·8–4·7 cm., ovate to narrowly lanceolate or rarely elliptic or elliptic-oblong, acuminate and frequently mucronate to acute or rarely obtuse at the apex, with margin acutely glandular-denticulate to crenulate-denticulate or more rarely entire, cuneate or angustate to rounded at the base, chartaceous to coriaceous, with lateral nerves and reticulate venation prominent below but not or scarcely so above; petiole (1)2–7(10) mm. long. Cymes monochasial after 1st to 3rd branching or pseudumbellate to fasciculate, with peduncle up to 21 mm. long or more rarely absent, in axil of foliage leaves; pedicels up to 13 mm. long or more rarely absent, articulated at the base or in the lower $\frac{2}{5}$; flowers c. 2–20 in each cyme or solitary, 3–5(6) mm. in diam. Sepals 5, unequal, 0·7–1·2 mm. long, suborbicular, rounded. Petals 5, green to white or white-tinged-red or deep red, 1·5–2·5 mm. long, broadly obovate to suborbicular, sessile or shortly unguiculate, entire. Stamens 5, with filaments 0·1–0·2 mm. long, flattened, united with outer margin of disk. Disk 5-angled, concave, with very short free margin. Ovary 3-locular, pyramidal, c. $\frac{1}{2}$-immersed in disk; style absent; stigma slightly 3-lobed. Capsule yellow to orange or red, 4–11 mm. long, 3-lobed with lobes rounded, smooth, ascending, or 2-lobed to obovoid or globose due to abortion of 1–2 loculi. Seeds 1–3, orange-brown to dark brown, completely enclosed by an orange aril.

In E. Africa from Kenya, Uganda and eastern Congo south to the Cape. In submontane forest and fringing forest, 1400–2400 m. in our area.

M. acuminata is a very variable species, especially in leaf-shape and inflorescence development. Specimens from western N. Rhodesia, for example, have well-developed peduncles and pedicels ("*M. bukobina*"), whereas in those from S. Rhodesia the flowers are sessile or almost so ("*M. amaniensis*"). The only variation in our area which seems distinct enough to warrant taxonomic recognition, however, is a small-leaved shrubby form from Mt. Mlanje.

Var. **acuminata**

Leaves ovate to narrowly lanceolate, (1·7)2·4–12·1 cm. long. Petals green to white or white-tinged-red.

N. Rhodesia. N: Chipalo, 5 km. N. of Luwingu, head of Lubansenshi R., fl. 21.v.1958, *Lawton* 373 (FHO; K). W: Mwinilunga Distr., Zambezi R., 6·4 km. N. of Kalene Hill Mission, fl. & fr. 20.ix.1952, *White* 3296 (BM; FHO; K; PRE). E: Nyika Plateau, Kangampande Mt., 2100 m., fl. 8.v.1952, *White* 2805 (BM; FHO; K). **S. Rhodesia.** E: Umtali Distr., Stapleford forest Reserve, 1680 m., fl. 16.vii.1955, *Chase* 5675 (BM; K; LISC; SRGH). **Nyasaland.** N: Nyika Plateau, between Chelinda and N. Rhodesian Government Rest House, 2100 m., st. 19.ix.1957, *McQueen* 96 (SRGH). S: Zomba Mt., 1800 m., near Cingwe's Hole, fr. 22.vi.1961 *Chapman* 1397 (FHO; K; SRGH). **Mozambique.** Z: Gúruè, R. Malema near Namuli peak, 1500 m., fl. 29.vi.1943, *Torre* 5605 (LISC). MS: Báruè, serra de Choa, 1400 m., fr. 8.ix.1943, *Torre* 5887 (LISC).

Throughout the range of the species.

Var. **uva-ursi** Brenan in Mem. N.Y. Bot. Gard. **8,** 3: 236 (1953) pro parte excl. specim. Transvaal. Type: Nyasaland, Mlanje Mt., Luchenya Plateau, 2200 m., *Brass* 16742 (BM; K, holotype; PRE; SRGH).

Leaves elliptic or elliptic-oblong, 1–3(3·7) cm. long. Petals deep red or pink-tinged-red.

Nyasaland. S: Mlanje Mt., Lower Ruo Plateau, 1800 m., fl. 8.viii.1956, *Newman & Whitmore* 400 (BM; LISC; SRGH).

Confined to Mlanje Mt., from 1800 m. to 2100 m.

10. **Maytenus chasei** N. Robson in Bol. Soc. Brot., Sér. 2, **39**: 26 (1965). Type: S. Rhodesia, Umtali Distr., Elephant Forest, *Chase* 5451 (BM, holotype; COI; K; PRE; SRGH).

Shrub or small tree 2–6 m. high, unarmed, glabrous except sometimes young shoots, without latex; branches 4-lined and finely puberulous or glabrous when young, becoming terete, longitudinally striate and glabrous. Leaves not fasciculate, petiolate; lamina dark green and glossy above, somewhat paler below, (5·5)6·5–13 × (2·5)3·2–7 cm., ovate to elliptic, acute or acuminate at the apex, with margin shallowly glandular-crenulate-denticulate, cuneate or shortly angustate to subrounded at the base, coriaceous, with c. 6–10 lateral nerves slightly prominent on both sides; petiole 4–9 mm. long. Cymes ± condensed, usually fasciculate, sessile or very shortly pedunculate, in axils of foliage leaves or bracts, pedicels 7–12 mm. long, articulated c. $\frac{1}{4}$–$\frac{5}{8}$ above the base; flowers c. 2–10 in each fascicle, rarely solitary, 6–7 mm. in diam. Sepals 5, unequal, 1·5–2 mm. long, ovate to subcircular, acute to rounded. Petals white to greenish-yellow, 2–2·5 mm. long, subcircular, unguiculate, with finely ciliolate crisped margin. Stamens 5, with filaments c. 0·5 mm. long, flattened, inserted within the outer margin of the disk. Disk shallowly 10-lobed, concave, with thin free margin. Ovary 2–3-locular, ovoid-pyramidal to subglobose, slightly immersed in disk; style elongate; stigma entire, capitate. Capsule green, c. 8–10 mm. long, obovoid to subglobose or shallowly 2–3-lobed, with lobes rounded and rugose-papillose, fleshy at first. Seeds orange, completely enclosed by an aril.

S. Rhodesia. E: Vumba Mts., Elephant Forest, 1710 m., fr. 3.vii.1955, *Chase* 5634 (BM; K; PRE; SRGH). **Mozambique.** MS: Quinta da Fronteira, near Penhalonga, 1140 m., fr. 16.vi.1962, *Wild* 5827 (BM; K; SRGH).

Known from only the Umtali and Melsetter Districts of S. Rhodesia and adjacent parts of Mozambique. Evergreen forest, 1140–1800 m.

M. chasei is related to *M. undata* (Thunb.) Blakelock and *M. acuminata* (L.f.) Loes., but differs from both in floral and fruit characters and from the latter in having no latex.

11. **Maytenus undata** (Thunb.) Blakelock in Kew Bull. **1956**: 237 (1956).—Keay & Blakelock in Keay F.W.T.A. ed. 2, **1,** 2: 624 (1958).—Wilczek, F.C.B. **9**: 118 (1960).—White, F.F.N.R.: 218 (1962). Type from S. Africa (Cape Prov.).

Celastrus undatus Thunb., Prodr. Pl. Cap.: 42 (1794).—Sond. in Harv. & Sond., F.C. **1**: 457 (1860). Type as above.

Celastrus ilicinus Burch., Trav. Int. S. Afr. **1**: 340 (1822). Type from S. Africa (Cape Prov.).

Celastrus lancifolius Thonn. apud. Schumach. in Kongel. Dansk Vid. Selsk.

Naturvid. & Math. Afh. **3**: 132 (1828).—Oliv., F.T.A. **1**: 364 (1868). Type from Ghana.

Celastrus luteolus Del. in Ann. Sci. Nat., Sér. 2, **20**: 90 (1843).—Oliv., tom. cit.: 363 (1868). Type from Ethiopia.

Celastrus laurifolius A. Rich., Tent. Fl. Abyss. **1**: 130 (1847).—Oliv., tom. cit.: 364 (1868). Type from Ethiopia.

Catha fasciculata Tul. in Ann. Sci. Nat., Sér. 4, **8**: 98 (1857). Syntypes from Madagascar and the Comoro Is.

Celastrus zeyheri Sond., tom. cit.: 456 (1860). Syntypes from S. Africa (Cape Prov.).

Celastrus huillensis Welw. ex Oliv., loc. cit.—Type from Angola.

Celastrus fasciculatus (Tul.) Boiv. ex Hoffm., Pl. Madag.: 12 (1881).—Grandidier, Hist. Nat. Madag., Bot. Madag., Bot. Atlas, **3**: t. 280 (1896). Syntypes as for *Catha fasciculata*.

Gymnosporia rehmannii Szyszyl., Polypet. Disc. Rehm.: 34 (1888).—Brenan, T.T.C.L.: 125 (1949). Type from Natal.

Gymnosporia undata (Thunb.) Szyszyl., loc. cit.—Davison in Bothalia, **2**: 296 (1927).—Burtt Davy, F.P.F.T. **2**: 448 (1932).—Hutch., Botanist in S. Afr.: 468 (1946). Type as for *Celastrus undatus*.

Gymnosporia zeyheri (Sond.) Szyszyl., op. cit.: 33 (1888).—Davison, tom. cit.: 294 (1927). Type as for *Celastrus zeyheri*.

Gymnosporia huillensis (Welw. ex Oliv.) Szyszyl., op. cit.: 35 (1888). Type as for *Celastrus huillensis*.

Gymnosporia lancifolia (Thonn.) Loes. in Engl., Bot. Jahrb. **17**: 548 (1893).—Brenan, T.T.C.L.: 125 (1949). Type as for *Celastrus lancifolius*.

Gymnosporia laurifolia (A. Rich.) Loes., loc. cit.—Brenan, loc. cit. Type as for *Celastrus laurifolius*.

Gymnosporia luteola (Del.) Loes., loc. cit.—Brenan, loc. cit. Type as for *Celastrus luteolus*.

Gymnosporia fasciculata (Tul.) Loes., op. cit. **19**: 232 (1895).—Burtt Davy, loc. cit.—Brenan, loc. cit. Syntypes as for *Catha fasciculata*.

Gymnosporia goetzeana Loes., op. cit. **30**: 344 (1901).—Brenan, loc. cit. Type from Tanganyika.

Celastrus albatus N. E. Br. in Kew Bull. **1906**: 16 (1906). Type from Natal.

Gymnosporia deflexa Sprague in Kew Bull. **1906**: 246 (1906).—Davison tom. cit.: 299 (1927).—Burtt Davy, loc. cit. Syntypes from the Transvaal.

Gymnosporia albata (N. E. Br.) Sim, For. & For. Fl. Col. Cap. Good Hope: 186 (1907). Type as for *Celastrus albatus*.

Gymnosporia ilicina (Burch.) Loes. in Engl., Pflanzenw. Afr. **3,** 2: 225 (1921).—Davison, tom. cit.: 296 (1927).—Hutch., op. cit.: 389 (1946). Type as for *Celastrus ilicinus*.

Gymnosporia peglerae Davison, tom. cit.: 298 (1927). Type from S. Africa (Cape Prov.).

Maytenus ilicina (Burch.) Loes. in Engl. & Prantl, Nat. Pflanzenfam., ed. 2, **20b**: 140 (1942). Type as for *Celastrus ilicinus*.

Maytenus fasciculata (Tul.) Loes., loc. cit.—Perrier, Fl. Madag., Celastrac.: 130 (1946). Type as for *Catha fasciculata*.

Maytenus lancifolia (Thonn.) Loes., loc. cit.—Exell & Mendonça, C.F.A. **2,** 1: 2 (1956). Type as for *Celastrus lancifolius*.

Maytenus huillensis (Welw. ex Oliv.) Loes., loc. cit.—Exell & Mendonça, loc. cit. Type as for *Celastrus huillensis*.

Maytenus zeyheri (Sond.) Loes., tom. cit.: 138 (1942). Type as for *Celastrus zeyheri*.

Gymnosporia maliensis Schnell in Bull. I.F.A.N. **15**: 96, t. 3 (1953). Syntypes from Mali.

Shrub or tree 1·5–10(12) m. high, much branched, sometimes scandent, unarmed, glabrous, without latex, very variable; branches ± markedly lined when young, becoming terete. Leaves not fasciculate, petiolate; lamina pale green or silver-green to dark green and sometimes glossy above, concolorous or sometimes paler below or whitish-glaucous below or on both surfaces, (1·1)3·2–13(20) × (0·9)1·2–7·5(11) cm., ovate or lanceolate or oblong to elliptic or oblanceolate or circular or oblate, acute to rounded at the apex, with margin glandular-denticulate to spinose-dentate or subentire, rarely reflexed or indurated, cuneate-angustate or rarely truncate to subcordate at the base, coriaceous to chartaceous, with 6–9 lateral nerves and reticulate venation varying in density and prominence; petiole 2–8(15) mm. long. Cymes fasciculate, sessile or rarely with peduncle up to 2 mm. long, in axil of foliage leaves; pedicels 2–9(15) mm. long, articulated at the base or in the lower $\frac{1}{3}$; flowers 2–c. 30 in each fascicle or rarely solitary, 4–8 mm. in diam., fragrant. Sepals 5, unequal, 0·7–1·5(2) mm. long, ovate to semicircular, rounded, sometimes tinged red. Petals white to cream or pale yellow, 1·5–3(5) mm. long, oblong or elliptic to obovate or oblanceolate, sessile or shortly unguiculate, with margin entire or eroded-denticulate, often ± crisped. Stamens 5, with filaments 0·7–2 mm. long, not or scarcely flattened, united with the base of the disk. Disk 5–10-lobed or subentire, flat or with thin free upturned margin or rarely somewhat convex. Ovary 3-locular, subglobose, c. $\frac{1}{3}$-immersed in the disk; style elongate or almost absent; stigma shortly 3-lobed. Capsule white or pinkish-red to yellow or orange, 4–7(10) mm. long, obovoid to subglobose or 3-gonous, with lobes rounded or carinate, smooth, not fleshy. Seeds orange-brown, completely or almost completely enclosed by an oblique aril.

Bechuanaland Prot. SE: Gaberones Distr., 25 km. N. of Lobatsi on Great North Road, fl. 15.xi.1948, *Hillary & Robertson* 561 (K; PRE). **N. Rhodesia.** N: Abercorn Distr., Mukoma stream, 27 km. from Abercorn, 1050 m., fr. 7.iv.1962, *Richards* 16309 (K; SRGH). W: Ndola Distr., Konkola R., fl. 13.vii.1952, *Holmes* 756 (K; SRGH). C: Serenje, fr. 24.ix.1961, *Fanshawe* 6711 (K). E: Fort Jameson to Lundazi, km. 130, st. 24.v.1952, *White* 2877 (FHO; K). **S. Rhodesia.** N: Satsi R., st. 26.ix.1937, *McGregor* 116/37 (BM; FHO). W: Matobo, Besna Kobila, 1350 m., fl. i.1954, *Miller* 2102 (K; LISC; SRGH). C: Rusape, Mona, fl. 9.i.1953, *Dehn* in GHS 43065 (K; SRGH). E: Umtali, 960 m., fl. 5.xii.1954, *Chase* 5344 (BM; COI; K; LISC; SRGH). S: Zimbabwe Ruins, fr. 1.vii.1930, *Hutchinson & Gillett* 3327 (BM; COI; K; LISC; SRGH). **Nyasaland.** N: Misuku Hills, Walindi Forest, 1650 m., fl. 12.i.1959, *Richards* 10614 (K; SRGH). C: Dedza, Ciwao Hill, st. 2.vii.1960, *Chapman* 799 (BM; K; SRGH). S: Shire Highlands, fl., *Adamson* 222 (E; K). **Mozambique.** MS: Cheringoma, Chiniziua, near Zuni village, fl. 11.iv.1957, *Gomes e Sousa* 4349 (COI; K; PRE). SS: near Macia, fl. 11.v.1944, *Torre* 6628 (LISC). LM: Goba, near Fonte dos Libombos, fl. 3.xi.1960, *Balsinhas* 179 (BM; K; LMJ; PRE; SRGH).

In W. Africa (Guinée, Togo, Ghana, N. Cameroons) and Angola, and from Eritrea, Somalia, Ethiopia and Sudan Republic southward to the Cape; also in Madagascar and the Comoro Is. Forest, fringing forest, woodland and evergreen scrub, c. 200–2100 m. in our area.

A very variable aggregate species in which the variation shows a certain amount of ecological and geographical correlation but no apparent morphological disjunctions. Thus in the rain-forests of W. Africa, Uganda, Tanganyika, etc., the leaves are large and the fruits pinkish-red to white (" *lancifolia* "). Eastward and southward the leaves become smaller and thicker (" *laurifolia* ", " *deflexa* " etc.) with pale yellow capsules. In drier regions in S. Rhodesia, Bechuanaland Prot. and S. Africa occur forms with still smaller leaves (" *undata* ", " *zeyheri* ") which are frequently ± densely glaucous (" *albata* ") and sometimes spinose-dentate (" *ilicina* ") and have yellow to orange fruits. Two extreme types from southern Africa have been treated as distinct species: *M. lucida* (L.) Loes., from the Cape region, with entire thick-margined leaves, and *M. procumbens* (L.f.) Loes., from the S.E. coastal region, with thick-margined spinulose-denticulate leaves.

12. **Maytenus procumbens** (L.f.) Loes. in Engl. & Prantl, Nat. Pflanzenfam., ed. 2, **20b**: 140 (1942).—Marais in Bothalia, **7**: 383 (1960). Type from S. Africa (Cape Prov.).

Celastrus procumbens L.f., Suppl. Pl.: 153 (1781).—Sond. in Harv. & Sond., F.C. **1**: 457 (1860). Type as above.

Gymnosporia procumbens (L.f.) Loes. in Engl., Bot. Jahrb. **39**: 169 (1906).—Davison in Bothalia, **2**: 294 (1927).—Hutch., Botanist in S. Afr.: 668 (1946). Type as above.

Shrub or small tree up to 6 m. high, sometimes bushy or scandent or drooping or ± procumbent (in S. Africa), unarmed, glabrous, without latex; branches lined when young, becoming terete. Leaves not fasciculate, petiolate; lamina pale green, concolorous, (1)3–6·6 × (1)1·3–3·1(4) cm., oblong or elliptic to obovate or oblanceolate, acute to obtuse or rounded at the apex, with margin reflexed, indurated, bearing 3–5 spinulose glandular denticles on each side or subentire, cuneate-angustate at the base, coriaceous, with c. 6 lateral nerves prominent on both sides or obsolete; petiole (1·5)2–4 mm. long. Cymes fasciculate, sessile, in axil of foliage leaves; pedicels 3–10 mm. long, articulated c. $\frac{1}{5}$–$\frac{1}{3}$ above the base; flowers c. 2–10 in each fascicle, 3–4 mm. in diam. Sepals 5, unequal, 1–1·5 mm. long, ovate, rounded. Petals white, 1–2(3) mm. long, oblong to elliptic, sessile, with margin subentire, ± crisped. Stamens 5, with filaments c. 1 mm. long, slightly flattened, inserted within the outer margin of the disk. Disk subentire, flat except for thin free upturned margin. Ovary 3-locular, subglobose, c. $\frac{1}{3}$-immersed in the disk; style almost absent; stigma capitate or very shortly 3-lobed. Capsule bright yellow to orange or orange-brown, 4–5 mm. long, obovoid or subglobose to 3-gonous or shallowly 3-lobed, with lobes rounded or keeled, smooth, dry. Seeds orange-brown, completely enclosed by an aril.

Mozambique. LM: Inhaca I., N. coast, Melville fishing village, fl. 17.vii.1957, *Mogg* 27308 (K; PRE; SRGH).

Also in Natal and E. Cape Prov. Woodland and scrub near the sea, on sandy soil, 0–150 m.

M. procumbens is a southern coastal derivative of the *M. undata* complex and is also closely related to *M. lucida* (L.) Loes. from the Cape region. It differs from the latter in having leaves with a denticulate to subentire (not entire) margin and from the rest of the *M. undata* complex by the combination of leaves with indurated margin and relatively few teeth and yellow to orange capsules. It may prove to be only subspecifically distinct from *M. undata*, in which case it would become the type subspecies.

3. PTEROCELASTRUS Meisn.

Pterocelastrus Meisn., Gen.: 68 (1837).

Asterocarpus Eckl. & Zeyh., Enum. Pl. Afr. Austr. Extratrop. **1**: 122 (1834–35?) non Necker ex Dumortier (1822) (" Astrocarpa ").

Trees or shrubs, without latex, glabrous; branches lined or angular. Leaves spiral, entire, petiolate; stipules free, small, thick, deciduous. Inflorescence pedunculate or rarely sessile, dichasial, ± numerous-flowered, simple, in axil of foliage leaves or rarely in axil of bracts forming " panicles "; bracts persistent. Flowers bisexual, pedicellate. Sepals 5, unequal, imbricate, ± united, with margin entire to subentire. Petals 5, white or greenish-white to cream or pale yellow, imbricate in bud, spreading. Disk inside stamens, single, ± shallowly concave, thin, not or scarcely lobed or crenulate. Stamens 5, with filaments thin, united with base of disk; anthers versatile, introrse, with separate thecae dehiscing longitudinally. Ovary superior, sessile, almost free or up to $\frac{1}{2}$-immersed in the disk, (2)3-locular, with 2 erect collateral ovules in each loculus; style ± short, cylindric, simple; stigma (2)3-branched. Fruit capsular, often waxy or fleshy, with horn- or wing-like emergences. Seeds 1–3, glossy, red-brown, almost covered by a thin yellow aril, with fleshy endosperm.

A genus of 4 species in south-eastern Africa. It is allied to the S. African *Maytenus oleoides* (Lam.) Loes. which has entire leaves, but its fruit is distinctive.

Pterocelastrus echinatus N.E.Br. in Kew Bull. **1906**: 16 (1906).—Davison in Bothalia, 2: 324 (1927).—Burtt Davy, F.P.F.T. **2**: 447 (1932).—Loes. in Engl. & Prantl, Nat. Pflanzenfam., ed. 2, **20b**: 157 (1942).—Hutch., Botanist in S. Afr.: 668 (1946). TAB. **77**. Type from Natal.

Pterocelastrus galpinii Loes. in Engl., Bot. Jahrb. **41**: 308 (1908); in Engl. & Prantl, loc. cit.—Davison, tom. cit.: 321 (1927).—Burtt Davy, loc. cit.—Burtt Davy & Hoyle, N.C.L.: 38 (1936).—Brenan in Mem. N.Y. Bot. Gard. **8**, 3: 238 (1953). Type from the Transvaal.

Pterocelastrus variabilis sensu Sim, For. Fl. Port. E. Afr.: 36 (1909).

Pterocelastrus rostratus sensu Bak. f. in Journ. Linn. Soc., Bot. **40**: 44 (1911).

Pterocelastrus rehmannii Davison, tom. cit.: 322, t. 13 fig. 2 (1927).—Loes. in Engl. & Prantl, loc. cit. Type from the Transvaal.

Gymnosporia nyasica Burtt Davy & Hutch. in Burtt Davy, tom. cit.: XXIII (1932). Type: Nyasaland, Mt. Mlanje, *Whyte* (K).

Shrub or tree (0·3)1–10·5 m. high (to 18 m. in S. Africa); branches angular and greyish when young, becoming vinous red and eventually terete with whitish lenticels. Leaves with lamina dark green to mid-green and glossy above, paler and dull beneath, (2)3·4–8·5 × (0·7)1·2–3·2(4·6) cm., ovate to lanceolate or oblong-elliptic, subacute or obtusely acuminate to obtuse or more rarely rounded at the apex, with margin entire, cuneate to angustate at the base, coriaceous, with c. 5–8 lateral nerves slightly prominent beneath but scarcely so above, without visible reticulate venation; petiole (2)4–8 mm. long. Cymes in axil of foliage leaves or bracts; peduncle 3–7 mm. long; flowers (3)7–19 in each dichasium, 3–4 mm. in diam., in corymbose or subglobose clusters. Sepals 0·5–1 mm. long (inner longer

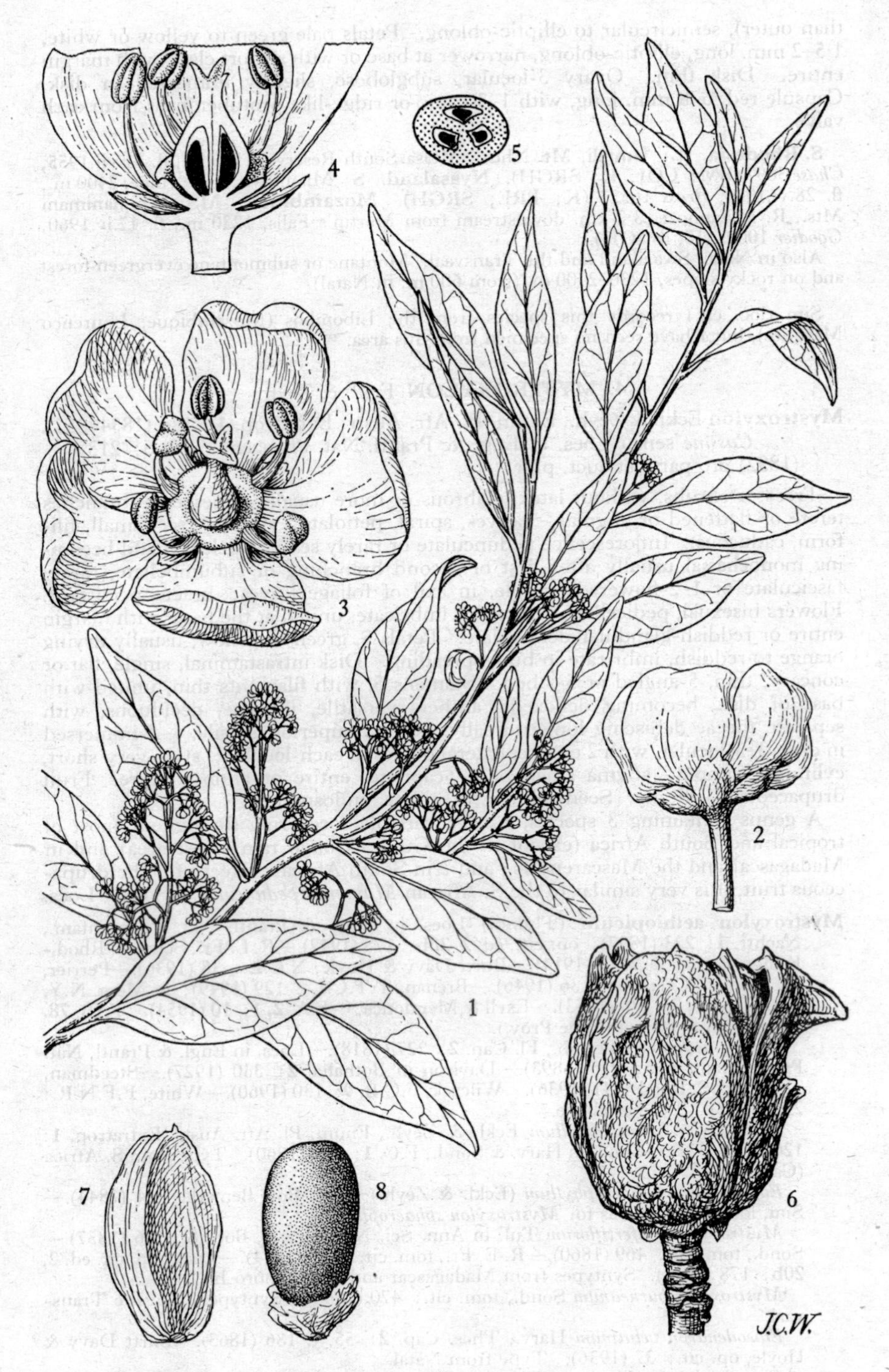

Tab. 77. PTEROCELASTRUS ECHINATUS. 1, flowering shoot ($\times \frac{2}{3}$); 2, flower bud ($\times 8$); 3, flower ($\times 12$); 4, vertical section of flower ($\times 12$); 5, transverse section of ovary ($\times 12$), all from *Newman & Whitmore* 134; 6, fruit ($\times 6$); 7, seed ($\times 6$); 8, seed with aril removed ($\times 6$), all from *Newman & Whitmore* 2991.

than outer), semicircular to elliptic-oblong. Petals pale green to yellow or white, 1·5–2 mm. long, elliptic-oblong, narrower at base or with a short claw, with margin entire. Disk thin. Ovary 3-locular, subglobose, slightly immersed in disk. Capsule red, 6–8 mm. long, with 1–3 conic or ridge-like protuberances from each valve.

S. Rhodesia. E: Umtali, Mt. Nhuri, Mtasa South Reserve, 1580 m., fl. 31.vii.1955, *Chase* 5697 (BM; COI; K; SRGH). **Nyasaland.** S: Mlanje Mt., SW. ridge, 2400 m., fl. 28.vi.1946, *Brass* 16523 (K; PRE; SRGH). **Mozambique.** MS: Chimanimani Mts., R. Mevumoze, 3·2 km. downstream from Martin's Falls, 1230 m., fl. 17.iv.1960, *Goodier* 1000a (K; SRGH).

Also in Natal, Swaziland and the Transvaal. Montane or submontane evergreen forest and on rocky slopes, 1230–2400 m. (from 600 m. in Natal).

Sim (loc. cit.) records this species from the Libombos (Mozambique, Lourenço Marques), but I have seen no specimen from this area.

4. MYSTROXYLON Eckl. & Zeyh.

Mystroxylon Eckl. & Zeyh., Enum. Pl. Afr. Austr. Extratrop. **1**: 125 (1834–35?).
Cassine sensu Loes. in Engl. & Prantl. Nat. Pflanzenfam. **3,** 5: 215 (1892) pro parte et auct. plur.

Trees or shrubs, without latex, glabrous or more usually pubescent; branches terete or flattened or angular. Leaves, spiral, petiolate; stipules free, small, filiform, caducous. Inflorescence pedunculate or rarely sessile, dichasial and becoming monochasial usually after first or second branching or subumbellate, rarely fasciculate or 1–2-flowered, simple, in axil of foliage leaves; bracts persistent. Flowers bisexual, pedicellate. Sepals 5, imbricate, united at the base, with margin entire or reddish-glandular-denticulate. Petals 5, green to yellow, usually drying orange to reddish, imbricate in bud, spreading. Disk intrastaminal, single, flat or concave, thin, 5-angled or 5-lobed. Stamens 5, with filaments thin, united with base of disk, becoming deflexed; anthers versatile, introrse, deciduous, with separate thecae dehiscing longitudinally. Ovary superior, sessile, $\frac{1}{2}$–$\frac{2}{3}$-immersed in disk, 2–3-locular, with 2 erect collateral ovules in each loculus; style very short, cylindric, simple; stigma small or $\pm$ capitate, entire or almost entire. Fruit drupaceous, $\pm$ fleshy. Seed solitary, exarillate, endospermic.

A genus containing 3 species, 1 very variable, occurring throughout most of tropical and South Africa (except in the Guinea-Congo rain forest area) and in Madagascar and the Mascarene Is., and 2 in South Africa. Except for the drupaceous fruit, it is very similar to the S. African *Maytenus peduncularis* (Sond.) Loes.

Mystroxylon aethiopicum (Thunb.) Loes. in Engl. & Prantl, Nat. Pflanzenfam., Nachtr. **1**: 223 (1897); op. cit., ed. 2, **20b**: 178 (1942).—R. E. Fr., Schwed. Rhod.-Kongo-Exped. **1**: 128 (1914).—Burtt Davy & Hoyle, N.C.L.: 38 (1936).—Perrier, Fl. Madag., Celastrac.: 36 (1946).—Brenan, T.T.C.L.: 129 (1949); in Mem. N.Y. Bot. Gard. **8,** 3: 238 (1953).—Exell & Mendonça, C.F.A. **2,** 1: 10 (1954). TAB. **78.** Type from S. Africa (Cape Prov.).

Cassine aethiopica Thunb., Fl. Cap. **2**: 227 (1818).—Loes. in Engl. & Prantl, Nat. Pflanzenfam. **3,** 5: 215 (1893).—Davison in Bothalia, **2**: 330 (1927).—Steedman, Trees etc. S. Rhod.: 41 (1936).—Wilczek, F.C.B. **9**: 130 (1960).—White, F.F.N.R.: 216 (1962). Type as above.

Mystroxylon sphaerophyllum Eckl. & Zeyh., Enum. Pl. Afr. Austr. Extratrop. **1**: 126 (1834–35?).—Sond. in Harv. & Sond., F.C. **1**: 470 (1860). Type from S. Africa (Cape Prov.).

Elaeodendron sphaerophyllum (Eckl. & Zeyh.) Presl, Bot. Bemerk.: 34 (1844).—Sim, loc. cit. Type as for *Mystroxylon sphaerophyllum.*

Mystroxylon confertiflorum Tul. in Ann. Sci. Nat., Sér. 4, Bot., **8**: 106 (1857).—Sond., tom. cit.: 469 (1860).—R. E. Fr., tom. cit.: 129 (1914).—Loes., op. cit. ed. 2, **20b**: 178 (1942). Syntypes from Madagascar and the Comoro Is.

Mystroxylon burkeanum Sond., tom. cit.: 470 (1860). Syntypes from the Transvaal.

Elaeodendron velutinum Harv., Thes. Cap. **2**: 55, t. 186 (1863).—Burtt Davy & Hoyle, op. cit.: 37 (1936). Type from Natal.

Elaeodendron aethiopicum (Thunb.) Oliv., F.T.A. **1**: 361 (1868).—Sim, For. Fl. Port. E. Afr.: 36 (1909). Type as for *Cassine aethiopica.*

Elaeodendron aethiopicum var. *pubescens* Oliv., loc. cit. Syntypes from Angola (Huila).

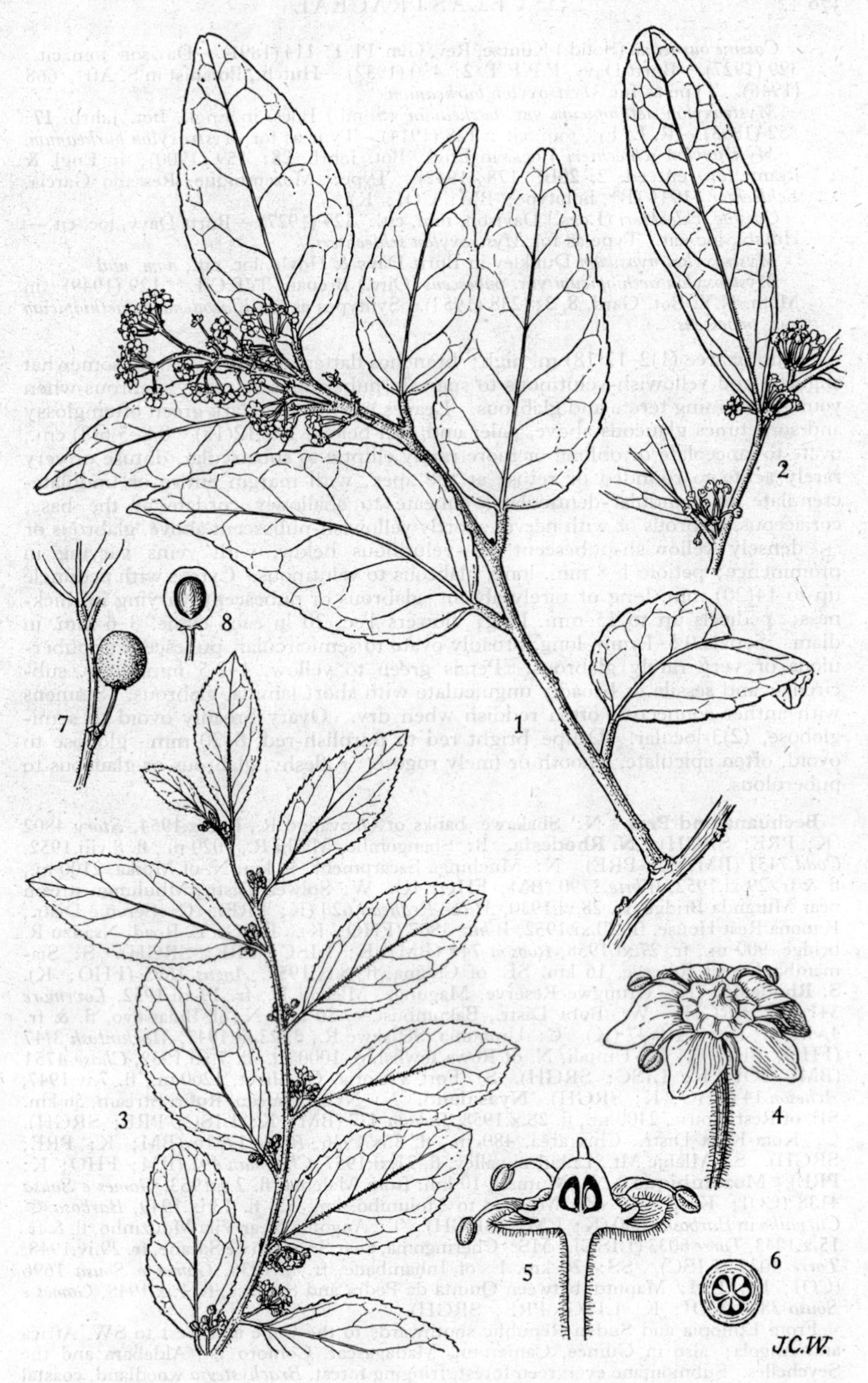

Tab. 78. MYSTROXYLON AETHIOPICUM. 1, flowering shoot ($\times\frac{2}{3}$) *Chase* 4224; 2, ibid. ($\times\frac{2}{3}$) *Brass* 17516; 3, ibid. ($\times\frac{2}{3}$) *Chase* 1916; 4, flower ($\times 4$) *Grout* 183; 5, vertical section of flower ($\times 6$); 6, ovary in transverse section ($\times 6$), both from *Chase* 4224; 7, fruit on branch ($\times\frac{2}{3}$); 8, vertical section of fruit ($\times\frac{2}{3}$), both from *Miller* 471.

Cassine burkeana (Sond.) Kuntze, Rev. Gen. Pl. **1**: 114 (1891).—Davison, tom. cit.: 329 (1927).—Burtt Davy, F.P.F.T. **2**: 450 (1932).—Hutch., Botanist in S. Afr.; 668 (1946). Type as for *Mystroxylon burkeanum*.

Mystroxylon aethiopicum var. *burkeanum* (Sond.) Loes. in Engl., Bot. Jahrb. **17**: 552 (1893).—R. E. Fr., tom. cit.: 128 (1914). Type as for *Mystroxylon burkeanum*.

Mystroxylon schlechteri Loes. in Engl., Bot. Jahrb. **28**: 159 (1900); in Engl. & Prantl, op. cit. ed. 2, **20b**: 178 (1942). Type: Mozambique, Ressano Garcia, *Schlechter* 11944 (B†, holotype; BM; COI; K).

Cassine schlechteri (Loes.) Davison, tom. cit.: 329 (1927).—Burtt Davy, loc. cit.—Hutch., loc. cit. Type as for *Mystroxylon schlechteri*.

Mystroxylon nyasicum Dunkley in Burtt Davy & Hoyle, loc. cit., *nom. nud.*

Mystroxylon aethiopicum var. *pubescens* (Oliv.) Brenan, T.T.C.L.: 129 (1949); in Mem. N.Y. Bot. Gard. **8**, 3: 238 (1953). Syntypes as for *Elaeodendron aethiopicum* var. *pubescens*.

Shrub or tree (1)2–12(18) m. high; branches flattened (or very rarely somewhat angular) and yellowish-velutinous to sparsely puberulous or rarely glabrous when young, becoming terete and glabrous. Leaves with lamina dark green often glossy and sometimes glaucous above, paler and dull below, 1·6–12(15) × 0·8–5·6(8) cm., ovate to lanceolate or oblong or more rarely elliptic to subcircular, obtuse or very rarely acute to rounded or retuse at the apex, with margin entire or undulate-crenulate to glandular-denticulate, cuneate to shallowly cordate at the base, coriaceous, glabrous or with nerves shortly yellowish-pubescent above, glabrous or ± densely yellowish-pubescent to -velutinous below, with veins varying in prominence; petiole 1–8 mm. long, glabrous to velutinous. Cymes with peduncle up to 14(20) mm. long or rarely absent, glabrous or pubescent, varying in thickness; pedicels up to 15 mm. long; flowers 1–c. 20 in each cyme, 3–6 mm. in diam. Sepals 0·5–1 mm. long, broadly ovate to semicircular, pubescent or puberulous or very rarely glabrous. Petals green to yellow, 1–2·5 mm. long, subcircular and sessile or broadly unguiculate with short lamina, glabrous. Stamens with anther-connective often reddish when dry. Ovary broadly ovoid to semi-globose, (2)3-locular. Drupe bright red to purplish-red, 8–20 mm., globose to ovoid, often apiculate, smooth or finely rugose, ± fleshy, glabrous or glaucous to puberulous.

Bechuanaland Prot. N: Shakawe, banks of Okovango R., fl. 3.x.1954, *Story* 4802 (K; PRE; SRGH). **N. Rhodesia.** B: Shangombo, Mashi R., 1020 m., fl. 8.viii.1952, *Codd* 7451 (BM; K; PRE). N: Muchinga Escarpment, 48 km. N. of Mpika, 2100 m., fl. & fr. 29.xi.1952, *White* 3790 (BM; FHO; K). W: Solwezi Distr., Mbulungu stream near Mutanda Bridge, fr. 28.vi.1930, *Milne-Redhead* 620 (K; PRE). C: Serenje Distr., Kanona Rest House, fr. 30.xi.1952, *White* 3807 (FHO; K). E: Gt. E. Road, Nsadzu R. bridge, 900 m., fr. 27.xi.1958, *Robson* 744 (BM; K; LISC; PRE; SRGH). S: Siamambo Forest Reserve, 16 km. SE. of Choma, fl. 8.ix.1957, *Angus* 1695 (FHO; K). **S. Rhodesia.** N: Urungwe Reserve, Magunge, Mlelchi R., fr. 19.xii.1952, *Lovemore* 341 (K; SRGH). W: Bubi Distr., Bakumbusi, c. 80 km. N. of Bulawayo, fl. & fr. 4.vi.1947, *Keay* A40/47 (K). C: Umvuma, Sebakwe R., fl. 23.iii.1947, *Mackintosh* 3/47 (FHO; SRGH). E: Umtali, N. of Rowa township, 1000 m., fl. 7.xii.1952, *Chase* 4751 (BM; COI; K; LISC; SRGH). S: Fort Victoria, Glenlivet, 1200 m., fl. 7.iv.1947, *Acheson* 14 (FHO; K; SRGH). **Nyasaland.** N: Nyika Plateau, Rufiri Stream, 56 km. SE. of Rest House, 2400 m., fl. 28.x.1958, *Robson* 437 (BM; K; LISC; PRE; SRGH). C: Kota-Kota Distr., Chia area, 480 m., fl. 3.ix.1946, *Brass* 17516 (BM; K; PRE; SRGH). S: Mlanje Mt., Lukulezi valley, fl. 21.xi.1957, *Chapman* 491 (BM; FHO; K; PRE). **Mozambique.** N: Mutuáli, 10 km. from Malema, fl. 2.x.1953, *Gomes e Sousa* 4138 (COI; K; PRE). Z: Milange to Molumbo, km. 55, fl. 13.ix.1949, *Barbosa & Carvalho* in *Barbosa* 4060 (K; LMJ; SRGH). T: Angónia, near Vila Mouzinho, fl. & fr. 15.x.1943, *Torre* 6033 (LISC). MS: Cheringoma, near Serração de Sarane, fr. 29.iv.1948, *Torre* 4033 (LISC). SS: 20 km. E. of Inhambane, fr. iii.1936, *Gomes e Sousa* 1696 (COI; K). LM: Maputo, between Quinta da Pedra and Santaca, fl. 1.ix.1948, *Gomes e Sousa* 3810 (COI; K; LISC; PRE; SRGH).

From Ethiopia and Sudan Republic southwards to the Cape and west to SW. Africa and Angola; also in Guinée, Cameroun, Madagascar, Comoro Is., Aldabara and the Seychelles. Submontane evergreen forest, fringing forest, *Brachystegia* woodland, coastal woodland or savanna woodland, 0–2100 m. in our area.

Mystroxylon aethiopicum is treated here in a broad sense. It is a very variable species and many of its forms have been given specific names. All these forms, however, appear to be interconnected by intermediates, so that it does not seem possible at present to recognise more than one species. Forms represented in our area include a large-leaved forest one (*Elaeodendron velutinum* Harv.), a narrow-leaved subglabrous south-eastern one (*M. schlechteri* Loes.) and a small-leaved few-flowered south-western one (*M. burkeanum* Sond.).

5. ALLOCASSINE N. Robson

Allocassine N. Robson in Bol. Soc. Brot., Sér. 2, **39**: 30 (1965).
Cassine L. subgen. *Eucassine* Loes. in Engl. & Prantl, Nat. Pflanzenfam. **3,** 5: 215 (1892) pro parte.

Shrubs, scrambling or climbing, without latex, glabrous; branches terete or flattened or quadrangular. Leaves alternate and foliar on young plants, alternate and bracteose on climbing shoots, opposite and foliar on lateral flowering shoots, petiolate; stipules free, small, triangular, deciduous. Inflorescence pedunculate, dichasial, few-flowered, ± condensed, simple or borne on specialised shoots forming " panicles ", in axil of foliage leaves; bracts persistent. Flowers bisexual, sessile. Sepals 5, imbricate, free, with margin reddish-glandular-fringed. Petals 5, cream or greenish-yellow to yellow, usually drying reddish, imbricate in bud, erect. Disk intrastaminal, single, ± deeply concave, thick to thin, not or scarcely lobed. Stamens 5, with filaments thin, united with base of disk; anthers versatile, introrse, with separate thecae dehiscing longitudinally. Ovary superior, sessile, almost free or up to c. $\frac{1}{3}$-immersed in the disk, 2-locular, with 2 erect collateral ovules in each loculus; style ± elongated, cylindric, simple; stigma 2-branched. Fruit baccate, fleshy. Seeds 2–3(4), exarillate, with endosperm.

A genus of 2 species in SE. Africa. Its nearest relatives appear to be *Mystroxylon* and *Polycardia* (from Madagascar), from both of which it differs *inter alia* by its baccate fruit, climbing habit and partially opposite leaves.

Allocassine laurifolia (Harv.) N. Robson in Bol. Soc. Brot., Sér. 2, **39**: 32 (1965). TAB. **79.** Lectotype from Natal.
Elaeodendron ? laurifolium Harv., Thes. Cap. **2**: 54, t. 185 (1863). Syntypes from Natal.
Cassine laurifolia (Harv.) Davison in Bothalia, **2**: 335 (1927). Syntypes as above.

Shrub, climbing or scrambling; branches ± flattened and grey when young, eventually becoming terete and dark purplish-red. Leaves on young and lateral shoots with lamina bluish-green and glossy on both surfaces, 6–14(18) × 2·5–6·5(8·5) cm., elliptic to oblong or ovate-oblong, rounded to retuse or rarely obtuse at the apex, with margin entire or remotely glandular-denticulate, cuneate to rounded at the base, coriaceous, with nerves and principal reticulate venation equally prominent on both sides, sometimes with very dense secondary reticulate venation prominent above; petiole 3–9 mm. long; stipules with fringed margin. Cymes simple, with peduncle (3)6–24 mm. long bearing a condensed dichasium with sessile or pedunculate lateral 1–3-flowered clusters, or rarely inflorescences compound with condensed cymes on specialised shoots forming " panicles "; pedicels absent; flowers (1)3–7 in each cyme, c. 3–4 mm. in diam. Sepals 2 mm. long, lanceolate, sometimes carinate, coriaceous, with membranous fimbriate margin. Petals greenish-yellow or yellow, 3 mm. long, narrowly oblong to oblanceolate, spreading above the erect sepals. Stamens with anthers c. 0·7 mm. long, oblong-lanceolate. Ovary narrowly conic. Berry orange-red, 18–25(28) mm. long, obovoid to subglobose, smooth, fleshy, ± glaucous. Seeds c. 15 mm. long, dark brown, curved, flattened-ellipsoid.

S. Rhodesia. E: Umtali, Eastlands, 900 m., fl. & fr. 25.xi.1951, *Chase* 4181 (BM; K; LISC; SRGH). **Mozambique.** MS: Manica, source of R. Revuè, 1290 m., st. 16.vi.1962, *Chase* 7766 (K; SRGH). LM: Namaacha, fl. & fr. 30.vi.1961, *Balsinhas* 503 (BM; K; LMJ).

Also in Natal. Evergreen forest, especially in gorges, 120–1600 m.

6. CATHA Forsk. ex Scop.

Catha Forsk. [Fl. Aegypt.-Arab.: 63 (1775) *nom. nud.*] ex Scop., Introd. Hist. Nat. Sist.: 228 (1777).

Trees or shrubs, without latex, glabrous. Leaves opposite on flowering shoots, alternate on vegetative ones, petiolate; stipules free or ± united interpetiolarly,

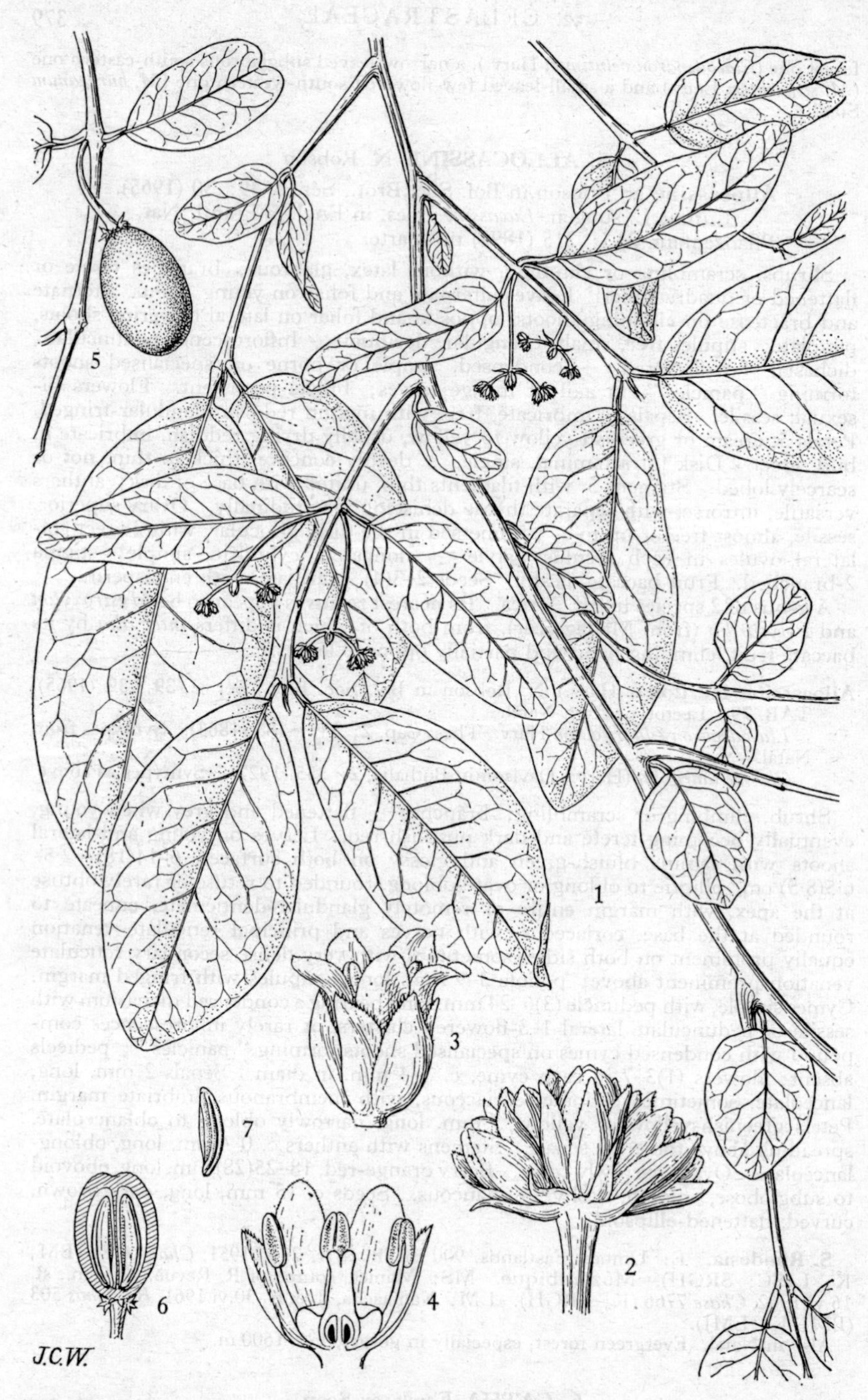

Tab. 79. ALLOCASSINE LAURIFOLIA. 1, flowering branch ($\times \frac{2}{3}$) *Chase* 5798; 2, partial inflorescence ($\times 6$); 3, flower ($\times 8$); 4, vertical section of flower ($\times 8$), all from *Chase* 4687; 5, fruit on branch ($\times \frac{2}{3}$); 6, vertical section of fruit ($\times \frac{2}{3}$); 7, seed ($\times \frac{2}{3}$), all from *Chase* 4181.

rapidly caducous. Inflorescence pedunculate, cymose, regularly dichasial, simple, axillary; bracts persistent. Flowers bisexual, pedicellate. Sepals 5, imbricate, united at the base. Petals 5, white or greenish, imbricate in bud. Disk intrastaminal, single, concave, thin, shallowly 5-lobed. Stamens 5, with filaments united with base of disk; anthers basifixed, with separate thecae dehiscing longitudinally; pollen simple. Ovary superior, sessile, 3-locular (sometimes incompletely so), with 2 erect collateral ovules per loculus; styles and stigmas 3, free. Fruit capsular, dehiscing loculicidally in 3 valves. Seeds 1–3, with a wing developed below the point of attachment.

A genus of 1 species, widespread in eastern tropical Africa.

Catha edulis (Vahl) Forsk. [Fl. Aegypt.-Arab.: cvii, 63 (1775)] ex Endl., Enchirid. Bot.: 575 (1841).—Hook. f. in Benth. & Hook., Gen. Pl. **1**: 361 (1862).—Oliv., F.T.A. **1**: 365 (1868).—Bak. f. in Journ. Linn. Soc., Bot. **40**: 44 (1911).—Eyles in Trans. Roy. Soc. S. Afr. **5**: 404 (1916).—Davison in Bothalia, **2**: 339 (1927).—Burtt Davy, F.P.F.T. **2**: 450 (1932).—Burtt Davy & Hoyle, N.C.L.: 37 (1936).—Loes. in Engl. & Prantl, Nat. Pflanzenfam., ed. 2, **20b**: 154 (1942).—Hutch., Botanist in S. Afr.: 668 (1946).—Brenan, T.T.C.L.: 123 (1949); in Mem. N.Y. Bot. Gard. **8**, 3: 236 (1953).—Paris & Moyse in Ann. Pharm. Fr. **15**: 89 (1957); in Trav. Lab. Mat. Med. **42**, 4: 1 (1958).—Wilczek, F.C.B. **9**: 126 (1960).—White, F.F.N.R.: 216 (1962). TAB. **80**. Type from Arabia.

Celastrus edulis Vahl, Symb. Bot. **1**: 21 (1790). Type as above.

Catha inermis Gmel. in L., Syst. Nat., ed. 13, **2**: 411 (1791). Type as above.

Methyscophyllum glaucum Eckl. & Zeyh., Enum. Pl. Afr. Austr. Extratrop. **1**: 152 (1834–35?).—Sond. in Harv. & Sond., F.C. **1**: 463 (1860). Syntypes from S. Africa (Cape Prov.).

Trigonotheca serrata Hochst. in Flora, **24**: 662 (1841). Type from Ethiopia.

Catha forskalii A. Rich., Fl. Abyss. Tent. **1**: 134, t. 30 (1847). Syntypes from Ethiopia.

Dillonia abyssinica Sacleux in Bull. Mus. Hist. Nat. Par., Sér. 2, **4**: 602 (1932). Type from Ethiopia.

Tree 2–15(25) m. high; stems pale grey-green and flattened when young, becoming vinous red, terete. Leaves with lamina dark green or greyish-green and glossy above, paler beneath, (3·7)5·5–11 × (0·8)1·5–4·5(6) cm., oblong to elliptic or obovate, acute to acuminate or more rarely obtuse at the apex, with margin markedly glandular-crenulate-denticulate, cuneate to angustate at the base, coriaceous or subcoriaceous, with densely reticulate venation more prominent below than above; petiole 3–10 mm. long; stipules c. 2 mm. long, triangular-acicular. Flowers ∞; peduncle 6–12 mm. long; bracts 0·5–1 mm. long, acicular to triangular. Sepals 0·5–0·7 mm. long, broadly ovate to semicircular, rounded, with margin ciliate-fimbriate. Petals 1–1·5 mm. long, elliptic-oblong, with margin ciliolate and paler in colour (at least when dry). Ovary broadly ovoid; styles short. Capsule red, 6–10 mm. long, narrowly oblong-3-gonous, pendulous. Seeds rugose-papillose, with a basal wing.

N. Rhodesia. N: Sunzu Hill, fl. 31.iii.1960, *Fanshawe* 5607 (K). W: Ndola, Lake Ishiku, fl. 16.iv.1957, *Fanshawe* 3183 (K; SRGH). **S. Rhodesia.** N: Mazoe, Citrus Estate, fl. 20.vi.1931, *Ford* in G.H.S. 5160 (PRE; SRGH). W: Matobo, Besner Kobila, 1435 m., fl. & fr., *Miller* 1827 (K; LISC; SRGH). C: Marandellas, path to Wedza Mt., fl. iii.1955, *Davies* 993 (COI; K; SRGH). E: Umtali, Commonage, 1160 m., fl. & fr. x.1943, *Chase* 37 (BM; PRE; SRGH). S: Fort Victoria, fl. 17.iv.1947, *Acheson* 20 (SRGH). **Nyasaland.** N: Misuku Hills, Mugesse Forest, fl. ix.1953, *Chapman* 160 (FHO; K). C: Nchisi Mt., 1350 m., fl. 1.viii.1946, *Brass* 17071 (BM; K; PRE; SRGH). S: Cholo, 1100 m., fr. 29.ix.1946, *Brass* 17873 (BM; K; PRE; SRGH). **Mozambique.** MS: Mossurize, near Catholic Mission, fl. 9.vi.1942, *Mendonça* 4268 (LISC).

In SW. Arabia and eastern Africa from Ethiopia to Cape Prov. In evergreen submontane or medium altitude forest, usually near the margins, or in woodland often on rocky hills, 1100–1435 m. (to 1800 m. in Uganda).

7. PLEUROSTYLIA Wight & Arn.

Pleurostylia Wight & Arn., Prodr. Fl. Penins. Ind. Or. **1**: 157 (1834).

Trees or shrubs, without latex, glabrous. Leaves opposite, petiolate; stipules very small, free, tardily deciduous. Inflorescence pedunculate, (1-) few-flowered,

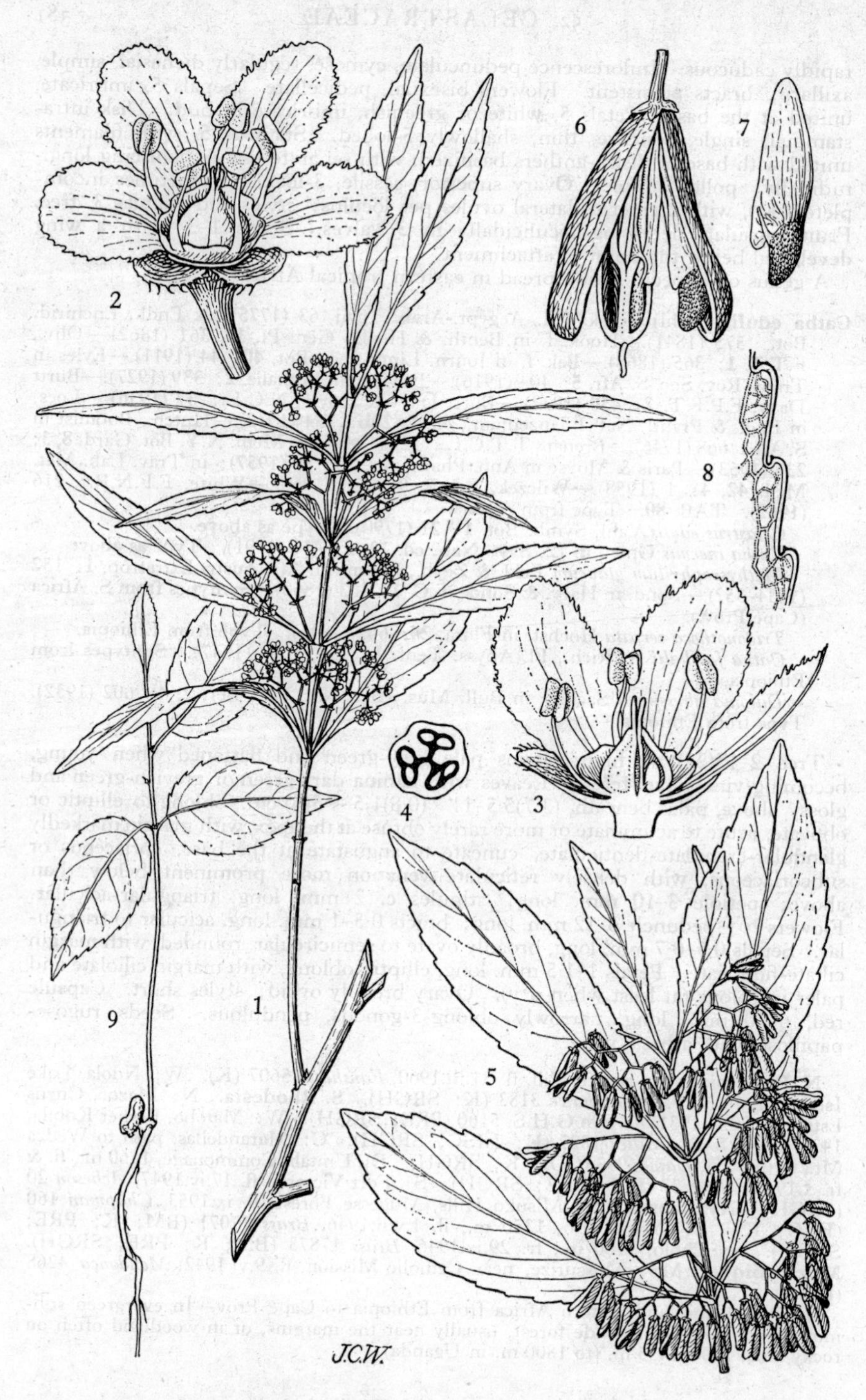

Tab. 80. CATHA EDULIS. 1, flowering shoot (× ⅔); 2, flower (×12); 3, vertical section of flower (×12); 4, transverse section of ovary (×12), all from *Eyles* 977; 5, fruiting shoot (× ⅔); 6, fruit (×4); 7, seed (×4); 8, leaf margin (× 2⅔), all from *Chase* 37; 9, seedling (× ⅔) *Swynnerton* 102.

cymose, usually regularly dichasial, simple, single or paired in leaf axil; bracts persistent. Flowers bisexual or rarely unisexual, pedicellate. Sepals (4)5, imbricate, united at the base. Petals (4)5, greenish, imbricate in bud. Disk intrastaminal, single, flat or concave, somewhat fleshy, (4)5-lobed. Stamens (4)5, with filaments united with base of disk; anthers sub-basifixed, introrse, with separate thecae dehiscing longitudinally. Ovary superior, sessile, 2-locular becoming 1-locular by abortion, with 2–8 erect ovules in 2 rows in the fertile loculus; style simple, sometimes very short, becoming lateral in fruit due to development of only 1 loculus; stigma unlobed. Fruit drupaceous, dry, 1-locular, with thin pericarp. Seeds 1(2), exarillate.

A genus of 3–4 species in tropical Africa, tropical S. Asia, Queensland and New Caledonia.

Leaves acute to acuminate, rarely glaucous; fruit green; branches becoming vinous red, not usually glaucous - - - - - - - - - - 1. *africana*
Leaves obtuse to rounded (in our area), ± glaucous; fruit white; branches becoming whitish - - - - - - - - - - - 2. *opposita*

1. **Pleurostylia africana** Loes. in Engl., Bot. Jahrb. **41**: 311 (1908); in Engl. & Prantl, Nat. Pflanzenfam., ed. 2, **20b**: 180 (1942).—Burtt Davy & Hoyle, N.C.L.: 45 (1936).—Hutch., Botanist in S. Afr.: 476 (1946).—Brenan, T.T.C.L.: 129 (1949).—Wilczek, F.C.B. **9**: 132 (1960).—White, F.F.N.R.: 219 (1962). TAB. **81** fig. A. Syntypes from Angola and Nyasaland, *Buchanan* 272 (B†, holotype; K).
Pleurostylia heynei var. *acutifolia* Suesseng. in Proc. & Trans. Rhod. Sci. Ass. **43**: 110 (1953). Type: S. Rhodesia, Marandellas, Willowkopje, *Dehn* 724 (M).

Tree or shrub 1–16(20) m. high, with rounded crown; stems pale grey-green and 4-lined when young, becoming vinous red, eventually terete. Leaf with lamina deep green or bright green and glossy above, concolorous or slightly paler beneath, 3·5–9·5(12) × 1–4 cm., elliptic to oblong-elliptic, acute to acuminate at the apex, with margin entire, angustate at the base, coriaceous or subcoriaceous, with reticulate venation equally prominent on both sides; petiole 3–8 mm. long; stipules c. 0·3 mm. long, triangular. Flowers 3–c. 15 in condensed rounded clusters; peduncle 1–4 mm. long or almost absent. Sepals 0·5–1 mm. long, broadly ovate to semicircular, rounded, with margin entire. Petals 1–1·5(2) mm. long, oblong, with margin finely ciliolate-denticulate. Disk with lobes truncate or shallowly emarginate. Ovary ovoid, slightly asymmetric; ovules 2(3). Fruit green, 7–8 mm. long, obovoid to ellipsoid. Seed 1, plano-convex.

N. Rhodesia. B: Balovale, 1050 m., fl. xii.1953, *Gilges* 315 (K; PRE; SRGH). N: Mpika, fl. 31.i.1955, *Fanshawe* 1910 (K). W: Mwinilunga Distr., c. 1·6 km. S. of Matonchi Farm, fr. 13.xi.1955, *Milne-Redhead* 4481 (BM; K). C: Lusaka South Fuel Reserve, st. 19.v.1934, *Miller* 27 (FHO). S: 1·6 km. above Victoria Falls, fr. 7.vii.1930, *Hutchinson & Gillett* 3424 (BM; K; SRGH). **S. Rhodesia.** N: Urungwe, Kariba, c. 600 m., fl. i.1959, *Goldsmith* 3/59 (BM; K; LISC; SRGH). C: Charter, Featherstone, fr. v.1951, *Mullin* 8/51 (K; LISC; SRGH). E: Umtali Golf Course, 1080 m., fl. 11.xii.1951, *Chase* 4257 (BM; K; LISC; SRGH). S: Buhera, 1050 m., fl. xi.1953, *Davies* 610 (K; SRGH). **Nyasaland.** C: near Salima, between Grand Beach and Lake Nyasa Hotels, fl. 16.ii.1959, *Robson* 1622 (BM; K; LISC; PRE; SRGH). S: Mlanje Mt., Chambe Peak, fr. 4.vi.1957, *Chapman* 376 (BM; FHO; K; PRE). **Mozambique.** Z: Gúruè, Lioma, fr. 1.vii.1943, *Torre* 5645 (LISC).

Also in Kenya, Uganda, Tanganyika, Katanga and Angola (Huila). On termite mounds and rocky hillsides and in deciduous woodland and forest margins on well-drained soil, 420–1100 m.

The S. African species, *P. capensis* (Turcz.) Loes., is closely allied to *P. africana* but has 6–8 ovules in each loculus.

2. **Pleurostylia opposita** (Wall.) Alston, Fl. Ceyl. **6**, Suppl.: 48 (1931).—Merrill & Metcalf in Lingn. Sci. Journ. **16**: 394 (1937).—Ding Hou in Fl. Males., Ser. 1, **6**: 288, t. 20 (1963). TAB. **81** fig. B. Type from India.
Celastrus oppositus Wall. in Roxb., Fl. Ind., ed. Carey & Wall., **2**: 398 (1824). Type as above.
Pleurostylia wightii Wight & Arn., Prodr. Fl. Penins. Ind. Or. **1**: 157 (1834).—Loes. in Engl. & Prantl, Nat. Pflanzenfam. ed. 2, **20b**: 180 (1942). Type from India.

Tree or shrub up to 15 m. high; stems pale grey-green and 4-lined when young, becoming whitish, eventually terete. Leaf with lamina grey-green, ± glaucous,

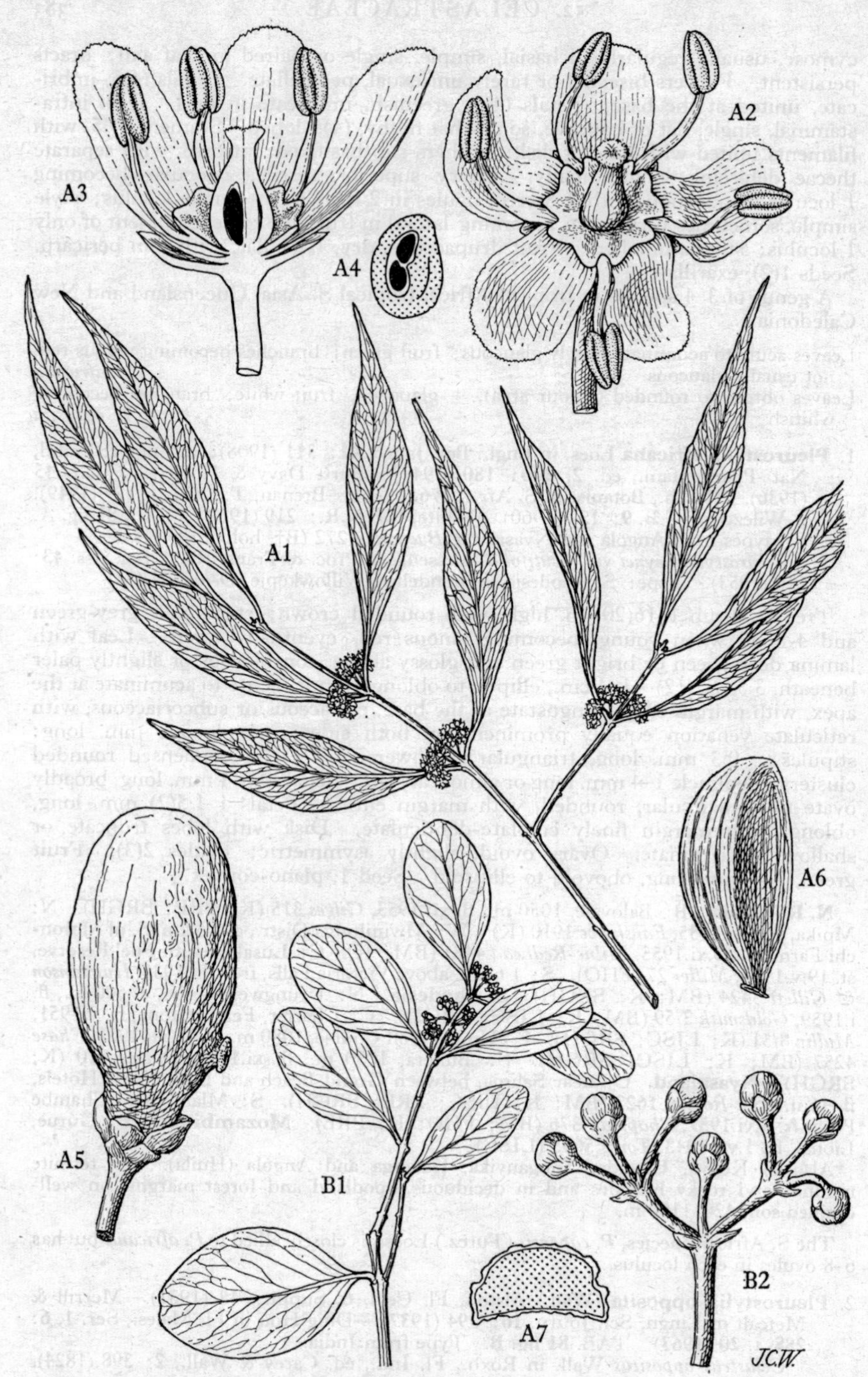

Tab. 81. A.—PLEUROSTYLIA AFRICANA. A1, flowering shoot (×⅔); A2, flower (×10); A3, vertical section of flower (×10); A4, transverse section of ovary (×12), all from *Goldsmith* 3/59; A5, fruit (×6); A6, seed (×6); A7, transverse section of seed (×6), all from *Chase* 3880. B.—PLEUROSTYLIA OPPOSITA. B1, flowering shoot (×⅔); B2, inflorescence (×6⅔), both from *Torre* 4656.

3–8 × 1·5–5·5 cm., ovate to elliptic-oblong or obovate, obtuse or obtusely acuminate to rounded at the apex, with margin entire, cuneate to angustate at the base, chartaceous, with reticulate venation equally prominent on both sides; petiole 2–5 mm. long; stipules very small, acicular-triangular. Flowers 1–7(20) in condensed rounded clusters; peduncle 1–3 mm. long. Sepals 0·3–0·5 mm., semicircular, rounded, with margin entire. Petals 1–1·5 mm. long, elliptic or broadly ovate, with entire margin. Disk with lobes rounded or shallowly emarginate. Ovary ovoid, slightly asymmetric; ovules 2. Fruit white, 5–7 mm. long, ellipsoid to obovoid-ellipsoid. Seeds 1(2).

Mozambique. Z: Pebane beach, fl. 24.x.1942, *Torre* 4656 (LISC).
Also in S. India, Ceylon, S. China and Malaysia to Queensland and New Caledonia. In dune scrub in our area.

The above specimen is the only one recorded for the African mainland; but it agrees very well with material from Ceylon and S. India apart from having more flowers on an average in each cyme. *P. pachyphloea* Tul. from Mauritius and (?) Madagascar is very closely allied to *P. opposita* but has smaller globose fruits.

8. ELAEODENDRON Jacq. f. ex Jacq.

Elaeodendron Jacq. f. ex Jacq., Ill. Pl. Rar. **1**: t. 48 (1782), 5 (1787); in Nov. Act. Helv. **1**: 36, t. 2 fig. 2 (1787).

Cassine subgen. *Elaeodendron* (Jacq. f. ex Jacq.) Loes. sect. *Euelaeodendron* Loes. in Engl. & Prantl, Nat. Pflanzenfam. **3,** 5: 215 (1892) pro parte excl. *C. crocea*, et auct. plur.

Trees or shrubs, without latex, glabrous. Leaves all opposite, or subopposite to alternate on young shoots, rarely heteromorphic, petiolate; stipules free, small, caducous. Inflorescence pedunculate to sessile, regularly dichasial, simple or borne on specialised shoots forming " panicles " in axil of foliage leaves or prophylls; bracts persistent. Flowers bisexual or unisexual, sometimes dioecious, pedicellate. Sepals 4–5, imbricate, united at the base, with margin entire or rarely finely serrulate. Petals 4–5, green to white or yellow, imbricate in bud, spreading. Disk intrastaminal, single, convex to flat, thick, entire or shallowly 4–5-lobed. Stamens 4–5, with filaments thin, united with base of disk; anthers versatile, introrse, with separate thecae dehiscing longitudinally. Ovary superior, sessile, ± immersed in disk, 2–3-locular, with (1)2 erect collateral ovules in each loculus; style short, pyramidal, simple; stigma small, entire or slightly 2–3-lobed. Fruit drupaceous, hard or ± fleshy. Seeds 1–2(3), exarillate, with endosperm.

A genus of c. 15 species in the Old World tropics, West Indies and C. America. It has been included in *Cassine* L., a S. African genus of 2 species which differs in several characters, e.g. strictly opposite leaves, sepals with glandular margins, white petals, thin concave disk, basifixed anthers and more elongated styles with free stigmas.

Inflorescence a simple dichasial cyme; flowers bisexual:
Flowers (except ovary) 5-merous; cymes in axil of foliage leaves (except sometimes in *E. orientale*); margin of juvenile leaves entire to denticulate:
Leaves homomorphic; fruit subglobose to obovoid or ellipsoid, rounded at the apex:
Plant a tree 2–20 m. high; fruit yellow; inflorescence-branches stout; flowers numerous (c. 40 or more) - - - - - - 1. *matabelicum*
Plant a shrub 1·5–4 m. high; fruit orange; inflorescence-branches slender; flowers few (c. 15–25) - - - - - - 2. *fruticosum*
Leaves heteromorphic (juvenile ones linear); fruit narrowly ovoid, acute to acuminate at the apex - - - - - - - - - - 6. *orientale*
Flowers (except ovary) 4-merous; cymes usually in axil of bract-like prophylls; margin of juvenile leaves spreading-spinose-dentate - - - - 3. *capense*
Inflorescence compound, paniculate (i.e. cymes in axil of bracts on specialised shoots); flowers bisexual or unisexual:
Flowers bisexual or, if unisexual, then staminodes stamen-like; fruit white; leaf-margin usually with ± spreading teeth - - - - - 4. *schlechteranum*
Flowers unisexual; staminodes petaloid; fruit pale yellow; leaf margin with ± incurved teeth - - - - - - - - 5. *buchananii*

1. **Elaeodendron matabelicum** Loes. in Engl., Bot. Jahrb. **40**: 61 (1907); in Engl. & Prantl, Nat. Pflanzenfam., ed. 2, **20b**: 174 (1942).—Eyles in Trans. Roy. Soc. S.

Afr. **5**: 404 (1916). TAB. **82** fig. A. Type: S. Rhodesia, near Matopos, *Engler* 2835 (B†).

Cassine matabelica (Loes.) Steedman, Trees etc. S. Rhod.: 41 (1933). Type as above.

Elaeodendron capense sensu O.B. Mill. in Journ. S. Afr. Bot. **18**: 48 (1952).

Cassine sp. 1.—White, F.F.N.R.: 216 (1962).

Tree 2–12(20) m. high; branches 4-lined and vinous red when young, becoming terete with prominent whitish lenticels and eventually greyish. Leaves with lamina yellowish-green to dark green above, greyish-green beneath, (3·5)4–10(13·2) ×1·5–4·5 cm., oblong to elliptic or lanceolate, acute or shortly acuminate to obtuse or apiculate at the apex, with margin obtusely glandular-denticulate or more rarely dentate or subentire, cuneate to angustate at the base, coriaceous, with densely reticulate yellowish venation more prominent above than below; petiole (5)7–18 mm. long. Cymes in axil of foliage leaves; peduncle 3–25 mm. long; flowers numerous in each dichasium, c. 4 mm. in diam., in subglobose clusters, bisexual. Sepals 5, c. 0·3 mm. long, semicircular, entire. Petals 5, green to yellow, 1–1·5 mm. long, oblong or elliptic-oblong, with margin paler, entire. Stamens 5. Disk slightly 5-angled, patelliform, puberulous. Ovary 3-locular, broadly ovoid-conic, c. $\frac{2}{3}$-immersed in disk. Drupe yellow or greenish-yellow, drying dark red, 13–18 mm. long, subglobose to ellipsoid or obovoid, smooth. Seed 1.

Bechuanaland Prot. N: Tati Distr., between Tsessebe and Ramaquebane, fl. ix.1948, *Miller* B919 (K; PRE). **N. Rhodesia.** S: Mazabuka Distr., Siamambo Forest Reserve, near Choma, fl. 2.vii.1952, *White* 3014 (BM; FHO; K; PRE). **S. Rhodesia.** N: Mazoe, Shamva, Chipoli Farm, 840–960 m., fl. vii–viii, *Moubray* in GHS 87776 (K; SRGH). W: Matopos Hills, fl. iii.1929, *Eyles* 6308 (K; SRGH). C: Hartley, Poole Farm, fl. 28.iii.1949, *Hornby* 2938 (PRE; SRGH). E: Umtali, Golf Course, fr. ii.1947, *Chase* 290 (BM; COI; SRGH). S: Gwanda, Koodoovale Motel, 900 m., fl. v.1958. *Davies* 2484 (K; LISC; SRGH). **Nyasaland.** C: Dowa Distr., near Lake Nyasa Hotel, 420 m., fl. 6.viii.1951, *Chase* 3881 (BM; SRGH). S: Zomba Distr., fl. i.1932, *Clements* 194 (FHO). **Mozambique.** T: Marávia, near Chicoa, fr. 25.ix.1942, *Mendonça* 418 (LISC). MS: Chimoio, mouth of R. Vanduzi, fl. & fr. 28.iv.1948, *Andrada* 1207 (LISC).

In S. Rhodesia and the regions immediately adjacent to it. In dry open deciduous woodland and wooded grassland, 425–1525 m.

2. **Elaeodendron fruticosum** N. Robson in Bol. Soc. Brot., Sér. 2, **39**: 39 (1965). Type: Mozambique, Gaza, Vila João Belo, *Torre* 3878 (LISC).

Shrub 1·5–4 m. high; branches 4-lined and vinous red when young, becoming terete and ± glaucous with whitish lenticels and eventually whitish. Leaves with lamina yellowish-green to grey-green above, greyish beneath, 4–8 ×1·7–3·5 cm., oblong to elliptic or oblanceolate, obtuse to rounded at the apex, with margin obtusely glandular-denticulate to subentire, cuneate to angustate at the base, coriaceous, with densely reticulate venation more prominent above than below; petiole 3–10 mm. long. Cymes in axil of foliage leaves; peduncle 8–12 mm. long; flowers c. 20–25 in each dichasium, c. 3·5 mm. in diam., in subglobose clusters, bisexual. Sepals 5, c. 0·2 mm. long, semicircular, entire. Petals 5, greenish, c. 1 mm. long, elliptic-oblong, with margin paler entire. Stamens 5. Disk slightly 5-angled, patelliform, finely papillose. Ovary 3-locular, broadly ovoid-conic, c. $\frac{2}{3}$-immersed in the disk. Drupe orange, drying dark red, 15–20 mm. long, subglobose to obovoid, smooth. Seed 1.

Mozambique. SS: Panda, fr. ix.1936, *Gomes e Sousa* 1871 (COI; K).

Known only from coastal regions of S. Mozambique. In dune scrub and dry woodland on sandy soil.

E. fruticosum appears to be a lowland derivative of *E. matabelicum*, from which it differs in habit, in fruit colour and in the size and number of flowers.

3. **Elaeodendron capense** Eckl. & Zeyh., Enum. Pl. Afr. Austr. Extratrop. **1**: 127 (1834–35?).—Grah. in Curt. Bot. Mag. **67**: t. 3835 (1841).—Sond. in Harv. & Sond., F.C. **1**: 468 (1860).—Bak. f. in Journ. Linn. Soc., Bot. **40**: 45 (1911).—Eyles in Trans. Roy. Soc. S. Afr. **5**: 404 (1916).—Marloth, Fl. S. Afr. **2**, 2: 154, t. 51 fig. E (1925).—Loes. in Engl. & Prantl, Nat. Pflanzenfam., ed. 2, **20b**: 173 (1942). Type from S. Africa (Cape Prov.).

Elaeodendron papillosum Hochst. in Flora, **27**: 305 (1844). Type from Natal.

Cassine papillosa (Hochst.) Kuntze, Rev. Gen. Pl. **1**: 114 (1891).—Davison in Bothalia, **2**: 334 (1927) pro parte excl. syn. *C. lacinulata* Loes. Type as for *Elaeodendron papillosum.*

Tree 2·5–10·5 m. high; branches 4-lined, flattened and pale grey-green to vinous red when young, becoming terete and grey-brown with prominent blackish lenticels. Leaves with lamina greyish-green above, concolorous, 5·3–11·5(15·8) × 2·1–4·4(6·7) cm., oblong to elliptic, acute or very shortly acuminate to obtuse at the apex, with margin indurated, shortly spinulose-denticulate to obtusely glandular-denticulate on flowering shoots, spreading-spinose-dentate on juvenile shoots, cuneate at the base, coriaceous, with densely to rather laxly reticulate venation equally prominent on both sides; petiole 5–11 mm. long. Cymes in axil of bract-like prophylls at base of current year's shoot, rarely also in axil of lowermost foliage leaf; peduncle (2)3–16 mm. long; flowers (1)3–7 in each dichasium, 4–5 mm. in diam., bisexual. Sepals 4, all c. 1 mm. long or inner pair shorter, ovate-oblong, rounded, entire. Petals 4, green to yellow, 1·5–2 mm. long, ovate to oblong, entire. Stamens 4. Disk slightly 4-lobed, with lobes shallowly emarginate, flat, glabrous. Ovary 2-locular, almost completely immersed in the disk except for the style. Drupe yellow, drying dark red, c. 20 mm. long, ovoid to broadly ovoid, smooth or with a few emergences (" papillae "). Seed 1.

S. Rhodesia. E: Umtali, Inyamatshira Mts., 1525 m., fr. 9.ix.1951, *Chase* 3944 (BM; COI; K; LISC; SRGH).

From S. Rhodesia to eastern Cape Prov.; introduced into St. Helena. Medium altitude evergreen forest, 1100–1980 m. in our area.

The above description refers to the S. Rhodesian specimens only. The Natal and Cape Prov. plants tend to have smaller flowers and narrower fruits; but, until fruiting material is collected in the Transvaal, it will not be possible to say whether or not the two populations are subspecifically distinct.

4. **Elaeodendron schlechteranum** (Loes.) Loes. in Engl. & Prantl, Nat. Pflanzenfam., Nachtr. **1**: 223 (1897); op. cit., ed. 2, **20b**: 173 (1942). Type: Mozambique, Tete, Boroma, *Menyhart* 2a (Z).

Cassine schlechterana Loes. in Bull. Herb. Boiss. **4**: 432 (1896).—Schinz, Pl. Menyharth.: 60 (1905).—Gomes e Sousa, Pl. Menyhart.: 78 (1936). Type as above.

Cassine lacinulata Loes., loc. cit. (1896).—Schinz, loc. cit.—Gomes e Sousa, loc. cit. Type: Mozambique, Boroma, R. Zambeze, *Menyhart* 1a (Z).

Elaeodendron lacinulatum (Loes.) Loes. in Engl. & Prantl, Nat. Pflanzenfam., Nachtr. **1**: 223 (1897). Type as for *Cassine lacinulata.*

Elaeodendron stuhlmannii Loes. in Engl., Bot. Jahrb. **28**: 156 (1900).—Brenan, T.T.C.L.: 124 (1949). Syntypes from Tanganyika.

Elaeodendron bussei Loes. in Engl., Bot. Jahrb. **41**: 309 (1908).—Brenan, tom. cit. 123 (1949). Type from Tanganyika.

Elaeodendron capense sensu Burtt Davy & Hoyle, N.C.L.: 37 (1936).

Elaeodendron papillosum sensu Brenan, tom. cit.: 123 (1949).

Cassine stuhlmannii (Loes.) Blakelock in Kew Bull. **1956**: 555 (1957). Syntypes as for *Elaeodendron stuhlmannii.*

Cassine buchananii sensu White, F.F.N.R.: 216 (1962) pro parte quoad specim. *Kirk.*

Shrub or tree 2–18 m. high; branches 4-lined or not, flattened to angular and pale grey to purplish when young, becoming terete and red-brown to grey-brown or whitish with prominent whitish lenticels. Leaves with lamina greyish-green and often ± glossy above, paler beneath, 3·5–12·6 × 1·8–7·2 cm., elliptic or oblong to oblanceolate or broadly obovate, obtuse or apiculate to rounded at the apex, with margin recurved or somewhat indurated, spreading- to erect- or rarely incurved-spinulose-glandular-dentate to -denticulate or subentire, cuneate to angustate at the base, coriaceous, with relatively laxly reticulate venation more prominent above than beneath; petiole 2–15 mm. long. Cymes in axils of bracts on specialised shoots arising from axils of bract-like prophylls or lowermost foliage leaves on current year's shoot, forming few-flowered paniculate inflorescences or often reduced to 1–3-flowered pseudocymes; peduncle (of cymes) up to c. 3 mm. long or absent; flowers 1–4 in each cyme, 4–6·5 mm. in diam., bisexual (or polygamous?). Sepals (4)5, c. 1 mm. long, broadly ovate to semicircular, rounded, entire or minutely serrulate. Petals (4)5, white or pale yellow, 2–3 mm. long, ovate to oblong, often shortly unguiculate, with raised longitudinal ridges often ending as a fringed appendage below 2 parallel depressions, entire. Stamens (4)5. Disk (4)5-lobed, flat or convex, glabrous. Ovary (2)3-locular, ovoid-conic, immersed in the disk. Drupe white, drying dark red or brown, 12–22 mm. long, ellipsoid to subglobose, smooth. Seed 1.

N. Rhodesia. C: Feira Distr., Katondwe, fl. 13.xi.1963, *Fanshawe* 8122 (K; NDO). S: highlands of Batoka country, fl. x.1860, *Kirk* (K). **S. Rhodesia.** N: Sebungwe, Chalala, Zambezi R. valley, 450 m., st. ix.1955, *Davies* 1487 (K; SRGH). S: Sabi-Lundi junction, Chitsa's Kraal, 240 m., st. 6.vi.1950, *Chase* 2286 (BM; K; SRGH). **Nyasaland.** Without precise locality, 600 m., st. 1.vi.1930?, *Carver* 38 (FHO). **Mozambique.** N: Mocímboa da Praia, Mechanga farm, 11° 15′ S; 40° 20′ E, fl. 6.iv.1961, *Gomes e Sousa* 4696 (COI; K; PRE). Z: between Ile and Mugeba, fr. 18.vi.1943, *Torre* 5506 (LISC). T: Zambeze valley, 48 km. above Tete, fr. 12.vi.1947, *Hornby* 2745 (K; PRE; SRGH). MS: Cheringoma, Chinizíua, between sawmills and R. Chinizíua bridge, fl. 29.ix.1958, *Gomes e Sousa* (K; PRE). SS: Gaza, Guijá, Malvernia road, near old and new roads to Mabalane, fl. 16.xi.1957, *Barbosa & Lemos* in *Barbosa* 8170 (COI; K; LISC; LMJ; SRGH). LM: Magude, Chobela, fr. 5.i.1948, *Torre* 7054 (LISC).

Also in Kenya and Tanganyika. Dry deciduous woodland especially on alluvium, fringing forest and on termite mounds; confined to lowlands and river valleys in our area, 6–600 m. (to c. 1300 m. in Tanganyika).

E. schlechteranum is a lowland species in our area; but it occurs at higher altitudes in central Tanganyika. This inland extension of the distribution may reach our area at Abercorn; but the only specimen (*Kafuli* 192 from Chinakila) is so poor that it is not possible to be certain that it belongs to this species and not to *E. buchananii*.

5. **Elaeodendron buchananii** (Loes.) Loes. in Engl. & Prantl, Nat. Pflanzenfam., Nachtr. **1**: 223 (1897).—Burtt Davy & Hoyle, N.C.L.: 37 (1936). TAB. **82** fig. B. Type: Nyasaland, *Buchanan* 710 (B†, holotype; BM; K).

Cassine buchananii Loes. in Engl., Bot. Jahrb. **17**: 551 (1893).—Keay & Blakelock in Keay, F.W.T.A. ed. 2, **1,** 2: 626 (1958).—Wilczek, F.C.B. **9**; 130 (1960).—White, F.F.N.R.: 216 (1962) pro parte excl. specim. *Kirk*. Type as above.

Elaeodendron afzelii Loes. in Engl., Bot. Jahrb. **28**: 157 (1900). Type from Sierra Leone.

Elaeodendron warneckei Loes. in Engl., Bot. Jahrb. **41**: 309 (1908).—Exell & Mendonça, C.F.A. **2,** 1: 11 (1956). Syntypes from Togo.

Elaeodendron keniense Loes. in Notizbl. Bot. Gart. Berl. **9**: 489 (1926). Type from Kenya.

Elaeodendron friesianum Loes., tom. cit.: 490 (1926). Type from Kenya.

Elaeodendron stolzii Loes. in Notizbl. Bot. Gart. Berl. **12**: 35 (1934).—Brenan, T.T.C.L.: 123 (1949). Type from Tanganyika.

Shrub or tree (1)4·5–12 m. high (to 30 m. in Uganda); branches 4-lined or not, flattened to quadrangular and pale grey when young, becoming terete and vinous-red to red-brown with prominent whitish lenticels. Leaves with lamina yellowish-green and dull above, paler green to grey-brown beneath, (5)6·5–14(17) × (2)2·7–8·1(10) cm., elliptic or lanceolate to obovate or oblanceolate, acute or shortly acuminate to obtuse or more rarely rounded at the apex, with margin indurated, glandular-crenulate-denticulate on flowering shoots, shortly appressed- or incurved-spinulose-denticulate on juvenile shoots, cuneate or rarely rounded at the base, coriaceous, with densely reticulate venation prominent beneath but not usually above; petiole 5–13 mm. long. Cymes in axils of bracts on specialised shoots arising from axils of bracteose prophylls or lowermost foliage leaves on current year's shoot, forming a paniculate multiflorous inflorescence; peduncle (of cymes) often indistinguishable from " panicle " branches; flowers 1–7(11) in each cyme, 3–5·5 mm. in diam., dioecious. Sepals 4–5, c. 1 mm. long, semicircular, rounded, entire or very finely serrulate. Petals 4–5, white or green to pale yellow, 1·5–2 mm. long in ♀ flowers, 2–2·5 mm. long in ♂ flowers, ovate to oblong, often shortly unguiculate, with raised longitudinal ridges, entire. ♂ flowers with 4–5 stamens; disk, slightly 4–5-lobed, flat, glabrous. ♀ flowers with or rarely without (4)5 petaloid staminodes; disk 4–5-lobed, annular, glabrous; ovary 2–3-locular, ovoid-conic, c. $\frac{1}{3}$-immersed in the disk. Drupe pale yellow, drying dark red or brown, 13–20 mm. long, ellipsoid to subglobose, smooth or with a few emergences. Seed 1.

N. Rhodesia. N: Abercorn Distr., Kalambo R., above Falls, fl. 21.x.1947, *Brenan* 8179 (FHO; K). W: Luanshya, fl. 28.vii.1954, *Fanshawe* 1403 (K; SRGH). **Nyasa-**

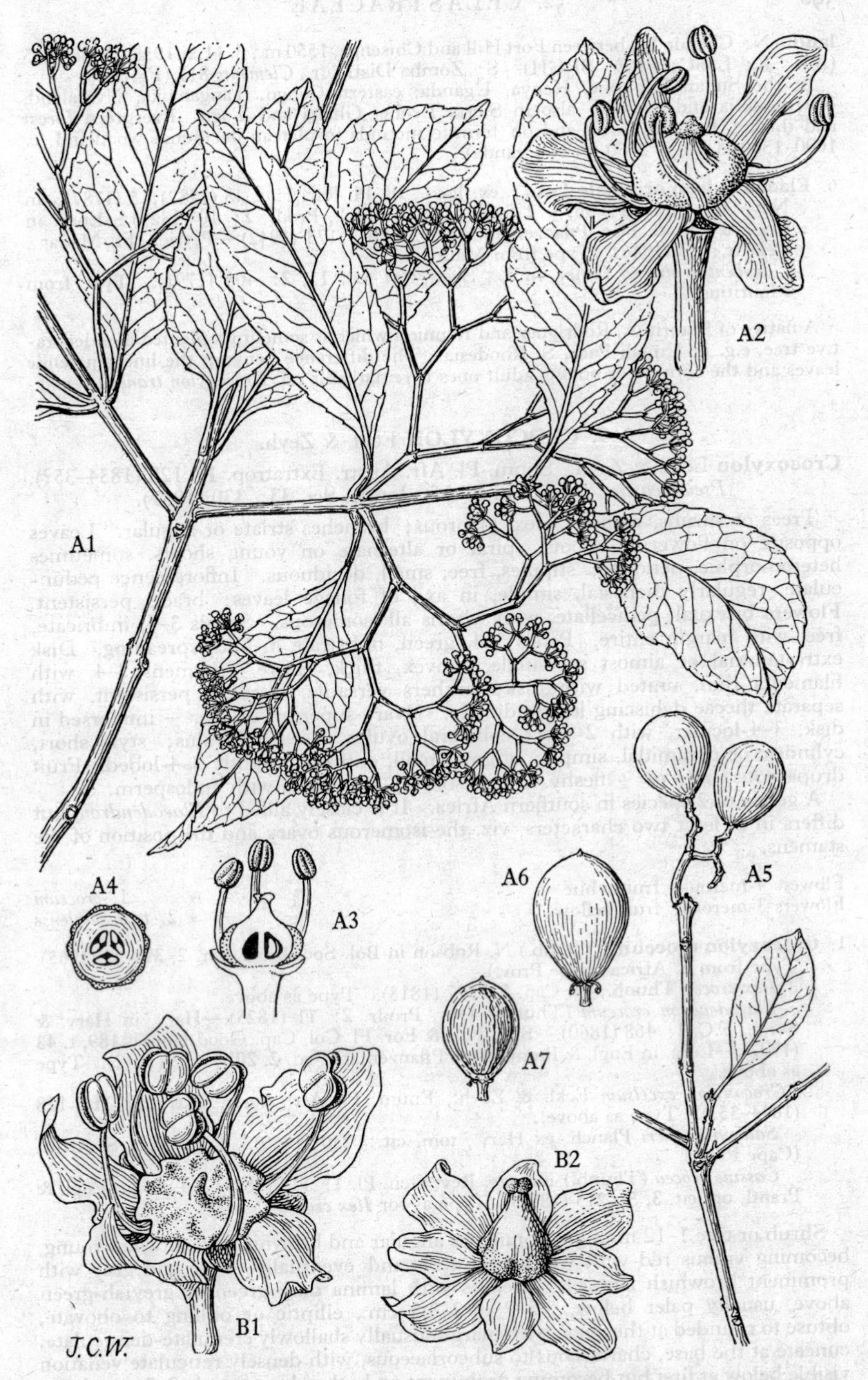

Tab. 82. A.—ELAEODENDRON MATABELICUM. A1, flowering shoot (×⅔); A2, flower (×10); A3, vertical section of flower (×6); A4, transverse section of ovary and disk (×6), all from *Chase* 290; A5, fruit on branch (×⅔); A6, fruit (×1); A7, fruit showing endocarp (×⅔), all from *Miller* 2145. B.—ELAEODENDRON BUCHANANII. B1, ♂ flower (×10) *Stolz* 2250; B2, ♀ flower (×10) *Prescott Decie* s.n.

land. N: Chendo R., between Fort Hill and Chisenga, 1550 m., fl. 11.xi.1958, *Robson* 555 (BM; K; LISC; PRE; SRGH). S: Zomba Distr., fr., *Clements* 811 (FHO).

In the Sudan Republic, Kenya, Uganda, eastern Congo, Tanganyika, Nyasaland, N. Rhodesia and Angola; also in Sierra Leone, Ghana and Togo. Evergreen forest and thicket, fringing forest and on termite mounds in plateau deciduous woodland, c. 1000–1500 m. (to c. 2250 m. in Uganda).

6. **Elaeodendron orientale** Jacq. f. ex. Jacq., Ill. Pl. Rar. **1**: t. 48 (1782); 5 (1787); in Nov. Act. Helv. **1**: 36, t. 2 fig. 2 (1787).—DC., Prodr. **2**: 10 (1825).—Loes. in Engl. & Prantl, Nat. Pflanzenfam., ed. 2, **20b**: 173 (1942).—Perrier, Fl. Madag., Celastr.: 55 (1946). Type from Mauritius.

Rubentia olivina Gmel. in L., Syst. Nat. ed. 13, **2**: 408 (1791). Type from Mauritius.

A native of Mauritius, Rodriguez and Réunion which is sometimes planted as a decorative tree, e.g. in Umtali Park, S. Rhodesia. The difference between the linear juvenile leaves and the obovate or oblong adult ones is remarkable (cf. *Crocoxylon transvaalense*).

9. CROCOXYLON Eckl. & Zeyh.

Crocoxylon Eckl. & Zeyh., Enum. Pl. Afr. Austr. Extratrop. **1**: 128 (1834–35?).
Pseudocassine Bredell in S. Afr. Journ. Sci. **33**: 330 (1937).

Trees or shrubs, without latex, glabrous; branches striate or angular. Leaves opposite on flowering shoots, spiral or alternate on young shoots, sometimes heteromorphic, petiolate; stipules, free, small, deciduous. Inflorescence pedunculate, regularly dichasial, simple, in axil of foliage leaves; bracts persistent. Flowers bisexual, pedicellate, with whorls all isomerous. Sepals 3–4, imbricate, free, with margin entire. Petals 3–4, green, imbricate in bud, spreading. Disk extrastaminal or almost so, single, convex, thick, entire. Stamens 3–4, with filaments thin, united with disk; anthers versatile, introrse, persistent, with separate thecae dehiscing longitudinally. Ovary superior, sessile, ± immersed in disk, 3–4-locular, with 2 erect collateral ovules in each loculus; style short, cylindric or pyramidal, simple; stigma small, entire or slightly 3–4-lobed. Fruit drupaceous, hard or ± fleshy. Seeds solitary, exarillate, with endosperm.

A genus of 2 species in southern Africa. It is closely allied to *Elaeodendron*, but differs in at least two characters, viz. the isomerous ovary and the position of the stamens.

Flowers 4-merous; fruit white - - - - - - - - - 1. *croceum*
Flowers 3-merous; fruit yellow - - - - - - - - 2. *transvaalense*

1. **Crocoxylon croceum** (Thunb.) N. Robson in Bol. Soc. Brot., Sér. 2, **39**: 41 (1965). Type from S. Africa (Cape Prov.).

Ilex crocea Thunb., Fl. Cap. **1**: 577 (1813). Type as above.

Elaeodendron croceum (Thunb.) DC., Prodr. **2**: 11 (1825).—Harv. in Harv. & Sond., F.C. **1**: 468 (1860).—Sim, For. & For. Fl. Col. Cap. Good Hope: 189, t. 43 (1907).—Loes. in Engl. & Prantl, Nat. Pflanzenfam., ed. 2, **20b**: 173 (1942). Type as above.

Crocoxylon excelsum Eckl. & Zeyh., Enum. Pl. Afr. Austr. Extratrop. **1**: 128 (1834–35?). Type as above.

Salacia zeyheri Planch. ex Harv., tom. cit.: 230 (1860). Type from S. Africa (Cape Prov.).

Cassine crocea (Thunb.) Kuntze, Rev. Gen. Pl. **1**: 114 (1891).—Loes. in Engl. & Prantl, op. cit. **3,** 5: 215 (1892). Type as for *Ilex crocea.*

Shrub or tree 1–12 m. high; branches angular and brownish-green when young, becoming vinous red with whitish lenticels and eventually terete and grey with prominent brownish lenticels. Leaves with lamina dark green or greyish-green above, usually paler below, 3·5–7 × 1·2–3·5 cm., elliptic or oblong to obovate, obtuse to rounded at the apex, with margin usually shallowly crenulate-denticulate, cuneate at the base, chartaceous to subcoriaceous, with densely reticulate venation visible below at first but becoming prominent on both sides; petiole 2–7 mm. long. Cymes with peduncle 4–10 mm. long; flowers 3–7 in each dichasium, 5–7 mm. in diam. Sepals 4, c. 1 mm. long, broadly ovate to subcircular. Petals 4, green, 2·5–3·5 mm. long, obovate, sometimes broadly unguiculate, with or without shallow longitudinal ridges and short fringed appendage at the base, entire.

Stamens 4. Ovary 4-locular, semiglobose, c. ½-immersed in disk. Drupe white, up to c. 25 mm. long, globose, smooth.

Mozambique. LM: Namaacha, Fonte de Goba, fl. 18.ix.1963, *Carvalho* 654 (K).
Also in S. Africa (Cape Prov.) near Port Elizabeth. Evergreen forest and near rivers.

The specimen from our area appears to belong to the same species as those from Cape Province, despite the large disjunction involved.

2. **Crocoxylon transvaalense** (Burtt Davy) N. Robson in Bol. Soc. Bot., Sér. 2, **39**: 41 (1965). TAB. **83.** Type from the Transvaal.
Hippocratea seineri Loes. ex Seiner in Engl., Bot. Jahrb. **46**: 44 (1911) *nom. nud.*
Salacia (?) *transvaalensis* Burtt Davy in Kew Bull. **1921**: 51 (1921); F.P.F.T. **2**: 451 (1932). Type as for *Crocoxylon transvaalense.*
Elaeodendron croceum var. *triandrum* Dinter in Fedde, Repert. **17**: 189 (1921) *nom. illegit.* Syntypes from SW. Africa.
Elaeodendron croceum var. *heterophyllum* Loes. in Notizbl. Bot. Gart. Berl. **12**: 35 (1934). Syntypes from SW. Africa.
Pseudocassine transvaalensis (Burtt Davy) Bredell in S. Afr. Journ. Sci. **33**: 330, tt. (1937).—O.B. Mill., B.C.L.: 35 (1948).—Pardy in Rhod. Agr. Journ. **53**: 631, photo (1956). Type as for *Crocoxylon transvaalense.*
Cassine sp.—O.B. Mill. in Journ. S. Afr. Bot. **18**: 48 (1952) pro parte quoad specim. *van Son* H28807 et *Miller* B/225.

Shrub or tree (1·5)2·5–13·5 m. high; branches angular or striate and grey-green when young, becoming vinous-red with whitish lenticels and eventually grey and terete. Leaves on young shoots with lamina up to 12·5 cm. long, linear to narrowly oblong-elliptic, with margin irregularly spreading-spinulose-dentate or -denticulate and reticulate venation prominent on both sides; leaves on mature shoots and short shoots apple-green or greyish green, usually concolorous, 2–7 × 0·8–3·1 cm., oblong to elliptic or obovate, obtuse to rounded or apiculate at the apex, with margin shallowly crenulate-denticulate to entire, cuneate to rounded at the base, coriaceous, with densely reticulate venation more prominent below than above; petiole 2·5–5 mm. long. Cymes subcorymbose, with peduncle 5–20 mm. long; flowers c. 20–30 in each dichasium, 4·5–6 mm. in diam. Sepals 3, c. 1 mm. long, broadly ovate to subcircular. Petals 3, green to cream, 2–2·5 mm. long, obovate, sometimes broadly unguiculate, with or without shallow longitudinal ridges and short fringed appendage at the base, entire. Stamens 3. Ovary 3-locular, semiglobose, c. ⅓–½-immersed in disk. Drupe yellow, drying dark red, 12–16(21) mm. long, globose to obovoid or rarely ellipsoid, smooth.

Bechuanaland Prot. N: Sherobe, fr. 10.vi.1930, *van Son* in Herb. Transv. Mus. 28807 (BM; K; PRE). SE: Serowe, Metsimesau, 1000 m., fr. x.1940, *Miller* B/224 (PRE). **N. Rhodesia.** S: Livingstone, bank of Zambezi R. near end of King's Mile, fr. 31.viii.1947, *Brenan* 7790 (FHO; K). **S. Rhodesia.** W: Bulawayo, Commonage, fl. i.1947, *Hodgson* 19/47 (K; SRGH). C: Hartley, Poole Farm, fl. 29.ii.1956, *Hornby* 3382 (K; SRGH). E: Umtali, Commonage, fl. & fr. 6.ii.1949, *Chase* 1217 (BM; COI; K; LISC; SRGH). S: Nuanetsi, Mwambe R., Ruware, 630 m., fl. xii.1955, *Davies* 1780 (K; SRGH). **Mozambique.** LM: Maputo, fl. 28.i.1948, *Gomes e Sousa* 3658 (COI; K; PRE; SRGH).
Also in Natal, Transvaal, SW. Africa and Angola (Huila). Open woodland, thickets and scrub, on sandy soil or termite heaps, up to 1400 m.

10. SALACIA L.

Salacia L., Mant. Pl. Alt.: 159, 293 (1771).

Lianes or shrubs or small trees, often with the upper branches scandent, rarely rhizomatous shrublets, with or without latex. Leaves opposite to alternate, petiolate; stipules small, free, deciduous, or absent. Inflorescence pedunculate or sessile, cymose, dichasial or monochasial or fasciculate, simple, without accessory branches, axillary or extra-axillary; bracts persistent. Flowers bisexual or functionally unisexual. Sepals (4)5(6), imbricate, free or ± united, the inner ones sometimes ± petaloid. Petals (4)5(6), green, white, yellow, orange, brownish or red, imbricate in bud. Disk extrastaminal, single, varying in shape, free or variously united with other floral parts. Stamens (2)3(4), with filaments often ± broadened below, free or united with the ovary or gynophore, becoming reflexed; anthers versatile, extrorse or apical, with thecae vertical to oblique or transverse,

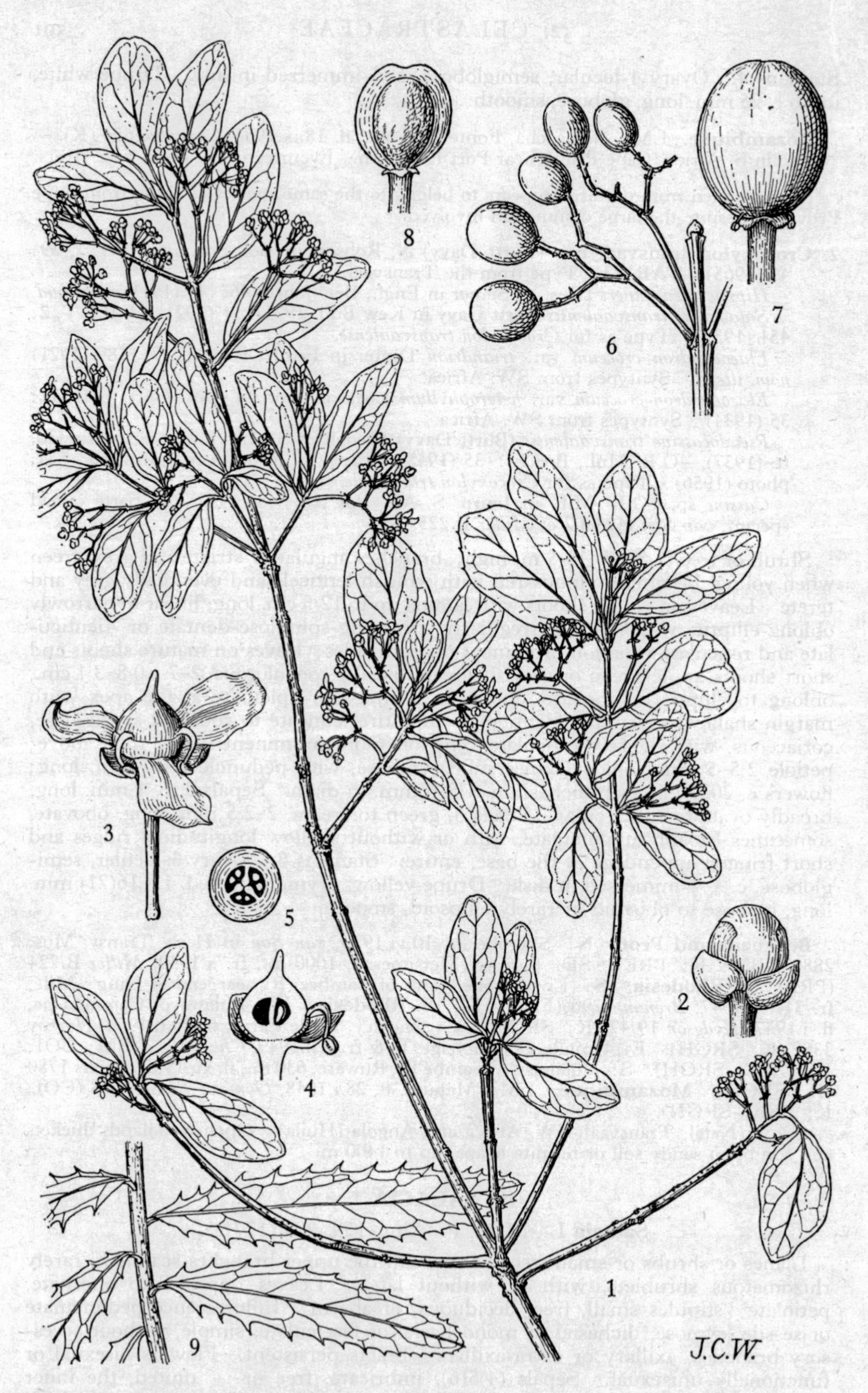

Tab. 83. CROCOXYLON TRANSVAALENSE. 1, flowering shoot (× ⅔); 2, flower bud (× 6); 3, flower (× 6); 4, vertical section of flower with petals removed (× 6); 5, transverse section of ovary (× 6), all from *Chase* 5958; 6, infructescence (× ⅔); 7, fruit (× 2); 8, fruit showing endocarp (× 1), all from *Chase* 1217; 9, shoot with juvenile leaves (× ⅔) *Chase* 1940.

free or ± confluent; pollen simple. Ovary superior, (2)3-locular, with 2–8 axile ovules in 1–2 rows per loculus, sessile or on a gynophore, usually immersed in the disk; style simple; stigma 3-lobed or entire. Fruit drupaceous, globose or ± elongated, with coriaceous exocarp and mucilaginous pulp (mesocarp), (2)3-locular, soon becoming 1-locular by evanescence of dissepiments, 1-c. 20-seeded. Seeds (pyrenes) ± irregular. Germination hypogeal (2 species). Chromosome number: 2n = 28 (8 species).

A genus of c. 150 species occurring throughout the tropics and subtropics of both hemispheres.

Stems, leaves and inflorescence without latex threads:
  Pedicels articulated above the base or, if at the base, then inflorescence pubescent; inflorescence branched:
    Stem, inflorescence branches and sepals glabrous; primary peduncle present; flowers bisexual, 12–15 mm. in diam. - - - - - - - 1. *bussei*
    Stem, inflorescence branches and sepals pubescent; primary peduncle absent; flowers functionally unisexual, 5–7 mm. in diam. - - - 2. *rhodesiaca*
  Pedicels articulated at the base; plant glabrous; inflorescence fasciculate, sessile or very shortly pedunculate:
    Plant a shrub, small tree or liane, 1–7 m. high or higher; petiole 3–8 mm. long or rarely shorter, but then with undulating margins:
      Leaf-lamina oblong to elliptic-oblong or obovate; petiole with markedly undulating margins; ovules (2)4–6 per loculus; young fruit ovoid-3-gonous or 3-crested 3. *erecta*
      Leaf-lamina lanceolate to oblong-lanceolate or ovate; petiole with straight or slightly undulating margins; ovules 2 per loculus; young fruit globose 4. *leptoclada*
    Plant a shrublet 30–60 cm. high; petiole 1–2 mm. long, with straight margins 5. *luebbertii*

Stems, leaves and inflorescence containing latex threads:
  Petals 2–7·5 mm. long; flower-buds 2–5 mm. long; flowers bisexual; disk convex, conic to cylindric; fruit usually over 2 cm. long:
    Flower-buds ovoid-conic to cylindric or narrowly obovoid, longer than broad:
      Stems smooth or becoming finely lenticellate-rugulose; buds ovoid to shortly conic:
        Stems flattened and without raised lines when young, becoming terete, purplish and eventually finely lenticellate-rugulose; disk conic-cylindric or shallowly cylindric with free spreading margin - - - - 6. *congolensis*
        Stems with paired raised lines when young, becoming quadrangular, yellow-brown, not lenticellate-rugulose; disk shallowly conic - 7. *orientalis*
      Stems becoming verruculose; flower-buds elongate-conic to narrowly obovoid:
        Inflorescence pedunculate, with primary peduncle up to 10 mm. long; flower-buds cylindric to narrowly obovoid, 3–5 mm. long; petals with markedly cucullate apex; fruit globose to ellipsoid, c. 12–20 seeded, sometimes with carpophore - - - - - - - - - - 8. *chlorantha*
        Inflorescence sessile or with peduncle up to 1 mm. long; flower-buds elongate-conic to cylindric, 2–3 mm. long; petals flat or slightly cucullate; fruit globose, c. 6-seeded, without carpophore - - - 9. *madagascariensis*
    Flower-buds globose or subglobose, broader than long;
      Leaves opposite or subopposite, except on climbing shoots; plant a shrub, small tree or liane; petals c. 2·5 mm. long. - - - - 10. *stuhlmanniana*
      Leaves all alternate; plant a shrub (rarely exceeding 1 m.) or rhizomatous shrublet; petals 3–4·5 mm. long - - - - - - - 11. *kraussii*
  Petals 1–2 mm. long; flower-buds 1–1·5 mm. long; flowers functionally unisexual; disk annular, flat or somewhat concave; fruit 1–2(2·5) cm. long:
    Stems not lined when young, becoming terete, yellowish-brown; sepal margin entire or subentire; inflorescence sessile, fasciculate, or with peduncle up to 1 mm. long 12. *pynaertii*
    Stems with paired raised lines when young, becoming angular or terete, deep reddish-purple; sepal margin ciliolate or denticulate (rarely subentire); inflorescence with primary peduncle 1–5 mm. long bearing fascicles or pseudumbels with 2–4 condensed monochasial branches - - - - - - 13. *elegans*

1. **Salacia bussei** Loes. in Engl., Bot. Jahrb. **44**: 176 (1910).—Brenan, T.T.C.L.: 248 (1949).—Wilczek, F.C.B. **9**: 201 (1960).—White, F.F.N.R.: 220 (1962). TAB. **84** fig. B. Syntypes from Tanganyika (Ungoni).

Salacia rehmanni var. *baumii* Loes. in Warb., Kunene-Samb.-Exped. Baum: 292 (1903). Type from Angola (Cubango).

Salacia baumii (Loes.) Exell & Mendonça, C.F.A. **2**, 1: 24 (1956). Type as above.

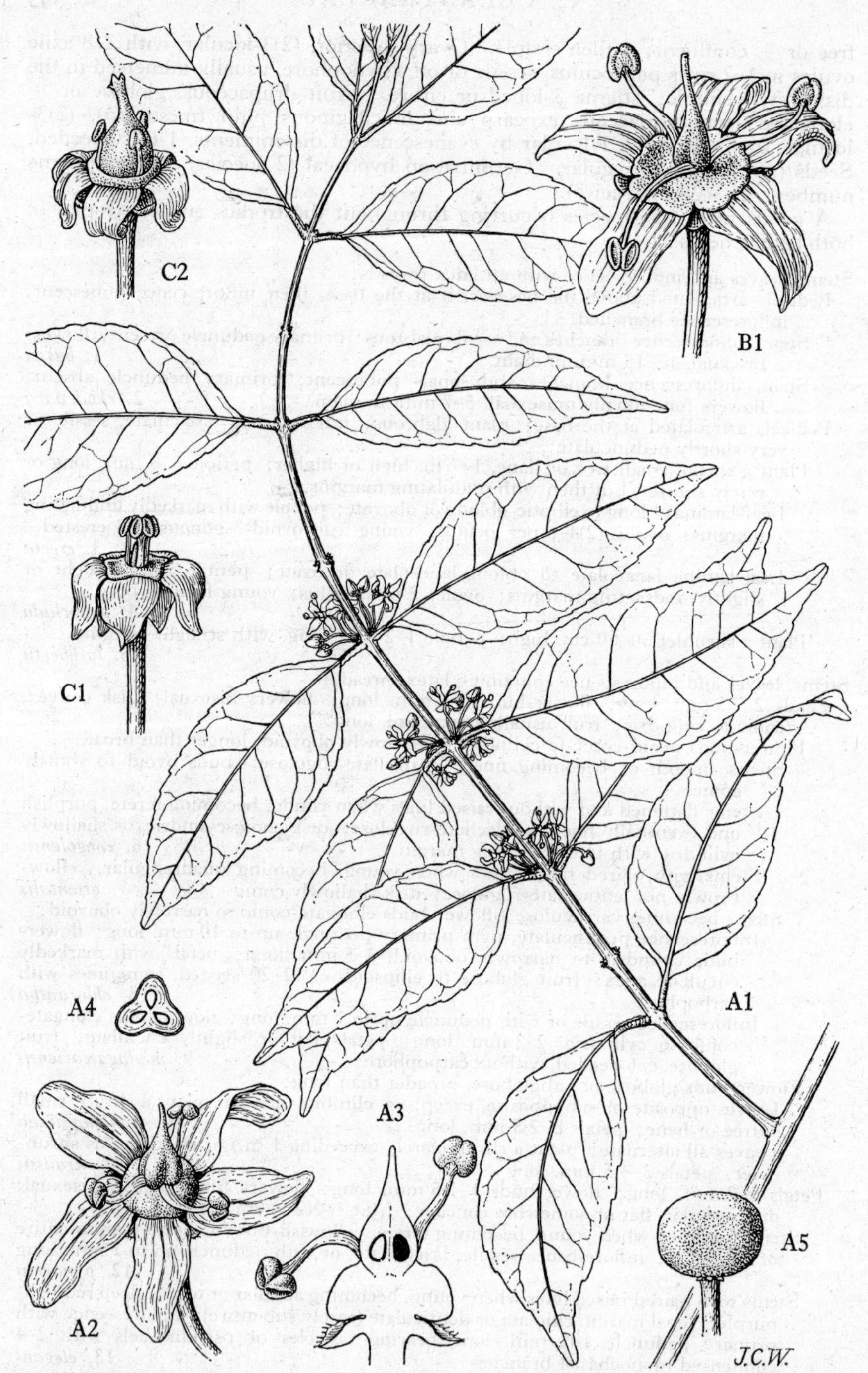

Tab. 84. A.—SALACIA LEPTOCLADA. A1, flowering shoot (×⅔); A2, flower (×6); A3, vertical section of flower (×8); A4, transverse section of ovary (×12), all from *Swynnerton* 1170; A5, fruit (×⅔) *Goldsmith* 80/62. B.—SALACIA BUSSEI. B1, flower (×6) *Fanshawe* 3673. C.—SALACIA ELEGANS. C1, ♂ flower (×8) *Angus* 519; C2, ♀ flower (×8) *Faulkner* 1860.

Shrubs, much branched, or shrublets with erect annual shoots from a woody ± creeping rootstock, 0·2–1 m. high (or higher and arborescent according to some collectors), sometimes forming colonies, without latex, glabrous; stems subterete or ± flattened, almost smooth or striate, green at first, becoming purple-brown. Leaves alternate or the upper ones subopposite or opposite, petiolate; lamina bright green to olive-green, concolorous or rather paler green below, shining above or with both surfaces dull, (4·5)5·2–9·5(11) × (1·6)2–5·1(6·5) cm., elliptic or oblong-elliptic to obovate or oblanceolate, acute to obtuse or rounded and shortly apiculate at the apex, with margin entire or ± deeply curved-dentate, cuneate or decurrent at the base, papyraceous to coriaceous, with (6)7–8(9) lateral nerves and densely reticulate venation more prominent below than above; petiole 1–4 mm. long, with entire margins; stipules absent. Flowers (1–2)3–15 in axillary dichasial or subumbellate cymes, bisexual, 12–15 mm. in diam.; buds 3–4 mm. long, ovoid-conic; peduncle 1–9(15) mm. long; pedicels 5–30 mm. long, articulated in the lower ½; bracts 0·5–1·5 mm. long, oblong to triangular, with margin entire to red-brown-scarious and denticulate, persistent. Sepals pale green with narrow whitish scarious entire or denticulate margin, 1–1·5 mm. long, unequal or subequal, ovate to oblong or semicircular, rounded, free. Petals pale green to yellowish, 4–7 mm. long, oblong to triangular, rounded, entire, united with the disk at the base. Disk dark to pale or yellowish green, shortly cylindric, round the bases of stamens and ovary, with margin broad and flat or concave, shallowly and ± irregularly 5-lobed. Stamens 3, with filaments up to 3 mm. long, slender, ± flattened; anthers pale to orange-yellow, dehiscing by 2 oblique or almost vertical clefts not confluent at the apex. Ovary ovoid-3-gonous or 3-lobed, with style narrow, elongated; stigma 3-lobed; ovules 2 per loculus. Fruit orange, 3·5–4 × 1·8–3·2 cm., subglobose to pyriform or irregular, apiculate, sometimes ridged, smooth, 1–3-seeded.

N. Rhodesia. B: Katuba valley, 1065 m., fl. 25.x.1953, *Gilges* 65 (K; PRE; SRGH). N: Abercorn, fl. 22.ix.1949, *Bullock* 1028 (K: LISC; SRGH).

In the Congo, southern Tanganyika, N. Rhodesia and Angola. In dry deciduous woodland or scrub, c. 700–1600m.

S. bussei is related to *S. nitida* (Benth.) N. E. Br. from the Guinea-Congo rainforest, but differs from it in habit and in its larger flowers with ± spreading disk.

S. bussei var. *grandifolia* Wilczek, from Bas Katanga, differs from the rest of *S. bussei* in having larger papyraceous leaves: it provides a morphological and geographical link with *S. nitida.*

2. **Salacia rhodesiaca** Blakelock in Kew Bull. **11**, 2: 244 (1956).—Wilczek, F.C.B. **9**: 192 (1960).—White, F.F.N.R.: 220 (1962). Type: N. Rhodesia, Abercorn, *Bullock* 2743 (EA; K, holotype; SRGH).

Shrub or small tree, (1·2)3–5(6) m. high, with rounded crown, without latex; bark smooth; stems ± flattened with paired raised lines at first, becoming terete and rugulose, densely chocolate-brown- to fawn-pubescent. Leaves opposite to subopposite or alternate, petiolate; lamina deep dull green above (drying grey-green), markedly paler below, (3·5)5–10·5(13·2) × (1·5)2·1–4·8(6·1) cm., oblong to elliptic or obovate, obtuse to acute or acuminate at the apex, with margin entire or shallowly crenulate-serrulate and slightly reflexed, cuneate to rounded or rarely shallowly cordate at the base, subcoriaceous, with (6)10–12(14) lateral nerves and densely reticulate venation more prominent below than above, sparsely pubescent towards the base below and on both sides of the midrib; petiole 2·5–6 mm. long, with entire margins, brownish-pubescent; stipules absent. Flowers 2–16 in fasciculate axillary subdichasial or monochasial cymes, functionally dioecious, 5–7 mm. in diam., sweetly scented; buds c. 2 mm. long, ovoid to subglobose; inflorescence branches densely red-brown-tomentose; primary peduncle absent, secondary peduncles fasciculate; pedicels 2–6 mm. long, articulated at the base or in the lower ½; bracts 0·5 mm. long, triangular, entire, red-brown, sparsely pubescent, persistent. Sepals brown, red-brown-tomentose outside, puberulous or glabrous within, c. 1 mm. long, subequal, ovate to oblong, obtuse or subacute, entire, free. Petals yellow or greenish-yellow to brownish, 1·5–2 mm. long, ovate to oblong, rounded, entire or with margin ciliolate, sometimes longitudinally ribbed, sparsely pubescent on both surfaces, free. Disk green, thick, convex,

5-angled, sometimes ± fluted, glabrous or puberulous. Stamens 3; in ♂ flowers with filaments slender, equalling the style, and anthers orange, large, fertile, glabrous or ± sparsely pubescent, dehiscing by 2 oblique clefts confluent at the apex; in ♀ flowers with filaments usually shorter than the anthers and anthers small, sterile, otherwise as in ♂ flowers. Ovary conic, glabrous or sparsely pubescent, immersed in the disk, with style narrow, elongated in both forms of flower; stigma slightly 3-lobed; ovules 2 per loculus. Fruit red to red-brown, sometimes glaucous, c. 1·5–2·3 × 1·5–2·7 cm., globose or dorsiventrally somewhat flattened, rugulose, 2–4 seeded. Seeds cylindric.

N. Rhodesia. N: midway between Fort Rosebery and Samfya Mission, fr. 16.viii.1952, *White* 3072 (FHO; K; PRE). W: Mwinilunga Distr., 0·8 km. S. of Matonchi Farm, ♀ fl. 7.ii.1938, *Milne-Redhead* 4462 (K; LISC; PRE; SRGH). C: 62 km. S. of Chitambo, ♂ fl. 7.vi.1931, *Stevenson* 318/31 (FHO).

Also in the Congo (Haut Katanga). In *Brachystegia* woodland or mixed dry deciduous woodland, 1250–1800 m.

S. rhodesiaca is very closely allied to *S. howesii* Hutch. & Moss from W. Africa, which differs *inter alia* in its longer inflorescences, less dense indumentum and usually larger leaves. Both species have dimorphic flowers of which the long-stamened (?♂) form is apparently much the rarer.

3. **Salacia erecta** (G. Don) Walp., Repert. **1**: 402 (1842).—Oliv., F.T.A. **1**: 377 (1868). —Loes. in Engl., Bot. Jahrb. **19**: 243 (1895); op. cit. **44**: 168 (1910).—R. E. Fr., Wiss. Ergebn. Schwed. Rhod.-Kongo-Exped. **1**: 129 (1914).—Exell & Mendonça, C.F.A. **2**, 1: 26 (1956) pro parte excl. specim. *Gossweiler* 4371, 13698, 14113.—Keay & Blakelock, F.W.T.A. ed. 2, **1**, 2: 633 (1958) pro parte excl. syn. *S. cornifolia*, *S. baumannii*, *S. elliotii*, (?) *S. prinoides* var. *liberica*.—Wilczek, F.C.B. **9**: 223 (1960). —Hallé in Mém. I.F.A.N. **64**: 196, t. 67 (1962). Type from Sierra Leone.
Calypso erecta G. Don, Gard. Dict. **1**: 629 (1831). Type as above.
Salacia alpestris A. Chev. [Expl. Bot. **1**: 133 (1920) *nom. nud.*] ex Hutch. & Dalz., F.W.T.A. **1**: 453 (1928). Type from Guinée.
Salacia sp. 1.—White, F.F.N.R.: 220 (1962).

Shrub, often with climbing branches, or liane, (1)2–7(15) m. high, sometimes forming thickets, without latex, glabrous; stems with paired raised lines or quadrangular, olive-green to purplish and rugulose-tuberculate at first, becoming subterete, purplish-grey and smooth or remaining rugulose. Leaves opposite to subopposite or alternate (on climbing shoots), petiolate; lamina dark green, paler below, glossy or more rarely rather dull on both surfaces, (3·3)4·4–10·8(15) × (1·2)2·1–4·5 cm., oblong or elliptic-oblong to obovate, acuminate at the apex with acumen long to short, obtuse or retuse, with margin shallowly rounded-denticulate, rarely subentire, cuneate to rounded at the base, chartaceous to softly coriaceous, with (6)7–10 lateral nerves and ± densely reticulate venation varying in prominence; petiole (2)3–4(6) mm. long, with margins markedly undulating; stipules small, asymmetric, ± ciliate-denticulate, caducous. Flowers 1–5, axillary, in sessile or very shortly pedunculate fascicles, bisexual, 4–7(9) mm. in diam.; buds 2–3 mm. long, ovoid to ovoid-cylindric; pedicels 3–5(7) mm. long, smooth or slightly rugulose, articulated at the base. Sepals green with pale subentire or irregularly denticulate margin, 0·4–1 mm. long, unequal, ovate to semicircular, rounded, free or shortly united. Petals greenish-yellow, 2–3(4) mm. long, oblong, rounded, entire. Disk brown, convex or convex-cylindric, without depressions, surrounding the ovary and wider than it. Stamens 3, with filaments slender; anthers dehiscing by 2 oblique clefts confluent at the apex. Ovary ovoid, 3-gonous or 3-lobed, with style narrowly pyramidal; stigma entire; ovules (2)4–6 per loculus. Fruit orange, ovoid-3-gonous and acute when young, becoming 3-crested and eventually globose, 1·3–3 cm. in diam., smooth or with a few tubercles, 1–3-seeded.

N. Rhodesia. N: Fort Rosebery Distr., Lake Bangweulu, near Samfya Mission, fr. 21.viii.1952, *White* 3112 (BM; FHO; PRE).

From Guinée and Sierra Leone to Uganda, Kenya, Tanganyika, the Congo, N. Rhodesia and Angola. In evergreen forest or thickets, 50–1700 m. outside our area.

S. erecta can be distinguished from its nearest relatives by the leaves with oblong to obovate acuminate lamina and petiole with markedly undulating margins and the pointed and 3-crested young fruits.

4. **Salacia leptoclada** Tul. in Ann. Sci. Nat., Sér. 4, Bot., **8**: 96 (1857).—Perrier, Fl. Madag., Hippocrat.: 9 (1946) pro parte quoad specim. Comorens. pro parte et Madag. et t. 1, fig. 1–2. TAB. **84** fig. A. Type from the Comoro Is. (Mayotte).

Salacia baumannii Loes. in Engl., Bot. Jahrb. **44**: 180 (1910).—Hallé in Mém. I.F.A.N. **64**: 194, t. 66 (1962). Type from Togo.

Salacia erecta sensu Keay & Blakelock in Keay, F.W.T.A. ed. 2, **1**, 2: 633 (1958) pro parte quoad syn. *S. baumannii*.

Salacia wardii Verdoorn in Bothalia, **8**: 114 (1962). Type from Natal.

Shrub or small tree or liane, 1–4(25) m. high, without latex, glabrous; stems with paired raised lines or rarely subterete, olive-green and smooth or slightly rugulose at first, becoming terete, purplish to grey-brown or brown and smooth or with ± prominent lenticels. Leaves opposite or alternate (on climbing shoots), petiolate; lamina dark green, paler below, glossy on both surfaces, 5–10·2(12·3) × (1·7)2·2–4(4·8) cm., lanceolate or oblong-lanceolate to ovate, acute or acutely to subobtusely acuminate at the apex, with margin entire or shallowly rounded-denticulate, cuneate at the base, chartaceous to softly coriaceous, with (7)9–11 lateral nerves and densely reticulate venation varying in prominence, more prominent below; petiole (3)4–6(8) mm. long, with margins straight or slightly undulating; stipules absent. Flowers 1–5(8), axillary, in sessile fascicles, bisexual, 5–7 mm. in diam.; buds 1·5–2·5(3) mm. long, broadly ovoid; pedicels 3–10 mm. long, smooth, articulated at the base. Sepals pale green with paler or reddish denticulate or ciliolate margin, c. 1 mm. long, subequal, semicircular, rounded, free. Petals yellow or greenish-yellow, 2–3 mm. long, oblong to elliptic or ovate and slightly unguiculate, rounded, entire or with margin partly ciliolate or denticulate. Disk greenish-yellow, cylindric or convex-cylindric, usually without depressions, surrounding or enveloping the ovary. Stamens 3, with filaments slender; anthers dehiscing by 2 oblique clefts confluent at the apex. Ovary subglobose or 3-lobed, with style narrowly pyramidal; stigma entire; ovules 2 per loculus. Fruit orange to orange-red, globose even when immature, 1·2–2(2·9) cm. in diam., smooth, 1(2–3)-seeded.

S. Rhodesia. E: Melsetter Distr., Nyamgamba R. valley, c. 610 m., fr. iii.1962, *Goldsmith* 80/62 (BM; SRGH). **Mozambique.** N: Nacala, near Fernão Veloso, fl. 27.x.1952, *Balsinhas* 5294 (BM; LISC). MS: Mossurize, Chicamboge Valley, 915 m., fl. 17.x.1906, *Swynnerton* 1170 (BM; K; SRGH). SS: between Chibuto and Gomes da Costa, fl. 14.xi.1957, *Barbosa & Lemos* in *Barbosa* 8114 (COI; LISC; LMJ; SRGH).

In E. Africa from Kenya to Natal, the Comoro Is. and Madagascar, also in Ivory Coast, Ghana and Togo. Evergreen or deciduous forest, 200–915 m.

S. leptoclada has been confused with *S. erecta*, but can be distinguished from it by the lanceolate to ovate acute leaves with petiole margins straight or almost so, the constantly 2-ovulate ovary loculi and the fruit which is round even when young. The distributions of these two species overlap in eastern Tanganyika, where some specimens with somewhat intermediate characters occur. The W. African specimens named *S. baumannii* Loes. appear to be conspecific with *S. leptoclada*.

5. **Salacia luebbertii** Loes. in Engl., Bot. Jahrb. **44**: 194 (1910). Type from SW. Africa.

Shrublet 30–60 cm. high, without latex, glabrous; stem with paired raised lines, olive-green and smooth at first, becoming quadrangular and reddish-purple. Leaves subopposite or alternate, petiolate; lamina deep green and glossy above, paler and duller below, 3–5·5 × (1)1·7–2·8 cm., lanceolate to oblong or elliptic, rounded at the apex, with margin entire, cuneate at the base, ± softly coriaceous, with c. 8–11 lateral nerves and densely reticulate venation slightly prominent below but not or scarcely so above; petiole 1–2 mm. long, with margins straight; stipules very small, eroded or fimbriate-ciliolate, persistent. Flowers solitary, axillary, bisexual, 6–8 mm. in diam.; buds c. 2 mm. long, depressed-globose; pedicels 5–8 mm. long, smooth, articulated at the base. Sepals green with pale or reddish ciliolate margin, c. 1 mm. long, unequal, semicircular, rounded, free. Petals yellowish-green, 2·5–3 mm. long, shortly unguiculate, with lamina broadly elliptic to circular, entire or with minutely denticulate margin. Disk yellowish-green, convex, with 5-lobed margin, surrounding the base of the ovary. Stamens 3, with filaments broad, flattened; anthers dehiscing by 2 oblique clefts confluent at the apex. Ovary ovoid-3-gonous, with style shortly pyramidal; stigma entire;

ovules 2 per loculus. Fruit pale orange-yellow to red when ripe, globose, 2·6–3 cm. in diam., smooth, 1(2–3)-seeded. Seeds flattened.

N. Rhodesia. S: Namwala, fl. & fr. 19.x.1959, *Fanshawe* 5246 (FHO; K).
Also in SW. Africa. In mixed woodland and thickets on Kalahari Sand.

S. luebbertii occurs in SW. Africa very near to Angola and the Bechuanaland Protectorate and is therefore very likely present in both these countries.

6. **Salacia congolensis** De Wild. & Dur. in Ann. Mus. Cong. Bot., Sér. 1, **1**: 85, t. 43 (June 1899); op. cit., Sér. 2, **1**: 16 (July 1899); Pl. Thonn.: 23, t. 20 (1900).—Wilczek, F.C.B. **9**: 196 (1960). Type from the Congo.
Salacia senegalensis sensu Oliv., F.T.A. **1**: 374 (1868) pro parte quoad specim. Smith.
Salacia demeusei sensu Exell & Mendonça, C.F.A. **2,** 1: 24 (1956) pro parte quoad specim. Smith.
Salacia pyriformis sensu Wilczek, tom. cit.: 198, t. 24 (1960) pro parte.—White, F.F.N.R.: 219 (1962) pro parte excl. specim. *White* 3344.

Shrub or liane, 1 m. high or higher, with latex, glabrous; stems flattened, sometimes 4-lined, glaucous-green and smooth at first, becoming blackish-purple, finely lenticullate-rugulose. Leaves opposite or subopposite (alternate on climbing shoots), petiolate; lamina dark green and ± glossy above, paler and dull below, (5)7·5–15(18) × (2)3·5–6·2(8) cm., oblong or elliptic-oblong to obovate, ± shortly acuminate at the apex, with margin shallowly rounded-denticulate, cuneate to rounded at the base, coriaceous, with c. 8–11 lateral nerves and densely reticulate venation more prominent below than above; petiole 5–10(13) mm. long, with margins straight; stipules very small, caducous. Flowers 2–9(18), bisexual, in axillary sessile or shortly pedunculate fascicles; buds 2–3 mm. long, ovoid-conic to subcylindric; peduncle absent or up to 2 mm. long; pedicels 4–8 mm. long, smooth, articulated at the base. Sepals yellow-green ? with pale margin, unequal, broadly ovate to subcircular, free, the outer ones c. 1 mm. long, subacute, with ciliolate-denticulate margin, the inner ones subpetaloid, c. 2 mm. long, rounded, with entire margin. Petals orange to yellow, 2–3·5 mm. long, ovate to elliptic-oblong, sessile or shortly unguiculate, rounded, entire or subentire. Disk cylindric or conic-cylindric, with top ± thickened and margin free and sometimes ± undulate or concave, surrounding the base of the ovary. Stamens 3, with filaments broadened at the base; anthers dehiscing by 2 oblique clefts sometimes confluent at the apex. Ovary ovoid-globose, with style c. 2 mm. long, slender; stigma entire, punctiform; ovules 2 per loculus. Fruit orange to reddish, globose, c. 2–3 cm. in diam., smooth to rugulose or somewhat muricate or with wart-like emergences, 2–3 seeded.

N. Rhodesia. N: Mporokoso Distr., Lake Mweru at Kafulwe, fl. 2.xi.1952, *Angus* 692 (BM; FHO; K).
Also in Angola (Zaire), the Congo, Central African Republic, Sudan and Ethiopia. Evergreen forest and fringing forest.

S. congolensis is closely related to *S. pyriformis* (Sabine) Steud., a forest species of the Guinea-Congo region which differs in having entire leaves, smooth stems, thicker pedicels, flower-buds globose or broader than long, a disk with a free lower margin and not usually contracted in the middle and smooth fruit with a triradiate crest.

7. **Salacia orientalis** N. Robson in Bol. Soc. Brot., Sér. 2, **39**: 44 (1965). Type: Mozambique, Mocímboa da Praia, *Andrada* 1348 (COI; LISC, holotype).
Salacia congolensis sensu Brenan, T.T.C.L.: 248 (1949) pro parte quoad specim. *Schlieben* 1300.

Shrub or liane, up to at least 3 m. high, with latex, glabrous; stems markedly quadrangular, pale brown to purplish-brown and smooth at first, becoming grey-brown to yellowish-brown and 4-lined or terete, without or with numerous small scarcely prominent lenticels. Leaves opposite or subopposite (alternate on climbing shoots), petiolate; lamina dull or somewhat glossy green and sometimes glaucous above, paler below, 6·7–10·4(12·6) × 2·1–4·2 cm., oblong to elliptic or obovate, obtuse to acuminate at the apex, with margin shallowly and ± remotely serrulate, cuneate at the base, subcoriaceous, with 7–10 lateral nerves and densely reticulate venation more prominent below than above; petiole 4–8 mm. long, with margins straight; stipules very small, caducous. Flowers 5–6(10), bisexual, in axillary sessile fascicles; buds c. 2 mm. long, ovoid-conic; pedicels 5–8 mm. long, smooth, articulated at the base. Sepals with pale margin, c. 0·8 mm. long, subequal or ± unequal, united at the base, broadly ovate, obtuse to rounded, with margin eroded-denticulate to ciliolate. Petals yellow to brownish-yellow with pale margin, 2·5–3 mm. long, oblong to elliptic-ovate, sessile, rounded, with eroded-denticulate margin, reflexed at anthesis. Disk shallowly conic with 5-angled free margin, surrounding the ovary. Stamens 3, with filaments scarcely broadened at the base; anthers dehiscing by 2 oblique clefts not confluent at the apex. Ovary ovoid-3-gonous, with style 2–3 mm. long, slender; stigma entire, punctiform; ovules 2 per loculus. Fruits (immature) 3-gonous or 3-crested, smooth, 2–3-seeded.

Mozambique. N: Mocímboa da Práia, road to Palma, fl. 15.ix.1948, *Andrada* 1348 (COI; LISC).

Also in SE. Tanganyika. *Brachystegia* or mixed damp woodland, to 700 m. in Tanganyika.

S. orientalis is closely related to *S. calypso* DC. from NE. Madagascar which differs in having entire leaves with a glossy upper surface and more numerous lateral nerves, entire sepals and petals, a deeper cylindric disk and usually purplish ± flattened young shoots with paired lines. It can be distinguished from *S. congolensis* by its pale quadrangular young shoots, shallower disk and relatively narrow leaves with shallowly serrulate margin.

8. **Salacia chlorantha** Oliv., F.T.A. **1**: 375 (1868). Type from Nigeria.
Salacia macrocarpa Welw. ex. Oliv., tom. cit.: 373 (1868) non Korthals (1848). Syntypes from Nigeria and Angola.

Subsp. **demeusei** (De Wild. & Dur.) Hallé [Thèses Fac. Sci. Univ. Par., Monogr. Hipp. Afr. Occ.: 222 (1958)] in Mém. I.F.A.N. **64**: 225 (1962) ("demeusii")*. Type from the Congo.
Salacia demeusei De Wild. & Dur. in Ann. Mus. Cong., Bot., Sér. 2, **1**: 11 (1900).—Exell & Mendonça, C.F.A. **2,** 1: 23 (1956). Type as for *Salacia chlorantha* subsp. *demeusei*.
Salacia macrocarpa var. *typica* Loes. in Engl., Bot. Jahrb. **44**: 166 (1910). Type from Angola.
Salacia senegalensis sensu Wilczek, F.C.B. **9**: 200 (1960).
Salacia sp. 2—White, F.F.N.R.: 220 (1962).

Shrub or liane up to 9 m. high or higher, with latex, glabrous; stems with paired lines, green, smooth and ± flattened at first, becoming terete or subangular, purple or reddish-purple and verruculose with numerous prominent lenticels. Leaves opposite or subopposite (alternate on climbing shoots), petiolate; lamina ± glossy above, paler below, (5)6·5–16·5 × 2·5–7 cm., oblong or elliptic-oblong to obovate, acuminate to caudate at the apex, with margin ± densely shallowly rounded-denticulate, cuneate to rounded at the base, chartaceous to subcoriaceous, with 8–10(12) lateral nerves and densely reticulate venation equally prominent on both sides; petiole 4–8(10) mm. long, with margins straight; stipules small, broad, irregularly denticulate, deciduous. Flowers ∞, bisexual, in axillary pedunculate fascicles; buds 3–5 mm. long, cylindric to narrowly obovoid; primary peduncle 1–10 mm. long, lenticellate-verruculose, sometimes dividing to form 2–4 secondary peduncles; pedicels 10–15 mm. long, smooth, articulated at the base. Sepals c. 1 mm. long, subequal, free, triangular-ovate, obtuse, with margin entire or rarely ciliolate-denticulate. Petals white or greenish-white, 3–5(7·5) mm. long, oblong, sessile, rounded, entire or subentire, spreading at anthesis. Disk deeply subcylindric, broadened at the base and usually at the apex, fluted, with 5-lobed lower and upper margins, partially surrounding the ovary. Stamens 3, with filaments slender, not broadened at the base; anthers dehiscing by 2 vertical clefts not confluent at the apex. Ovary narrowly ovoid-3-gonous, on a short gynophore, with style 1–1·5(2) mm. long, conic, 3-lobed in section; stigma entire, punctiform; ovules 6–8 per loculus. Fruit orange, 3–8·5 cm. long, ellipsoid to globose with a 3-radiate apical crest, sometimes narrowed at the base to form a carpophore, smooth or with a few pointed tubercles, c. 12–20-seeded.

* Although Hallé's thesis was duplicated and some copies were distributed, it was not effectively published and has no nomenclatural status.

N. Rhodesia. W: Solwezi R. gorge near Solwezi, fr. 10.ix.1952, *Angus* 394 (BM; COI; FHO; K; PRE).

From Cameroun and the Congo to N. Rhodesia and Angola. Evergreen forests and fringing forests.

9. **Salacia madagascariensis** (Lam.) DC., Prodr. **1**: 570 (1824).—Perrier, Fl. Madag., Hippocrat.: 3 (1946) pro parte. Type from Madagascar.

Hippocratea madagascariensis Lam., Tabl. Encycl. Méth. Bot. **1**: 101 (1791). Type as above.

Hippocratea verticillata var. *madagascariensis* (Lam.) Pers., Syn. Pl. **1**: 41 (1805). Type as above.

Tonsella madagascariensis (Lam.) Vahl, Enum. Pl. **2**: 32 (1806). Type as above.

Hippocratea senegalensis var. *madagascariensis* (Lam.) Poir. in Lam., Encycl. Méth., Suppl. **1**: 607 (1810). Type as above.

Salacia simtata Loes. in Engl., Bot. Jahrb. **44**: 182 (1910).—Pflanzenw. Afr. **3**, 2: 244, t. 120 (1921).—Chiov., Fl. Somal. **2**: 135, t. 90 (non t. 89) (1932).—Brenan, T.T.C.L.: 248 (1949). Syntypes from Tanganyika.

Shrub or liane, up to at least 6 m. high, with latex, glabrous; stems ± quadrangular, green and smooth at first, becoming terete, purplish and eventually grey, verruculose with numerous prominent lenticels. Leaves opposite or subopposite (alternate on climbing shoots), petiolate; lamina glossy on both sides, paler below, 4–10·5 × 2·1–4·7(5·8) cm., oblong to elliptic, acuminate at the apex, with margin densely serrulate to subentire, cuneate to decurrent at the base, chartaceous to subcoriaceous, with 7–10 lateral nerves and densely reticulate venation more prominent below than above; petiole 5–9 mm. long, with margins straight; stipules small, acicular-triangular, deciduous. Flowers 2–12 (rarely solitary), bisexual, in axillary sessile or shortly pedunculate fascicles; buds 2–3 mm. long, cylindric to elongate-conic; peduncles absent or up to 1 mm. long; pedicels 5–10 mm. long, smooth, articulated at the base. Sepals c. 0·5–0·8 mm. long, subequal, united at the base, broadly ovate to triangular-ovate, acute to obtuse or rounded, with margin eroded-denticulate to subentire. Petals greenish-yellow to deep yellow, (2·5)3–4 mm. long, oblong to ovate, sessile or very shortly unguiculate, rounded, entire, reflexed at anthesis. Disk cylindric or deeply conic, sometimes fluted, with 5-lobed lower margin, surrounding the ovary. Stamens 3, with filaments scarcely broadened at the base; anthers dehiscing by 2 vertical or oblique clefts not confluent at the apex. Ovary ovoid-3-gonous, with style 1–2 mm. long, slender; stigma entire, punctiform; ovules 2(3–4) per loculus. Fruit orange to pinkish, 1·8–3 cm. in diam., globose, finely rugulose, c. 6-seeded.

Mozambique. Z: Murroa to Namuera, 3·2 km., fl. 2.x.1949, *Barbosa & Carvalho* in *Barbosa* 4270 (K; LMJ). SS: Vilanculos, Mapinhana, Mabote road, fl. 1.ix.1944, *Mendonça* 1939 (LISC).

Coastal regions of E. Africa from S. Somalia to Mozambique and in Madagascar. Littoral scrub, coastal evergreen forest or deciduous woodland, often on sandy soils, 0–600 m. (in Tanganyika).

S. madagascariensis has been confused with *S. senegalensis* (Lam.) DC., a West African species with markedly discolorous leaves which turn reddish-brown beneath when dry, a disk usually with thickened upper margin (i.e. hour-glass-shaped rather than cylindric or conic) and free lower margin, an ovary with 3–5 ovules per loculus and smooth more elongated fruits. It is also related to two Madagascar plants, *S. dentata* Bak., which has more deeply dentate leaves, brownish stems and globose buds, and *S. madagascariensis* (Lam.) DC. forma *minimifolia* Perrier, which has smaller thicker entire markedly discolorous leaves and is probably specifically distinct.

10. **Salacia stuhlmanniana** Loes. in Engl., Bot. Jahrb. **19**: 241 (1894); op. cit. **44**: 179 (1910).—Brenan, T.T.C.L.: 248 (1949). Type from Zanzibar.

Salacia pyriformis var. *obtusa* Oliv., F.T.A. **1**: 375 (1868).—Loes., op. cit. **44**: 179 (1910). Type: Mozambique, Chupanga, R. Zambeze, *Kirk* (K).

Salacia pyriformis sensu Sim, For. Fl. Port. E. Afr.: 37 (1909).

Salacia livingstonii Loes., op. cit. **44**: 178 (1910). Type: Mozambique, " Zambezi region? ", *Kirk* (K).

Salacia somalensis Chiov., Fl. Somala, **2**: 137, t. 91 (1932). Type from S. Somalia.

Shrub or small tree or liane, with latex, glabrous; stems with faint paired lines, yellow-green, smooth and flattened at first, becoming terete, greyish-brown or reddish, with lenticels scarcely visible except on young shoots. Leaves opposite or subopposite (alternate on climbing shoots), petiolate; lamina glossy above, paler and dull below, 5·3–11(15) × 3–6·4(9) cm., broadly elliptic to oblong-elliptic or obovate, obtuse or shortly acuminate to rounded or retuse at the apex, with margin entire or very shallowly undulate-denticulate, cuneate at the base, coriaceous, with 6–8(10) lateral nerves and densely reticulate venation prominent below but not or less so above; petiole 6–12 mm. long, with margins straight; stipules c. 1 mm. long, narrowly triangular, subentire, caducous. Flowers c. 8–16, bisexual, in axillary sessile fascicles; buds c. 2 mm. long, subglobose; pedicels 4–8 mm. long, smooth, articulated at the base. Sepals unequal, outer 2 very small, inner 3 c. 1 mm. long, free, broadly ovate to semicircular, rounded or irregular, with margin thin, ciliolate-denticulate to subentire. Petals greenish-yellow, c. 2·5 mm. long, oblong to obovate, sessile, rounded or irregular, subentire or ± irregularly ciliolate-fimbriate, ± reflexed at anthesis. Disk conic, surrounding the ovary. Stamens 3(4), with filaments broadened at the base; anthers dehiscing by 2 widely oblique confluent clefts (almost forming a straight line). Ovary ovoid-3-gonous, with style c. 1 mm. long, conic; stigma entire, punctiform; ovules 2(3) per loculus. Fruit orange, 2–2·5 cm. in diam., globose, smooth to rugose or covered with wart-like emergences, ± markedly 3-crested.

Mozambique. N: mouth of R. Msalu, fl. ix.1911, *Allen* 59 (K). Z: Chamo, R. Chire, fl. viii.1859, *Kirk* (K). MS: Marromeu, near Lacerdónia, bank of R. Zambeze, fl. 10.ix.1944, *Mendonça* 2025 (LISC).

Coastal regions from S. Somalia to the Zambezi R. and also in the Usambaras, Ulugurus and Iringa Distr. (Tanganyika). Deciduous woodland or scrub, 0–280 m.

S. stuhlmanniana appears to be a derivative of *S. pyriformis* (Sabine) Steud., a plant of the Guinea-Congo region differing in size of various parts (e.g. flowers, leaves) and in length of pedicels. The inland Tanganyika plants are somewhat intermediate, so that *S. stuhlmanniana* may eventually prove to be only subspecifically distinct. Hallé (1962) includes the W. African *S. lomensis* Loes. in *S. stuhlmanniana*, but to me the former appears to be more closely related to *S. congolensis*.

11. **Salacia kraussii** (Harv.) Harv. in Harv. & Sond., F.C. **1**: 230 (1860).—Loes. in Engl., Bot. Jahrb. **44**: 177 (1910). Type from Natal.
Diplesthes kraussii Harv. in Hook., Lond. Journ. Bot. **1**: 19 (1842). Type as above.
Salacia alternifolia Hochst. in Flora, **27**: 306 (1844). Type as above.

Shrub up to 1(2–3) m. high (sometimes climbing?) or a shrublet with annual shoots from a rhizome with chrome yellow layer beneath the bark, with latex, glabrous; stems purple-red, ridged and flattened at first, becoming striate or angular and eventually terete, greyish, with numerous lenticels slightly prominent on older shoots. Leaves alternate, petiolate; lamina glossy above, paler and dull below, 3·7–9·1 × 1·1–5·5 cm., oblong to elliptic or obovate, acute to obtuse or rounded at the apex, with margin entire or undulate-denticulate towards the apex, cuneate at the base, coriaceous, with 5–9 lateral nerves and densely reticulate venation prominent below but scarcely so above; petiole 1–4 mm. long, stout, with margins entire; stipules c. 1 mm. long, acicular, denticulate, caducous. Flowers c. 2–16 (rarely solitary), bisexual, in axillary sessile or shortly pedunculate fascicles; buds 2·5–3 mm. long, subglobose; peduncle absent or up to 2 mm. long pedicels 4–10 mm. long, smooth or almost so, articulated at the base. Sepals unequal, outer 2 c. 1 mm. long, inner up to 2·5 mm. long, with innermost often subpetaloid, free, semicircular, rounded, with margin papyraceous, entire. Petals 5–7, yellow or greenish-yellow, 3–4·5 mm. long, obovate, sessile or shortly unguiculate, rounded, entire or eroded-denticulate, spreading at anthesis. Disk conic-cylindric, surrounding the ovary. Stamens 3, with filaments broadened at the base; anthers dehiscing by 2 oblique clefts confluent at the apex. Ovary ovoid-3-gonous, with style 1·5–2 mm. long, slender; stigma entire, small, punctiform; ovules 2 per loculus. Fruit orange to reddish-orange, 3·5–5 cm. in diam., globose, rugulose, 2–3-seeded.

Mozambique. SS: Inharrime, Ponta Zavora, fl. 16.x.1947, *Barbosa & Lemos* in *Barbosa* 8073 (COI; K; LISC; LMJ). LM: Maputo, Bela Vista, Santaca, fl. 19.ix.1947, *Gomes e Sousa* 3612 (COI; K; LISC).

Also in Natal. Dune scrub, thickets and open deciduous woodland near the coast, 0–90 m. (to 200 m. in Natal).

S. kraussii is very closely allied to *S. gerrardii* Harv., a forest climber from Natal which differs in habit and in its quadrangular young shoots and leaves which are opposite to subopposite (except in climbing shoots), acuminate and usually more markedly dentate.

12. **Salacia pynaertii** De Wild. in Ann. Mus. Cong., Bot., Sér. 5, **2**: 295 (1908).—T. & H. Dur., Syll. Fl. Cong.: 103 (1909) pro parte quoad specim. *Pynaert* 1705. Type from the Congo (Eala).

Salacia erecta sensu Exell & Mendonça, C.F.A. **2,** 1: 26 (1956) pro parte quoad specim. *Gossweiler* 13698, 14113.

Salacia elegans var. *pynaertii* (De Wild.) Wilczek, F.C.B. **9**: 196 (1960). Type as above.

Salacia pyriformis sensu White, F.F.N.R.: 219 (1962) pro parte quoad specim. *White* 3344.

Liane up to 6 m. high or higher, with latex, glabrous; stems flattened, not lined, pale green and smooth at first, becoming terete and yellowish-brown, smooth, with scarcely prominent lenticels. Leaves opposite to subopposite (alternate on climbing shoots), petiolate; lamina ± glossy above or on both surfaces, paler below, 6·4–11·8(13) × 2·9–5·9 cm., oblong or elliptic-oblong to obovate, ± shortly and usually obliquely acuminate at the apex, with margin finely denticulate to subentire, cuneate or decurrent at the base, subcoriaceous, with c. 6–8 lateral nerves and densely reticulate venation more prominent below than above; petiole 4–7 mm. long, with margins straight; stipules very small, caducous. Flowers c. 8–18, functionally unisexual, in axillary, sessile or shortly pedunculate fascicles, rarely also in terminal fascicles; buds c. 1·5 mm. long, ovoid; peduncles up to 1 mm. long (up to c. 5 mm. in Congo specimens); pedicels 3–5 mm. long, smooth, articulated at the base. Sepals with pale margin, c. 0·5 mm. long, semicircular, rounded, ± unequal, united at the base, with margin entire or subentire. Petals greenish-yellow, with pale margin, 1·5–2 mm. long, ovate to oblong, shortly unguiculate, rounded, entire. Disk annular, flat or rather concave, surrounding the base of the ovary. Stamens 3, with filaments slender, longer than the style (♂ forms) or much shorter (♀ forms); anthers apiculate and sterile in ♀ forms, dehiscing by 2 longitudinal clefts. Ovary ovoid-3-gonous, with style c. 0·5 mm. long; stigma entire, 3-lobed (♀ forms) or punctiform (♂ forms); ovules 2 per loculus. Fruit orange-pink with small whitish spots, globose, 1–2 cm. in diam., smooth, 1–3-seeded.

N. Rhodesia. W: Mwinilunga Distr., Zambezi R. 6·4 km. N. of Kalene Hill Mission, fr. 23.ix.1952, *White* 3344 (BM; COI; FHO; K; PRE).

Also in Angola (Lunda), the Congo, Cameroun and Nigeria. Evergreen and fringing forest.

S. pynaertii has been confused with *S. elegans*, but can be easily distinguished by several characters, e.g. stems not lined, becoming yellowish-brown, not purple; leaves subcoriaceous, not papyraceous, usually with finely denticulate margin and fewer lateral nerves; inflorescence sessile or very shortly pedunculate; sepals entire or subentire, not ciliolate or denticulate; styles not varying in length.

13. **Salacia elegans** Welw. ex Oliv., F.T.A. **1**: 373 (1868).—Loes. in Engl., Bot. Jahrb. **44**: 172 (1910).—R. E. Fr., Wiss. Ergebn. Schwed. Rhod.-Kongo-Exped. **1**: 129 (1914).—Exell & Mendonça, C.F.A. **2,** 1: 26 (1956).—Wilczek, F.C.B. **9**: 195 (1960).—Hallé in Mém. I.F.A.N. **64**: 203 (1962). TAB. **84** fig. C. Type from Angola (Pungo Andongo).

Salacia floribunda Tul. in Ann. Sci. Nat., Sér. 4, **8**: 97 (1857) non Wight (1840).—Loes., tom. cit.: 171 (1910).—White, F.F.N.R.: 219 (1962). Type from the Comoro Is.

Salacia zanzibarensis Vatke ex Loes. in Engl. & Prantl, Nat. Pflanzenfam. **3,** 5: 230 (1892) *nom. nud.*—Brenan, T.T.C.L.: 248 (1949). Syntypes from Zanzibar.

Salacia floribunda forma *mombassensis* Loes., op. cit. **19**: 240 (1894); op. cit. **44**: 171 (1910).—Chiov., Fl. Somala, **2**: 135, t. 89 (non t. 90) (1932). Syntypes from Kenya and Zanzibar.

Salacia floribunda forma *kumbensis* Loes., op. cit. **19**: 240 (1894). Syntypes from the Comoro Is. and Zanzibar.

Salacia floribunda forma *amaniensis* Loes., op. cit. **44**: 171 (1910). Syntypes from Tanganyika.

Salacia floribunda forma *subintegra* Loes., op. cit. **44**: 172 (1910). Type from Tanganyika (Usambaras).

Salacia semlikiensis De Wild., Pl. Bequaert, **2**: 73 (1923).—Robyns, Fl. Parc Nat. Alb. **1**: 510 (1948). Type from the Congo.

Salacia semlikiensis var. *subcordatifolia* De Wild., tom. cit.: 75 (1923). Type from the Congo.

Salacia leptoclada sensu Perrier, Fl. Madag., Hippocrat.: 9 (1946) pro parte quoad specim. Comorens. pro parte et t. 1, fig. 3–4.

Shrub, often with climbing branches, or liane, 1–6(40) m. high, with latex, glabrous; stems flattened, with paired raised lines, pale green and smooth at first, becoming terete or angular (climbing shoots), deep reddish-purple with numerous small whitish scarcely prominent lenticels. Leaves opposite to subopposite (alternate on climbing shoots), petiolate; lamina glossy above or on both sides, paler below, 3·5–10(18) × 1·5–5(8) cm., oblong or elliptic to obovate, shortly acuminate to caudate at the apex, with margin regularly crenate-dentate to subentire, cuneate to rounded at the base, papyraceous or rarely subcoriaceous, with c. 8–14 lateral nerves and densely reticulate venation more prominent below than above; petiole (2)4–6(12) mm. long, with margins straight; stipules very small, caducous. Flowers (1)3–∞, usually functionally unisexual, in axillary pedunculate or subsessile fascicles or pseudumbels with 2–4 condensed monochasial branches; buds 1–1·5 mm. long, globose to shortly cylindric; peduncle 1–5 mm. long, ± slender; pedicels 2–6 mm. long, smooth, articulated at the base. Sepals with pale margin, c. 0·5 mm. long, broadly ovate to semicircular, rounded, ± unequal, free, with margin ciliolate to denticulate or rarely subentire. Petals greenish yellow to orange-yellow, 1–2 mm. long, ovate to oblong, sessile, rounded, entire. Disk annular, flat, surrounding the ovary. Stamens 3, with filaments slender or slightly wider at the base, longer than the style (♂) or as long (☿) or much shorter (♀); anthers ± coherent round the stigma in ♂ forms, small and often apiculate in ♀ forms, dehiscing by 2 longitudinal clefts. Ovary ovoid-3-gonous, with style up to 1 mm. long, ⅓ as long as the stamens; stigma 3-lobed (♀ and ☿ forms) or punctiform (♂ forms); ovules 2 per loculus. Fruit orange to red, globose, or dorsiventrally compressed, 1–2·5 cm. in diam., smooth or rugulose, 1–2-seeded.

N. Rhodesia. N: Fort Rosebery Distr., Lake Bangweulu at Samfya, fl. 21.viii.1952, *Angus* 262 (BM; COI; FHO; K; PRE). W: Mwinilunga Distr., Zambezi R. 6·4 km. N. of Kalene Hill Mission, fl. 22.ix.1952, *Angus* 519 (BM; COI; FHO; K; PRE). **Nyasaland.** C: between Lake Nyasa and Senga Bay Hotel, 480 m., fl. 17.ii.1959, *Robson* 1638 (BM; K; LISC; PRE; SRGH). **Mozambique.** N: between Mossuril and Cabeceira, fl. 1884–5, *Carvalho* (COI). Z: Mocuba, Namagoa Estate, fl. 24.ix.1949, *Faulkner* K486 (COI; K; PRE; SRGH). SS: Gaza, Chipenhe, Régulo Chiconela, Chirindzeni Forest, fr. 13.x.1957, *Barbosa & Lemos* in *Barbosa* 8033 (K; LISC; LMJ).

From Guinée to the Congo, Uganda, Kenya and S. Somalia and southward to Mozambique, Nyasaland, N. Rhodesia and Angola; also in the Comoro Is. Evergreen and fringing forest and dense deciduous woodland, 0–600 m. (to 1500 m. in Uganda).

S. elegans has several forms with recognisable geographical distributions but without any morphological discontinuity. Thus, the large-leaved f. *subintegra* occurs in the Congo and the Usambaras; the western and central forms have smaller oblong regularly dentate leaves with long-pedunculate inflorescences; and the eastern forms (e.g. f. *mombassensis*) tend to have broader less deeply dentate leaves and inflorescences with short peduncles.

11. HIPPOCRATEA L.

Hippocratea L., Sp. Pl. **2**: 1191 (1753); Gen. Pl. ed. 5: 498 (1754).

Hippocrateaceae tribu *Hippocrateae* Hallé in Mém. I.F.A.N. **64**: 50 (1962).

Lianes or ± scandent shrubs or small trees, without latex (except in sect. *Hippocratea*). Leaves opposite or subopposite, petiolate; stipules free or ± united interpetiolarly, usually deciduous, with opposite pairs sometimes connected by ridges. Inflorescence pedunculate, cymose, dichasial or submonochasial, simple or compound, with or without accessory branches, axillary; bracts persistent. Flowers bisexual. Sepals (4)5, imbricate, free or ± united. Petals (4)5(6), green to white or yellow, often with paler margin, imbricate or valvate in bud. Disk extrastaminal, single or double, varying in shape, free or variously united with other floral parts. Stamens (2)3(4), with filaments often ± broadened below, free or united at the base or sometimes united with the ovary or gynophore; anthers versatile, apical or extrorse, with the thecae ± confluent, dehiscing transversely; pollen simple or rarely in polyads (sect. *Hippocratea*), not in tetrads. Ovary superior, 3-locular, with 2–20 axile ovules in 2 rows per loculus, sessile or on a gynophore, sometimes immersed in the disk; style simple or absent; stigmas 3, free or ± united, sometimes sessile or very small. Fruit of 3 capsular mericarps united at the base or rarely (not in African species) for most of their length; mericarps dorsiventrally flattened or rarely biconvex, dehiscing by the median suture. Seeds with long stalks, rarely sessile; stalks usually expanded into a wing with the raphe forming a submedian or rarely marginal vein and the thickened integuments forming a vein along the other margin. Germination epigeal or hypogeal. Chromosome number: 2n = 56 (3 species).

A genus of 90–100 species occurring throughout the tropics and subtropics of both hemispheres. *H. clematoides* Loes. has been recorded from Nyasaland on the basis of one specimen (*Burtt Davy* 22407 (BM)). This was wrongly labelled and was actually collected in Pemba I., E. Africa.

Reasons for rejecting segregate genera from *Hippocratea* are discussed in Bol. Soc. Brot., Sér. 2, **39**: 46 (1965).

Style present; stigma 3-lobed or entire; disk continuous; inflorescence a simple or compound cyme, with or without accessory branches; mericarps 2–∞ -seeded:
- Inflorescence without accessory branches; ovules 4–20 per loculus; seeds 2–∞ per mericarp; cotyledons united:
  - Petals imbricate-cucullate in bud; secondary wood cylindric, regular in section; whole plant glabrous:
    - Flower-bud cylindric; petals not unguiculate, 3·5–6 mm. long; disk cylindric, united with an androgynophore - - - - - 1. *goetzei*
    - Flower-bud globose; petals unguiculate or, if not, then not over 1·5 mm. long; disk concave to shallowly truncate-conic, not united with an androgynophore:
      - Petals unguiculate, with lamina ovate to semicircular or subcircular; leaves with reticulate venation ± prominent on both surfaces:
        - Flowers 6–8 mm. in diam.; petals 2–3 mm. long; sepals and petals not obviously veined:
          - Petals pale green to yellow with paler margin; leaves entire 2. *ritschardii*
          - Petals dark green with paler margin; leaves with margin obtusely dentate 3. *delagoensis*
        - Flowers 3–4·5 mm. in diam.; petals 1–1·5 mm. long; sepals and usually petals obviously dark-green-veined - - - - 4. *volkensii*
      - Petals not unguiculate, elliptic to oblong; leaves without obvious reticulate venation - - - - - - 5. *longipetiolata*
  - Petals valvate or slightly imbricate in bud, not cucullate; secondary wood fluted, undulate in section; at least the inflorescence pubescent:
    - Stem, leaves and mericarps glabrous; petals wholly glabrous or puberulous outside 6. *africana*
    - Stem pubescent; leaves and mericarps pubescent to glabrescent; petals densely pubescent or puberulous on both sides:
      - Disk broad with a 5-angled margin, adnate to the base of the stamen-filaments; mericarps broadly obovate to oblong (l/b = 1·2–1·6) - 7. *crenata*
      - Disk narrow with a circular margin, not adnate to the base of the stamen-filaments; mericarps lanceolate to elliptic-oblong (l/b = 1·8–3) 8. *apocynoides*
- Inflorescence with accessory branches; ovules 2 per loculus; seeds 2 per mericarp; cotyledons free:
  - Inflorescence simple; petals glabrous; mature stems glabrous or, if pubescent, then 4-lined:
    - Petals plane; stems terete, rugulose-lenticellate; leaves coriaceous or subcoriaceous, shining above, glabrous - - - - - - 9. *parvifolia*
    - Petals incurved or inrolled; stems ± quadrangular, smooth; leaves chartaceous, with both surfaces dull, glabrous or pubescent:
      - Stem, leaves and inflorescence glabrous; petals yellowish; leaves usually concolorous; petioles (4)8–12 mm. long - - - - 10. *indica*
      - Stem, leaves, inflorescence-branches and sepals pubescent; petals greenish; leaves ± markedly discolorous; petioles 2–7(9) mm. long - 11. *parviflora*
  - Inflorescence compound; petals puberulous outside; mature stems pubescent, terete - - - - - - - - - - 12. *buchananii*

Style absent; stigma 3-lobed, each lobe extending over a stamen; disk interrupted, each lobe enclosing a stamen; inflorescence a simple cyme without accessory branches; mericarps 2-seeded - - - - - - - - - 13. *pallens*

1. **Hippocratea goetzei** Loes. in Engl., Bot. Jahrb. **30**: 346 (1901).—R. E. & T. C. E. Fr. in Notizbl. Bot. Gart. Berl. **9**: 323 (1925).—Brenan, T.T.C.L.: 246 (1949); in Mem. N.Y. Bot. Gard. **8,** 3: 239 (1953).—Cufod. in Senckenb. Biol. **43**: 315 (1962). TAB. **85** fig. B. Type from Tanganyika (Kinga Mts.).
   Hippocratea scheffleri Loes. in Engl., Bot. Jahrb. **34**: 115 (1904).—Brenan, T.T.C.L.: 247 (1949). Type from Tanganyika (Usambaras).
   Simirestis goetzei (Loes.) Hallé [in Bull. Mus. Hist. Nat. Par., Sér. 2, **30**: 465 (1958) *sine basion.*] ex Wilczek, F.C.B. **9**: 162 (1960). Type as for *Hippocratea goetzei.*

Liane or rarely shrub, up to at least 15 m. high, glabrous; stems green and 4-angled when young or climbing, becoming vinous-red and eventually red-brown, 4-lined, with whitish lenticels becoming ± elongated and protuberant. Leaf-lamina dark green or greyish-green, paler beneath, (5)6–11·5(13) × 2·5–6·5(8·3) cm., ovate to oblong or elliptic, rounded to obtuse or obtusely acuminate at the apex, with margin glandular-crenulate or -crenulate-denticulate to subentire, cuneate to rounded at the base, papyraceous to subcoriaceous, with (6)7–9(10) lateral nerves and reticulate venation prominent beneath or on both sides; petiole 5–14 mm. long; stipules c. 0·5 mm. long, triangular-acicular, eventually deciduous, not united. Flowers c. 3–15 in lax simple axillary dichasia; buds 2·5–5 mm. long, cylindric, rounded; pedicels (2)3–7 mm. long; bracts 1–1·5 mm. long, triangular, rounded at apex, with margin entire or ciliolate, persistent. Sepals dark to yellow-green with paler ciliolate-denticulate margin, c. 1 mm. long, subequal, broadly ovate-triangular to semicircular, rounded. Petals dark to yellow-green with paler margin, (3·5)4·5–6 mm. long, not unguiculate, oblanceolate to oblong-spathulate or oblong, rounded, imbricate in bud. Disk simple, elongated, cylindric, united with the androgynophore. Stamens 3, with filaments long, ± narrow; anthers extrorse, subcircular, 1-thecal. Ovary with style elongated; stigmas partially united; ovules 6–7 per loculus. Mericarps and seeds as yet unknown.

Nyasaland. N: Misuku Hills, Walindi Forest, 2000 m., fl. 12.xi.1958, *Robson* 590 (BM; K; LISC; PRE; SRGH). S: Cholo Mt., 1300 m., fl. 20.ix.1946, *Brass* 17684 (BM; K; PRE; SRGH).

Also in Ethiopia, Uganda, Kenya, Tanganyika and eastern Congo (Kivu). In evergreen rainforest, 1300–2000 m. (to 3000 m. in East Africa).

Easily distinguished from the other species in our area by the elongated buds, long petals and disk-covered androgynophore. Related to *Simirestis dewildemaniana* Hallé from the Guinea-Congo region, which has smaller more numerous flowers with a much shorter androgynophore.

2. **Hippocratea ritschardii** (Wilczek) N. Robson in Bol. Soc. Brot., Sér. 2, **39**: 49 (1965). Type from the Congo (Haut Katanga).
   Loeseneriella ritschardii Wilczek in Bull. Jard. Bot. Brux. **26**: 406 (1956). Type as above.
   Simirestis ritschardii (Wilczek) Hallé [in Bull. Mus. Hist. Nat. Par., Sér. 2, **30**: 465 (1958) *sine basion.*] Wilczek, F.C.B. **9**: 168 (1960). Type as above.
   Hippocratea sp. 2.—White, F.F.N.R.: 218 (1962).

Liane, up to at least 3 m., glabrous; stems pale green and terete or 4-lined when young, becoming purplish-brown, terete, with dense whitish sometimes laterally elongated lenticels. Leaf-lamina bright green, concolorous, 3–10 × 1·5–4 cm., oblong to elliptic or ovate-lanceolate, obtuse to acute or obtusely acuminate at the apex, with margin entire, cuneate to rounded at the base, subcoriaceous, with 7–10 lateral nerves and densely reticulate venation almost equally prominent on both sides; petiole 5–10 mm. long; stipules c. 0·5 mm. long, laciniate, deciduous, not united. Flowers c. 10–40 in lax simple axillary dichasia; buds c. 1·5 mm. long, globose; pedicels 1·5–3·5(5) mm. long; bracts c. 0·5 mm. long, triangular, acute to obtuse, with margin eroded or subentire, persistent. Sepals green with paler ciliolate-denticulate margin, 0·5–1 mm. long, subequal, semicircular, rounded. Petals pale green to yellow, with paler eroded-ciliolate margin, 2·5–3 mm. long, unguiculate, with claw long and lamina semicircular, rounded, cucullate, imbricate in bud. Disk single, convex with margin slightly 5-angled, flat or concave. Stamens 3, with filaments short, broad; anthers extrorse, subcircular, 2-thecous. Ovary with style elongated; stigmas completely united; ovules 6–8 per loculus.

Mericarps (immature) olive-brown, 3·8–4·3 × 1·9 cm., flattened, ovate-oblong, acute to obtuse at the apex, striate. Seeds as yet unknown.

N. Rhodesia. W: Mwinilunga Distr., Kalene Hill, fl. 26.ix.1952, *Holmes* 914 (FHO).
Known from only 3 collections, the one cited above and two from Katanga. In dry rocky or sandy soils.

3. **Hippocratea delagoensis** Loes. in Engl., Bot. Jahrb. **34**: 119 (1904). TAB. **85** fig. A. Type: Mozambique, Lourenço Marques, *Schlechter* 11517 (B†, holotype; BM; COI; K; PRE).
Hippocratea delagoensis Sim, For. Fl. Port. E. Afr.: 37, t. 30B (1909) *nom. illegit.* Type: Mozambique, Delagoa Bay, *Sim* 6390 (?).

Shrub, scandent, up to at least 6 m. high, glabrous; stems olive-green and 4-lined when young, becoming vinous-red and eventually red-brown, terete, with ± prominent round whitish lenticels. Leaf-lamina dark green or yellow-green, slightly paler beneath or concolorous, 2·7–6·9(7·4) × (1·5)2·2–3·7 cm., elliptic to oblong or subcircular, obtuse to rounded at the apex, with margin shallowly obtusely glandular-dentate, cuneate at the base, subcoriaceous, with 5–7 lateral nerves and reticulate venation prominent on both sides; petiole 6–8 mm. long; stipules c. 1·5–2 mm. long, acicular, caducous, not united. Flowers (2)3–15(19) in ± lax simple axillary dichasia; buda 2–3 mm. long, globose; pedicels (2)3–5(6) mm. long; bracts c. 1 mm. long, triangular, subacute to apiculate, with margin entire, persistent. Sepals dark green with paler ciliolate-denticulate margin, c. 0·7 mm. long, subequal, broadly ovate to semicircular, rounded. Petals dark green with paler minutely ciliolate margin, 2–3 mm. long, unguiculate, with claw short and lamina ovate-cordate to subcircular, rounded, cucullate, imbricate in bud. Disk single, shallowly truncate-conic, with margin flattened and subentire or slightly 5-lobed. Stamens 3, with filaments short, broad; anthers extrorse, subcircular, 1-thecous. Ovary with style elongated; stigmas united; ovules c. 8 per loculus. Mericarps green, 4·5–5 × 2·3–2·8 cm., flattened, obovate to oblong-obovate, rounded to truncate at the apex, c. 8-seeded, smooth. Seeds winged, with veins marginal and submedian.

Mozambique. SS: Chibuto, Manjacaze road, fl. 12.x.1957, *Barbosa* 8007 (COI; K; LISC; LMJ). LM: near Lourenço Marques, Costa do Sol, fr. 8.v.1946, *Gomes e Sousa* 3447 (COI; K; LISC; SRGH).
Also in Natal (Zululand). In open woodland and dune scrub, 0–120 m.

H. delagoensis is sometimes confused with *H. longipetiolata* Oliv., but is easily distinguished from it by the leaves with prominent reticulate venation and usually shorter petioles and the larger flowers with broader clawed petals and a flat disk.

4. **Hippocratea volkensii** Loes. in Engl., Bot. Jahrb. **19**: 237 (1894); op. cit. **34**: 114 (1904).—Brenan, T.T.C.L.: 247 (1949). Syntypes from Tanganyika (Usambara).
Hippocratea sp. 1—White, F.F.N.R.: 217 (1962).

Shrub or small tree, usually scandent or scrambling, up to 5 m. high, glabrous; stems grey-green to olive-green and 4-lined when young, becoming reddish-brown and eventually greyish, terete, with numerous ± prominent whitish lenticels. Leaf-lamina dark green or rarely yellowish-green, concolorous, (3·2)3·8–9(11·2) × (1·5)2·3–3·6(5) cm., oblong to elliptic (rarely obovate), obtuse to acute or acuminate at the apex, with margin obtusely denticulate, usually with each tooth bearing an acicular gland, cuneate at the base, membranous to thinly subcoriaceous, with 5–7 lateral nerves and reticulate venation prominent on both sides; petiole 5–10(14) mm. long; stipules c. 1·5 mm. long, lanceolate to filiform, entire or ± laciniate, eventually deciduous, not united. Flowers 8–40 in rather lax simple subcorymbose axillary dichasia; buds 1–1·5 mm. long, globose; pedicels 1–2·5 mm. long; bracts 0·5–1 mm. long, triangular, acute, with margin entire or ± eroded, persistent. Sepals yellow, veined green, with ciliolate-denticulate margin, 0·5–1 mm. long, subequal, broadly ovate to semicircular, rounded. Petals pale yellow or greenish-yellow, usually veined green, with ciliolate margin, 1–1·5 mm. long, unguiculate, with claw short and lamina ovate-oblong to semicircular, rounded, cucullate, imbricate in bud. Disk single, concave, annular or slightly 5-lobed. Stamens 3, with filaments narrow, scarcely broadened towards the base; anthers extrorse, oblate, 2-thecous. Ovary with style elongated; stigmas united; ovules

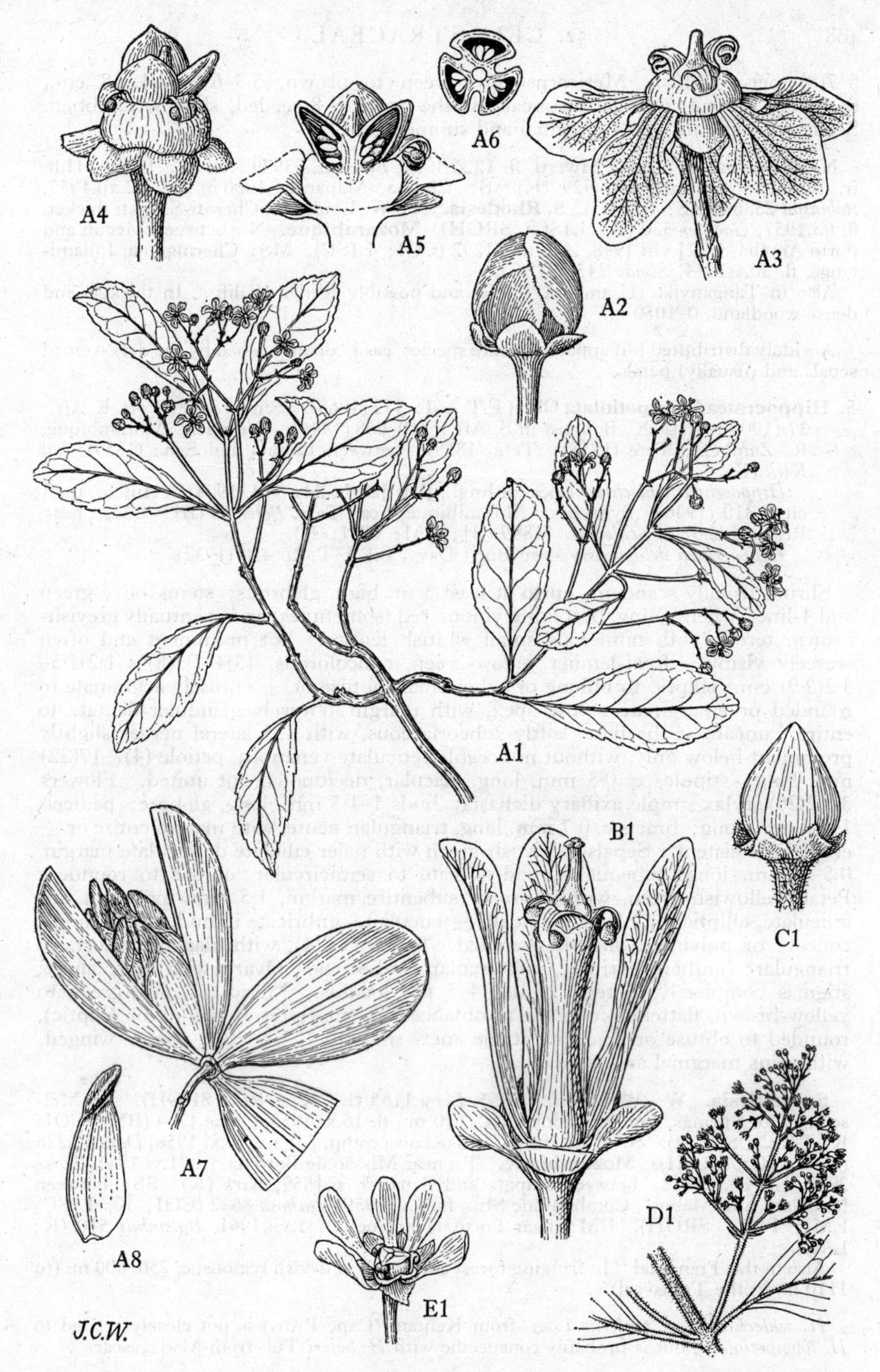

Tab. 85. A.—HIPPOCRATEA DELAGOENSIS. A1, flowering shoot (×⅔); A2, flower bud (×6); A3, flower (×6); A4, developing fruit (×6); A5, vertical section of developing fruit (×6); A6, transverse section of developing fruit (×6), all from *Pimenta* 27801; A7, fruit with ½ mericarp removed (×⅔); A8, seed (×⅔), both from *Gomes e Sousa* 3447. B.—HIPPOCRATEA GOETZEI. B1, flower with 2 petals removed (×6) Stolz 367. C.—HIPPOCRATEA CRENATA. C1, flower bud (×6) *Drummond & Rutherford-Smith* 7593. D.—HIPPOCRATEA BUCHANANII. D1, inflorescence (×⅔) *Chase* 1666. E.—HIPPOCRATEA PALLENS. E1, flower (×6) *Vaughan* 2719.

6–7(8) per loculus. Mericarps olive-green to brown, 5·3–6·1 × 2·4–2·8 cm., flattened, oblong, undulate-truncate at the apex, 5–8-seeded, shallowly ribbed. Seeds winged with veins marginal and submedian.

N. Rhodesia. N: Lake Mweru, fl. 12.xi.1957, *Fanshawe* 3929 (K). C: Broken Hill, fr. 16.viii.1964, *Mutimushi* 929 (K). S: Choma, Mapanza, 1050 m., fl. 22.xii.1957, *Robinson* 2536 (PRE; SRGH). **S. Rhodesia.** N: W. Urungwe, Cheroti-Sanyati thicket, fl. 6.i.1957, *Goodier* 530 (K; LISC; SRGH). **Mozambique.** N: between Mecufi and Porto Amélia, fr. 21.viii.1948, *Andrada* 1292 (COI; LISC). MS: Cheringoma, Inhamitanga, fl. 30.x.1944, *Simão* 245 (LISC).

Also in Tanganyika (Usambara, Ufipa) and possibly Kenya (Kilifi). In thickets and dense woodland, 0–1050 m.

A widely distributed but apparently rare species, easily distinguished by the dark-veined sepals and (usually) petals.

5. **Hippocratea longipetiolata** Oliv., F.T.A. **1**: 372 (1868).—Sim, For. Fl. Port. E. Afr.: 37 (1909).—Hutch., Botanist in S. Afr.: 310, 668 (1946). Syntypes: Mozambique, R. Zambeze, Tete (1858); Tete (1859); between Lupata and Sena (1859); all *Kirk* (K).

Hippocratea schlechteri Loes. in Engl., Bot. Jahrb. **34**: 114 (1904).—Hutch., tom. cit.: 310 (1946). Syntypes: Mozambique, near Tete, *Peters* 8 (B†; BM); near Ressano Garcia, *Schlechter* 11899 (B†; BM; COI; K).

Hippocratea delagoensis sensu Burtt Davy, F.P.F.T. **2**: 450 (1932).

Shrub, usually scandent, up to at least 5 m. high, glabrous; stems olive-green and 4-lined when young, becoming vinous-red (sometimes) and eventually greyish-brown, terete, with numerous small whitish lenticels not prominent and often scarcely visible. Leaf-lamina yellow-green, concolorous, (3)4–7·5(8) × (1·2)1·5–3·2(3·9) cm., elliptic to oblong or subcircular, obtuse or ± obtusely acuminate to rounded or emarginate at the apex, with margin obtusely glandular-dentate to entire, cuneate at the base, softly subcoriaceous, with 4–5 lateral nerves slightly prominent below only, without noticeable reticulate venation; petiole (4)7–17(22) mm. long; stipules c. 0·5 mm. long, acicular, deciduous, not united. Flowers 3–12(26) in lax simple axillary dichasia; buds 1–1·5 mm. long, globose; pedicels 1·5–3 mm. long; bracts c. 0·7 mm. long, triangular, acute, with margin entire or ± eroded, persistent. Sepals yellowish-green with paler ciliolate denticulate margin, 0·5–1 mm. long, subequal, broadly ovate to semicircular, obtuse to rounded. Petals yellowish-green, with entire or subentire margin, 1·5 mm. long, not unguiculate, elliptic to oblong, rounded, ± cucullate, imbricate in bud. Disk single, concave or pulvinate, slightly 5-angled. Stamens 3(4), with filaments short, ± triangular; anthers extrorse, subcircular, 1-thecous. Ovary with style short; stigmas completely united; ovules 4–5 per loculus. Mericarps olive-green to yellow-brown, flattened, obovate to oblanceolate (more rarely oblong or elliptic), rounded to obtuse or apiculate at the apex, smooth, 1–5-seeded. Seeds winged, with veins marginal and submedian.

S. Rhodesia. W: Wankie, fl. i.1955, *Levy* 1165 (E; K; PRE; SRGH). E: Melsetter, Hot Springs, near Dokodoko Mt., 610 m., fl. 16.xi.1953, *Chase* 4709 (BM; COI; K; LISC; SRGH). S: Nuanetsi R., Chikadziwa camp, 335 m., fl. xi.1956, *Davies* 2226 (K; LISC; SRGH). **Mozambique.** T: near Missão de Boroma, fr. 21.vi.1941, *Torre* 2918 (LISC). MS?: between Lupata and Sena, fr. ii.1859, *Kirk* (K). SS: between Mabalane and Mabote, Combomune Stn., fr. 5.vi.1959, *Barbosa* 8632 (COI; K; LISC; LMJ; PRE; SRGH). LM: near Porto Henrique, fr. 1.vii.1961, *Balsinhas* 510 (K; LMJ).

Also in the Transvaal. In fringing forest and scrub in dryish regions, c. 250–600 m. (to 1110 m. in the Transvaal).

H. schlechteri var. *peglerae* Loes. from Kentani (Cape Prov.) is not closely related to *H. longipetiolata*, but is probably conspecific with *H. bojeri* Tul. from Madagascar.

6. **Hippocratea africana** (Willd.) Loes. in Engl., Bot. Jahrb. **44**: 197 (1910); in Engl., Pflanzenw. Afr. **3**, 2: 240 (1921).—Keay & Blakelock, F.W.T.A. ed. 2, **2**: 628 (1958).—Cufod. in Bull. Jard. Bot. Brux. **28**, Suppl.: 483 (1958).—White, F.F.N.R.: 216 (1962). Type from Ghana.

Tonsella africana Willd. in L., Sp. Pl. ed. 4, **1**: 194 (1797). Type as above.

Salacia africana (Willd.) DC., Prodr. **1**: 570 (1824). Type as above.

Calypso africana (Willd.) G. Don, Gen. Syst. **1**: 629 (1831). Type as above.

Loeseneriella africana (Willd.) Wilczek [ex Hallé, Thèses Fac. Sci. Univ. Par., Monogr. Hipp. Afr. Occ.: 99 (1958)] F.C.B. **9**: 154 (1960).—Hallé in Mém. I.F.A.N. **64**: 104 (1962). Type as above.

Liane or scandent shrub, 2–15 m. high; stems pale green or olive green, 4-lined or subterete and glabrous or slightly puberulous when young, becoming terete and reddish to purple-brown, with numerous lenticels varying in prominence. Leaf-lamina greyish-green to yellowish-green, sometimes ± shining above, concolorous, 3–10(20) × 2–5·5(10) cm., ovate to oblong or elliptic-oblong, rounded to obtuse or obtusely acuminate (rarely acutely acuminate) at the apex, with margin obtusely glandular-denticulate to entire, cuneate to rounded at the base, papyraceous to coriaceous, glabrous, with (5)6–8(9) lateral nerves and reticulate venation varying in density and prominence, equally prominent on both sides; petiole (2)3–7(10) mm. long, glabrous; stipules 1·5–2 mm. long, linear-lanceolate, entire, pubescent, deciduous, the opposite pairs not united but sometimes connected by ridges. Flowers 3–∞ in simple axillary dichasial sometimes corymbose cymes with pubescent or glabrescent branches; buds 2–7 mm. long, subglobose to elongate-conic; pedicels 1–6 mm. long, pubescent; bracts 0·5–1 mm. long, triangular, acute, entire, pubescent, persistent. Sepals greenish, pubescent, convex, sometimes rugose outside, c. 1 mm. long, subequal, ovate-triangular, obtuse or rounded, free. Petals greenish, 3–6 mm. long, lanceolate-triangular, acute or subacute, often with subapical cusp, plane or with ± undulate margin, wholly glabrous or puberulous outside, slightly imbricate in bud. Disk single, ± undulate or 5-lobed, concave (free or united to the petals) or ± convex (i.e. surrounding the ovary) with or without a free margin. Stamens 3, with filaments broad or narrow at the base, glabrous; anthers extrorse, circular, 2-thecous, with orange-red pollen. Ovary glabrous, with style elongated, glabrous; stigmas united; ovules 6–14 per loculus. Mericarps green or brownish-green, 3·5–5·5 × 2·3–3·9 cm., oblong or elliptic to oblanceolate or obovate, rounded or retuse at the apex, glabrous, 6–14-seeded, striate or smooth. Seeds winged, with veins marginal and intramarginal to submedian, the thickened marginal one surrounding the embryo or not.

From Senegal eastward to Ethiopia and south to the Transvaal, S. Rhodesia, Bechuanaland Prot. and the Caprivi Strip. In dry or damp evergreen forest, fringing forest or woodland on rocky ground.

A very variable species which in this concept comprises several taxa with distinct but overlapping distributions. The numerous intermediate forms between these taxa, however, appear to preclude their treatment as subspecies or species, at least until a detailed study is made of the whole group. In our area the specimens from northern and lowland areas tend to have a ± convex disk with a flat or upturned margin (i.e. they agree with *H. obtusifolia* var. *fischerana* Loes.), whereas in those from further south and inland the disk tends to be without such a margin (as in var. *richardiana* (Cambess.) N. Robson); but the other characters said by Wilczek (loc. cit.) to be correlated do not appear to be so in our area. It therefore appears to be impossible to recognise var. *fischerana* as distinct.

Var. **richardiana** (Cambess.) N. Robson in Bol. Soc. Brot., Sér. 2, **39**: 52 (1965). Type from Senegal.

Hippocratea richardiana Cambess. in St.-Hil., Fl. Bras. Merid. **2**: 102 (1829).—Guill. & Perr. in Guill., Perr. & Rich., Fl. Senegamb. Tent **1**: 112, t. 26 (1831).—Hutch. & Moss., F.W.T.A. **1**: 449 (1928).—Robyns, Fl. Parc Nat. Alb. **1**: 509 (1948).—Schnell, I.F.A.N. Ic. Pl. Afr. **1**: 12 (1953).—Exell & Mendonça, C.F.A. 2, **1**: 22 (1954). Type as above.

Hippocratea obtusifolia sensu Oliv., F.T.A. **1**: 369 (1868).—Gibbs in Journ. Linn. Soc. **37**: 436 (1905).—Bak. f. in Journ. Linn. Soc., Bot. **40**: 45 (1911).—R. E. Fr., Schwed. Rhod.-Kongo-Exped.: 129 (1914).—Eyles in Trans. Roy. Soc. S. Afr. **5**: 405 (1916).—Burtt Davy, F.P.F.T. **2**: 450 (1932).—Hutch., Botanist in S. Afr.: 473, t. (1946).—O. B. Mill., B.C.L.: 36 (1948).—Brenan, T.T.C.L.: 246 (1949).

Hippocratea obtusifolia var. *richardiana* (Cambess.) Loes. in Engl., Bot. Jahrb. **19**: 236 (1894); op. cit. **34**: 108 (1904); in Engl. & Prantl, Nat. Pflanzenfam. ed. 2, **20b**: 213, fig. 67 (1942).—Brenan, op. cit.: 247 (1949). Type as above.

Hippocratea obtusifolia var. *fischerana* Loes. in Engl., op. cit. **19**: 237 (1894); op. cit. **34**: 108 (1904).—Brenan, loc. cit. Syntypes from Tanganyika.

Hippocratea obtusifolia var. *eminiana* Loes. in Engl., op. cit. **19**: 237 (1894); op. cit. **34**: 108 (1904).—Brenan, loc. cit. Syntypes from Tanganyika.

Hippocratea c.f. *obtusifolia* Schinz, Pl. Menyharth.: 61 (1905).—Gomes e Sousa, Pl. Menyhart.: 78 (1936).
Hippocratea cymosa sensu Eyles, loc. cit.
Hippocratea nitida Obermeyer in Ann. Transv. Mus. **17**: 210 (1937).—O.B. Mill. in Journ. S. Afr. Bot. **18**: 49 (1952). Type from the Transvaal.
Loeseneriella africana var. *richardiana* (Cambess.) [Hallé, Thèses Fac. Sci. Univ. Par., Monogr. Hipp. Afr. Occ.: 100 (1958)] Wilczek, F.C.B. **9**: 155 (1960).—Hallé in Mém. I.F.A.N. **64**: 104, t. 38 (1962). Type as for *Hippocratea richardiana*.
Loeseneriella africana var. *fischerana* (Loes.) Wilczek, tom. cit.: 156 (1960). Syntypes as for *H. obtusifolia* var. *fischerana*.

Leaf-lamina oblong to elliptic-oblong, rounded to obtuse or obtusely acuminate at the apex, with margin shallowly glandular-crenulate-denticulate to entire, usually subcoriaceous to coriaceous. Flower-buds 3·5–7 mm. long, ± elongate-conic. Sepals not rugose outside. Petals 4–6 mm. long. Disk ± convex, with or without a free margin. Stamen-filaments broad or narrow at the base. Mericarps oblanceolate to obovate. Seeds with a thickened marginal vein ± surrounding the embryo.

Caprivi Strip: east of Cuando R., 945 m., fl. x.1945, *Curson* 917 (PRE). **Bechuanaland Prot.** N: Kabulabula, Chobe R., fr. vii.1930, *van Son* in Herb. Transv. Mus. 28849 (BM; COI; K; PRE; SRGH). **N. Rhodesia.** B: Balovale, near Chavuma, fl. 13.x.1952, *White* 3491 (BM; COI; FHO; K; PRE). N: Chinsali Distr., Mbesuma, edge of Chambeshi Flats, fl. 11.x.1960, *Robinson* 3958 (K; SRGH). W; Kitwe, fl. & fr. 6.vii.1955, *Fanshawe* 2363 (K; LISC; SRGH). E; Lutembwe R. gorge, east of Machinji Hills, 900 m., fl. 14.x.1958, *Robson* 105 (BM; K; LISC; PRE; SRGH). S: Livingstone Distr., Katombora, Zambezi R., fl. 6.i.1953, *Angus* 1110 (BM; K; PRE). **S. Rhodesia.** N: Mkoto Res., Nyangomba R., fl. 1.x.1948, *Wild* 2686 (K; SRGH). W: Kandahar Is., Victoria Falls, 900 m., fl. 3.ix.1955, *Chase* 5772 (BM; COI; SRGH). E: Umtali, fr. 11.vi.1947, *Chase* 373 (BM; COI; K; PRE; SRGH). S: Ndanga, Nyamakasana R. near Chiredzi R. confluence, fl. 3.xii.1959, *Goodier*, 707 (K; SRGH). **Nyasaland.** S: Katunga, Shire Valley, fl. v.1888, *Scott* (K). **Mozambique.** N: between Montepuez and Namuno, fl. 3.ix.1948, *Barbosa* 1990 (LISC). Z: between Mopeia and Marral, fl. 15.x.1941, *Torre* 3663 (LISC). T: Mutarara, Lago Lifumba, fr. 20.vi.1949, *Andrada* 1614 (COI; LISC). MS: Chemba, near Marínguè, Vila Paiva road, fl. 2.x.1944, *Mendonça* 2366 (LISC). SS: Guijá, Aldeia da Barragem, R. Limpopo, fl. 16.xi.1957, *Barbosa & Lemos* in *Barbosa* 8147 (COI; K; LISC; LMJ). LM: Maputo, fl. 1.ii.1947, *Hornby* 2545 (K; PRE; SRGH).
Distribution as for the species. In fringing forest or woodland or rocky ground. Up to 1250 m. in our area.

7. **Hippocratea crenata** (Klotzsch) K. Schum. & Loes. in Engl. & Prantl, Nat. Pflanzenfam. **3,** 5: 228 (1893).—Loes. in Engl., Bot. Jahrb. **19**: 237 (1894); op. cit. **34**: 108, t. 1, figs. M–N (1904).—Burtt Davy & Hoyle, N.C.L.: 46 (1936).—Brenan, T.T.C.L.: 246 (1949).—Cufod. in Bull. Jard. Bot. Brux. **28,** Suppl.: 484 (1958). TAB. **85** fig. C. Type: Mozambique, Tete, Rios de Sena, *Peters* (B†).
Gymnema crenata Klotzsch in Peters, Reise Mossamb. Bot. **1**: 273 (1862). Type as above.
Hippocratea kirkii Oliv., F.T.A. **1**: 370 (1868).—Sim, For. Fl. Port. E. Afr.: 37 (1909).—Gomes e Sousa, Pl. Menyhart.: 78 (1936). Type: Mozambique, R. Zambeze, Tete, *Kirk* (K).
Loeseneriella crenata (Klotzsch) Wilczek, F.C.B. **9**: 152 (1960). Type as for *Gymnema crenata*.

Shrub, often scandent or scrambling, 2–4 m. high; stems grey, striate and densely red-brown-pubescent when young, becoming terete and greyish-pubescent, without noticeable lenticels. Leaf-lamina dark green and ± shining above, paler below, 1·5–5·4(6·3) × 1–2·8 cm., elliptic to oblong, obtuse to rounded at the apex, with margin obtusely glandular denticulate to subentire, narrowly to broadly cuneate at the base, subcoriaceous, reddish-brown- or greyish-pubescent near base (especially on midrib and margins), otherwise glabrous, with 5–7(9) lateral nerves and reticulate venation more prominent below than above or prominent below only; petiole 3–5 mm. long, reddish-brown- or greyish-pubescent; stipules c. 0·7 mm. long, narrowly triangular, entire, reddish-brown- or greyish-pubescent, not united or connected by transverse ridges. Flowers (1)3–5(8) in rather dense shortly pedunculate simple terminal and axillary dichasia; buds 2·5–3·5 mm. long, conic; pedicels 2–3 mm. long, pubescent; bracts c. 0·7–1 mm. long, lanceolate-triangular, acute, persistent. Sepals greenish, reddish-brown- or greyish-pubescent, c. 0·7–1 mm. long, subequal, triangular, acute, united in lower ⅓. Petals greenish, 2·5–4 mm. long, unguiculate, with claw short, glabrous within, and lamina lanceolate-triangular, acute, plane, densely reddish-brown-pubescent outside, greyish-puberulous within, valvate in bud. Disk broad with free concave 5-angled margin and cylindric inner part adnate to the base of the stamen-filaments. Stamens 3, with filaments broad, sparsely pubescent; anthers extrorse, transversely broadly elliptic, 2-thecous. Ovary densely pubescent, with style elongated, glabrous; stigmas completely united; ovules 4–5 per loculus. Mericarps greenish- to reddish-brown, (3)3·5–4·6 × 2·3–3 cm., flattened, broadly obovate to oblong, truncate or emarginate at the apex, sparsely pubescent, 4–5-seeded, striate. Seeds winged, with veins marginal and submedian.

N. Rhodesia. S: Lusitu, fr. 25.ix.1959, *Fanshawe* 5222 (K). **S. Rhodesia.** E: Melsetter Distr., E. bank of Sabi R. opposite Chitsa's village, 240 m., fl. 11.vi.1950, *Chase* 2422 (BM; COI; K; SRGH). S: Ndanga Distr., Bikita road, N. Sangwe Reserve, 420 m., fl. i.1960, *Farrell* 137 (K; SRGH). **Nyasaland.** S?: without precise locality, fl. 1891, *Buchanan* 1256 (BM; K). **Mozambique.** T: between Tete and Boroma, fr. 5.v.1948, *Mendonça* 4085 (LISC). SS: Mabalane to Mapai, 8 km., fr. 4.vi.1959, *Barbosa* 8618 (COI; K; LISC; PRE; SRGH).

Also in SE. Kenya and the Transvaal and recorded from Ethiopia. Confined in our area to the basins of the Zambezi, Sabi and Limpopo rivers, in dry bush and rocky places, up to c. 450 m.

H. rubiginosa Perrier, from Madagascar, is a close relative of *H. crenata* and may not be specifically distinct from it.

I have seen no specimens of *H. crenata* from Ethiopia, the Congo or Tanganyika. The records from the last two countries are based on small-leaved forms of *H. apocynoides*.

8. **Hippocratea apocynoides** Welw. ex Oliv., F.T.A. **1**: 368 (1868).—Loes. in Engl., Bot. Jahrb. **34**: 108 (1904).—Exell & Mendonça, C.F.A. **2**, 1: 20, t. 4 fig. A (1954). Type from Angola (Cuanza Norte).

Hippocratea bruneelii De Wild. in Ann. Mus. Cong., Bot. Sér. 5, **2**: 292 (1908). Type from the Congo.

Hippocratea apocynoides var. *typica* forma *australis* Loes. in Mildbr., Wiss. Ergebn. Deutsch. Z. Afr. Exped. 1907–08, **2**: 468 (1912). Type as for *H. apocynoides.*

Hippocratea apocynoides var. *mildbraedii* Loes., loc. cit. Type from Rwanda.

Hippocratea gossweileri Exell in Journ. of Bot. **65**, Suppl. Polypet.: 77 (1927).—Exell & Mendonça, loc. cit. Type from Angola (Cabinda).

Hippocratea crenata sensu Brenan, T.T.C.L.: 246 (1949).

Loeseneriella crenata sensu Wilczek, F.C.B. **9**: 152 (1960) pro parte quoad syn. *Hippocratea apocynoides* var. *mildbraedii* et specim. cit.

Shrub or more usually liane, up to 25 m. high. Stems purple-brown, striate and densely red-brown-pubescent when young, becoming terete, without noticeable lenticels. Leaf-lamina dull green above, paler below, (2·4)4–13(17) × (1·6)2·2–6·9(8·7) cm., elliptic to oblong or obovate, very shortly obtusely acuminate to rounded at the apex, with the margin entire or very shallowly serrulate or crenulate, broadly cuneate to subcordate at the base, papyraceous to subcoriaceous, ± sparsely pubescent below when young, glabrescent, with c. 7 lateral nerves and venation more prominent below than above; petiole 3–7 mm. long; stipules c. 1 mm. long, linear-lanceolate, entire, pubescent, ± persistent, not usually united by transverse ridges. Flowers 10–∞, in rather dense or lax pedunculate simple or compound terminal and axillary dichasia; buds 2·5–3 mm. long, conic to acicular; pedicels 0·5–2 mm. long, pubescent; bracts 0·5–2 mm. long, triangular, acute, eventually deciduous. Sepals reddish-brown-pubescent, c. 0·5–1 mm. long, subequal, triangular, acute, free. Petals greenish-cream, 1·5–3·5 mm. long, ovate-triangular to triangular-acicular, densely red-brown- to fawn-pubescent outside, red-brown- to grey-puberulous within, valvate in bud. Disk narrow, with a free concave circular margin, not adnate to the base of the stamen-filaments. Stamens 3, with filaments broad, glabrous; anthers extrorse, oblate, 2-thecous. Ovary pubescent, with style short; stigmas completely united; ovules 4–10 per loculus. Mericarps dull brown, c. 5–6 × 2–2·5 cm., flattened, elliptic-oblong to lanceolate, obtuse to rounded or slightly emarginate at the apex, sparsely puberulous, 4–6-seeded, striate. Seeds winged, with veins marginal and submedian.

From Guinée to the Central African Republic, Uganda and Tanganyika and south to Angola, the Congo and N. Rhodesia. In swamp forest and fringing forest.

H. apocynoides comprises two subspecies with overlapping distributions, one in the Congo basin and Angola (subsp. *apocynoides*) with a reduced form in the eastern Congo, Rwanda and Tanganyika (var. *mildbraedii*) and the other in W. Africa and round the north and east margins of the Congo basin (subsp. *guineensis*). Only the latter subspecies occurs in our area.

Subsp. **guineensis** (Hutch. & Moss) N. Robson in Bol. Soc. Brot., Sér. 2, **39**: 55 (1965). Type from Nigeria.

Hippocratea apocynoides var. *typica* forma *borealis* Loes. in Mildbr., loc. cit. Type from the Congo.

Hippocratea guineensis Hutch. & Moss, F.W.T.A. **1**: 449 (1928); in Kew Bull. **1929**: 20 (1929). Type as for *H. apocynoides* subsp. *guineensis*.

Loesneriella guineensis (Hutch. & Moss) Hallé [Thèses Fac. Sci. Univ. Par., Monogr. Hippocrat. Afr. Occ.: 114 (1958)] in Mém. I.F.A.N. **64**: 120, t. 44 (1962). Type as above.

Inflorescence a simple dichasium. Petals usually short (1·5–2·5 mm. long), but up to 3 mm. long in N. Rhodesian specimens.

N. Rhodesia. N: Fort Rosebery, fl. 22.v.1964, *Fanshawe* 8668 (K; NDO).

From Guinée to Caméroun and in the Central African Republic, the Congo (Kivu and Katanga), Uganda and N. Rhodesia.

It is not possible to base the separation of subsp. *guineensis* and subsp. *apocynoides* on petal-shape or -length, but a division based on simple or compound dichasia gives an almost complete separation.

Wilczek, in F.C.B. **9**: 152 (1960), included both subspecies under the *later* epithet, *guineensis*. His combination under *Loeseneriella* is illegitimate as he cited no basionym.

9. **Hippocratea parvifolia** Oliv., F.T.A. **1**: 368 (1868).—Loes. in Engl., Bot. Jahrb. **34**: 105 (1904).—Exell & Mendonça, C.F.A. 2, 1: 15 (1954). Type from Angola (Huila).

Hippocratea busseana Loes., loc. cit. Type from Tanganyika (Songea).

Hippocratea kageraensis Loes. in Mildbr., Wiss. Ergebn. Deutsch. Z. Afr. Exped. 1907–1908, 2: 467, t. 60 (1912).—Brenan, T.T.C.L.: 246 (1949). Type from Tanganyika (Bukoba).

Hippocratea buchholzii sensu Exell & Mendonça, tom. cit.: 16 (1956) pro parte quoad specim. *Nolde* 358 et *Milne-Redhead* 4243.

Reissantia parvifolia (Oliv.) Hallé [Thèses Fac. Sci. Univ. Par., Monogr. Hipp. Afr. Occ.: 88 (1958); in Bull. Mus. Hist. Nat. Par., Sér. 2, **30**: 466 (1958) *sine basion.*—Wilczek, F.C.B. **9**: 147 (1960)] in Mém. I.F.A.N. **64**: 92 (1962). Type as for *Hippocratea parvifolia*.

Reissantia parvifolia var. *kageraensis* (Loes.) Hallé in Mém. I.F.A.N. **64**: 93 (1962). Type as for *Hippocratea kageraensis*.

Hippocratea indica sensu White, F.F.N.R.: 217 (1962) pro parte quoad specim. *Bullock* 2634, *White* 2890, 3017.

Shrub or small tree, often scandent or scrambling, 0·3–4·5 m. high, or liane up to c. 10 m. high, glabrous; stems greyish-green to purple (if climbing), smooth and slightly 4-lined when young, soon becoming terete and reddish-purple to grey-brown and ± rugose owing to numerous lenticels. Leaf-lamina 2·7–10(13) × 1·1–4·4 cm., oblong to broadly or narrowly oblong-elliptic or ovate, acute to obtuse or acuminate at the apex, with margin entire or shallowly glandular-denticulate, narrowly cuneate to rounded at the base, coriaceous, deep green and shining above, paler below, with midrib and 6–9 lateral nerves and reticulate venation more prominent below than above or plane; petiole 1–9 mm. long; stipules c. 0·3 mm. long, triangular to bifid or irregularly dentate, the opposite pairs not united. Flowers ∞ in lax simple axillary dichasia with accessory branches; peduncles and inflorescence-branches terete or flattened or slightly 4-lined; buds c. 1 mm. long, ovoid-cylindric; pedicels 0·2–0·5(1) mm. long; bracts 0·5–1(2) mm. long, triangular, acute, eroded-denticulate, persistent. Sepals yellowish-green (purplish-brown with pale margin when dry), c. 0·4 mm. long, equal, ± semicircular, obtuse, rounded, with margin irregularly ciliolate, united in lower $\frac{1}{3}$–$\frac{1}{2}$. Petals yellowish-green to cream (reddish-purple with pale margin when dry), 0·5–1(1·7) mm. long, elliptic-oblong, rounded, entire, with margins plane, imbricate in bud, spreading at anthesis. Disk single, subcupular, shallowly 3-lobed, surrounding the ovary. Stamens 3, with filaments short, narrow; anthers extrorse, circular, 1-thecous. Ovary with short style; stigmas completely united; ovules 2 per loculus. Mericarps purplish-brown, 3–5·1(5·8) × 1·4–2·1 cm., flattened, oblong to narrowly obovate, obtuse to truncate or apiculate at the apex, 2-seeded, almost smooth. Seeds winged, with veins marginal and submedian.

N. Rhodesia. N: Abercorn, road to Isoko Village, 1190 m., fl. 31.i.1955, *Richards* 4301 (K; SRGH). C: Kapiri Mposhi, fl. 22.i.1955, *Fanshawe* 1817 (K). E: Petauke Distr., Great East Road, 16 km. west to Kachalolo Rest House, st. 27.v.1952 *White* 2890 (K). S: Mazabuka Distr., Siamambo Forest Reserve, near Choma, fr. 24.vii.1952, *White* 3017 (BM; K). **S. Rhodesia.** N: Sebungwe, Binga Hill, c. 520 m., fl. 10.xi.1958, *Phipps* 1419 (K; LISC; SRGH). W: Wankie, fl. iii.1931, *Pardy* in GHS 4637 (SRGH). S: Beitbridge, Chiturupazi, fl. 25.ii.1961, *Wild* 5404 (COI; K; PRE; SRGH). **Nyasaland.** N: Karonga Distr., Fort Hill, 1400 m., fr. 1960 *Young* 195 (K; SRGH). C: Kota-Kota, 365 m., fl. 16.ii.1944, *Benson* 324 (PRE). **Mozambique.** N: Nampula, fl. 14.ii.1937, *Torre* 1354 (COI; LISC). T: Boroma, Sisito Camp, Ulere Station, st. 9.vii.1950, *Chase* 2644 (BM; K; SRGH). SS: Guijà, between Caniçado and Papai, fl. 6.v.1944, *Torre* 6583 (LISC; PRE).

From Mali to Ethiopia and southward to Mozambique, the Transvaal, S. Rhodesia and Angola; in dry woodland, often on rocky outcrops or sand, 275–1400 m. in our area (to 1900 m. in Angola).

H. parvifolia is very variable in habit as well as in leaf-size and -shape. There appears to be no break in variation between forms with large or entire leaves and the much smaller denticulate leaves of typical *H. parvifolia*, so that *H. busseana* and *H. kageraensis* cannot be recognised as distinct even at the varietal level.

10. **Hippocratea indica** Willd. in L., Sp. Pl. ed. 4, **1**: 193 (1797).—Roxb., Pl. Corom. **2**: 16, t. 130 (1799).—Oliv., F.T.A. **1**: 368 (1868).—Ficalho, Pl. Ut. Afr. Port.: 119 (1884).—Loes. in Engl., Bot. Jahrb. **34**: 106 (1904); in Fedde, Repert. **49**: 227 (1940).—Sim, For. Fl. Port. E. Afr.: 36 (1909).—Burtt Davy & Hoyle, N.C.L.: 46 (1936).—Brenan, T.T.C.L.: 246 (1949).—Blakelock & Keay, F.W.T.A. ed. 2, **1**, 2: 627 (1954).—White, F.F.N.R.: 217 (1962) pro parte quoad specim. *White* 2397, *Angus* 1083. Type from India (Coromandel).

Hippocratea indica var. *β* Oliv., loc. cit. Type: Mozambique, R. Rovuma, *Kirk* (K).

Hippocratea cf. *indica*.—Schinz, Pl. Menyharth.: 60 (1905).—Gomes e Sousa, Pl. Menyhart.: 78 (1936).

Hippocratea loesenerana Hutch. & Moss, F.W.T.A. **1**: 450 (1928); in Kew Bull. **1929**: 21 (1929).—Robyns, Fl. Parc Nat. Alb. **1**: 506 (1948).—O.B. Mill., B.C.L.: 36 (1948); in Journ. S. Afr. Bot. **18**: 49 (1952).—Exell & Mendonça, C.F.A. **2**, 1: 18 (1954). Type from Sierra Leone.

Hippocratea indica forma *longepetiolata* Loes. in Fedde, Repert. **49**: 227 (1940). Type: Mozambique, M'Kumba, *Tiesler* 88 (B†).

Pristimera indica (Willd.) A. C. Sm. in Amer. Journ. Bot. **28**: 438 (1941). Type as for *Hippocratea indica*.

Reissantia indica (Willd.) Hallé [in Bull. Mus. Hist. Nat. Par., Sér. 2, **30**: 466 (1958) *sine basion.*] in Mém. I.F.A.N. **64**: 85 (1962). Type as for *Hippocratea indica*.

Reissantia indica var. *loesenerana* (Hutch. & Moss) Hallé [Thèses Fac. Sci. Univ. Par., Monogr. Hipp. Afr. Occ.: 82 (1958)] ex Wilczek, F.C.B. **9**: 146 (1960).—Hallé in Mém. I.F.A.N. **64**: 85, t. 32 (1962). Type as for *Hippocratea loesenerana*.

Shrub, often scandent or scrambling, 1–5 m. high, or liane up to c. 12 m. high, more rarely a small tree, glabrous; stems greyish- to yellowish-green, smooth and 4-lined when young, becoming obtusely 4-lobed or terete and reddish-purple when mature, smooth, without lenticels. Leaf-lamina greyish- or yellowish-green, concolorous or slightly paler and sometimes with brownish nervation below, 4–11·7(15) × (2)3·2–7(10) cm., oblong or elliptic to obovate or subcircular, obtuse to acuminate at the apex, with margin denticulate to subentire, cuneate, ± decurrent at the base, chartaceous, with midrib and 5–7 lateral nerves ± prominent below but the venation otherwise plane; petiole (4)8–12 mm. long; stipules c. 0·5–0·7 mm. long, triangular to semicircular, laciniate or irregularly dentate, not united. Flowers ∞ in lax to rather dense simple axillary dichasia with accessory branches; peduncles and inflorescence-branches quadrangular, sometimes glandular along angles; buds c. 0·5 mm. long, ovoid to globose; pedicels 0·5–1 mm. long; bracts c. 0·5–0·7 mm. long, triangular, acute, eroded-denticulate, persistent. Sepals reddish-brown, 0·7–1 mm. long, equal, oblong to ovate-lanceolate, acute or subacute, with margin irregularly ciliolate, united in lower $\frac{1}{3}$–$\frac{2}{3}$. Petals yellowish to cream, drying reddish-brown, 1–2 mm. long, linear-lanceolate, acute to subacute, entire, with margins inrolled, imbricate in bud, erect or ± spreading but straight or incurved towards the apex at anthesis. Disk single, shallowly 3-lobed, surrounding the ovary. Stamens 3, with filaments short, narrow; anthers extrorse, circular, 1-thecous. Ovary with a short style; stigmas completely united; ovules 2 per loculus. Mericarps pale green, 2·8–4(4·5) × 0·9–1·5 cm., flattened, narrowly oblong to oblanceolate, rounded at the apex, 2-seeded, striate. Seeds winged, with veins marginal and submedian.

Caprivi Strip. Mpilili I., c. 900 m., fl. 13.i.1949, *Killick & Leistner* 3345 (K; PRE; SRGH). **Bechuanaland Prot.** N: Kazungula, fl. iv.1936, *Miller* B128 (BM; FHO; K). **N. Rhodesia.** S: Choma, Mapanza, 1050 m., fl. 26.i.1958, *Robinson* 2752 (K; PRE; SRGH). **S. Rhodesia.** N: Sebungwe Distr., c. 64 km. N. of Gokwe, fr. 8.vii.1952, *Vincent* 2 (SRGH). W; Shangani R., fl., *Davies* in GHS 32074 (K; SRGH). S: Ndanga Distr., Chiredzi R., fr. 14.x.1951, *Thompson* 114/51 (SRGH). **Nyasaland.** N: Rumpi Distr., near Njakwa, fr. 30.iv.1952, *White* 2540 (FHO; K). S: Farringdon road, near Fort Johnston, fl. 14.iii.1955, *E. M. & W.* 872 (BM; LISC; SRGH). **Mozambique.** N: Nampula, fl. 3.iii.1936, *Torre* 839 (COI; LISC). T: Boroma, near Máguè, fr. 23.vii.1950, *Chase* 2688 (BM; K; LISC; SRGH). MS: Reserve Florestal do Mucheve, Chibabava, fr. 20.vii.1963, *Carvalho* 635 (K). SS: between Chibuto and Chongoene, fr. 30.vii.1947, *Pedro & Pedrogão* 2097 (PRE).

From Senegal to Cameroun and Fernando Po, Central African Republic, the Congo, Tanganyika and Angola, and occurring south of our area in the Transvaal; also in tropical Asia from India and Ceylon to the Philippines and Timor; in damp or dryish forest or woodland or on river banks, often in stony ground, 210–1170 m.

Although most specimens from our area have the long-petioled obovate leaf with short broad acumen characteristic of forma *longepetiolata*, the transitional forms to the typical northern form are so numerous as to make separation of these forms impossible. Likewise there appear to be no constant differences between the African and Asian forms of this species. On the other hand, unlike Wilczek (loc. cit.), I think that *Reissantia astericantha* Hallé is quite distinct from *H. indica*.

11. **Hippocratea parviflora** N. E. Br. in Kew Bull. **1909**: 99 (1909).—O.B. Mill., B.C.L.: 36 (1948) (" parvifolia "); in Journ. S. Afr. Bot. **18**: 49 (1952). Type: Bechuanaland Prot., Kwebe Hills, *Lugard* 180 (K).

Hippocratea buchananii forma *dolichocarpa* Loes. in Engl., Bot. Jahrb. **34**: 106 (1904).—Brenan T.T.C.L.: 246 (1949). Type from Tanganyika.

Hippocratea hirtiuscula Dunkley in Kew Bull. **1934**: 185 (1934). Type N. Rhodesia, Bombwe Forest, *Martin* 354 (FHO, holotype; K).

Hippocratea sp.—O.B. Mill., B.C.L.: 35 (1948); in Journ. S. Afr. Bot. **18**: 49 (1952) pro parte quoad *Pole Evans* 4203.—Brenan, loc. cit.

Hippocratea indica var. *parviflora* (N. E. Br.) Blakelock in Kew Bull. **11**, 3: 556 (1957).—White, F.F.N.R.: 217 (1962). Type as for *H. parviflora*.

Hippocratea indica sensu White, F.F.N.R.: 217 (1962) pro parte quoad specim. Trapnell.

Liane up to at least 4·5 m. high; stem green, 4-lined and finely pubescent when young, becoming reddish-brown or purplish-brown, persistently 4-lined or 4-angled and pubescent, without lenticels or with numerous small faint ones visible only on the older branches. Leaf-lamina deep green above, pale green below, sometimes with brownish nervation, 4·8–8·2(12) × 1·9–4(6) cm., oblong to oblong-elliptic or obovate, obtuse or more rarely rounded or shortly acuminate at the apex, with margin glandular-denticulate, cuneate at the base, chartaceous, softly pubescent on midrib and nerves and sparsely on rest of lamina or rarely subglabrous, with midrib and 5–6(7) lateral nerves slightly prominent below but plane or almost so above; petiole 2–7(9) mm. long, softly pubescent; stipules c. 0·5–1 mm. long, linear-filamentous with broad base, reddish-brown, pubescent, not united. Flowers ∞ in ± lax simple axillary dichasia with accessory branches; peduncles and inflorescence-branches quadrangular, pubescent; buds c. 0·8 mm. long, cylindric; pedicels 0·5–2 mm. long; bracts c. 0·7 mm. long, triangular, entire, pubescent. Sepals greenish with paler margin, c. 0·3 mm. long, equal, ovate-triangular to semicircular, obtuse to rounded, entire, pubescent, united in the lower $\frac{1}{3}$–$\frac{1}{2}$. Petals greenish to brownish, 0·8–1 mm. long, narrowly oblong, rounded, entire, with margin ± incurved, glabrous, imbricate in bud, suberect or

± spreading at anthesis. Disk single, shallowly 3-lobed, surrounding the ovary. Stamens 3, with filaments short, narrow; anthers extrorse, circular, 1-thecous. Ovary with short style; stigmas completely united; ovules 2 per loculus. Mericarps pale green or yellow-green, 3·7–5·2(8) × 1·4–1·9 cm., flattened, narrowly oblong to oblanceolate, obtuse to rounded at the apex, 2-seeded, striate. Seeds winged, with veins marginal and submedian.

Caprivi Strip. 4·8 km. S. of Katima Mulilo on Ngoma road, 915 m., fl. 22.xii.1958, *Killick & Leistner* 3021 (K; PRE; SRGH). **Bechuanaland Prot.** N: near Kasane, fr. 12.vii.1937, *Pole Evans* 4203 (K; PRE). **N. Rhodesia.** B: Sesheke, S. of Namena Forest, Machili-Mwandi (old road), 17·6 km., fl. 21.xii.1952, *Angus* 989 (BM; K). C: Lusaka, fl. 6.xii.1957, *Fanshawe* 4112 (FHO; K). S: Mazabuka Distr., Pemba Forest Reserve, 4·8 km. S. of Pemba, fr. 17.i.1952, *White* 1940 (FHO; K). **S. Rhodesia.** W: Shangani Distr., Gwampa Forest Reserve, fr. ii.1955, *Goldsmith* 75/55 (LISC; SRGH). **Mozambique.** MS: Madanda Forests, c. 120 m., fl. 5.xii.1906, *Swynnerton* 1235 (BM).

Also in central Tanganyika and SW. Africa; dry forest, woodland or thickets, often on Kalahari sand, 120–915 m.

H. parviflora has been confused with *H. buchananii*, but differs in several well-marked characters e.g. glabrous petals, shorter sepals, simple inflorescence, leaf-shape and -margin, persistently 4-lined branches. It is more closely related to *H. indica*, but can be distinguished from the latter by the pubescent stem, leaves etc., smaller flowers with greenish (not yellowish) petals, usually larger fruits and markedly discolorous leaves.

12. **Hippocratea buchananii** Loes. in Engl., Bot. Jahrb. **19**: 235 (1894); op. cit. **34**: 106 (1904).—Eyles in Trans. Roy. Soc. S. Afr. **5**: 405 (1916).—Burtt Davy & Hoyle, N.C.L.: 46 (1936).—Brenan, T.T.C.L.: 245 (1949).—White, F.F.N.R.: 217 (1962). TAB. **85** fig. D. Syntypes from Tanganyika (Usambaras) and Nyasaland, *Buchanan* 248 (B†; K).

Hippocratea menyharthii Schinz, Pl. Menyharth.: 60 (1905).—Gomes e Sousa, Pl. Menyhart.: 78 (1936). Type: Mozambique, Boroma, *Menyhart* 692 (K; Z, holotype).

Reissantia buchananii (Loes.) Hallé [in Bull. Mus. Hist. Nat. Par., Sér. 2, 30: 466 (1958) *sine basion.*] in Mém. I.F.A.N. **64**: 84 (1962). Syntypes as for *Hippocratea buchananii*.

Shrub or small tree, often scandent or scrambling, 1·5–7·5 m. high; stems greenish-brown, 4-lined and densely fawn-pubescent when young, becoming terete, reddish- to greyish-brown, persistently pubescent, with numerous small lenticels. Leaf-lamina deep green to yellowish-green above, paler below, sometimes with brownish nervation, 4·6–9·5(15) × 2·5–5·8(6·5) cm., rhombic to oblong-elliptic or obovate, obtuse to acute or shortly acuminate (rarely rounded) at the apex, with margin shallowly glandular-crenulate-serrulate to entire, cuneate at the base, chartaceous to membranous, softly fawn-pubescent on midrib and nerves and sparsely on rest of lamina, with midrib and c. 5 lateral nerves slightly prominent below but plane or almost so above; petiole 4–10 mm. long, softly fawn-pubescent; stipules c. 0·5 mm. long, triangular, entire, reddish-brown, fawn-pubescent, not united. Flowers ∞, sweetly scented, in lax to rather dense compound axillary submonochasial cymes with accessory branches; peduncles and inflorescence branches terete, densely fawn-pubescent; buds c. 1 mm. long, cylindric-ellipsoid; pedicels 1–1·5 mm. long; bracts 1–3 mm. long, linear, entire, fawn-pubescent outside, red-brown and glabrous within. Sepals greenish, 0·5–0·8 mm. long, equal, oblong-linear, rounded, entire, fawn-pubescent outside, glabrous within, free almost to the base. Petals yellowish-brown, c. 1 mm. long, narrowly oblong, rounded, entire, with margin incurved, fawn-puberulous outside, glabrous within, imbricate in bud, ± spreading at anthesis. Disk single, shallowly 3-lobed, surrounding the ovary. Stamens 3, with filaments short, narrow; anthers extrorse, circular, 1-thecous. Ovary with short style; stigmas completely united; ovules 2 per loculus. Mericarps pale green to yellowish, 4–5 × 1·6–2·2 cm., flattened, obovate to oblong, rounded to obtuse or sometimes acuminate and ± bicornute at the apex, 2-seeded, striate. Seeds winged, with veins marginal and submedian.

Bechuanaland Prot. N: Kasane, Chobe swamps, fr. 28.ix.1949, *Pole Evans* 4617 (PRE). **N. Rhodesia.** B: Nangweshi, 1035 m., fr. 23.vii.1952, *Codd* 7155 (BM; K; PRE; SRGH). E: Luangwa R., 520 m., *E. M. & W.* 1193 (BM; LISC; SRGH). S: Mazabuka Distr., slopes of Kariba Hills, fl. 4.xii.1957, *Goodier* 436 (K; LISC; SRGH).

S. Rhodesia. N: Urungwe Distr., Sunde Gorge, 760 m., fl. 26.xi.1953, *Wild* 4274 (K; LISC; SRGH). W: Wankie Distr., Deka R., fl. 25.xii.1952, *Lovemore* 346 (SRGH). E: Melsetter Distr., near Hot Springs, fl. & fr. 29.xii.1948, *Chase* 1462 (BM; COI; K; SRGH). S: Nuanetsi Distr., between Chikadziwa and Chilonja, 335 m., fl. xi.1956, *Davies* 2214 (K; LISC; SRGH). **Nyasaland.** S?: without precise locality, fl. 1891, *Buchanan* 248 (K). **Mozambique.** T: Msusa to Máguè, 16 km., fr. 27.vii.1950, *Chase* 2809 (BM; COI; K; SRGH). Ms: Marromeu, Lacerdónia, 8.v.1942, *Torre* 4110 (LISC).

Also in Tanganyika. In *Colophospermum mopane* woodland or dry scrub in river valleys, 215–1035 m. in our area.

13. **Hippocratea pallens** Planch. ex Oliv., F.T.A. **1**: 367 (1868).—Loes. in Engl., Bot. Jahrb. **34**: 103, t. 1 fig. A (1904).—Brenan, T.T.C.L.: 246 (1949).—Keay & Blakelock, F.W.T.A. ed. 2, **1**, 2: 627 (1958).—White, F.F.N.R.: 217 (1962). TAB. **85** fig. E. Syntypes from Sierra Leone and Angola.

Hippocratea buchholzii Loes. in Engl., Bot. Jahrb. **19**: 234 (1894); op. cit. **34**: 103 (1904).—Exell & Mendonça, C.F.A. **2**, 1: 16 (1954) pro parte excl. specim. *Nolde* 358 et *Milne-Redhead* 4243. Type from Cameroun.

Hippocratea verdickii De Wild., Ann. Mus. Cong., Bot., Sér. 4, **1**, 3: 208 (1903). Type from the Congo (Katanga).

Hippocratea chariensis A. Chev., Étud. Fl. Afr. Centr. **1**: 57 (1913) *in synon.*

Hippocratea pallens var. *keniensis* Loes. apud R.E. & T.C.E. Fr. in Notizbl. Bot. Gart. Berl. **9**: 323 (1925). Type from Kenya.

Hippocratea oliverana Hutch. & Moss., F.W.T.A. **1**: 449 (1928) pro parte excl. specim. *Warnecke* 157, 302. Syntypes from Sierre Leone, Ghana and Nigeria.

Apodostigma pallens (Planch. ex Oliv.) Wilczek in Bull. Jard. Bot. Brux. **26**: 403, t. 2 (1956); F.C.B. **9**: 144, t. 18 (1960).—Hallé in Mém. I.F.A.N. **64**: 95, tt. 34–37 (1962). Syntypes as for *Hippocratea pallens.*

Liane or scandent shrub, to over 5 m. high (over 30 m. in W. Africa), glabrous; stems greyish-brown, smooth and slightly 4-lined or terete when young, soon becoming terete, without lenticels. Leaf-lamina ± deep green above, paler below, often ± glaucous, 3·5–9·2 × 1·7–3·6 cm. in our area (to 12 × 6·6 cm. in Cameroun), oblong or elliptic-ovate or obovate, acuminate at the apex, with margin entire (in our area) to undulate or glandular-serrulate, cuneate to truncate and decurrent at the base, thin to subcoriaceous, with midrib and 4–6 lateral nerves plane or slightly prominent on both sides, and reticulate venation often visible below; petiole 5–12(14) mm. long; stipules 0·2–0·5 mm. long, broadly triangular to semi-lunar, fimbriate, not united. Flowers ∞, scented, in lax or ± dense simple axillary dichasia; peduncles and inflorescence branches quadrangular; buds c. 0·6 mm. long, globose; pedicels 0·2–1(2·5) mm. long; bracts 0·5–0·8 mm. long, triangular, acute, entire or eroded-denticulate, with fimbriate basal auricles, persistent. Sepals greenish, c. 0·5 mm. long, unequal, ovate to oblong, obtuse to rounded, with margin irregularly ciliolate or subentire, free or united in the lower $\frac{1}{3}$. Petals green or white to yellow or orange, 1–1·5 mm. long, obovate to oblong-oblanceolate, rounded, entire, with margins plane, imbricate in bud, spreading or suberect and grouped 2 : 2 : 1 over the stamens at anthesis. Disk single, deeply 3-lobed, each almost separate concave lobe enclosing a stamen. Stamens 3, with filaments short, broad; anthers extrorse, circular, 1-thecous. Ovary with 3 sessile divergent stigmas, each curved down over a stamen; ovules 2 per loculus. Mericarps green tinged purplish-brown, c. 3 × 1 cm., flattened, narrowly oblong, obtuse to rounded at the apex, 2-seeded, almost smooth. Seeds winged, with veins marginal and submedian.

N. Rhodesia. W: Mwinilunga Distr., Lunga R. at the Boma, fl. 11.vii.1955, *Holmes* 1157 (K). **Mozambique.** N: Mechanga, c. 5 km. N. of Mocímboa da Praia, fl. 6.iv.1961, *Gomes e Sousa* 4695 (COI). Z: Bajone, between Namuera and Murroa, fr. 2.x.1949, *Barbosa & Carvalho* in *Barbosa* 4286 (K; L.MJ). MS: Buzi, Madanda Forest, c. 120 m., fl. 5.xii.1906, *Swynnerton* 1234 (BM; K; SRGH). SS: Chibuto-Alto Changane, 12 km., fl. 12.ii.1959, *Barbosa & Lemos* in *Barbosa* 8397 (COI; K; LMJ; PRE; SRGH).

From Senegal to Ethiopia and southward to N. Rhodesia and Angola; also in the east from Kenya to Mozambique; in lowland rain forest, dense deciduous woodland and thickets, 0–c. 1500 m.

The concept of *H. pallens* adopted here is that of Hallé (loc. cit.) and Wilczek (loc. cit.), both of whom agree that the large thick-leaved forms with large inflorescences are linked with the typical form by intermediates. It may be possible to recognise some infraspecific taxa, in which case the plants in Mozambique (with thin relatively narrow leaves showing markedly reticulate venation below) could be treated as a subspecies.

12. CAMPYLOSTEMON Welw. ex Hook. f.

Campylostemon Welw. ex Hook. f. in Benth. & Hook., Gen. Pl. **1**: 998 (1862).
Hippocrateaceae tribu *Campylostemonae* Hallé in Mém. I.F.A.N. **64**: 132 (1962).

Lianes or ± scandent shrubs, with latex. Leaves opposite, petiolate; stipule; free, caducous. Inflorescence pedunculate, cymose, dichasial, simple, without axillary branches, axillary and sometimes terminal; bracts persistent. Flowers bisexual. Sepals 5, imbricate, free or ± united. Petals (4)5, imbricate in buds Disk absent. Stamens 3–5, with filaments united at the base, erect or incurved. anthers versatile, apical or introrse (? rarely extrorse), with thecae ± confluent, dehiscing transversely; pollen in tetrads. Ovary superior, 3-locular, with 5–16 axile ovules per loculus, sessile; style absent; stigmas 3, free, divergent. Fruit of 3 capsular mericarps united at the base; mericarps dorsiventrally flattened, dehiscing by the median suture. Seeds with long winged stalks as in *Hippocratea*. Germination epigeal (1 species). Chromosome number: 2n = 56 (1 species).

A genus of c. 10 species in tropical Africa. The species placed in *Bequaertia* and *Tristemonanthus* agree with typical *Campylostemon* in so many ways that differences in the number of stamens and their posture (erect or incurved) do not seem sufficient to warrant generic segregation. I have therefore included the characters of these genera in the above description of *Campylostemon*.

Campylostemon angolense Oliv. in Journ. Linn. Soc., Bot. **10**: 44 (1867); F.T.A. **1**: 366 (1868).—Loes. in Notizbl. Bot. Gart. Berl. **13**: 574 (1937); in Engl. & Prantl, Nat. Pflanzenfam. ed. 2, **20b**: 112 (1942).—Exell & Mendonça, C.F.A. **2**, 1: 11 (1954) pro parte excl. specim. *Gossweiler* 644, 6230, 9266.—Keay & Blakelock, F.W.T.A. ed. 2, **1**, 2: 626 (1958) pro parte excl. syn. *C. warneckeanum* et *C. kennedyi*.—Hallé, Thèses Fac. Sci. Univ. Par., Monogr. Hipp. Afr. Occ.: 131 (1958); in Mém. I.F.A.N. **64**: 138 (1962).—Wilczek, F.C.B. **9**: 179 (1960). TAB. **86.** Type from Angola (Cazengo).

Campylostemon duchesnei De Wild. & Dur., Bull. Soc. Roy. Bot. Belg. **39**: 57 (1900); in Ann. Mus. Cong., Bot., Sér. 1, **1**: 141, t. 71 (1900). Type from the Congo (Kasai).

Hippocratea chevalieri Hutch. & Moss, F.W.T.A. **1**: 449 (1928). Type from Ivory Coast.

Liane or scandent shrub up to 24 m. high, glabrous; stems greenish-purple to reddish-purple, smooth and narrowly 4-winged when young, becoming rugose-lenticellate, terete and eventually greyish. Leaf-lamina deep green above, somewhat paler below, (3)4–9(14·5) × 1·6–5 cm., elliptic to oblong or ovate-lanceolate, narrowly acuminate to subcaudate at the apex, with margin shallowly to rather deeply crenulate-serrulate, cuneate to rounded at the base, papyraceous to chartaceous, with midrib prominent on both sides, 6–7 lateral nerves and reticulate venation not or scarcely prominent; petiole 4–9 mm. long; stipules c. 1·5 mm. long, lanceolate, fimbriate. Flowers c. 10–25, slightly scented, in corymbose axillary dichasia 1·5–3(3·5) cm. long; peduncles and inflorescence-branches 4-winged, smooth; buds 1–1·5 mm. long, subglobose; pedicels 2–4 mm. long, smooth or rugulose; lower bracts (2)3–4 mm. long, linear to narrowly triangular, upper bracts c. 1 mm. long, ovate-triangular, acute, ciliate. Sepals c. 1 mm. long, broadly ovate, rounded, with margin eroded-denticulate, united at the base. Petals white to yellow turning yellow to orange, 2–3·2 mm. long, oblong to elliptic-oblong, rounded, entire. Stamens 5, with filaments united at the base; anthers reddish, introrse, 4-lobed. Ovary with 3 very short sessile divergent stigmas; ovules 5–9 per loculus. Mericarps 4–6 × 2·2–4 cm., ovate, 4–9-seeded. Seeds winged, with veins marginal and submedian.

N. Rhodesia. N: Lake Mweru, fl. 12.xi.1957, *Fanshawe* 3925 (K; LISC; SRGH).

In Liberia, Ivory Coast, Congo, Uganda, N. Rhodesia and Angola; in fringing forest and " muteshi " thicket.

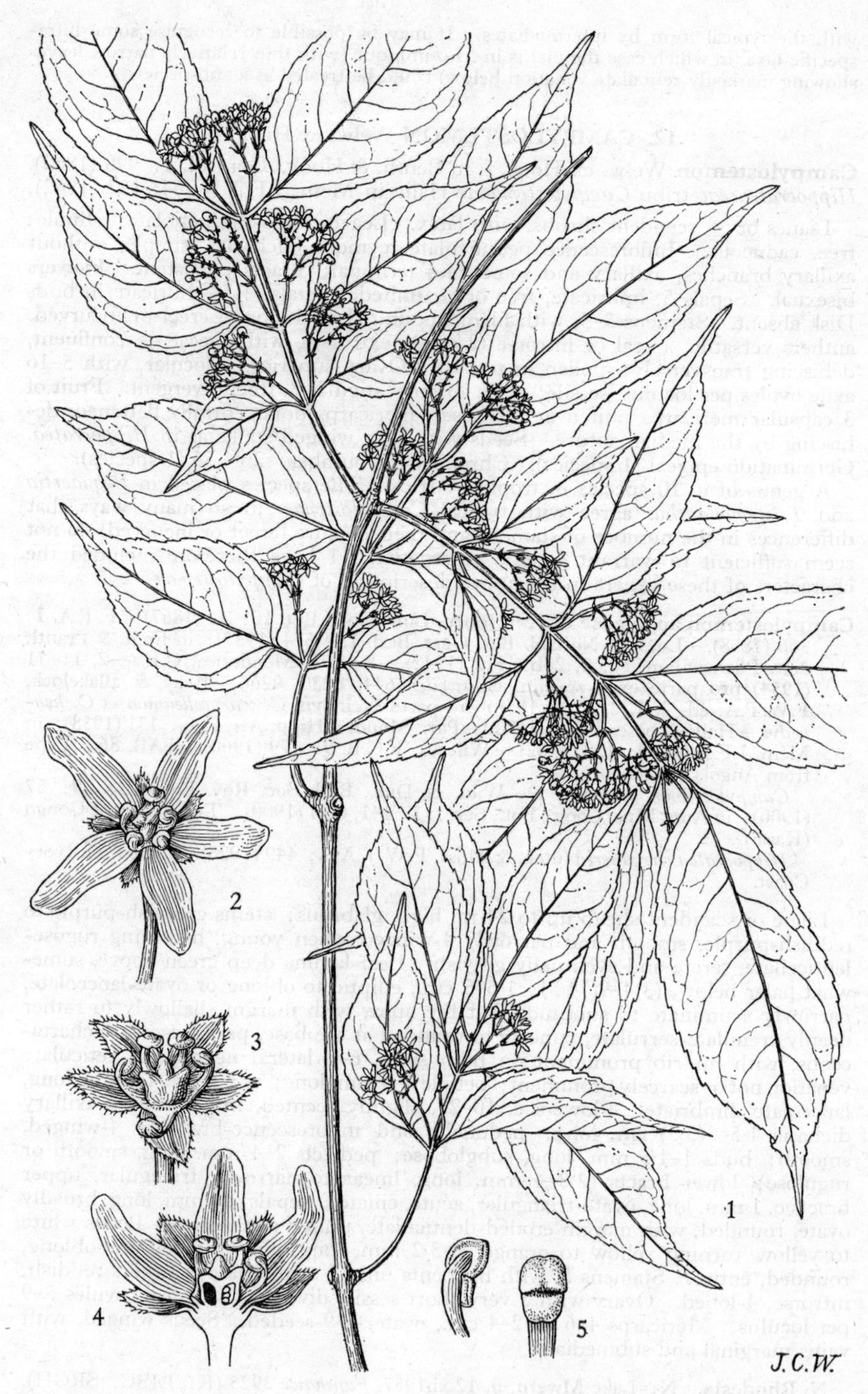

Tab. 86. CAMPYLOSTEMON ANGOLENSE. 1, flowering shoot ($\times \frac{2}{3}$); 2, flower ($\times 6$); 3, flower with petals removed ($\times 10$); 4, vertical section of flower ($\times 10$); 5, upper part of stamen (2 views) ($\times 20$), all from *Welwitsch* 1331.

53. RHAMNACEAE

By R. B. Drummond

Trees, shrubs, shrublets or lianes, glabrous or with simple hairs; branches rarely with coiled tendrils; leaves alternate or rarely opposite, simple, entire to toothed, petiolate, penninerved or 3–5-nerved from the base; stipules present, rarely interpetiolar, sometimes spinescent. Flowers often in axillary cymes or umbels (rarely solitary), or in racemes arranged in terminal panicles or thyrses, bisexual (rarely unisexual), actinomorphic. Receptacle flattish to obconic or hemispherical. Sepals (4)5, valvate. Petals (4)5 or absent, usually smaller than the sepals and unguiculate, cucullate, closely surrounding the stamens. Stamens (4)5, antipetalous; filaments free; anthers 2-thecous (rarely 1-thecous), introrse, dehiscing longitudinally. Disk usually present and well developed, intrastaminal, perigynous, very variable in shape, large, filling the receptacle or cup-shaped with free margins, or lining the receptacle; ovary syncarpous, sessile, free or immersed in the disk, superior, subinferior or inferior, 2–4-locular; style entire or 2–4-lobed; ovules solitary in each loculus, erect, anatropous. Fruit a drupe or septicidal capsule or schizocarp, (1)2–3(4)-locular, sometimeswinged. Seeds 1 in each loculus; embryo large, straight; endosperm usually copious.

Maesopsis eminii Engl. a native of the Sudan and Uganda, extending westwards to the Caméroun Republic and southwards to Angola, is planted for forestry purposes at Ndola. *Hovenia dulcis* Thunb., a native of China, is grown in gardens in S. Rhodesia.

Leaves alternate or, if subopposite, stipules not interpetiolar:
  Ovary superior or subinferior; branches without coiled tendrils:
    Plants armed with stipular or axillary spines:
      Leaves penninerved; spines not stipular - - - - - - - **4. Scutia**
      Leaves 3–5-nerved from the base; branches often with stipular spines **1. Ziziphus**
    Plants unarmed or branches spine-tipped:
      Fruit a fleshy drupe:
        Leaves 3–5-nerved from base - - - - - - - **1. Ziziphus**
        Leaves penninerved:
          Fruit 2-locular, ellipsoid, leaves entire; flowers 5-merous - **2. Berchemia**
          Fruit 2–4-locular; leaves usually crenate or serrate, if entire then flowers 4-merous - - - - - - **3. Rhamnus**
      Fruit dry, capsular:
        Leaf-lamina 3–7·5 cm. long, 3–5-nerved at base, toothed - **5. Colubrina**
        Leaf-lamina under 1·5 cm. long, entire, not several-nerved from the base **6. Phylica**
  Ovary inferior; fruit a capsule or schizocarp; branches often with coiled tendrils:
    Leaves toothed; schizocarp 3-winged or 3-angled - - - - **7. Gouania**
    Leaves entire; capsule obovoid or globose - - - - - - **8. Helinus**
Leaves opposite, stipules interpetiolar - - - - - - - **9. Lasiodiscus**

1. ZIZIPHUS Mill.

Ziziphus Mill., Gard. Dict. Abridg. ed. **3** (1754).

Trees, shrubs or shrublets. Leaves alternate, petiolate; lamina with margin dentate to serrulate, often markedly asymmetric at the base, with 3–5 nerves from the base; stipules frequently spinescent. Cymes axillary or rarely terminal, sessile or pedunculate. Flowers bisexual, pedicellate. Receptacle obconic. Sepals 5. Petals 5 (or absent), cucullate. Stamens 5, inserted under the edge of the disk. Disk flat, covering the receptacle, 5–10-lobed or rarely entire, with margin free. Ovary immersed in the disk and adnate to the receptacle, 2(4)-locular; ovules 1 in each loculus; style 2(4)-lobed. Fruit a drupe with fleshy exocarp and woody endocarp, (1)2(4)-seeded. Seeds with a thin shining testa.

Branches (some at least) with paired stipular spines:
  Shrublet not more than 60 cm. high - - - - - - - 4. *zeyherana*

Trees or shrubs:
Leaf-bases cuneate to rounded, subequal or only slightly asymmetric 1. *mauritiana*
Leaf-bases markedly asymmetric:
Leaf-lamina with nerves impressed above, always glabrous above and persistently tomentose below - - - - - - - - - - 2. *abyssinica*
Leaf-lamina with nerves not markedly impressed above, glabrous or pubescent above, glabrescent or with a coarse pale-brown tomentum below 3. *mucronata*
Branches without paired stipular spines:
Leaf-bases markedly asymmetric - - - - - - - - - 3. *mucronata*
Leaf-bases not markedly asymmetric:
Cymes sessile; style 3-fid; ovary 3-locular; fruit globose - - 5. *rivularis*
Cymes pedunculate; style 2-fid; ovary 2-locular; fruit ovoid - 6. *pubescens*

1. **Ziziphus mauritiana** Lam., Encycl. Méth., Bot. **3**: 319 (1789).—Brenan, T.T.C.L.: 469 (1949).—Suesseng. in Engl. & Prantl, Nat. Pflanzenfam. ed. 2: **20d**: 124 (1953) pro parte.—Verdcourt in Bull. Jard. Bot. Brux. **27**: 354 (1957).—Keay, F.W.T.A ed. 2, **1**, 2: 668 (1958).—Evrard, F.C.B. **9**: 440 (1960).—Dale & Greenway, Kenya Trees and Shrubs: 394 (1961).—White, F.F.N.R.: 228 (1962). TAB. **87** fig. C. Type from Mauritius.
Rhamnus jujuba L., Sp. Pl. **1**: 194 (1753). Type from India.
Ziziphus jujuba (L.) Lam., Encycl. Méth., Bot. **3**: 318 (1789) non Mill. (1768).—Hemsl. in Oliv., F.T.A. **1**: 379 (1868) pro parte.—Sim, For. Fl. Port. E. Afr.: 35 (1909) pro parte excl. var. *nemoralis*.—Perrier, Fl. Madag., Rhamnac.: 11 (1950). Type as above.

Shrub or tree up to 15 m. tall; bark greyish; branchlets densely white-tomentellous when young. Leaf-lamina 3–8 × 1·5–5 cm., elliptic to broadly elliptic, apex rounded or subacute to emarginate, margin minutely serrulate, base cuneate to rounded, subequal or only slightly asymmetric, 3-nerved from the base almost to the apex, upper surface glabrous, lower surface covered with a persistent white tomentum; petiole up to 10 mm. long; stipules mostly spinescent, one hooked and one straight at each node or both hooked or only one or neither developed into a spine. Cymes tomentose, sessile or very shortly pedunculate; pedicels 2–8 mm. long. Sepals 2 mm. long, deltate, pubescent outside. Petals 2 mm. long, unguiculate with expanded portion circular, 1·5 mm. in diam. Stamens with filaments up to 2 mm. long. Disk 3 mm. in diam. Ovary immersed in the disk, 2-locular; style 2-fid, 1 mm. long. Fruit edible, up to 2 cm. in diam., 2-seeded. Seeds compressed.

N. Rhodesia. E: Petauke Distr., Luangwa R., fl. & fr. 17.iv.1952, *White* 2700 (BM; FHO; K; NDO). S: Chirundu, fl. 26.ii.1953, *Wild* 4050 (K; SRGH). **S. Rhodesia.** N: Darwin Distr., Mkumbura R., fl. 23.i.1960, *Phipps* 2393 (K; SRGH). C: Salisbury, Mabelreign, st. i.1965, *Wild* 6770 (SRGH). **Nyasaland.** S: Port Herald Distr., fl. 19.ii.1960, *Phipps* 2561 (K; SRGH). **Mozambique.** N: Lúrio, fl. & fr. 28.x.1953, *Balsinhas* 33 (BM; K; LISC; LMJ). Z: Quelimane, st. 1908, *Sim* 21204 (PRE). T: between Chicoa and Fíngoè, fl. 26.vi.1949, *Barbosa & Carvalho* 3282 (K; LMJ). MS: Chemba, Chiou, fl. & fr. 13.iv.1960, *Lemos & Macuácua* 98 (BM; K; LISC; LMJ; PRE; SRGH). SS: Inhambane, fl. & fr. xii.1936, *Gomes e Sousa* 1944 (COI; K; LISC).

Widely distributed in the Old World tropics and subtropics; in tropical Africa from Senegal to Ethiopia and south to the Congo and Tanganyika. In our area it is not native but widely naturalized, especially in the Zambezi Valley. Cultivated over most of its range for its edible fruit.

2. **Ziziphus abyssinica** Hochst. ex A. Rich., Tent. Fl. Abyss. **1**: 136 (1847).—Brenan, T.T.C.L.: 469 (1949); in Mem. N.Y. Bot. Gard. **8**, 3: 239 (1953).—Exell & Mendonça, C.F.A. **2**, 1: 29 (1954).—Verdcourt in Bull. Jard. Bot. Brux. **27**: 353 (1957).—Keay, F.W.T.A. ed. 2, **1**, 2: 669 (1958).—Evrard, F.C.B. **9**: 440 (1960) pro parte.—Dale & Greenway, Kenya Trees and Shrubs: 393 (1961).—White, F.F.N.R.: 228 (1962). TAB. **87** fig. B. Type from Ethiopia.
Ziziphus jujuba sensu Hemsl. in Oliv., F.T.A. **1**: 379 (1868) pro parte quoad syn. *Z. abyssinica*.—Eyles in Trans. Roy. Soc. S. Afr. **5**: 406 (1916) pro parte quoad specim. Rogers.
Ziziphus jujuba var. *nemoralis* Sim, For. Fl. Port. E. Afr.: 35 (1909). Type: Mozambique, *Sim*.
Ziziphus mauritiana sensu Wild, Guide Fl. Vict. Falls: 151 (1952).—Suesseng. in Engl. & Prantl, Nat. Pflanzenfam. ed. 2, **20d**: 124 (1953) pro parte.

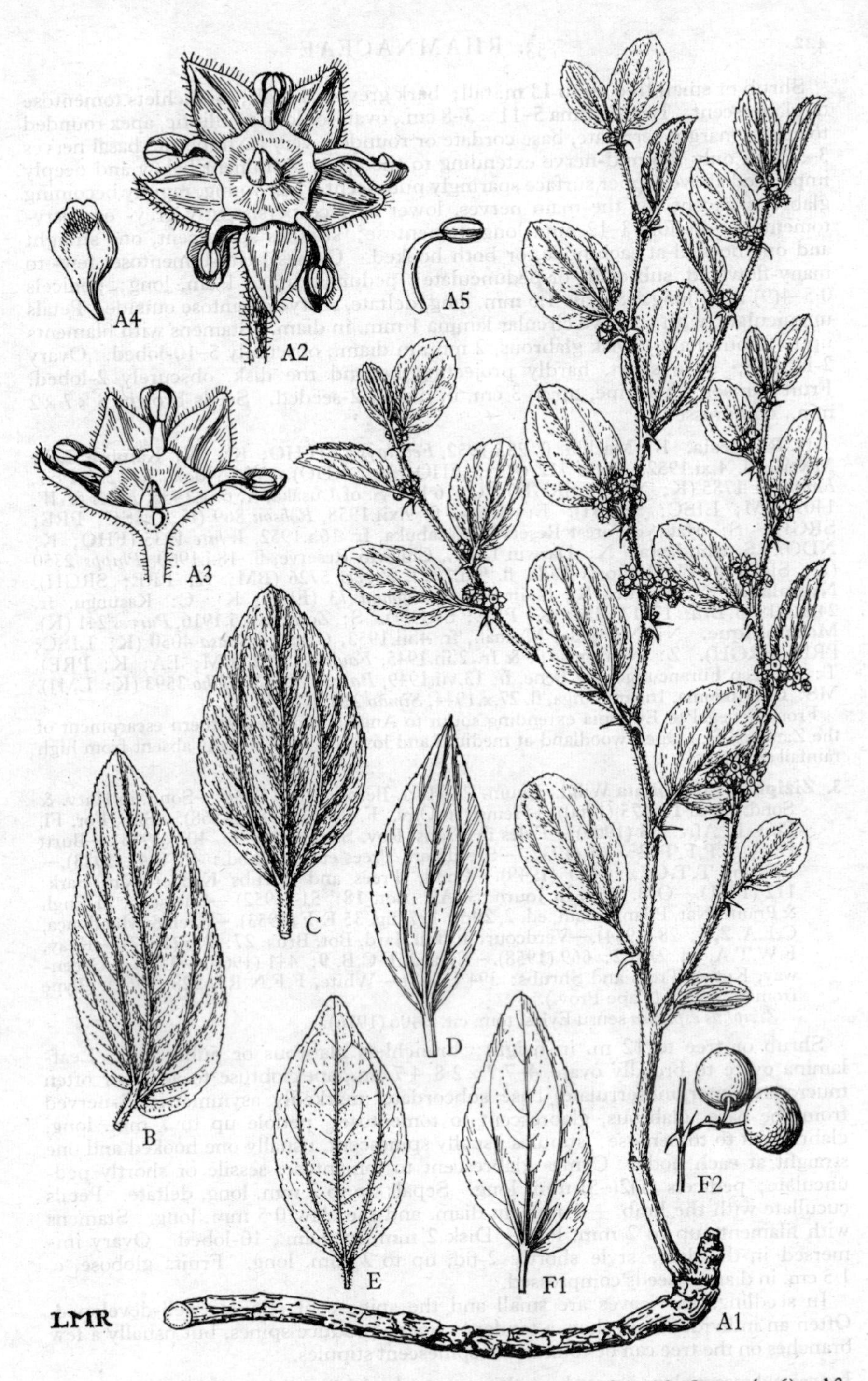

Tab. 87. A.—ZIZIPHUS ZEYHERANA. A1, flowering shoot ($\times \frac{2}{3}$); A2, flower ($\times 6$); A3, vertical section of flower ($\times 6$); A4, petal ($\times 8$); A5, stamen ($\times 8$). A1 from *Drummond* 4923, A2–A5 from *Story* 4826. B.—ZIZIPHUS ABYSSINICA, leaf ($\times \frac{2}{3}$) *Richards* 4779. C.—ZIZIPHUS MAURITIANA, leaf ($\times \frac{2}{3}$) *Barbosa & Carvalho* 3253. D.—ZIZIPHUS RIVULARIS, leaf ($\times \frac{2}{3}$) *van der Schijff* 3974. E.—ZIZIPHUS PUBESCENS SUBSP. PUBESCENS, leaf ($\times \frac{2}{3}$) *White* 2404. F.—ZIZIPHUS MUCRONATA SUBSP. MUCRONATA. F1, leaf ($\times \frac{2}{3}$) *Myres* 730; F2, portion of fruiting branch ($\times \frac{2}{3}$) *Gomes e Sousa* 3696.

Shrub or small tree up to 13 m. tall; bark greyish, rough; branchlets tomentose or glabrescent. Leaf-lamina 5–11 × 3–8 cm., ovate-oblong to elliptic, apex rounded to acute, margin serrulate, base cordate or rounded, very asymmetric, basal nerves 3–5 with only the mid-nerve extending to the apex, prominent below and deeply impressed above, upper surface sparingly pubescent when young, rapidly becoming glabrous except for the main nerves, lower surface persistently grey- or rusty-tomentose; petiole 3–12 mm. long, tomentose; stipules spinescent, one straight and one hooked at each node, or both hooked. Cymes rusty-tomentose, few- to many-flowered, subsessile or pedunculate; peduncle up to 1 cm. long; pedicels 0·5–4(9) mm. long. Sepals 1·5 mm. long, deltate, rusty-tomentose outside. Petals unguiculate, with concave circular lamina 1 mm. in diam. Stamens with filaments up to 2 mm. long. Disk glabrous, 2 mm. in diam., obscurely 5–10-lobed. Ovary 2-locular; style short, hardly projecting beyond the disk, obscurely 2-lobed. Fruit crimson when ripe, up to 3 cm. in diam., 2-seeded. Seeds brown, 8 × 7 × 2 mm., compressed.

N. Rhodesia. B: Machili, fl. 28.i.1952, *Fewdays* 11 (FHO; K). N: Kafulwe, Lake Mweru, fl. 4.xi.1952, *Angus* 712 (BM; FHO; K; NDO). W: Ndola, fl. 12.xi.1957, *Fanshawe* 1785 (K; NDO; SRGH). C: 16 km. S. of Lusaka, fr. 6.iv.1955, *E.M. & W.* 1409 (BM; LISC; SRGH). E: Sasare, fl. 9.xii.1958, *Robson* 869 (K; LISC; PRE; SRGH). S: Magoye Forest Reserve, Mazabuka, fr. 16.i.1952, *White* 1935 (FHO; K; NDO). **S. Rhodesia.** N: Darwin Distr., Chiswiti Reserve, fl. 18.i.1960, *Phipps* 2350 (K; SRGH). W: Victoria Falls, fl. 9.ii.1912, *Rogers* 5726 (BM; K; PRE; SRGH). **Nyasaland.** N: Rukuru R., fl. iii.1953, *Chapman* 73 (FHO; K). C: Kasungu, fr. 24.viii.1946, *Brass* 17413 (BM; K; PRE; SRGH). S: Zomba, fl. i.1916, *Purves* 241 (K). **Mozambique.** N: Nampula, Mutuáli, fr. 4.iii.1953, *Gomes e Sousa* 4050 (K; LISC; PRE; SRGH). Z: Namagoa, fl. & fr. 2.iii.1945, *Faulkner* P62 (BM; EA; K; PRE). T: between Furancungo and Bene, fr. 13.vii.1949, *Barbosa & Carvalho* 3593 (K; LMJ). MS: Cheringoma, Inhamitanga, fl. 27.x.1944, *Simão* 216 (LISC).

From Senegal to Ethiopia extending south to Angola and the southern escarpment of the Zambezi R. Open woodland at medium and low altitudes, usually absent from high rainfall areas.

3. **Ziziphus mucronata** Willd., Enum. Pl. Hort. Berol.: 251 (1809).—Sond. in Harv. & Sond., F.C. **1**: 475 (1860).—Hemsl. in Oliv., F.T.A. **1**: 380 (1868).—Sim, For. Fl. Port. E. Afr.: 85 (1909).—Eyles in Trans. Roy. Soc. S. Afr. **5**: 407 (1916).—Burtt Davy, F.P.F.T. **2**: 469 (1932).—Steedman, Trees etc. S. Rhod.: 46, t. 45 (1933).—Brenan, T.T.C.L.: 470 (1949).—Codd, Trees and Shrubs Kruger Nat. Park: 112 (1951).—O.B. Mill. in Journ. S. Afr. Bot. **18**: 51 (1952).—Suesseng. in Engl. & Prantl, Nat. Pflanzenfam, ed. 2, **20d**: 127, fig. 35 E-F (1953).—Exell & Mendonça, C.F.A. **2,** 1: 28 (1954).—Verdcourt in Bull. Jard. Bot. Brux. **27**: 355 (1957).—Keay, F.W.T.A. ed. 2, **1,** 2: 669 (1958).—Evrard, F.C.B. **9**: 441 (1960).—Dale & Greenway, Kenya Trees and Shrubs: 394 (1961).—White, F.F.N.R.: 228 (1962). Type from S. Africa (Cape Prov.).

Ziziphus espinosa sensu Eyles, tom. cit.: 406 (1916).

Shrub or tree to 12 m. in height; branchlets glabrous or pubescent. Leaf-lamina ovate to broadly ovate, 4–7·7 × 2·8–4·7 cm., apex obtuse to acute or often mucronate, margin serrulate, base subcordate, markedly asymmetric, 3-nerved from the base, glabrous, glabrescent to tomentose; petiole up to 7 mm. long, glabrescent to tomentose; stipules usually spinescent, usually one hooked and one straight at each node. Cymes glabrescent to tomentose, sessile or shortly pedunculate; pedicels (1)2(–5) mm. long. Sepals up to 2 mm. long, deltate. Petals cucullate with the limb ± 1 mm. in diam. and the claw 0·5 mm. long. Stamens with filaments up to 2 mm. long. Disk 2 mm. in diam., 10-lobed. Ovary immersed in the disk; style shortly 2-fid, up to 2 mm. long. Fruits globose, c. 1·5 cm. in diam. Seeds compressed.

In seedlings the leaves are small and the spinescent stipules well-developed. Often an individual will show a tendency not to produce spines, but usually a few branches on the tree can be found with spinescent stipules.

Leaves pubescent along nerves beneath, more rarely glabrous - - subsp. *mucronata*
Leaves with dense coarse pale-brown tomentum especially below, not becoming glabrous with age - - - - - - - - - - - subsp. *rhodesica*

Subsp. **mucronata.** TAB. **87** fig. F.

Bechuanaland Prot. SE: N. of Molepolole, st. 11.vi.1956, *Story* 4873 (K; PRE). **N. Rhodesia.** B: Sesheke, fl. 21.xii.1952, *Angus* 999 (K; FHO; NDO). S: Kazungula, fl. 4.i.1957, *Gilges* 702 (SRGH). **S. Rhodesia.** W: Lupane, fl. 17.xii.1952, *Whellan* 692 (K; SRGH). C: 24 km. S. of Selukwe, 24.xii.1959, *Leach* 9663 (SRGH). E: E. bank of Sabi R., Sabi-Lundi junction, 10.vi.1950, *Wild* 3489 (BM; SRGH). S: Chipinda Pools, Lundi R., fl. 10.i.1960, *Goodier* 766 (K; SRGH). **Nyasaland.** S: Port Herald, between Muona and Shire R., 20.iii.1960, *Phipps* 2595 (K; SRGH). **Mozambique.** N: between Meconta and Nampula, st. 6.v.1948, *Pedro & Pedrógão* 3178 (K). Z: between Megaza and Aguas Quentes, fl. 14.vi.1949, *Barbosa & Carvalho* 3763 (LMJ). T: 17·6 km. from Changara on Mtoko road, fr. 28.ix.1948, *Wild* 2663 (K; SRGH). MS: Bimba, Lower Buzi, 5.xii.1906, *Swynnerton* 1407 (BM; SRGH). SS: Bilene, 9.vi.1960, *Lemos & Balsinhas* 55 (K; LISC; LMJ; PRE; SRGH). LM: Maputo, fl. 17.i.1949, *Gomes e Sousa* 3944 (COI; K; PRE).

Widespread in drier areas from Senegal to Ethiopia and Arabia in the north to the Cape Province of S. Africa in the south. Common in open woodland in lower rainfall areas.

Subsp. **rhodesica** R. B. Drummond in Bol. Soc. Brot., Sér. 2, **39**: 57 (1965). Type: N. Rhodesia, Choma, fl. 27.x.1955, *Bainbridge* 169/55 (FHO; K; SRGH, holotype).
Ziziphus abyssinica sensu O.B. Mill. in Journ. S. Afr. Bot. **18**: 51 (1952).—Palgrave, Trees of Central Afr.: 371 cum photogr. et tab. (1956).
Ziziphus mauritiana sensu Suesseng. in Proc. & Trans. Rhod. Sci. Ass. **43**: 110 (1951).—Wild, S. Rhod. Bot. Dict.: 139 (1955).
Ziziphus mucronata var. A.—White, F.F.N.R.: 228 (1962).

Caprivi Strip. Linyanti area, fl. 28.xii.1958, *Killick & Leistner* 3159 (PRE; SRGH). **Bechuanaland Prot.** N: Chobe R., st. 11.vii.1937, *Erens* 387 (K; PRE). **N. Rhodesia.** N: Abercorn Distr., confluence of Lufubu R. and Mulugu R., fl. 6.x.1956, *Richards* 6375 (K; SRGH). W: S. of Matonchi Farm, fl. 16.x.1937, *Milne-Redhead* 2812 (BM; K; PRE). C: Mt. Makulu, fl. 17.xi.1956, *Angus* 1447 (BM; EA; FHO; K; PRE). S: Southern Mazabuka, fl. 13.xii.1952, *Angus* 928 (FHO; K). **S. Rhodesia.** N: Urungwe, Zwipani, fl. 27.x.1957, *Goodier* 339 (K; SRGH). W: Bulawayo, fl. xi.1903, *Eyles* 1209 (K; SRGH). C: Salisbury, fl. x.1917, *Eyles* 865 (BM; K; SRGH). E: Inyanga, fl. 14.xi.1958, *West* 3769 (K; SRGH). S: Chidumo Clinic, Ndanga, st. i.1959, *Farrell* 28 (SRGH).

Almost confined to our area, not having been recorded S. of the Limpopo and only just reaching S. Tanganyika and Katanga. Replaces *Z. mucronata* in higher-rainfall areas in *Brachystegia* woodland; especially common on termite mounds.

4. **Ziziphus zeyherana** Sond. in Harv. & Sond., F.C. **1**: 476 (1860).—Burtt Davy, F.P.F.T. **2**: 470 (1932).—Hopkins, Bacon & Gyde, Comm. Veld Fl.: 68 cum fig. (1940).—O.B. Mill. in Journ. S. Afr. Bot. **18**: 51 (1952) pro parte. TAB. **87** fig. A. Type from the Transvaal.
Ziziphus helvola Sond., loc. cit.—Burtt Davy, loc. cit. Type from the Transvaal.
Ziziphus jujuba var. *nana* Engl. in Sitz. Königl. Preuss. Akad. Wiss. Berl. **52**: 890 (1906).—Eyles in Trans. Roy. Soc. S. Afr. **5**: 407 (1916). Type: S. Rhodesia, Mashonaland, *Engler* (B, holotype†).

Shrublet up to 60 cm. tall; stems annual from a creeping rhizome; branchlets pubescent. Leaf-lamina ovate to broadly ovate, 2–5 × 1–4 cm., apex acute, rounded, truncate, retuse, or mucronate, margin serrulate, base cuneate, rounded, or subcordate, asymmetric to subequal, 3-nerved from base to near apex, tertiary venation reticulate, glabrous or glabrescent above, pubescent below; petiole 1–9 mm. long, pubescent; stipules spinescent, one hooked and one straight at each node or both hooked. Cymes pubescent, few-flowered; peduncle up to 5 mm. long; pedicels up to 5 mm. long. Sepals 2 mm. long, deltate, pubescent outside. Petals 2 mm. long, cucullate, unguiculate. Stamens with filaments up to 2 mm. long. Disk 2·5 mm. in diam., obscurely 5-lobed. Ovary immersed in the disk, 2-locular; style 2-fid, up to 2 mm. long. Fruits up to 7 mm. in diam., globose, 2-seeded. Seeds up to 5 × 5 mm., compressed.

Bechuanaland Prot. SE: without precise locality, fr. i.1946, *Miller* 403 (PRE). **S. Rhodesia.** W: Figtree, fl. 13.x.1954, *Story* 4826 (K; PRE). C: Salisbury, fl. 23.x.1955, *Drummond* 4923 (K; LISC; LMJ; SRGH).

Also in S. Africa. Often on termite mounds, in grassland and woodland from c. 1250–1500 m.

5. **Ziziphus rivularis** Codd in Bothalia, **7**: 31 (1958). TAB. **87** fig. D. Type from the Transvaal.

Shrub or small tree up to 7 m. tall; bark grey, smooth; branchlets densely puberulous with greyish curled hairs. Leaf-lamina 3·5–7 × 1·2–3·4 cm., narrowly elliptic-ovate to elliptic-ovate, apex acute, margin serrulate, base rounded, not markedly asymmetric, upper surface somewhat glossy, very prominently 3-nerved below from the base to the apex, with prominent reticulate venation on both sides, sparingly pubescent especially on the main nerves above and below but becoming glabrous with age; petiole 4–10 mm. long, pubescent; stipules up to 1 mm. long, not spinescent. Cymes sessile, of (1)3–5 flowers; pedicels 1–1·5 mm., elongating to up to 3 mm. long in fruit, pubescent. Sepals 1·5 mm. long, deltate, pubescent outside. Petals up to 1 mm. long, obovate, unguiculate. Stamens with filaments 1 mm. long; anthers medifixed. Disk obscurely 5-lobed. Ovary 3-locular; style 3-fid, c. 1 mm. long. Fruit up to 7 mm. in diam., globose, 3-seeded. Seeds up to 6 mm. long, somewhat compressed.

Mozambique. LM: between Umbeluzi and Boane, fr. 22.iv.1914, *Torre* 6503 (LISC; PRE).

Also in the northern and eastern Transvaal. Banks of streams and rocky watercourses.

6. **Ziziphus pubescens** Oliv., in Trans. Linn. Soc., Ser. 2, **2**: 330 (1887).—Brenan, T.T.C.L., 470 (1949).—Suesseng. in Engl. & Prantl, Nat. Pflanzenfam. ed. 2, **20d**: 130 (1953).—Verdcourt in Bull. Jard. Bot. Brux. **27**: 355 (1957).—Evrard, F.C.B. **9**: 442, t. 45 (1960).—Dale & Greenway, Kenya Trees and Shrubs: 395 (1961).—White, F.F.N.R.: 228 (1962). Type from Tanganyika.

Small tree up to 10(20) m. tall; young branchlets pubescent later glabrescent. Leaf-lamina 3·8–8 × 1·7–4 cm., ovate, elliptic or narrowly ovate, apex rounded to acute or subacuminate, base rounded to cuneate often slightly asymmetric, margin serrulate, 3-nerved from the base, greyish-puberulous to glabrous above, greyish-pubescent or glabrous except for main veins below; petiole 2–6(9) mm., pubescent; stipules narrowly ovate to subulate, up to 3·5 mm. long. Cymes subsessile or with a peduncle up to 4 mm. long, tomentose; pedicels up to 3 mm. long, tomentose. Sepals 1·5 mm. long, dentate, pubescent outside. Petals spathulate, 1·5 mm. long and 0·5 to 1 mm. wide in expanded portion. Stamens with filaments 2 mm. long. Disk 5-lobed, up to 2 mm. in diam. Ovary 2-locular; style 2-fid, 1 mm. long. Fruit ovoid, 10 × 8 mm. Seeds reddish-brown, 6 × 4 mm., plano-convex.

Leaf-lamina pubescent above and below - - - - subsp. *pubescens*
Leaf-lamina glabrous above and below (except sometimes with a few hairs on the main nerves below) - - - - - - - - - subsp. *glabra*

Subsp. **pubescens.** TAB. **87** fig. E.

N. Rhodesia. E: Luangwa R., st. 18.iv.1952, *White* 2404 (FHO; NDO). **S. Rhodesia.** S: Ndanga, Chitsa's Kraal, fr. 5.vi.1950, *Chase* 2288 (SRGH). **Nyasaland.** S: Chikwawa Boma, fr. 22.v.1963, *Chapman* 2073 (SRGH). **Mozambique.** MS: Nhamacolongo, fr. 6.vii.1947, *Simão* 1385 (LM; SRGH).

Also from the Sudan, Kenya, Uganda, Tanganyika and the Congo. In riverine vegetation.

Subsp. **glabra** R. B. Drummond in Bol. Soc. Brot., Sér. 2, **39**: 59 (1965).

Type: Mozambique, Sul do Save, Guijá, fl. 16.xi.1957, *Barbosa & Lemos* 8154 (COI; LISC, holotype). SS: between Mabalane and Mapai, fr. vi.1959, *Barbosa & Lemos* 8620 (COI; K; LISC; LMJ; PRE; SRGH). LM: Magude, fl. 29.xi.1944, *Mendonça* 3140 (LISC).

Endemic in southern Mozambique. Coastal or sub-coastal forest.

2. BERCHEMIA Neck. ex DC.

Berchemia Neck. ex DC., Prodr. **2**: 22 (1825) *nom. conserv. propos.*

Shrubs or trees. Leaves alternate or opposite or subopposite, petiolate; lamina entire, penninerved. Cymes axillary (or terminal). Stipules small, not spinescent. Flowers bisexual, pedicellate. Receptacle flattish. Sepals 5. Petals 5, unguiculate, cucullate. Stamens 5, inserted under the edge of the disk. Disk cupular or swollen and enveloping the ovary. Ovary free or immersed in the disk, 2-locular; ovules 1 in each loculus. Style 2-fid or notched. Fruit a drupe with fleshy exocarp and woody endocarp. Seeds 2.

Ovary free from disk; pedicels long and slender, 4–14 mm. long; leaf-lamina 1·2–4 × 0·6–2·5 cm. - - - - - - - - - - - - 1. *zeyheri*

Ovary immersed in disk; pedicels shorter and stouter, 9 mm. long in fruit; leaf-lamina 5–11 × 3–6 cm. - - - - - - - - - - - 2. *discolor*

1. **Berchemia zeyheri** (Sond.) Grubov in Act. Inst. Bot. Acad. Sci. URSS, Ser. 1, **8**: 374 (1949). TAB. **88** fig. B. Type from the Transvaal.
Rhamnus zeyheri Sond. in Harv. & Sond., F.C. **1**: 477 (1860).—Sim, For. Fl. Port. E. Afr.: 36, t. 26 (1909).—Burtt Davy, F.P.F.T. **2**: 470 (1932).—Steedman, Trees etc. S. Rhod.: 45 (1933).—Codd, Trees and Shrubs Kruger Nat. Park: 112 (1951).—O.B. Mill. in Journ. S. Afr. Bot. **18**: 51 (1952).—Suesseng. in Engl. & Prantl, Nat. Pflanzenfam. ed. 2, **20d**: 63 (1953).—Cardoso, Mad. Moçamb. **5**: *Rhamnus zeyheri* (1960). Type as above.
Rhamnus sp.—Eyles in Trans. Roy. Soc. S. Afr. **5**: 407 (1916).
Phyllogeiton zeyheri (Sond.) Suesseng. in Mitt. Bot. Staatssamml. Münch. **1**: 181 (1953). Type as for *Berchemia zeyheri*.

Tree 6–8(12) m. high, with rough grey bark; branches spreading; young branchlets puberulous. Leaves usually opposite or subopposite, glabrous; lamina 12–40(60) × 6–25 mm., elliptic to ovate, entire; midrib prominent beneath; lateral nerves 5–6, extending to the leaf-margin and prominent beneath; petiole 1–3 mm. long, channelled above. Flowers 1–4(6) in leaf-axils; pedicels slender, 4–14 mm. long. Flower bud globose, 2 mm. in diam. Sepals 2 mm. long, deltate, glabrous. Petals 2 mm. long, obovate, cucullate, often emarginate. Stamens 1·5–2 mm. long. Disk cupular, not enveloping the ovary, margin notched above insertion of filaments. Ovary free from disk; style 1·5–3 mm. long, 2-fid. Drupe 6–9 × 3 mm., ellipsoid, with yellow edible flesh surrounding the 2-seeded kernel.

Bechuanaland Prot. SE: Ootsi, fl. xi.1940, *Miller* 232 (PRE). **S. Rhodesia.** W: Matopos Research Station, fl. 30.xi.1951, *Plowes* 1366 (K; LISC; SRGH). C: Gwelo, fr. xi.1959, *Davies* 2638 (K; SRGH). **Mozambique.** LM: Santaca, fl. 23.xi.1948, *Gomes e Sousa* 3878 (COI; EA; K; LISC; PRE; SRGH).
Also in Natal, Swaziland and the Transvaal. In open woodland. Red Ivory; Pau-rosa. The heartwood is close-grained and pinkish and is used for ornaments and curios.

2. **Berchemia discolor** (Klotzsch) Hemsl. in Oliv., F.T.A. **1**: 381 (1868).—Sim, For. Fl. Port. E. Afr.: 35 (1909).—Bak. f. in Journ. Linn. Soc., Bot. **40**: 45 (1911).—Eyles in Trans. Roy. Soc. S. Afr. **5**: 407 (1916).—Burtt Davy, F.P.F.T. **2**: 470 (1932).—Brenan, T.T.C.L.: 416 (1949).—Codd, Trees and Shrubs Kruger Nat. Park: 111 (1951).—O.B. Mill. in Journ. S. Afr. Bot. **18**: 50 (1952).—White, F.F.N.R.: 227 (1962). TAB. **88** fig. A. Type: Mozambique, Sena, *Peters* (B, holotype†).
Scutia discolor Klotzsch in Peters, Reise Mossamb. Bot. **1**: 110, t. 21 (1861). Type as above.
Phyllogeiton discolor (Klotzsch) Herzog in Beih. Bot. Zentralbl. **15**: 169 (1903).—Suesseng. in Engl. & Prantl, Nat. Pflanzenfam. ed. 2, **20d**: 140 (1953).—Exell & Mendonça, C.F.A. **2**, 1: 30 (1954).—Palgrave, Trees of Central Afr.: 367 cum photogr. et tab. (1956).—Verdcourt in Bull. Jard. Bot. Brux. **27**: 356 (1957).—Dale & Greenway, Kenya Trees and Shrubs: 390 (1961). Type as above.

Tree up to 20 m. or more tall with dense rounded crown; bark very rough and tending to exfoliate in large pieces; branchlets pubescent or glabrous. Leaves alternate towards base of shoots, opposite to subopposite distally; lamina 5–11 × 3–6·6 cm., ovate or obovate-elliptic, midrib and 5–8 pairs of secondary nerves extending to leaf margin very prominent beneath and tertiary venation prominent and subparallel in adult leaves, glabrous or sparingly pubescent especially along the nerves above and below; petiole 5–15 mm. long, glabrous or pubescent. Flowers subfasciculate in axillary 1–10-flowered cymes. Pedicels up to 9 mm. long. Flower-bud 3 mm. in diam., glabrous. Sepals deltate, 2 mm. long, glabrous. Petals 1·5–2 mm. long. Stamens 1·5–2 mm. long. Disk enveloping the ovary. Ovary 2-locular, 1 seed per loculus; styles either 0·5 mm. long and notched at the top or 1·5 mm. long and markedly 2-fid. Drupe yellow, up to 2 × 0·8 cm., ellipsoid, edible. Seeds c. 10 × 4 mm., compressed.

Bechuanaland Prot. N: Serondela, Chobe R., fr. 27.vii.1950, *Robertson & Elffers* 57 (K; PRE). **N. Rhodesia.** B: Sesheke, fr. 29.xii.1952, *Angus* 1071 (BM; FHO; K; NDO). N: Kasama, Chibutubutu, fr. 5.ii.1961, *Mutimushi* (NDO). E: Fort Jameson, fl. 29.i.1961, *Grout* 248 (NDO). S: Bombwe, fr. 1933, *Martin* 630/33 (BM; K; FHO; NDO). **S. Rhodesia.** N: Chirundu, fl. 8.i.1958, *Goodier* 540 (K; LISC; SRGH). W: Victoria Falls, fr. ii.1960, *Armitage* (SRGH). C: Hartley, fr. v.1912, *Bell* in GHS

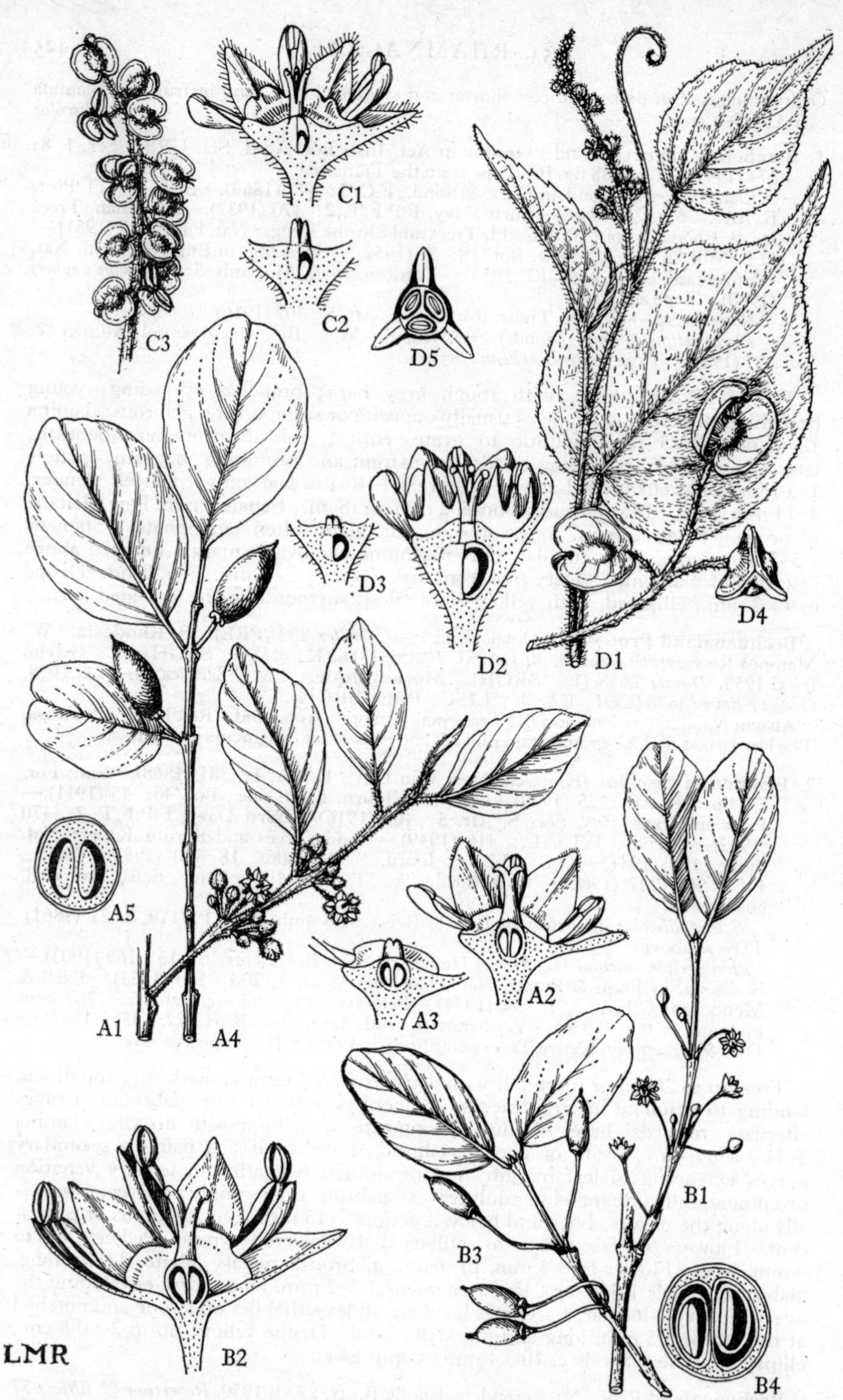

Tab. 88. A.—BERCHEMIA DISCOLOR. A1, flowering branch ($\times \frac{2}{3}$); A2, vertical section of flower ($\times 4$), both from *Lugard* 33; A3, vertical section of flower ($\times 4$) *Moubray* in GHS 92979; A4, fruiting branch ($\times \frac{2}{3}$); A5, transverse section of fruit ($\times 2$), both from *Phipps* 2461. B.—BERCHEMIA ZEYHERI. B1, flowering branch ($\times \frac{2}{3}$); B2, vertical section of flower ($\times 8$), both from *Gomes e Sousa* 3878; B3, fruiting branch ($\times \frac{2}{3}$); B4, transverse section of fruit ($\times 4$), both from *Barbosa & Lemos* 7391. C.—GOUANIA LONGISPICATA. C1 and C2, vertical sections of flower ($\times 6$); C3, portion of infructescence ($\times \frac{2}{3}$), all from *Swynnerton* 96. D.—GOUANIA SCANDENS. D1, flowering branch ($\times \frac{2}{3}$); D2 and D3, vertical sections of flower ($\times 8$); D4, portion of infructescence ($\times \frac{2}{3}$); D5, transverse section of fruit ($\times 2$). D1 and D2 from *Migeod* 393, D3 from *Schlieben* 5733, D4 and D5 from *Kirk* 227.

2095 (K; SRGH). E: Sabi R., fl. 9.xi.1906, *Swynnerton* 1215 (BM; K). S: Chipinda Pools, fl. 28.xi.1959, *Goodier* 680 (K; LISC; SRGH). **Nyasaland.** N: Kayuni, *Lewis* 88 (FHO). S. Chikwawa, fr. 1937, *Townsend* 134 (FHO). **Mozambique.** N: Nampula, fl. 6.xii.1935, *Torre* 758 (COI; LISC). Z: between Mocuba and Milange, fr. 16.iii.1943, *Torre* 4950 (LISC). T: between Chetima and Tete, fr. 1.vii.1949, *Barbosa & Carvalho* 3424 (K; LMJ). MS: Vila Machado, st. 17.iv.1948, *Mendonça* 3999 (LISC). SS: Guijá, st. 23.vi.1947, *Pedro & Pedrógão* 1188 (K; PRE; SRGH). LM: Magude, fl. 29.xi.1944, *Mendonça* 3119 (LISC).

Eastern Africa from the Sudan and Ethiopia south to the Transvaal and across to Angola and SW. Africa. Also in Arabia. In dry woodland and riverine vegetation in lower rainfall areas mainly below 1000 m.

Recent authors, notably Suessenguth, have supported the separation of *Phyllogeiton* as a separate genus. The chief character used to distinguish *Phyllogeiton* has been that it has the flowers in axillary subfasciculate cymes, whereas *Berchemia* has the flowers mostly in paniculate thyrses. This character is not of sufficient fundamental importance to be of generic value. Other characters used to separate *Phyllogeiton* have been opposite leaves and an ovary immersed in the disk. But both our species sometimes have alternate leaves and *Berchemia zeyheri*, which is without doubt congeneric with *B. discolor*, has an ovary which is not immersed in the disk.

3. RHAMNUS L.

Rhamnus L., Sp. Pl. **1**: 193 (1753); Gen. Pl. ed. 5: 89 (1754).

Small trees or shrubs. Leaves alternate or opposite or fasciculate, petiolate; lamina penninerved, margin often toothed. Stipules small, soon deciduous. Inflorescence an axillary fascicle (in our area), less often flowers solitary. Flowers usually bisexual. Receptacle urceolate or flat. Sepals 4–5, valvate. Petals 4–5 or absent, cucullate or flat. Stamens 4–5. Disk thin, lining the receptacle. Ovary superior, 2–4-locular; ovules 1 per loculus; style simple or 2–3(4)-fid. Fruit obovoid, drupaceous with a fleshy or somewhat woody endocarp.

Leaves with petiole 3–10 mm. long; lamina elliptic or elliptic-oblong, 3–10 × 1·5–4 cm.; flowers 5-merous - - - - - - - - - - - 1. *prinoides*
Leaves with petiole 1–3 mm. long; lamina narrowly elliptic, elliptic, narrowly obovate or obovate, 1·2–2·5 × 0·7–1·5 cm.; flowers 4-merous - - - - 2. *staddo*

1. **Rhamnus prinoides** L'Hérit., Sert. Angl.: 6, t. 9 (1788).—Sond. in Harv. & Sond., F.C. **1**: 477 (1860).—Hemsl. in Oliv., F.T.A. **1**: 382 (1868).—Bak. f. in Journ. Linn. Soc., Bot. **40**: 45 (1911).—Eyles in Trans. Roy. Soc. S. Afr. **5**: 407 (1916).—Burtt Davy, F.P.F.T. 2: 470 (1932).—Brenan, T.T.C.L.: 468 (1949); in Mem. N.Y. Bot. Gard. **8**, 3: 239 (1953).—Suesseng. in Proc. & Trans. Rhod. Sci. Ass. **43**: 110 (1951); in Engl. & Prantl, Nat. Pflanzenfam., ed. 2, **20d**: 65 (1953).—Exell & Mendonça, C.F.A. **2**, 1: 31 (1954).—Pardy in Rhod. Agr. Journ. **53**: 965 cum photogr. (1956).—Verdcourt in Bull. Jard. Bot. Brux. **27**: 358 (1957).—Keay, F.W.T.A. ed. 2, **1**, 2: 670 (1958).—Evrard, F.C.B. **9**: 433 (1960).—Dale & Greenway, Kenya Trees and Shrubs: 391 (1961).—White, F.F.N.R.: 227 (1963). TAB. **89** fig. B. Type from S. Africa (Cape Prov.).

Shrub or small tree, sometimes scrambling; bark blackish, often with many lenticels; young branches pubescent, glabrescent with age. Leaves alternate; lamina 3–10 × 1·5–4 cm., elliptic or oblong-elliptic, apex acute or acutely acuminate, margin serrulate, base cuneate or rounded, glabrous, shining, with nerves impressed above, pubescent on nerves below, tertiary venation conspicuous; petiole 3–10 mm. long, pubescent; stipules subulate, up to 5 mm. long. Inflorescence a sessile axillary 2–10-flowered fascicle; pedicels 2–7 mm. long, puberulous. Receptacle puberulous. Sepals 5, deltate, c. 2 mm. long. Petals 5, small, spathulate, hyaline or absent. Stamen-filaments <1 mm. long. Disk 2·5 mm. in diam. Ovary 3(4)-locular; ovules 1 in each loculus; style 2 mm. long, 3(4)-fid. Fruit turning through red to purple-black when ripe, c. 5 mm. in diam., slightly fleshy. Seeds 3–4, obconic.

N. Rhodesia. N: Kawimbe, fl. & fr. 12.x.1956, *Richards* 6425 (K). W: Solwezi, fl. 12.ix.1952, *White* 3227 (BM; FHO; K; NDO; PRE). C: Broken Hill, fl. & fr. 8.xi.1909, *Rogers* 8639 (K; PRE; SRGH). E: Nyika Plateau, st. 17.xii.1956, *White* 2759A (FHO). **S. Rhodesia.** N: near Mazoe Dam, fl. 2.ix.1960, *Rutherford-Smith* 38 (SRGH). C: Beatrice, fl. 14.ix.1923, *Eyles* 4532 (K; SRGH). E: Chirinda, fl. 22.x.1947, *Wild* 2123 (K; SRGH). **Nyasaland.** N: Nyika Plateau, fl. 15.xi.1958, *Robson & Fanshawe* 633

(K; LISC; PRE; SRGH). C: Chongoni Forest Reserve, st. 5.v.1960, *Chapman* 716 (K; SRGH). S: Mlanje, Luchenya Plateau, fl. & fr. 26.vi.1946, *Brass* 16439 (BM: K; PRE; SRGH). **Mozambique.** MS: Tsetsera, fr. 10.ii.1955, *E.M. & W.* 339 (BM; LISC; SRGH).

Widespread at higher elevations from Ethiopia south to Angola and the Cape Prov. of S. Africa. Fringing forest and forest margins.

2. **Rhamnus staddo** A. Rich., Tent. Fl. Abyss. **1**: 138 (1847).—Hemsl. in Oliv., F.T.A. **1**: 382 (1868).—Evrard, F.C.B. **9**: 434 (1960).—Dale & Greenway, Kenya Trees and Shrubs: 391 (1961). TAB. **89** fig. C. Type from Ethiopia.

Rhamnus holstii Engl. in Abh. Königl. Preuss. Akad. Wiss. Berl. **1894**: 69 (1894).—Brenan, T.T.C.L.: 468 (1949).—Verdcourt in Bull. Jard. Bot. Brux. **27**: 355 (1957). Type from Tanganyika.

Rhamnus rhodesicus Suesseng. in Mitt. Bot. Staatssamml. Münch. **1**: 181 (1953). Type: S. Rhodesia, Rusape, *Dehn* R40/52 (K; LISC; M, holotype; SRGH).

Shrub or tree to 7 m. tall; branches puberulous or glabrous, frequently spine-tipped; bark black. Leaves crowded at the end of short shoots; lamina 1·2–2·5 (3·5) × 0·7–1·5 cm., narrowly elliptic, elliptic, narrowly obovate or obovate, rounded and often emarginate at the apex, margin entire or obscurely denticulate, cuneate at the base, shining above, dull beneath, secondary nerves in 4–5 pairs, sparingly puberulous above; petiole 1–3 mm. long; stipules minute, ciliate. Inflorescences of few-flowered axillary fascicles. Flowers 4-merous, with pedicels 1–3 mm. long lengthening to 8–10 mm. long in fruit. Receptacle 4-angled. Sepals 1·5 mm. long, deltate. Petals 1 mm. long, shorter than the stamens, hyaline. Stamens with filaments 1 mm. long. Ovary 3-locular; 1 ovule per loculus; style short, 3-fid. Fruit 5 mm. in diam., somewhat fleshy. Seeds 3.

S. Rhodesia. C: Gwelo, fl. 9.x.1958, *Loveridge* 231 (SRGH). E: Inyanga, Alicedale Farm, 7.vi.1951, *Chase* 3897 (SRGH).

A species showing a remarkable discontinuous distribution, not having been found between northern Tanganyika and the Inyanga District of S. Rhodesia. On termite mounds and on granite hills.

4. SCUTIA Commers. ex Brongn.

Scutia Commers. ex Brongn. in Ann. Sci. Nat. **10**: 362 (1827) *nom. conserv.*

Shrubs or lianes, unarmed or armed with straight or hooked axillary spines. Leaves usually opposite, oval to oblong, entire or denticulate, glabrous, penninerved; stipules very small, deciduous. Inflorescence an axillary fascicle or umbel. Flowers bisexual. Receptacle turbinate. Sepals 5. Petals 5, inserted at the back of the margin of the disk, shortly unguiculate, emarginate, cucullate. Stamens 5, ± equal in length to the petals. Disk lining the receptacle. Ovary superior, 2(4)-locular; ovules 1 in each loculus; style 2(4)-lobed. Fruit an ovoid or subglobose somewhat fleshy drupe. Seeds 2–4, each surrounded by a thin but tough endocarp.

Scutia myrtina (Burm. f.) Kurz in Journ. As. Soc. Beng. **44**, 2: 168 (1875).—Brenan, T.T.C.L.: 469 (1949).—Perrier, Fl. Madag., Rhamnac.: 4 (1950).—Verdcourt in Bull. Jard. Bot. Brux. **27**: 358 (1957).—Evrard, F.C.B. **9**: 431 (1960).—Dale & Greenway, Kenya Trees and Shrubs: 391 (1961).—White, F.F.N.R.: 227 (1962). TAB. **89** fig. A. Type from India.

Rhamnus myrtina Burm. f., Fl. Ind.: 60 (1768). Type as above.

Scutia commersonii Brongn. in Ann. Sci. Nat. **10**: 363 (1827).—Sond. in Harv. & Sond., F.C. **1**: 477 (1860). Type from Madagascar.

Scutia indica var. *oblongifolia* Engl., Bot. Jahrb. **19**, Beibl. 47: 37 (1894). Type from Tanganyika.

Scutia myrtina var. *oblongifolia* (Engl.) Brenan in Mem. N.Y. Bot. Gard. **8**, 3: 239 (1953).—Evrard, F.C.B. **9**: 432 (1960). Type as above.

Shrub or tree, usually somewhat scandent; branchlets glabrous or sparingly puberulous. Prickles up to 9 mm. long, recurved, in leaf-axils. Leaves opposite or subopposite; lamina very variable in size and shape, 2–6 × 1·5–4 cm., ovate or elliptic, apex acute to rounded or retuse, apiculate, margin entire to crenate, base rounded to cuneate, penninerved with 5–8 secondary nerves on either side of the midrib, discolorous in dry state, glabrous; petiole 3–10 mm. long, glabrous or puberulous. Cymes condensed; peduncle up to 7 mm. long. Flowers sessile or

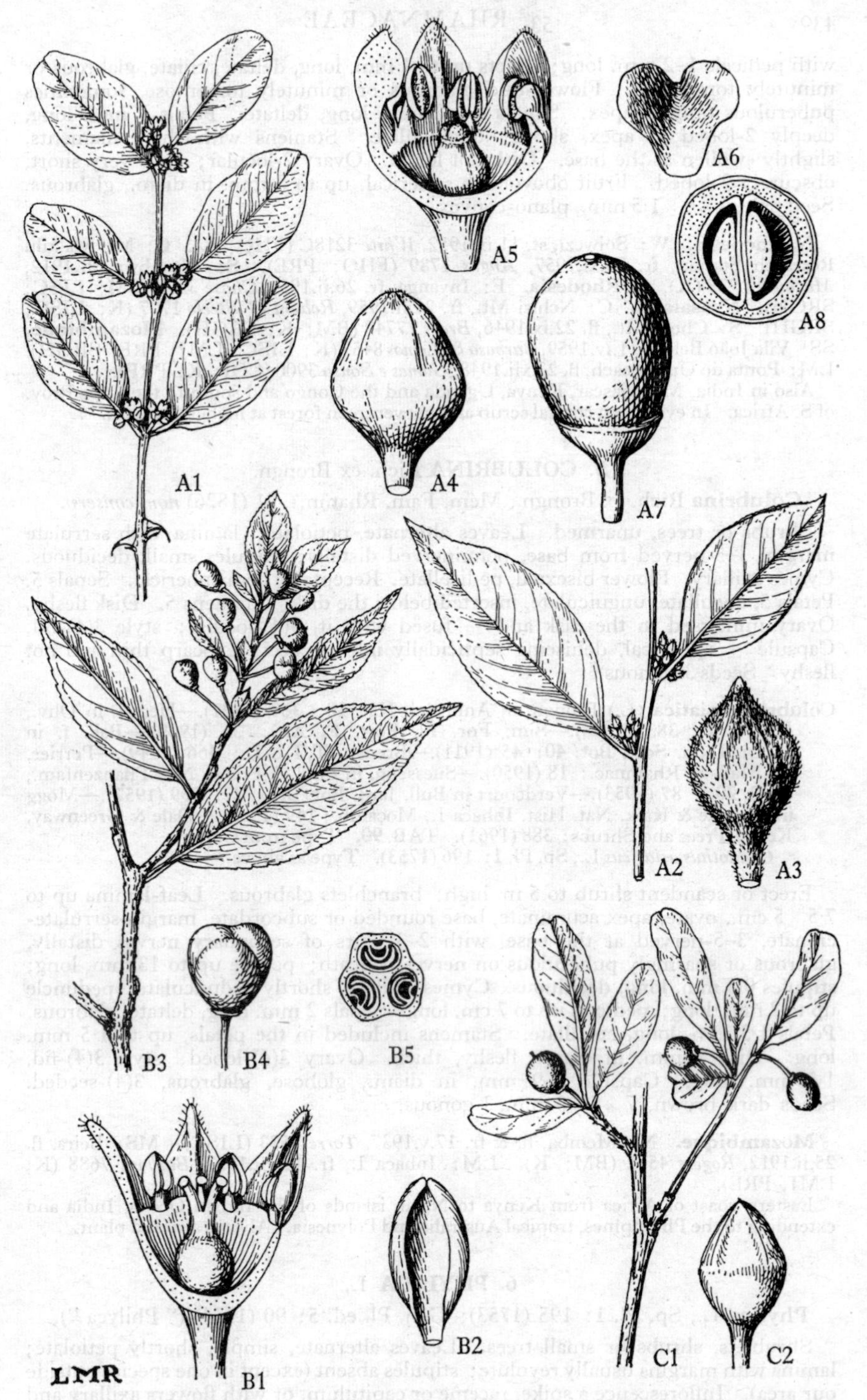

Tab. 89. A.—SCUTIA MYRTINA. A1 and A2, flowering branchlets (×⅔); A3 and A4, flower-buds (×8); A5, vertical section of flower (×8); A6, petal (×16); A7, fruit (×4); A8, transverse section of fruit; A1 and A4–A6 from *Barbosa & Lemos* 7984, A2 and A3 from *Brass* 17742, A7 and A8 from *Barbosa & Lemos* 8453. B.—RHAMNUS PRINOIDES. B1, vertical section of flower (×8); B2, petal (×21), both from *White* 3227; B3, fruiting branchlet (×⅔); B4, fruit (×2); B5, transverse section of fruit (×2). B3–B5 from *Fanshawe* 2936. C.—RHAMNUS STADDO. C1, portion of fruiting branch (×⅔); C2, flower-bud (×8); both from *Dehn* in GHS 40237.

with pedicels 1–2 mm. long; bracts up to 1 mm. long, deltate, ciliate, glabrous or minutely tomentose. Flower-bud glabrous or minutely tomentose, sometimes puberulous only at apex. Sepals 1·5–2 mm. long, deltate. Petals 1 mm. long, deeply 2-lobed at apex, shortly unguiculate. Stamens with short filaments, slightly swollen at the base. Disk not lobed. Ovary 2-locular; style very short, obscurely 2-lobed. Fruit obovoid to spherical, up to 8 mm. in diam., glabrous. Seed pale, 6 × 6 × 1·5 mm., plano-convex.

N. Rhodesia. W: Solwezi, st. 11.ix.1952, *White* 3218C (FHO; K). C: Mt. Makulu Research Station, fr. 20.xi.1957, *Angus* 1789 (FHO; PRE). S: Mumbwa, fl. 1912, *Macaulay* 925 (K). **S. Rhodesia.** E: Inyanga, fr. 26.ii.1951, *Chase* 3578 (BM; LISC; SRGH). **Nyasaland.** C: Nchisi Mt., fr. 21.ii.1959, *Robson & Steele* 1707 (K; LISC; SRGH). S: Cholo Mt., fl. 22.ix.1946, *Brass* 17742 (BM; K; SRGH). **Mozambique.** SS: Vila João Belo, fr. 1.iv.1959, *Barbosa & Lemos* 8453 (K; LISC; LMJ; PRE; SRGH). LM: Ponta do Ouro beach, fl. 23.xii.1948, *Gomes e Sousa* 3900 (COI; K; PRE).

Also in India, Madagascar, Kenya, Uganda and the Congo and south to the Cape Prov. of S. Africa. In evergreen coastal scrub and in evergreen forest at higher altitudes.

5. COLUBRINA Rich. ex Brongn.

Colubrina Rich. ex Brongn., Mém. Fam. Rhamn.: 61 (1826) *nom conserv.*

Shrubs or trees, unarmed. Leaves alternate, petiolate; lamina with serrulate margin, 3–5-nerved from base, penninerved distally; stipules small, deciduous. Cymes axillary. Flower bisexual, pedicellate. Receptacle hemispherical. Sepals 5. Petals 5, cucullate, unguiculate, inserted below the disk. Stamens 5. Disk fleshy. Ovary immersed in the disk and ± fused with it, 3(4)-locular; style 3(4)-fid. Capsule ± spherical, dehiscing septicidally into 3 cocci; exocarp thin and not fleshy. Seeds 3-gonous.

Colubrina asiatica (L.) Brongn, in Ann. Sci. Nat. **10**: 369 (1827).—Hemsl. in Oliv., F.T.A. **1**: 383 (1868).—Sim, For. Fl. Port. E. Afr.: 36 (1909).—Bak. f. in Journ. Linn. Soc., Bot. **40**: 45 (1911).—Brenan, T.T.C.L.: 466 (1949).—Perrier, Fl. Madag., Rhamnac.: 18 (1950).—Suesseng. in Engl. & Prantl, Nat. Pflanzenfam., ed. 2, **20d**: 87 (1953).—Verdcourt in Bull. Jard. Bot. Brux. **27**: 359 (1957).—Mogg in Macnae & Kalk, Nat. Hist. Inhaca I., Moçamb.: 149 (1958).—Dale & Greenway, Kenya Trees and Shrubs: 388 (1961). TAB. **90.** Type from Ceylon.
Ceanothus asiaticus L., Sp. Pl. **1**: 196 (1753). Type as above.

Erect or scandent shrub to 5 m. high; branchlets glabrous. Leaf-lamina up to 7·5 × 5 cm., ovate, apex acuminate, base rounded or subcordate, margin serrulate-crenate, 3–5-nerved at the base, with 2–3 pairs of secondary nerves distally, glabrous or sparingly puberulous on nerves beneath; petiole up to 13 mm. long; stipules 0·5 mm. long, deciduous. Cymes axillary, shortly pedunculate; peduncle up to 2 mm. long; pedicels up to 7 cm. long. Sepals 2 mm. long, deltate, glabrous. Petals 1·5 mm. long, cucullate. Stamens included in the petals, up to 1·5 mm. long. Disk 3 mm. in diam., fleshy, thick. Ovary 3(4)-lobed; style 3(4)-fid, 1·5 mm. long. Capsule 6–9 mm. in diam., globose, glabrous, 3(4)-seeded. Seeds dark brown, 7 × 5 × 3 mm., 3-gonous.

Mozambique. N: Memba, fl. & fr. 17.v.1937, *Torre* 1533 (LISC). MS: Beira, fl. 25.ii.1912, *Rogers* 4552 (BM; K). LM: Inhaca I., fr. 12.vii.1957, *Barbosa* 7688 (K; LMJ; PRE).

Eastern coast of Africa from Kenya to Natal, islands of the Indian Ocean, India and extending to the Philippines, tropical Australia and Polynesia. Always a strand plant.

6. PHYLICA L.

Phylica L., Sp. Pl. **1**: 195 (1753); Gen. Pl. ed. 5: 90 (1754) (" Philyca ").

Shrublets, shrubs or small trees. Leaves alternate, simple, shortly petiolate; lamina with margins usually revolute; stipules absent (except in one species outside our area). Inflorescence a spike, raceme or capitulum, or with flowers axillary and solitary. Flowers bisexual. Receptacle campanulate, urceolate, cylindric, obconic or terete, 5-angled. Sepals 5, erect or spreading, variously shaped, often with a prominent median vein within. Petals 5 (or absent), small, cucullate, unguiculate. Stamens 5, inserted below the petals; anthers 1–2-thecous. Disk usually distinct

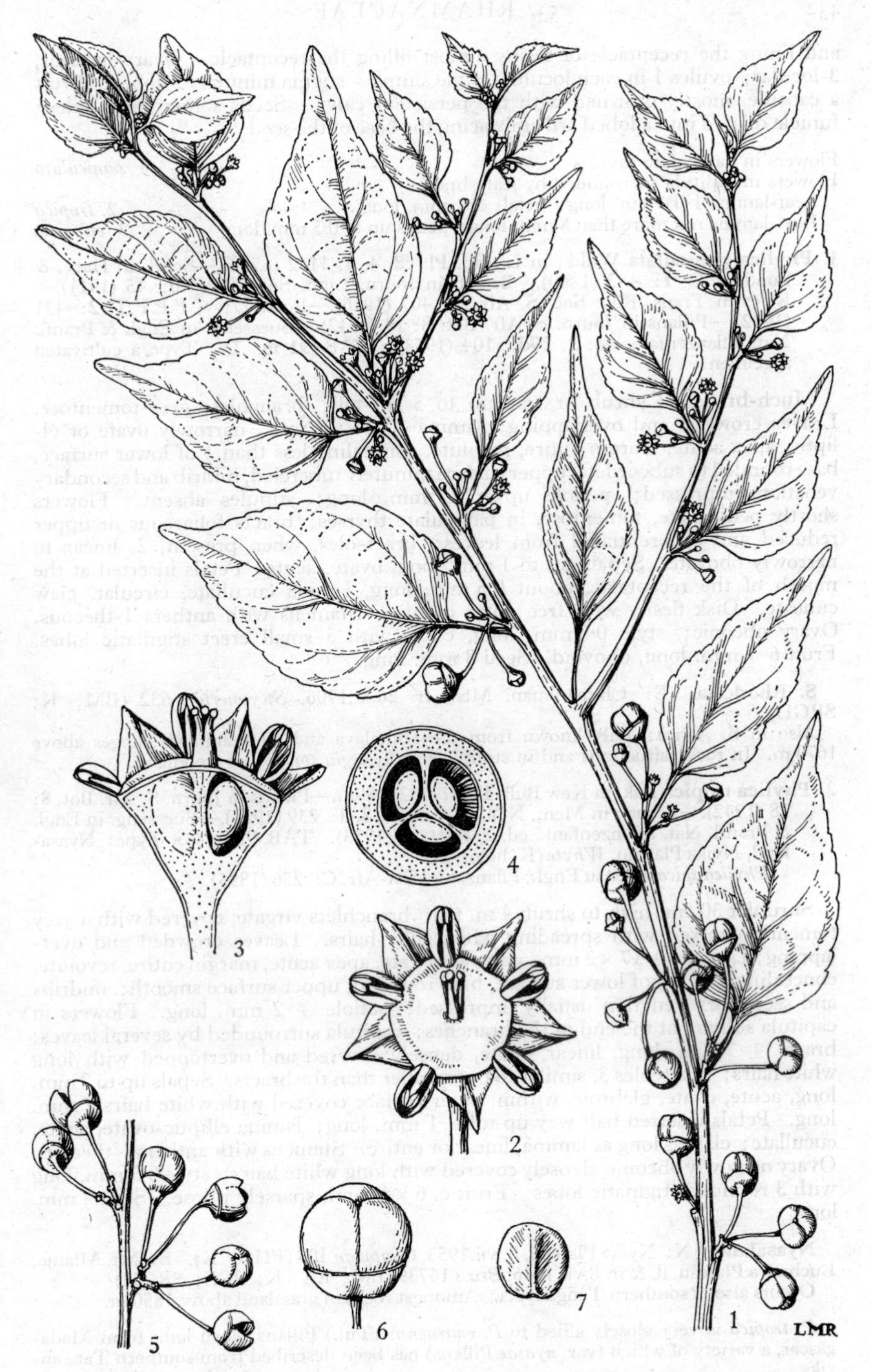

Tab. 90. COLUBRINA ASIATICA. 1, flowering and fruiting branch ($\times \frac{2}{3}$); 2, flower ($\times 8$); 3, vertical section of flower ($\times 8$); 4, transverse section of ovary ($\times 8$); 5, portion of fruiting branch ($\times \frac{2}{3}$); 6, fruit ($\times 2$); 7, seed ($\times 2$). 1–4 from *Drummond & Hemsley* 3252, 5–7 from *Volkens* 161.

and lining the receptacle or fleshy almost filling the receptacle. Ovary inferior, 3-locular; ovules 1 in each loculus; style simple; stigma minutely 3-lobed. Fruit a capsule, mostly crowned with the persistent calyx. Seeds somewhat 3-sided; funicle dilated into a lobed aril embracing the base of the seed.

Flowers in paniculate thyrses - - - - - - - - 1. *paniculata*
Flowers in capitula surrounded by leafy bracts:
Leaf-lamina 1–1·7 cm. long; petals c. 1 mm. long - - - - 2. *tropica*
Leaf-lamina not more than 8 mm. long; petals up to 0·5 mm. long - 3. *ericoides*

1. **Phylica paniculata** Willd., in L., Sp. Pl. ed. 4, **1,** 1112 (1798).—Sond. in Harv. & Sond., F.C. **1**: 482 (1860).—Bak. f. in Journ. Linn. Soc., Bot. **40**: 45 (1911).—Eyles in Trans. Roy. Soc. S. Afr. **5**: 407 (1916).—Burtt Davy, F.P.F.T. **2**: 471 (1932).—Pillans in Journ. S. Afr. Bot. **8**: 19 (1942).—Suesseng. in Engl. & Prantl, Nat. Pflanzenfam. ed. 2, **20d**: 104 (1953). TAB. **91** fig. D. Type a cultivated specimen.

Much-branched shrub or tree up to 5 m. tall; branchlets grey-tomentose. Leaves crowded and overlapping; lamina 10–15 × 3 mm., narrowly ovate or elliptic, apex acute, margin entire, revolute, concealing less than ½ of lower surface, base rounded to subcordate, upper surface minutely tubercled, midrib and secondary venation impressed; petiole up to 2 mm. long; stipules absent. Flowers shortly pedicellate, tomentose, in paniculate thyrses; bracts foliaceous or upper reduced and differentiated from leaves; bracteoles, when present, 2, linear to narrowly obovate. Sepals up to 1 mm. long, ovate, acute. Petals inserted at the mouth of the receptacle, about 0·5 mm. long, lamina cucullate, circular, claw cuneate. Disk fleshy with free outer margin. Stamens with anthers 1-thecous. Ovary obconic; style 0·5 mm. long, conic with 3 small erect stigmatic lobes. Fruit 6–7 mm. long, obovoid. Seed 3 mm. long.

S. Rhodesia. E: Chimanimani Mts., fr. 26.ix.1906, *Swynnerton* 632 (BM; K; SRGH).
Also in S. Africa; only known from the Himalaya and Chimanimani ranges above 1670 m. In rocky situations and in stunted *Brachystegia spiciformis* woodland.

2. **Phylica tropica** Bak. in Kew Bull. **1898**: 302 (1898).—Pillans in Journ. S. Afr. Bot. **8**: 28 (1942).—Brenan in Mem. N.Y. Bot. Gard. **8,** 3: 239 (1953).—Suesseng. in Engl. & Prantl, Nat. Pflanzenfam., ed. 2, **20d**: 97 (1953). TAB. **91** fig C. Type: Nyasaland, Nyika Plateau, *Whyte* (K, holotype).
Phylica spicata sensu Engl., Pflanzenw. Ost-Afr. **C**: 256 (1895).

Shrublet 30 cm. high to shrub 4 m. tall; branchlets virgate, covered with a grey tomentum mixed with spreading white silky hairs. Leaves crowded and overlapping; lamina 10–17 × 2 mm., narrowly ovate, apex acute, margin entire, revolute, concealing about ½ of lower surface, base rounded, upper surface smooth; midribs and secondary venation usually impressed; petiole ± 2 mm. long. Flowers in capitula solitary at the end of the branches; capitula surrounded by several leaves; bracts ± 7 mm. long, linear, acute, densely covered and overtopped with long white hairs; bracteoles 3, similar to but smaller than the bracts. Sepals up to 2 mm. long, acute, ovate, glabrous within, outer surface covered with white hairs 2 mm. long. Petals inserted half way up tube, 1 mm. long; lamina elliptic-ovate, acute, cucullate; claw as long as lamina, linear or entire. Stamens with anthers 1-thecous. Ovary narrowly obconic, densely covered with long white hairs; style 0·5 mm. long with 3 rounded stigmatic lobes. Fruit c. 6 × 4 mm., sparsely pilose. Seed 4 mm. long.

Nyasaland. N: Nyika Plateau, fl. vii.1953, *Chapman* 103 (FHO; K). S: Mt. Mlanje, Luchenya Plateau, fl. & fr. 8.vii.1946, *Brass* 16739 (BM; EA; K; PRE; SRGH).
Occurs also in southern Tanganyika. Amongst rocks in grassland above 1850 m.

P. tropica is very closely allied to *P. emirnensis* (Tul.) Pillans sensu lato, from Madagascar, a variety of which (var. *nyasae* Pillans) has been described from southern Tanganyika.

3. **Phylica ericoides** L., Sp. Pl. **1**: 195 (1753).—Sond. in Harv. & Sond., F.C. **1**: 499 (1860).—Pillans in Journ. S. Afr. Bot. **8**: 85 (1942).—Suesseng. in Engl. & Prantl, Nat. Pflanzenfam., ed. 2, **20d**: 106 (1953); in Mitt. Bot. Staatssamml. Münch. **1**: 341 (1953). TAB. **91** fig. E. Type from S. Africa (Cape Prov.).

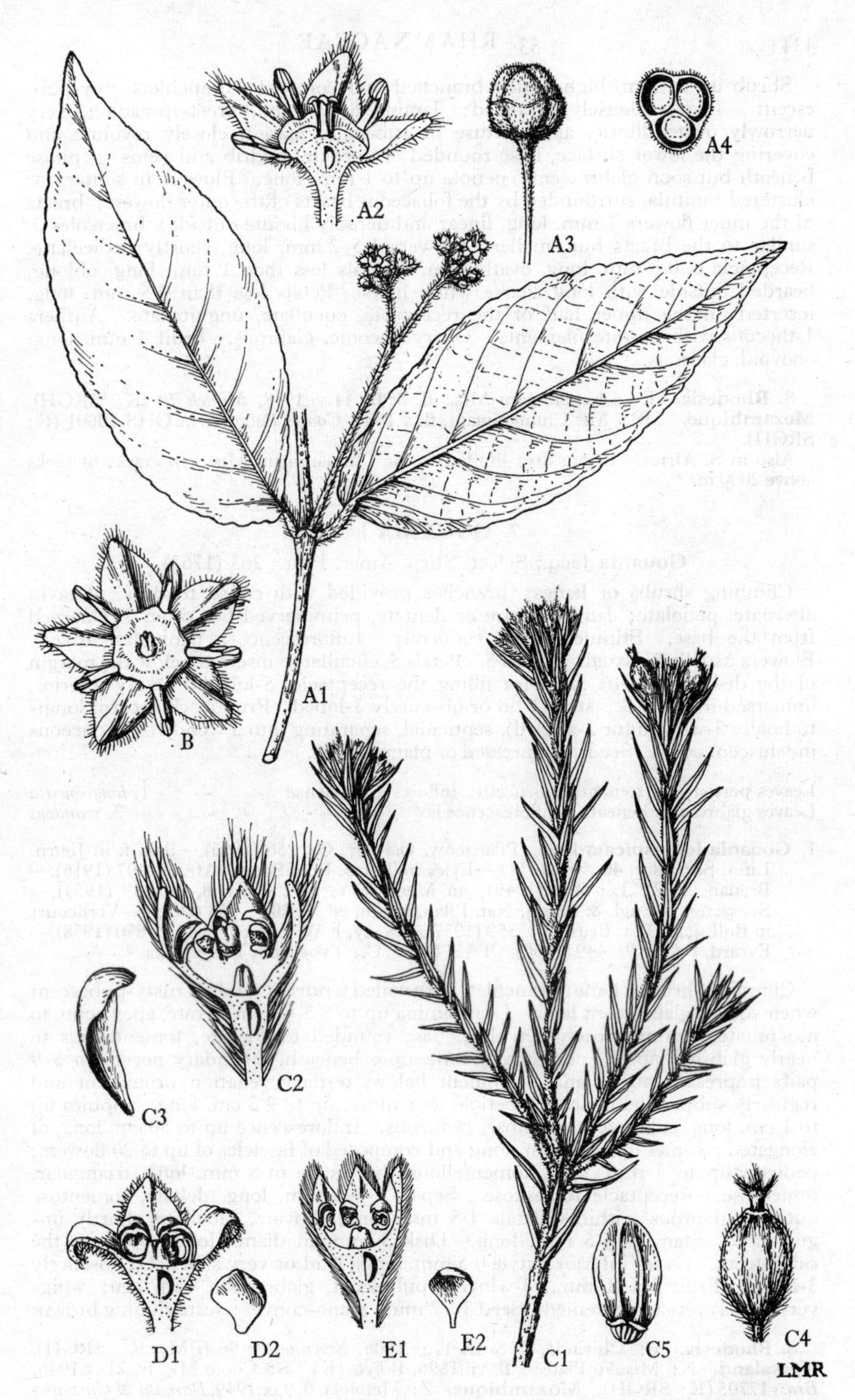

Tab. 91. A.—LASIODISCUS MILDBRAEDII. A1, portion of flowering branch (×⅔); A2, vertical section of flower (×4); A3, fruit (×⅔); A4, transverse section of fruit (×⅔). A1 and A2 from *Gomes e Sousa* 4444, A3 and A4 from *Swynnerton* 1038. B.—LASIODISCUS USAMBARENSIS, flower (×4) *Swynnerton* 121. C.—PHYLICA TROPICA. C1, flowering and fruiting branch (×⅔); C2, vertical section of flower (×6); C3, petal (×16); C4, fruit (×2); C5, seed (×4). C1–C3 from *Brass* 16652, C4 and C5 from *Brass* 16739. D.—PHYLICA PANICULATA. D1, vertical section of flower (×10); D2, petal (×16), both from *Swynnerton* 632. E.—PHYLICA ERICOIDES. E1, vertical section of flower (×6); E2, petal (×16), both from *Phipps* 648.

Shrub up to 1 m. high, much branched and compact; branchlets grey-pubescent. Leaves densely arranged; lamina 8 × 3 mm., erect-spreading, very narrowly ovate-elliptic, apex obtuse or subacute, margin closely revolute and covering the lower surface, base rounded or cordate, midrib and veins ± pilose beneath but soon glabrescent; petiole up to 1 mm. long. Flowers in solitary or clustered capitula, surrounded by the foliaceous bracts of the outer flowers; bracts of the inner flowers 1 mm. long, linear and densely hirsute outside; bracteoles 2, similar to the bracts but smaller. Flowers 1·5–2 mm. long, shortly pedicellate. Receptacle c. 0·5 mm. long, cyathiform. Sepals less than 1 mm. long, oblong, bearded outside with long coarse white hairs. Petals less than 0·5 mm. long, inserted on the upper half of the receptacle, cucullate, unguiculate. Anthers 1-thecous with minute filaments. Ovary obconic, glabrous. Fruit 4 mm. long, obovoid, glabrous.

S. Rhodesia. E: Chimanimani Mts., fl. & fr. 11.vi.1948, *Munch* 70 (K; SRGH). **Mozambique.** MS: Mt. Chimanimani, fl. v.1956, *Coates Palgrave* in GHS 70601 (K; SRGH).

Also in S. Africa. In our area limited to the Chimanimani Mts. In cracks of rocks above 2000 m.

7. GOUANIA Jacq.

Gouania Jacq., Select. Stirp. Amer. Hist.: 263 (1763).

Climbing shrubs or lianes; branches provided with coiled tendrils. Leaves alternate, petiolate; lamina entire or dentate, penninerved but often 3–5-nerved from the base. Stipules small, caducous. Inflorescence a paniculate thyrse. Flowers usually bisexual. Sepals 5. Petals 5, cucullate; inserted below the margin of the disk. Stamens 5. Disk filling the receptacle, 5-lobed. Ovary inferior, immersed in the disk; style 3-fid or obscurely 3-lobed. Fruit a schizocárp, longitudinally 3-winged (or 3-angled), septicidal, separating into 3 woody or coriaceous indehiscent cocci. Seeds compressed or planoconvex.

Leaves persistently tomentose beneath; inflorescence dense - - - 1. *longispicata*
Leaves glabrescent beneath; inflorescence lax - - - - - - 2. *scandens*

1. **Gouania longispicata** Engl., Pflanzenw. Ost-Afr. **C**: 256 (1895).—Bak. f. in Journ. Linn. Soc., Bot. **40**: 45 (1911).—Eyles in Trans. Roy. Soc. S. Afr. **5**: 407 (1916).—Brenan, T.T.C.L.: 466 (1949); in Mem. N.Y. Bot. Gard. **8**, 3: 239 (1953).—Suesseng. in Engl. & Prantl, Nat. Pflanzenfam. ed. 2, **20d**: 168 (1953).—Verdcourt in Bull. Jard. Bot. Brux. **27**: 359 (1957).—Keay, F.W.T.A. ed. 2, **1,** 2: 670 (1958).—Evrard, F.C.B. **9**: 449 (1960). TAB. **88** fig. C. Type from Tanganyika.

Climbing shrub or liane; branchlets with coiled tendrils; tendrils rusty-pubescent when young, glabrescent later. Leaf-lamina up to 8·5 × 7 cm., ovate, apex acute to mucronate, margin closely serrulate, base rounded to cordate, tomentellous to nearly glabrous above, persistently tomentose beneath, secondary nerves in 5–9 pairs impressed above and prominent below, tertiary venation prominent and regularly subparallel beneath; petiole tomentose, up to 2·5 cm. long; stipules up to 1 cm. long, subulate, tomentose, caducous. Inflorescence up to 50 cm. long, of elongated racemes up to 20 cm. long and composed of fascicles of up to 20 flowers; pedicels up to 5 mm. long, tomentellous; bracts up to 8 mm. long, triangular, tomentose. Receptacle tomentose. Sepals 1·5–2 mm. long, deltate, tomentose outside, glabrous within. Petals 1·5 mm. long, obovate, cucullate, shortly unguiculate. Stamens 1·5 mm. long. Disk 3 mm. in diam., lobed opposite the calyx-teeth. Ovary inferior: style 0·5 mm. long, 3-fid or very short and obscurely 3-lobed. Fruit 6 × 11 mm., 3-winged, pubescent, glabrescent with age; wings very loosely reticulate-veined. Seed 3 × 2 mm., plano-convex; testa shining brown.

S. Rhodesia. E: Chirinda, fl. & fr. 17.iv.1906, *Swynnerton* 96 (BM; K; SRGH). **Nyasaland.** N: Misuku Plateau, fl. vii.1896, *Whyte* (K). S: Cholo Mt., fr. 21.ix.1946, *Brass* 17705 (K; SRGH). **Mozambique.** Z: Metolola, fl. 9.ix.1949, *Barbosa & Carvalho* 4005 (K; LMJ; PRE). MS: Mossurize, between Espungabera and Gogoi, fl. 12.vi.1942, *Torre* 4312 (LISC).

Also in Nigeria, Congo and the Sudan, south through Kenya, Uganda and Tanganyika reaching its southern limit in our area. Forest edges and clearings.

2. **Gouania scandens** (Gaertn.) R.B. Drummond, comb. nov. TAB. **88** fig. D. Lectotype an illustration of a plant collected in Mauritius.
Retinaria scandens Gaertn., Fruct. **2**: 187, t. 120 fig. 4 (1791). Type as above.
Gouania retinaria DC., Prodr. **2**: 40 (1825). Type as above.
Gouania longipetala sensu Hemsl. in Oliv., F.T.A. **1**: 383 (1863) quoad specim. Kirk.
Gouania tiliifolia sensu Bak., Fl. Maurit. & Seychelles: 52 (1877) pro parte.—Brenan, T.T.C.L.: 467 (1949).—Suesseng. in Engl. & Prantl, Nat. Pflanzenfam. ed. 2, **20d**: 168 (1953) pro parte quoad syn. *G. retinaria*.
Gouania mozambicensis M.L. Green in Kew Bull. **1916**: 199 (1916).—Suesseng., loc. cit. Type: Mozambique, Chupanga, *Kirk* (K; holotype).

Climbing shrub; branchlets and coiled tendrils pilose when young, glabrescent. Leaf-lamina up to 7·3 × 5·8 cm., ovate, apex acute or shortly acuminate, margin crenate, base rounded to cordate, pilose on the nerves above and below at least when young, secondary nerves in 6–8 pairs, tertiary venation tending to be subparallel below; petiole 2–4 cm. long, pubescent; stipules 4 mm. long, subulate. Inflorescences up to 12 cm. long; inflorescence-branches 6 mm. long; pedicels 2–3 mm. long; bracts 2 mm. long, ovate. Receptacle pilose. Sepals 2 mm. long, deltate, glabrous. Petals 1·5 mm. long, obovate, cucullate, shortly unguiculate. Stamens 1·5 mm. long. Disk 3 mm. in diam., with triangular lobes opposite the calyx-teeth. Ovary inferior; style 1·5 mm. long, 3-lobed. Fruit up to 1·7 × 1·8 cm., wings prominently reticulate. Seeds 5 × 4 × 2 mm., planoconvex; testa shining greyish-brown.

Mozambique. N: Nampula, fl. & fr. 19.iii.1937, *Torre* 1313 (COI). T: 51 km. SW of Zóbuè, fr. 1.v.1960, *Leach & Brunton* 9880 (K; LISC; SRGH). MS: Inhaminga, fr. 25.v.1948, *Mendonça* 4375 (LISC).

Also from SE. Tanganyika and from Mauritius and other islands in the Indian Ocean. In coastal scrub and riverine vegetation.

In Pflanzenw. Ost-Afr. **C**: 256 (1895), Engler records *Gouania pannigera* Tul. from Gorongosa in Mozambique. This species is treated by Perrier de la Bâthie in the Flore de Madagascar as a subspecies of *G. mauritiana* Lam. This is the only record from the mainland of Africa and confirmation of its occurrence is desirable. The specimen on which the record is based has not been traced.

8. HELINUS E. Mey. ex Endl.

Helinus E. Mey. ex Endl., Gen.: 1102 (1840) *nom. conserv.*

Shrublets or climbing shrubs; branches with coiled tendrils. Leaves alternate, petiolate; lamina penninerved, margin entire; stipules subulate, caducous. Inflorescence a 1-several-flowered pedunculate axillary umbel. Flowers bisexual, pedicellate. Receptacle obconic, adnate to the ovary. Sepals 5. Petals 5, cucullate, inserted at the margin of the disk. Stamens 5, equalling the petals, inserted at the margin of the disk. Disk flat, filling the receptacle. Ovary inferior, 3-locular; ovules 1 in each loculus; style 3-fid or obscurely 3-lobed (sometimes heterostylous). Fruit an obovoid or globose capsule, splitting into 3 cocci which at length dehisce. Seeds black, plano-convex.

Shrublet under 1 m. tall; coiled tendrils usually absent; leaves narrowly ovate 1. *spartioides*
Shrubs usually over 1 m. tall, climbing with coiled tendrils; leaves ovate or broadly ovate to circular:
Tendrils and flowers glabrous; capsules smooth and glabrous - - 2. *integrifolius*
Tendrils and flowers hairy; capsules densely tuberculate and pubescent 3. *mystacinus*

1. **Helinus spartioides** (Engl.) Schinz ex Engl., Pflanzenw. Afr. **3,** 2: 316 (1921).—Suesseng. in Engl. & Prantl, Nat. Pflanzenfam. ed. 2, **20d**: 172 (1953). Type from S. Africa (Cape Prov.).
Marlothia spartioides Engl., Bot. Jahrb. **10**: 39, t. 5 (1889).—O.B. Mill. in Journ. S. Afr. Bot. **18**: 51 (1952). Type as above.

Shrublet up to 1 m. tall; branchlets virgate, green, very rarely with coiled tendrils, sparingly puberulous or glabrous. Leaves petiolate; lamina 20–40 × 4–5 mm., narrowly ovate, apex acute, margin entire, base rounded to cuneate, midrib prominent beneath with 3 pairs of secondary nerves, nerves not visible on upper surface, sparingly puberulous; the upper leaves smaller and narrower; petiole 1·5–5 mm.

long, sparingly puberulous; stipules up to 2 mm. long, ovate, puberulous, purplish-brown. Inflorescence 1–3-flowered; pedicels 3–5 mm. long, glabrous; bracts ovate, 1–2 mm. long, purplish-brown. Sepals 2·5–3 mm. long, ovate, glabrous; receptacle 1–1·5 mm. long. Petals 2 mm. long, ovate, cucullate, unguiculate. Stamens 2 mm. long, inserted at the edge of the disk. Disk 5-lobed, covering the ovary. Ovary 2–3-locular; 1 ovule in each loculus; style 3-fid, with spreading branches. Fruit 6 mm. in diam., obovoid. Seeds 4·5 × 3·5 mm., plano-convex; testa shining black.

Bechuanaland Prot. SW: 13 km. S. of Kang, fl. 18.ii.1960, *Wild* 5013 (BM; K; SRGH). SE: 6 km. N. of Murumush, fl. & fr. 17.ii.1960, *Wild* 5006 (K; SRGH).

Also in SW. Africa and Cape Prov. In arid areas with grassland and scattered bushes.

2. **Helinus integrifolius** (Lam.) Kuntze, Rev. Gen. Pl. **1**: 120 (1891).—Exell & Mendonça, C.F.A. **2**, 1: 31, t. 6 (1954).—Evrard, F.C.B. **9**: 450 (1960).—White, F.F.N.R.: 434 (1962). TAB. **92** fig. A. Type a cultivated specimen of uncertain origin.

Gouania integrifolia Lam., Encycl. Méth., Bot. **3**: 5 (1789). Type as above.

Willemetia scandens Eckl. & Zeyh., Enum. Pl. Afr. Austr. Extratrop. **1**: 130 (1834–35?). Type from S. Africa (Cape Prov.).

Helinus scandens (Eckl. & Zeyh.) A. Rich., Tent. Fl. Abyss. **1**: 139 (1847).—Brenan, T.T.C.L.: 467 (1949).—Suesseng. in Engl. & Prantl, Nat. Pflanzenfam. ed. 2, **20d**: 172 (1953).—Verdcourt in Bull. Jard. Bot. Brux. **27**: 360 (1957).—White, F.F.N.R.: 227 (1962). Type as above.

Helinus ovatus E. Mey. ex Sond. in Harv. & Sond., F.C. **1**: 479 (1860).—Hemsl. in Oliv., F.T.A. **1**: 384 (1868).—Eyles in Trans. Roy. Soc. S. Afr. **5**: 407 (1916).—Burtt Davy, F.P.F.T. **2**: 471 (1932).—Steedman, Trees etc. S. Rhod.: 45 (1933).—Wild, Guide Fl. Vict. Falls: 151 (1952). Type from S. Africa (Cape Prov.).

Helinus mystacinus sensu O.B. Mill. in Journ. S. Afr. Bot. **18**: 51 (1952).

Climbing shrub to c. 6 m.; branchlets glabrous or glabrescent, provided with glabrous coiled tendrils. Leaves petiolate; lamina 1·5–3·5 cm. long, ovate to broadly ovate or circular, apex rounded or mucronate, margin entire, base rounded to subcordate, penninerved, with 5–6 pairs of secondary nerves; lower surface puberulous, upper surface glabrous; petiole 5–7 mm. long, puberulous; stipules subulate, 2–5 mm. long. Flowers in axillary pedunculate umbels; peduncle up to 2 cm. long; pedicels up to 6 mm. long, slender. Sepals 2 mm. long, ovate-deltate, glabrous. Petals 1·5–2 mm. long, obovate, cucullate, unguiculate. Stamens 1·5–2 mm. long. Disk 1–1·5 mm. in diam., annular. Ovary immersed in the disk; style 2 mm. long, with 3 spreading branches, or less than 1 mm. long and obscurely 3-lobed. Capsule obovoid, glabrous. Seeds 3-angled or plano-convex; testa shining black.

Caprivi Strip. Singalamwe, fl. 1.i.1959, *Killick* 3247 (K; PRE; SRGH). **Bechuanaland Prot.** N: Macloutsie, fr. 15.iv.1957, *de Beer* in GHS 81507 (K; SRGH). **N. Rhodesia.** W: Luanshya, fl. 7.iii.1957, *Fanshawe* 3024 (K; NDO). C: Makalaikwa, 64 km. W. of Lusaka, fl. & fr. 7.iv.1960, *Angus* 2207 (K; NDO; SRGH). S: Mapanza, fl. 10.ii.1957, *Robinson* 2131 (EA; K; NDO; SRGH). **S. Rhodesia.** N: Mtoko Distr., Kafefe R., fl. i.1953, *Phelps* 19 (SRGH). W: Matopos, fl. iii.1918, *Eyles* 971 (BM; K; PRE; SRGH). C: Salisbury, Twentydales, fl. 1.i.1946, *Wild* 615 (SRGH). S: Ndanga, fr. 24.v.1959, *Noel* 1971 (SRGH). **Mozambique.** N: Nampula, fl. & fr. 13.iv.1937, *Torre* 1544 (COI; LISC). Z: near Maganja da Costa, fl. & fr. 27.ix.1949, *Barbosa & Carvalho* 4226 (K). MS: Manica, Mavita, fl. & fr. 26.iv.1948, *Barbosa* 1573 (LISC). SS: Macia, 31.iii.1959, *Barbosa & Lemos* 8425 (COI; K; LISC; LMJ; SRGH). LM: Goba, fl. 31.iii.1945, *Sousa* 135 (LISC; PRE).

In eastern Africa from Kenya south to the Cape Prov. of S. Africa, also in Angola and the Congo. Widespread in savanna and savanna woodland especially in rocky places.

3. **Helinus mystacinus** (Ait.) E. Mey. ex Steud., Nom. Bot. ed. 2, **1**: 742 (1840).—Hemsl. in Oliv., F.T.A. **1**: 385 (1868).—Bak. f. in Journ. Linn. Soc., Bot. **40**: 45 (1911).—Eyles in Trans. Roy. Soc. S. Afr. **5**: 407 (1916).—Brenan, T.T.C.L.: 467 (1949).—Suesseng. in Engl. & Prantl, Nat. Pflanzenfam. ed. 2, **20d**: 172 (1953). Verdcourt in Bull. Jard. Bot. Brux. **27**: 360 (1957).—Evrard, F.C.B. **9**: 452 (1960).—White, F.F.N.R.: 227, t. 39 (1962). TAB. **92** fig. B. Type a cultivated specimen, probably from Ethiopia.

Rhamnus mystacinus Ait., Hort. Kew. **1**: 266 (1789). Type as above.

Climbing shrub; branchlets pubescent, provided with pubescent coiled tendrils. Leaves petiolate; lamina 2–5 × 1·3–5 cm., ovate, broadly ovate or circular, apex

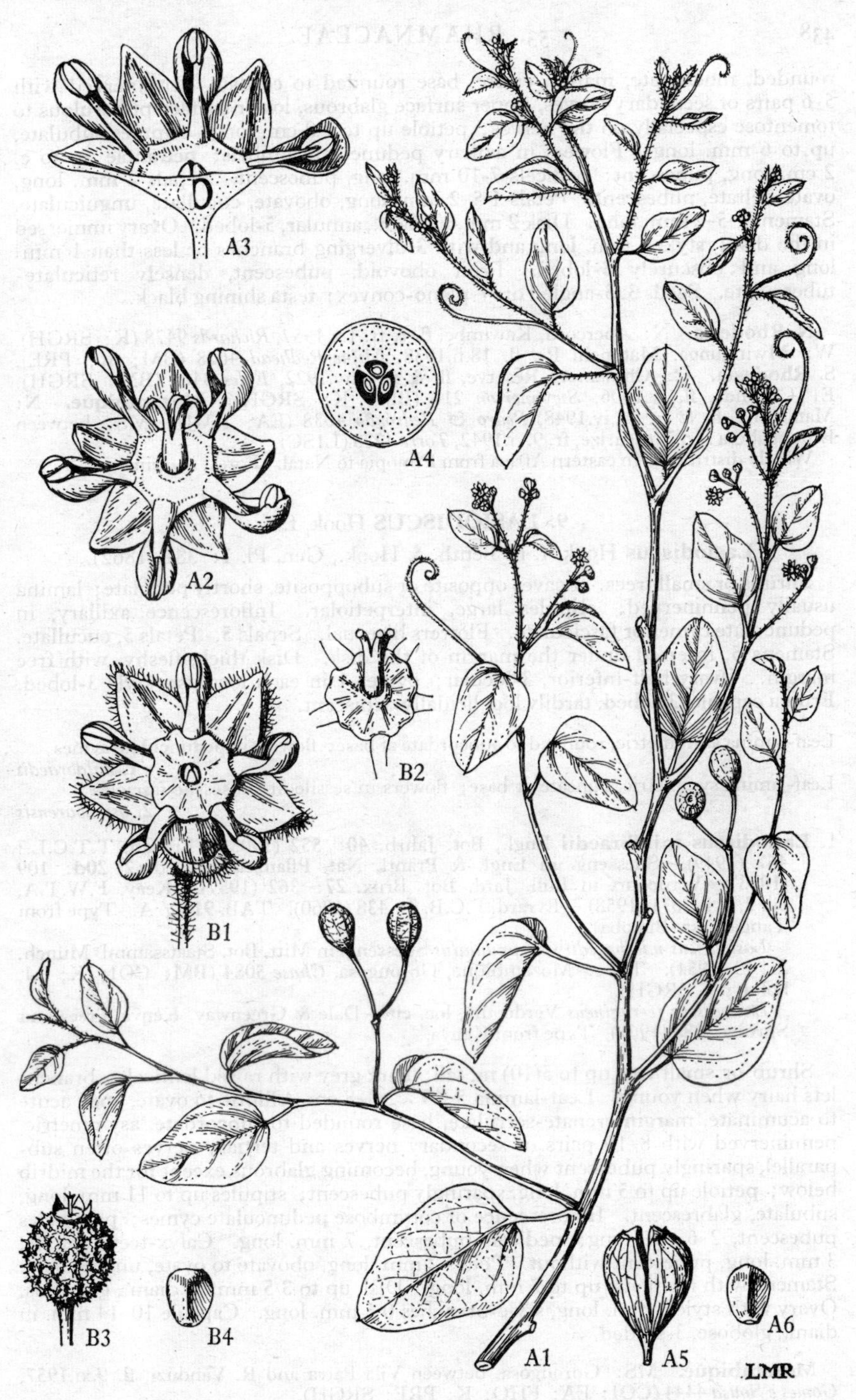

Tab. 92. A.—HELINUS INTEGRIFOLIUS. A1, flowering and fruiting branch (×⅔); A2, flower (×6); A3, vertical section of flower (×6); A4, transverse section of ovary (×6); A5, fruit (×2); A6, seed (×2). A1, A3 and A4 from *Barbosa & Carvalho* 4226, A2, A5 and A6 from *Lemos & Balsinhas* 43. B.—HELINUS MYSTACINUS. B1, flower (×6) *Richards* 697; B2, disk and style (×6) *Richards* 5568; B3, fruit (×2); B4, seed (×2), both from *Sturgeon* in GHS 18198.

rounded, mucronate, margin entire, base rounded to cordate, penninerved with 5–6 pairs of secondary nerves, upper surface glabrous, lower surface puberulous to tomentose especially on the nerves; petiole up to 1·8 cm. long; stipules subulate, up to 6 mm. long. Flowers in axillary pedunculate umbels; peduncle up to c. 2 cm. long, pubescent; pedicels 7-10 mm. long, pubescent. Sepals 2 mm. long, ovate-deltate, pubescent. Petals 1·5–2 mm. long, obovate, cucullate, unguiculate. Stamens 1·5–2 mm. long. Disk 2 mm. in diam., annular, 5-lobed. Ovary immersed in the disk; style 2 mm. long and with 3 diverging branches or less than 1 mm. long and obscurely 3-lobed. Fruit obovoid, pubescent, densely reticulate-tuberculate. Seeds 3, 3-angled or ± plano-convex; testa shining black.

N. Rhodesia. N: Abercorn, Kawimbe, fl. & fr. 1.v.1957, *Richards* 9478 (K; SRGH). W: Mwinilunga, Matonchi R., fl. 18.ii.1938, *Milne-Redhead* 4628 (BM; K; PRE). **S. Rhodesia.** C: Chinamora Reserve, fl. & fr. 5.iv.1922, *Eyles* 3405 (BM; SRGH). E: Chirinda, fl. iv.1906, *Swynnerton* 217 (BM; K; SRGH). **Mozambique.** N: Maniamba, fl. & fr. 28.iv.1948, *Pedro & Pedrógão* 4038 (EA; LMJ). MS: between Espungabera and Mossurize, fr. 9.vi.1942, *Torre* 4273 (LISC).

Widely distributed in eastern Africa from Ethiopia to Natal. Forest margins.

9. LASIODISCUS Hook. f.

Lasiodiscus Hook. f. in Benth. & Hook., Gen. Pl. **1**: 381 (1862).

Shrubs or small trees. Leaves opposite or subopposite, shortly petiolate; lamina usually penninerved; stipules large, interpetiolar. Inflorescence axillary, in pedunculate cymes or fasciculate. Flowers bisexual. Sepals 5. Petals 5, cucullate. Stamens 5, inserted under the margin of the disk. Disk thick, fleshy, with free margin. Ovary half-inferior, 3-locular; ovules 1 in each loculus; style 3-lobed. Fruit a capsule, 3-lobed, tardily loculicidally dehiscent.

Leaf-lamina asymmetric, rounded to subcordate at base; flowers in pedunculate cymes
1. *mildbraedii*

Leaf-lamina symmetric, cuneate at base; flowers in sessile or subsessile fascicles
2. *usambarensis*

1. **Lasiodiscus mildbraedii** Engl., Bot. Jahrb. **40**: 552 (1908).—Brenan, T.T.C.L.: 467 (1949).—Suesseng. in Engl. & Prantl, Nat. Pflanzenfam., ed. 2, **20d**: 109 (1953).—Verdcourt in Bull. Jard. Bot. Brux. **27**: 362 (1957).—Keay, F.W.T.A, ed. 2, **1,** 2: 671 (1958).—Evrard, F.C.B. **9**: 438 (1960). TAB. **91** fig. A. Type from Tanganyika (Bukoba).

Lasiodiscus mildbraedii var. *undulatus* Suesseng. in Mitt. Bot. Staatssamml. Münch. **2**: 40 (1954). Type: Mozambique, Gorongosa, *Chase* 5084 (BM; COI; K; M, holotype; SRGH).

Lasiodiscus ferrugineus Verdcourt, loc. cit.—Dale & Greenway, Kenya Trees and Shrubs: 388 (1961). Type from Kenya.

Shrub or small tree up to 5(10) m. tall; bark grey with raised lenticels; branchlets hairy when young. Leaf-lamina 5–19 × 2·3–8 cm., elliptic to ovate, apex acute to acuminate, margin crenate-serrulate, base rounded to subcordate, asymmetric, penninerved with 8–10 pairs of secondary nerves and tertiary nerves often subparallel, sparingly pubescent when young, becoming glabrous except for the midrib below; petiole up to 5 mm. long, sparingly pubescent; stipules up to 11 mm. long, subulate, glabrescent. Inflorescence of corymbose pedunculate cymes; peduncles pubescent, 2–6 cm. long; pedicels pubescent, 7 mm. long. Calyx-teeth deltate, 3 mm. long, pubescent without. Petals 2 mm. long, obovate to ovate, unguiculate. Stamens with filaments up to 2 mm. long. Disk up to 3·5 mm. in diam., glabrous. Ovary with style 2 mm. long, style-branches 0·5 mm. long. Capsule 10–14 mm. in diam., globose, 3-seeded.

Mozambique. MS: Gorongosa, between Vila Paiva and R. Vandúzi, fl. 9.xi.1957, *Gomes e Sousa* 4444 (COI; EA; FHO; K; PRE; SRGH).

From the Ivory Coast to the Sudan and Kenya in the north and to the Congo, Tanganyika and Mozambique in the south. In our area restricted to forest in the coastal areas of Mozambique.

2. **Lasiodiscus usambarensis** Engl., Bot. Jahrb. **40,** 551 (1908).—Brenan, T.T.C.L.: 467 (1949).—Suesseng. in Engl. & Prantl, Nat. Pflanzenfam., ed. 2, **20d**: 109 (1953).

—Verdcourt in Bull. Jard. Bot. Brux. **27**: 361 (1957). TAB. **91** fig. B. Type from Tanganyika.

Lasiodiscus holtzii sensu Bak. f. in Journ. Linn. Soc., Bot. **40**: 46 (1911).—Eyles in Trans. Roy. Soc. S. Afr. **5**: 407 (1916).

Tree to 6·5 m. tall; bark grey with raised lenticels; branchlets sparsely pubescent when young, later glabrescent. Leaf-lamina up to 4·5(6) × 14(17) cm., elliptic to ovate, apex acute to acuminate, margin crenate-serrulate, base cuneate, penninerved with 8–9 pairs of secondary nerves impressed above and prominent below, tertiary venation prominent beneath, sparsely hairy on nerves beneath or glabrous; petiole 3–9 mm. long, pubescent; stipules 5–9 mm. long, subulate, pubescent. Flowers in sessile or subsessile axillary fascicles. Pedicels 3–7 mm. long, pubescent. Sepals up to 2 mm. long, deltate, ovate, pubescent without. Petals 1·2 mm. long, obovate, cucullate, shortly unguiculate. Stamens with filaments 1 mm. long. Disk 2 mm. in diam., fleshy. Ovary inferior; style 0·5 mm. long, 3-lobed. Capsule up to 9 mm. in diam., persistently minutely tomentellous. Seeds 7 × 5 mm., compressed.

S. Rhodesia. E: Chirinda, fl. & fr. 12.x.1947, *Wild* 2218 (K; SRGH).
Also in eastern and southern Tanganyika. An understorey tree of evergreen forest.

54. VITACEAE

By H. Wild & R. B. Drummond

Erect trailing or climbing perennial herbs, climbing shrubs or rarely small trees. Leaves alternate, simple or digitately compound or rarely pedate, margin variously toothed or rarely entire; stipules petiolar; tendrils present or absent, leaf-opposed or arising from the peduncle. Flowers actinomorphic, usually bisexual. Calyx subentire or 4–6-lobed. Petals 4–6, free, valvate. Stamens 4–6, opposite the petals; filaments free; anthers 2-locular, medifixed with longitudinal dehiscence. Disk intrastaminal, annular or of separate glands. Ovary superior, 2-locular; ovules 2 in each loculus; style short; stigma subulate to capitate. Fruit baccate, with 1–4 seeds. Seeds with copious sometimes ruminate endosperm.

Petals 5–6:
- Inflorescence usually with tendrils; style ± conical and very short; disk with vertical furrows - - - - - - - - - **1. Ampelocissus**
- Inflorescence never with tendrils; style slender, readily visible; disk without furrows **2. Rhoicissus**

Petals 4:
- Flower-bud cylindric or lageniform with apex often ± inflated, constricted at or near the middle; disk divided into 4 glands; leaves usually compound **4. Cyphostemma**
- Flower-bud globose or conical, not constricted at the middle; disk annular:
  - Flower-bud conical to ovoid; disk thick; leaves usually simple, occasionally digitately compound; cymes usually leaf-opposed; fruit usually 1 (2)-seeded - **3. Cissus**
  - Flower-bud depressed-globose; disk thin; leaves 3-foliolate or pedate; cymes usually axillary; fruit (2) 4-seeded - - - - - - **5. Cayratia**

In addition to the above genera the Grape Vine (*Vitis vinifera* L.) is frequently cultivated and is being grown on a commercial scale in S. Rhodesia. Virginia Creeper (*Parthenocissus quinquefolia* (L.) Planch.), is commonly cultivated in gardens in our area.

1. AMPELOCISSUS Planch.

Ampelocissus Planch. in La Vigne Amér. **8**: 372 (1884); in A. & C. DC., Mon. Phan. **5**, 2: 368 (1887) *nom. conserv.*
Botria Lour., Fl. Cochinch.: 153 (1790).

Erect perennials, climbers or lianes; tendrils present, arising from the peduncle, rarely absent. Leaves simple, entire or lobed, or digitately 3–5-foliolate; leaflets sessile or petiolulate; stipules deltate, inconspicuous. Inflorescences of condensed cymes with flowers in capitate heads or cymes more lax and paniculate. Calyx entire or ± lobed. Petals 5. Disk ± entire, clasping the ovary. Style simple, very short; stigma not wider than the style. Seeds usually 4, flattened-ellipsoid, with a strongly marked longitudinal keel on one side and a median furrow or pit on the other, variously rugose.

An examination of the material of *Ampelocissus* from our area reveals that the character of polygamo-monoecious flowers, often used in keys to separate this genus from *Rhoicissus*, is of no value, at least in this Flora.

Leaves digitate:
 Leaves quite glabrous, membranous; inflorescence rather lax - - 1. *multistriata*
 Leaves hairy (usually with a ± villous indumentum at least below), often somewhat coriaceous; inflorescence densely capitate or of dense cymes:
  Plant ± erect, usually without tendrils - - - 2. *obtusata* subsp. *obtusata*
  Plant scrambling or climbing, usually with tendrils - 2. *obtusata* subsp. *kirkiana*
Leaves simple, broadly ovate to circular, entire or shallowly 3–5-lobed (sometimes more deeply so) - - - - - - - - - - - 3. *africana*

1. **Ampelocissus multistriata** (Bak.) Planch. in A. & C. DC., Mon. Phan. **5**, 2: 398 (1887).—Keay, F.W.T.A. ed. 2, **1**, 2: 682 (1958).—Willems, F.C.B. **9**: 556 (1960). Type from Senegambia.
 Vitis pentaphylla Guill. & Perr. in Guill., Perr. & Rich., Fl. Senegamb. Tent. **1**: 135, t.33 (1832) non Thunb. (1784).—Gilg & Brandt in Engl., Bot. Jahrb. **46**: 427 (1911).—Suesseng. in Engl. & Prantl, Nat. Pflanzenfam. ed. 2, **20d**: 301 (1953). Type as above.
 Vitis multistriata Bak. in Oliv., F.T.A. **1**: 410 (1868). Type as above.
 Ampelocissus sarcantha Gilg & Brandt, tom. cit.: 428 (1911).—Brenan, T.T.C.L.: 25 (1949). Syntypes from Tanganyika and Mozambique, Tete, Boruma, *Menyhart* 717 (B†, type; Z).

Liane or climber with cylindric hollow glabrous finely striate stems. Leaves (3) 5-foliolate or very occasionally simple and 5-lobed; petiole up to 13 cm. long, glabrous, finely striate; leaflets subsessile (or petiolule of terminal leaflet up to c. 5 mm. long); leaflet-lamina up to 22 × 11 cm., membranous, the lateral ones somewhat smaller than the terminal one, terminal leaflet narrowly obovate, lateral ones elliptic with the basal pair often asymmetric, apices acuminate, margins strongly serrate or serrate-crenate, bases cuneate, glabrous on both sides, paler below. Inflorescence of rather lax much-branched paniculate cymes; peduncle up to 11 cm. long, glabrous, tendril branching from peduncle c. 1·5 cm. below the base of the inflorescence-rhachis; rhachis and all branches of inflorescence puberulous; pedicels up to 2 mm. long, puberulous. Calyx 1 mm. in diam., salver-shaped, glabrous, entire or undulate. Petals wine-coloured (or cream), 1 mm. long, oblong, glabrous. Stamens shorter than the corolla lobes with wine-coloured filaments and yellow anthers. Disk wine-coloured, cupular and almost hiding the pentagonal ovary. Fruit c. 1 cm. long, ellipsoid. Seeds dark brown, up to 9 × 4 mm., oblong-ellipsoid.

N. Rhodesia. E: between Changwe and Luangwa R., fl. 16.xii.1958, *Robson* 966 (K; SRGH). **Nyasaland.** C: Dedza Distr., Mua Livulezi Forest, fr. 19.iii.1955, *E.M. & W.* 1053 (BM; SRGH). **Mozambique.** N; Nampula., fl. 23.i.1936, *Torre* 861 (COI; LISC). Z: Mocuba, Namagoa, fl. & fr. 1945, *Faulkner* 9 (EA; PRE). MS: Chimoio, Gondola, fl. & fr. 12.ii.1948, *Garcia* 171 (LISC).
Also in Senegal, Gambia, Mali Republic, Guinée Republic, Nigeria, Cameroun, Ubangi-Shari, Chad, Sudan and Tanganyika. A climber or liane in forest, riverine forest or bamboo brakes.

2. **Ampelocissus obtusata** (Welw. ex Bak.) Planch. in La Vigne Amér. **1885**: 48 (1885); in A. & C. DC., Mon. Phan. **5**, 2: 401 (1887).—Gilg & Brandt in Engl., Bot. Jahrb. **46**: 432 (1911).—Eyles in Trans. Roy. Soc. S. Afr. **5**: 408 (1916).—Suesseng. in Engl. & Prantl, Nat. Pflanzenfam. ed. 2, **20d**: 304 (1953).—Exell & Mendonça, C.F.A. **2**, 1: 37 (1954).—Willems, F.C.B. **9**: 560 (1960). Type from Angola.
 Vitis obtusata Welw. ex Bak. in Oliv., F.T.A. **1**: 414 (1868). Type as above.
 Ampelocissus aesculifolia Gilg & Brandt, tom. cit.: 433 (1911).—Gilg & Fr. in R.E.Fr., Schwed. Rhod.-Kongo-Exped. **1**: 133 (1914).—Brenan, T.T.C.L.: 25 (1949).—Suesseng., loc. cit.—White, F.F.N.R.: 230 (1962). Syntypes from Tanganyika and Nyasaland, *Buchanan* 199 (BM; K).

An erect perennial up to c. 2 m. tall but with its branches often eventually twining, or with scrambling or climbing branches from the beginning but even then not climbing usually to more than c. 2 m. high. Branches rather coarsely striate, densely ferruginous- or cinnamon-villous, later becoming greyish-villous or finally glabrescent. Leaves 3–5-foliolate, somewhat coriaceous; petiole up to c. 17 cm. long, villous, at least when young; leaflet-lamina up to c. 22 × 10 cm., terminal leaflet obovate to narrowly obovate or occasionally broadly elliptic, apex rounded or subacute, cuneate at the base, with a petiolule up to c. 2·5 cm. long, lateral leaflets somewhat smaller, ± asymmetric, more broadly cuneate, or subtruncate at the base, sessile or with petiolules almost as long as the terminal leaflets, leaflet margins irregularly serrate or serrate-crenate, greyish- or cinnamon-villous above when young, later glabrescent, more densely villous below. Inflorescence of dense capitate-globose cymes, heads simple or compound or sometimes heads only moderately dense; peduncle up to c. 8 cm. long, often villous when young; tendrils villous, branching from the base of the inflorescence, or just below, or apparently sometimes absent; rhachis and branches of inflorescence usually villous. Calyx c. 1·5 mm. in diam., cyathiform, entire or shallowly lobed, glabrous to villous. Petals deep red, c. 1·5 mm. long, oblong, glabrous. Stamens $\frac{2}{3}$ the length of the petals; filaments red; anthers yellow. Disk deep red, enclosing c. $\frac{2}{3}$ of the strongly ribbed ovary. Fruit red, c. 1 cm. in diam. Seeds brown, c. 7 × 5 mm., ellipsoid, longitudinal keel not much raised, with an elliptic pit and radiating ridges on the other side.

Also in Angola, Tanganyika, the Congo and the Transvaal.

Subsp. **obtusata.**

Plant ± erect, usually without tendrils; leaflet-lamina 7–22 cm. long, apex ± rounded; lateral leaflets subsessile; petiolules short.

N. Rhodesia. N: Chilongowelo, fl. 24.xii.1956, *Richards* 7344 (K). W: Mwinilunga, fl. 13.ii.1938, *Milne-Redhead* 2900 A (K). **Nyasaland.** N: Karonga, fl. 3.ii.1953, *Williamson* 139 (BM). S: Shire, fl. xii.1893, *Scott Elliot* 8556 (BM; K). **Mozambique.** N: Maniamba, fl. 25.i.1935 *Torre* 746 (COI; K; LISC).

Also in the Congo, Angola and Tanganyika. Open woodland.

Material from Nyasaland, the Northern Division of N. Rhodesia, Tanganyika and Mozambique has larger and more acute leaflets (as in the syntypes of *A. aesculifolia*) than the material from Angola, the Congo and the Western Division of N. Rhodesia (as in the type of *A. obtusata*).

Subsp. **kirkiana** (Planch.) Wild & Drummond in Kirkia, **3**: 16 (1963). Type: drawing 145 by Kirk (K) made from a specimen growing by the R. Luabo, Mozambique.

Ampelocissus kirkiana Planch. in A. & C. DC., Mon. Phan. **5**, 2: 403 (1887).—Gilg & Brandt, tom. cit.: 435 (1911).—Bullock in Kew Bull. **3**: 187 (1948).—Suesseng, loc. cit. Type as above.

Ampelocissus pulchra Gilg in Engl., Pflanzenw. Ost-Afr. **C**: 257 (1895).—Gilg & Brandt, tom. cit.: 434, fig. 1F (1911).—Gilg & Fr., loc. cit.—Brenan, tom. cit.: 26 (1949).—Suesseng. loc. cit. Type from Tanganyika.

Ampelocissus elisabethvilleana De Wild. in Bull. Jard. Bot. Brux. **4**: 365 (1914).—Willems, F.C.B. **9**: 562 (1960). Type from the Congo.

Ampelocissus venenosa De Wild., tom. cit.: 364 (1914). Type from the Congo.

Ampelocissus rhodesica Suesseng. in Proc. & Trans. Rhod. Sci. Ass. **43**: 110 (1951); in Engl. & Prantl, loc. cit. Type: S. Rhodesia, Marandellas, *Dehn* 527 (M, holotype; SRGH).

Plant scrambling or climbing, usually with tendrils; leaflet-lamina 4–c. 11 cm. long, apex acute or obtuse; lateral leaflets with petiolules up to 5 cm. long.

Bechuanaland Prot. N: Ngamiland., fl. xii.1930, *Curson* 733 (PRE). **N. Rhodesia.** B: Mongu, Lake Kande, fl. 11.xi.1959, *Drummond & Cookson* 6345 (SRGH). N: Abercorn, fl. 12.i.1955, *Richards* 4053 (K). W: Kitwe, fr. 26.i.1954, *Fanshawe* 717 (K; NDO). C: Serenje, fl. 30.x.1956, *White* 3809 (FHO). S: Bombwe, fr. 27.ii.1933, *Martin* 598 (FHO). **S. Rhodesia.** N: Sebungwe, Kariangwe, fl. 13.xi.1956, *Lovemore* 494 (SRGH). W: Nyamandhlovu, fl. x.1930, *Pardy* in GHS 4974 (SRGH). C: Salisbury, Twentydales, fl. 31.xi.1951, *Wild* 3673 (SRGH). E: Umtali, fl. 11.xii.1953, *Chase* 5153 (BM; SRGH). **Nyasaland.** S: Ncheu Distr., Lower Kirk Range, fr. 17.iii.1955, *E.M. & W.* 968 (BM; SRGH). **Mozambique.** Z: Mocuba, Namagoa, fl. 14.xi.1948, *Faulkner* 385 (COI; K). MS: Chimoio, fr. 25.ii.1948, *Garcia* 154 (LISC). SS: Vila de João Belo, fl. 17.xi.1957, *Barbosa & Lemos* in *Barbosa* 8098 (COI; LISC; LMJ).

Also in the Congo, Angola (Moxico), and the Transvaal. Woodland and often on rocky outcrops.

A. pulchra represents a form with globose-capitate inflorescences but this character, particularly in specimens with young inflorescences, occurs occasionally throughout the range of subsp. *kirkiana*. It seems of little significance. The ranges of the two subspecies overlap considerably in N. Rhodesia and in this territory intermediates are fairly common e.g., *Macaulay* 1004 (K) from Mumbwa, *White* 3597 (FHO) from Kawambwa and *White* 1949 (FHO) from Sesheke.

3. **Ampelocissus africana** (Lour.) Merr. in Trans. Amer. Phil. Soc., n.s., **24**: 253 (1935).—Suesseng. in Engl. & Prantl, Nat. Pflanzenfam. ed. 2, **20d**: 302 (1953).—Wild & Drummond in Kirkia, **3**: 17 (1963). TAB. **93**. Type from Zanzibar.
Botria africana Lour., Fl. Cochinch. **1**: 154 (1790). Type as above.
Vitis mossambicensis Klotzsch in Peters, Reise Mossamb. Bot. **1**: 180 (1861).—Bak. in Oliv., F.T.A. **1**: 397 (1868). Type: Mozambique, Zambesiland, *Peters* (B†, holotype).
Vitis asarifolia Bak., tom. cit.: 396 (1868).—Planch. in A. & C. DC., Mon. Phan. **5**, 2: 393 (1887). Syntypes from Sudan, Zanzibar and Mozambique (or Tanganyika), R. Rovuma, *Meller* (K).
Vitis grantii Bak., tom. cit.: 400 (1868). Type from Uganda.
Ampelocissus mossambicensis (Klotzsch) Planch. in La Vigne Amér. **1885**: 49 (1885); in A. & C. DC., Mon. Phan. **5**, 2: 392 (1887).—Bak. f. in Journ. Linn. Soc., Bot. **40**: 46 (1911).—Gilg & Brandt in Engl., Bot. Jahrb. **46**: 431 (1911).—Eyles in Trans. Roy. Soc. S. Afr. **5**: 408 (1916). Type as above.
Ampelocissus asarifolia (Bak.) Planch. in La Vigne Amér. **1885**: 29 (1885); in A. & C. DC., Mon. Phan. **5**, 2: 393 (1887). Syntypes as above.
Ampelocissus grantii Bak., tom. cit.: 400 (1868).—Planch. in La Vigne Amér. **1885**: 32 (1885); in A. & C. DC., Mon. Phan. **5**, 2: 394 (1887).—Gilg & Brandt, tom. cit.: 428, fig. 1D, 2A–F (1911).—Gilg & Fr. in R.E.Fr., Schwed. Rhod.-Kongo-Exped. **1**: 133 (1914).—Eyles, loc. cit.—Brenan, T.T.C.L.: 26 (1949).—Keay, F.W.T.A. ed. 2, **1**, 2: 682 (1958).—Willems, F.C.B. **9**: 560 (1960).—White, F.F.N.R.: 230 (1962). Type as above.

Climber or liane, or sometimes prostrate; branches densely pubescent when young or entirely glabrous, striate. Leaves simple; petiole up to 13 cm. long, pubescent or glabrous; leaf-lamina up to 20 cm. in diam., from very broadly ovate to circular, entire or shallowly 3–5-lobed, or more rarely deeply lobed, apices of lobes rounded or acute, margins serrate or crenate, base cordate with a rounded sinus c. 2 cm. deep, pubescent above when young or glabrous, varying from tomentellous (when young) to pubescent or glabrous below. Inflorescence of moderately dense or rather lax cymes; penduncle up to c. 6 cm. long, from subtomentose to glabrescent; tendril arising from the base of the inflorescence-rhachis or below; pedicels less than 1 mm. long. Flowers in subcapitate clusters. Calyx c. 1·5 mm. in diam., ± salver-shaped or cyathiform, glabrous, margin entire or shallowly lobed. Petals reddish, up to 1·5 mm. long, oblong, glabrous. Stamens c. ½ the length of the petals. Disk reddish, c. ½ the length of the ovary, cupular. Fruit subglobose, c. 1 cm. in diam. Seeds shining brown, c. 7 × 4 mm., ellipsoid.

Caprivi Strip: 9 km. from Katima Mulilo on road to Ngoma, fl. 5.i.1959, *Killick & Leistner* 3302 (PRE; SRGH). **Bechuanaland Prot.** N: Ngamiland, fl., *Curson* 398 (PRE). **N. Rhodesia.** B: Sesheke, fl. 28.xii.1952, *Angus* 1054 (FHO). N: Kasaba Game Reserve, st. 17.xi.1959, *McCallum Webster* 632 (K). W: Ndola, fl. 30.i.1954, *Fanshawe* 743 (EA; NDO; SRGH). S: Bombwe, fl. 25.xi.1932, *Martin* 393/32) (FHO; NDO). **S. Rhodesia.** N: Chiswiti Reserve, Darwin Distr., fl. 22.i.1960, *Phipps* 2384 (K; LISC; PRE; SRGH). W: Nyamandhlovu, fl. 15.ii.1956, *Plowes* 1932 (SRGH). C: Goromonzi, fl. 7.xii.1951, *Wild* 3695 (SRGH). E: Inyanga, Rhodes Estate, 16–20.xi.1931, *Brain* 7357 (SRGH). S: Ndanga, Mtilikwe R., fl. 26.i.1949, *Wild* 2771 (SRGH). **Nyasaland.** S: Fort Johnston, fr. 14.iii.1955, *E.M. & W.* 863 (BM; SRGH). **Mozambique.** N: Porto Amélia, fl. 14.iii.1959, *Myre & Macedo* 3520 (LM; SRGH). Z: Mocuba, Namagoa, fl. 3.i.1949, *Faulkner* 386 (COI; K). T: Angónia, fr. 31.v.1943, *Viana* IA (LISC; PRE). MS: Chimoio, Bandula, fr. 27.ii.1948, *Garcia* 406 (LISC).

Also in Guinée Republic, Ivory Coast, Dahomey, Nigeria, Cameroun, Ubangi-Shari, Chad, Sudan, Tanganyika, Zanzibar and Kenya. Climber in woodland or at forest edges.

Ampelocissus mossambicensis represents the most hairy form of this species, which shows considerable variation in its amount of indumentum, one form being quite glabrous in its vegetative parts.

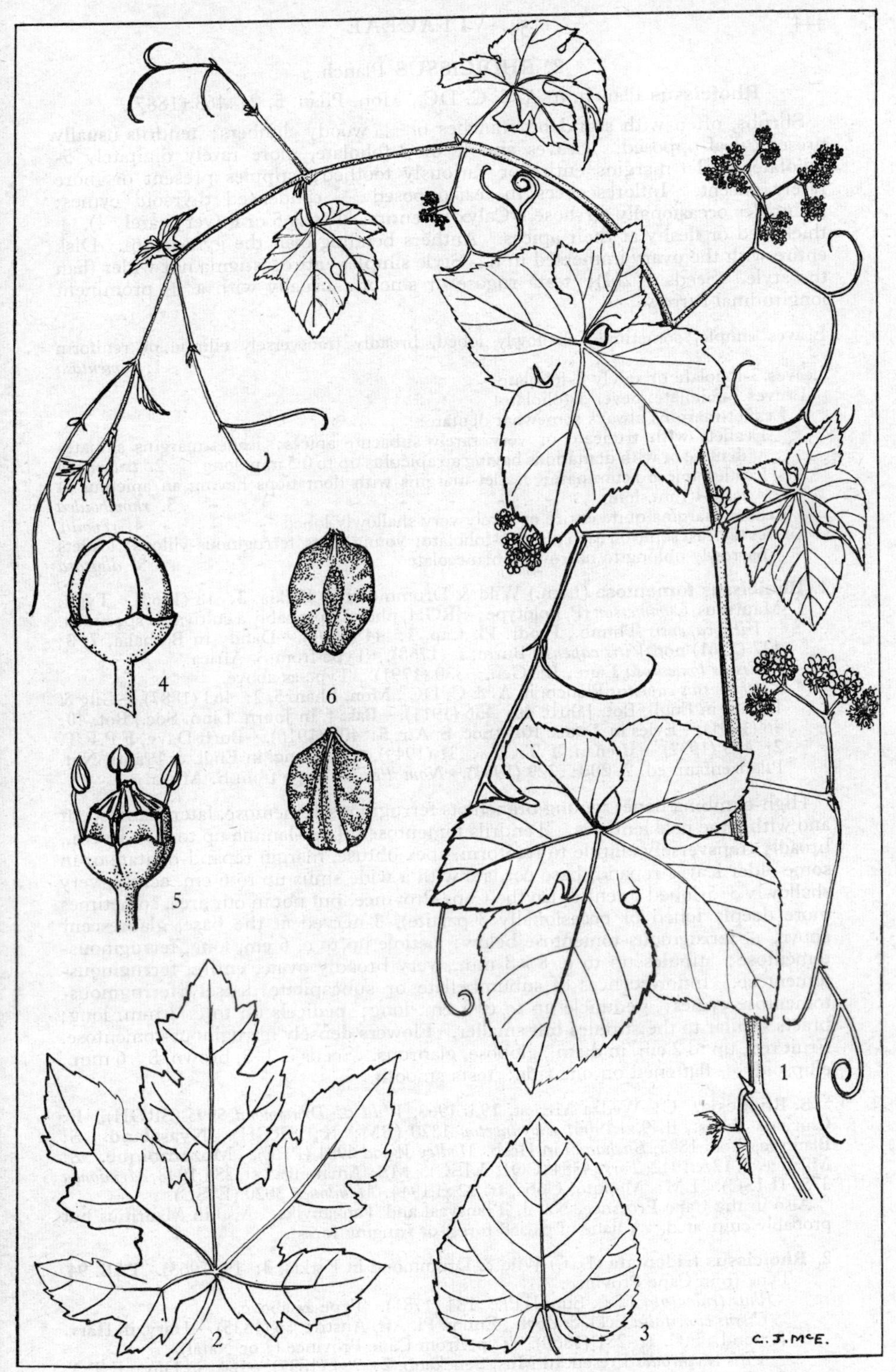

Tab. 93. AMPELOCISSUS AFRICANA. 1, flowering branch (×½) *Brain* 7340; 2, lobed leaf (×⅓) *Myre & Macedo* 3531; 3, simple leaf (×⅓) *Wild* 2269; 4, flower-bud (×8) *Richards* 7259; 5, flower with petals removed (×8) *Richards* 1951; 6, seeds (×2) *Germain* 5993.

2. RHOICISSUS Planch.

Rhoicissus Planch. in A. & C. DC., Mon. Phan. **5,** 2: 463 (1887).

Shrubs, often with scandent branches or ± woody climbers; tendrils usually present, leaf-opposed. Leaves simple or 3-foliolate, more rarely digitately 5-foliolate, leaflet-margins entire or variously toothed; stipules present or more rarely absent. Inflorescences in leaf-opposed ± condensed thyrsoid cymes; peduncles occasionally cirrhose. Calyx ± entire. Petals 5 or 6 (very rarely 4), ± thickened or fleshy at their apices. Anthers bending over the gynoecium. Disk entire with the ovary immersed in it. Style simple, entire; stigma not wider than the style. Seeds 1–2(4); testa rugose or smooth, usually with a ± prominent longitudinal furrow.

Leaves simple, sometimes shallowly lobed, broadly transversely elliptic or reniform - 1. *tomentosa*
Leaves 3-foliolate or rarely 5-foliolate:
  Leaves 3-foliolate, never 5-foliolate:
    Leaflet-margins always somewhat dentate:
      Leaflets with truncate or very rarely subacute apices; leaflet-margins sinuate-dentate or with dentations having an apiculus up to 0·5 mm. long 2. *tridentata*
      Leaflet-apices acuminate; leaflet-margins with dentations having an apiculus at least 1 mm. long - - - - - - - - - 3. *rhomboidea*
    Leaflet-margins quite entire or rarely very shallowly lobed - - - 4. *revoilii*
  Leaves (at least some on each plant) 5-foliolate; young parts ferruginous-villous; leaflets narrowly oblong to narrowly oblanceolate - - - - - 5. *digitata*

1. **Rhoicissus tomentosa** (Lam.) Wild & Drummond in Kirkia, **3**: 18 (1963). Type: Mauritius, *Commerson* (P, holotype; SRGH, photo.), probably a cultivated specimen.
Vitis capensis Thunb., Prodr. Pl. Cap. **1**: 44 (1794).—Dandy in Bothalia, **7,** 3: 427 (1961) non *Vitis capensis* Burm. f. (1768). Type from S. Africa.
Cissus tomentosa Lam., Ill. Gen.: 330 (1791). Type as above.
Rhoicissus capensis Planch. in A. & C. DC., Mon. Phan. **5,** 2: 463 (1887).—Gilg & Brandt in Engl., Bot. Jahrb. **46**: 436 (1911).—Bak. f. in Journ. Linn. Soc., Bot. **40**: 46 (1911).—Eyles in Trans. Roy. Soc. S. Afr. **5**: 408 (1916).—Burtt Davy, F.P.F.T. **2**: 473 (1932).—Brenan, T.T.C.L.: 31 (1949).—Suesseng. in Engl. & Prantl, Nat. Pflanzenfam. ed. 2, **20d**: 329 (1953). *Nom illegit.* Type from S. Africa.

High-climbing liane; young branchlets ferruginous-tomentose, later glabrescent and with large pale lenticels. Tendrils tomentose. Leaf-lamina up to 20 × 16 cm., broadly transversely elliptic to reniform, apex obtuse, margin repand-dentate or in some older leaves repand, base cordate with a wide sinus up to 6 cm. across, very shallowly 3–5-lobed or entire (in the Cape Province, but not in our area, sometimes more deeply lobed or occasionally 3-partite), 3-nerved at the base, glabrescent above, ± ferruginous-tomentose below; petiole up to c. 6 cm. long, ferruginous-tomentose; stipules up to c. 8 × 8 mm., very broadly ovate, entire, ferruginous-tomentose. Inflorescences of subumbellate or subcapitate densely ferruginous-tomentose cymes; peduncle up to c. 2 cm. long; pedicels up to c. 2 mm. long; bracts similar to the stipules but smaller. Flowers densely ferruginous-tomentose. Fruit red, up to 2 cm. in diam., globose, glabrous. Seeds c. 1–3, brown, 8 × 6 mm., ellipsoid, ± flattened on one side; testa smooth.

S. Rhodesia. C: Wedza Mt., st. 19.ii.1963, *Wild & Drummond* 5995 (SRGH). E: Chirinda Forest, fl. 9.xi.1906, *Swynnerton* 1370 (BM; K; SRGH). **Nyasaland.** S: Blantyre, fl. xii.1895, *Buchanan* in Herb. *Medley Wood* 6981 (PRE). **Mozambique.** Z: Milange, fl. 12.x.1942, *Torre* 4584 (COI; LISC). MS: Amatongas, st. 28.i.1948, *Mendonça* 3755 (LISC). LM: Maputo, Goba, fr. 22.xi.1944, *Mendonça* 3020 (LISC).
Also in the Cape Province, Natal, Transvaal and Tanganyika. Also in Mauritius but probably cultivated. A liane of closed forest or fringing forest.

2. **Rhoicissus tridentata** (L. f.) Wild & Drummond in Kirkia, **3**: 19 (1963). TAB. **94.** Type from Cape Province.
Rhus tridentata L. f., Suppl. Pl.: 184 (1781). Type as above.
Cissus cuneifolia Eckl. & Zeyh., Enum. Pl. Afr. Austr.: 56 (1835).—Harv. in Harv. & Sond., F.C. **1**: 251 (1860). Type from Cape Province (? or Natal).
Vitis erythrodes Fresen. in Mus. Senckenb. **2**: 284 (1837).—Bak. in Oliv., F.T.A. **1**: 401 (1868). Type from Ethiopia.
Rhoicissus cuneifolia (Eckl. & Zeyh.) Planch. in A. & C. DC., Mon. Phan. **5,** 2: 466 (1887).—Bak. f. in Journ. Linn. Soc., Bot. **40**: 47 (1911).—Eyles in Trans. Roy.

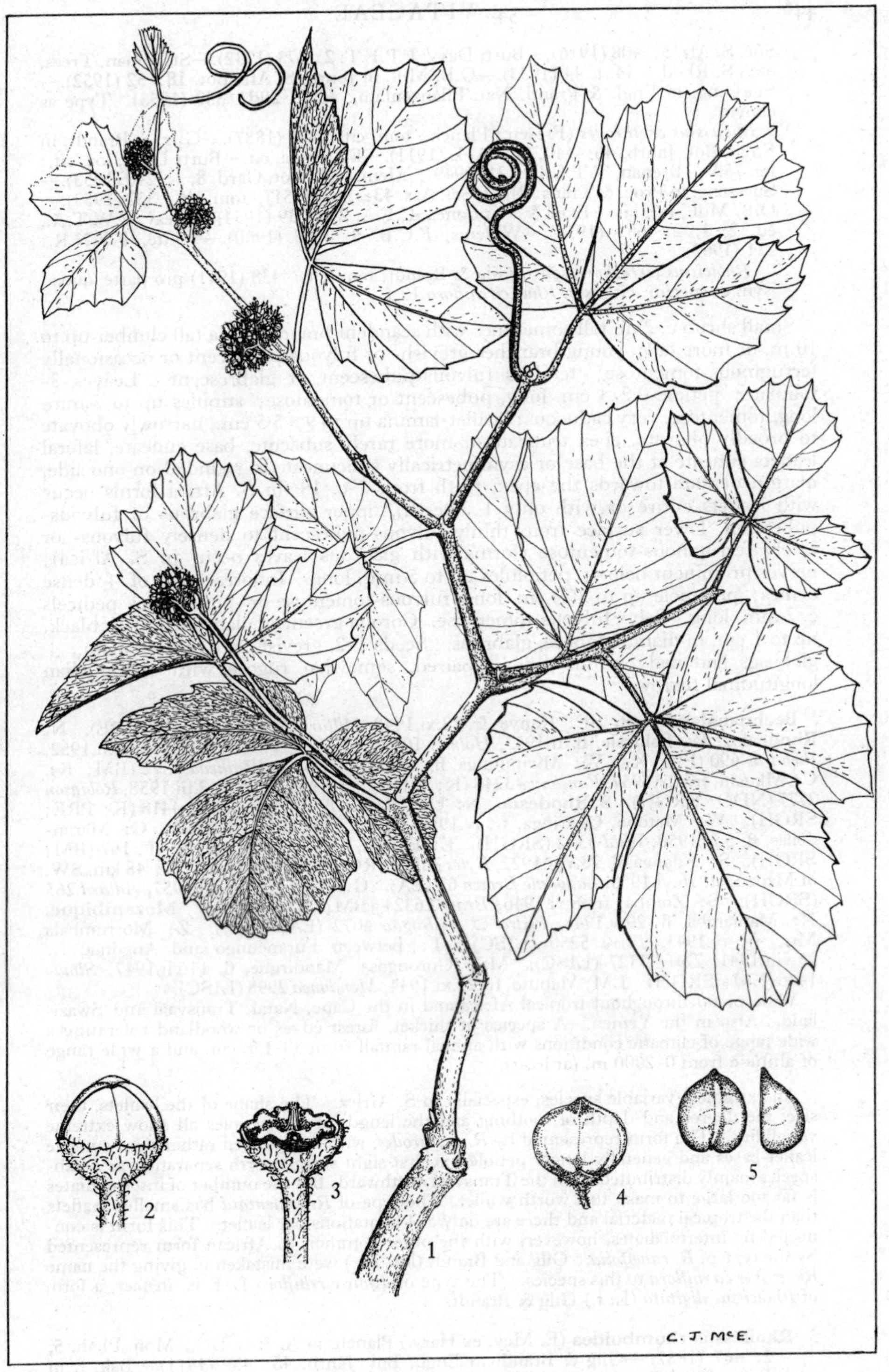

Tab. 94. RHOICISSUS TRIDENTATA. 1, flowering branch (×½) *Holmes* 1059; 2, flower-bud (×8) *Robinson* 2777; 3, flower with petals and stamens fallen (×8) *Wild* 3312; 4, fruit (×⅘) *West* 2143; 5, seeds (×2) *West* 2143.

Soc. S. Afr. **5**: 408 (1916).—Burtt Davy, F.P.F.T. **2**: 473 (1932).—Steedman, Trees, etc., S. Rhod.: 44, t. 44 (1933).—O.B. Mill. in Journ. S. Afr. Bot. **18**: 52 (1952).—Suesseng. in Engl. & Prantl, Nat. Pflanzenfam., ed. 2, **20d**: 330 (1953). Type as above.

Rhoicissus erythrodes (Fresen.) Planch., tom. cit.: 465 (1887).—Gilg. & Brandt, in Engl., Bot. Jahrb. **46**: 440, fig. 3A–L (1911).—Eyles, loc. cit.—Burtt Davy, loc. cit.: fig. 75A.—Brenan, T.T.C.L.: 31 (1949); Mem. N.Y. Bot. Gard. **8**, 3: 240 (1953).—Suesseng. in Proc. & Trans. Rhod. Sci. Ass. **43**: 111 (1951); tom. cit.: 332 (1953).—O.B. Mill., loc. cit.—Exell & Mendonça, C.F.A. **2**, 1: 39 (1954).—Keay, F.W.T.A., ed. 2, **1**, 2: 681 (1958).—Willems, F.C.B. **9**: 565 (1960).—White, F.F.N.R.: 231 (1962).

Rhoicissus cirrhiflora sensu Gilg & Brandt, tom. cit.: 438 (1911) pro parte quoad syn. et specim. cit. excl. *Rhus cirrhiflora* L. f.

Small shrub c. 2 m. tall sometimes with scandent branches or a tall climber up to 10 m. or more tall; young branches greyish- or fulvous-pubescent or occasionally ferruginous-tomentose; tendrils fulvous-pubescent or glabrescent. Leaves 3-foliolate; petiole 0·2–4 cm. long, pubescent or tomentose; stipules up to 2 mm. long, tomentose, very caducous; leaflet-lamina up to 9 × 5·5 cm., narrowly obovate to broadly obovate, apex truncate or more rarely subacute, base cuneate, lateral leaflets cuneate at the base or asymmetrically subcuneate to rounded on one side, margin dentate towards the apex, teeth from 3–c. 18 (in S. Africa forms occur with leaflets entire or with only 1–3 teeth), upper surface glabrous or fulvous-pubescent, lower surface from thinly fulvous-pubescent to densely fulvous- or rarely ferruginous-tomentose (forms with glabrous leaves occur in S. Africa); nerves prominent below; petiolules up to 5 mm. long. Inflorescences of ± dense cymes; peduncle up to 2(3) cm. long, fulvous-tomentose or -pubescent; pedicels c. 2 mm. long. Calyx fulvous-tomentose. Corolla greenish, glabrous. Fruit black, up to 1 cm. in diam., globose, glabrous. Seeds 1–2, greyish, c. 6 mm. in diam., ± globose, flattened on one side if paired, somewhat rugose with one median longitudinal furrow.

Bechuanaland Prot. SE: Kanye, fr. 13.xi.1948, *Hillary & Robertson* 495 (PRE). **N. Rhodesia.** B: Balovale, fl. ii.1953, *Holmes* 1059 (FHO). N: Abercorn, fl. 2.ii.1952, *Richards* 690 (EA; K). W: Mwinilunga, fl. 29.xii.1937, *Milne-Redhead* 3872 (BM; K). C: Mkushi, fr. 2.v.1957, *Fanshawe* 3248 (K; NDO). S: Mapanza, fl. 2.iii.1958, *Robinson* 2777 (NDO; SRGH). **S. Rhodesia.** N: Umvukwes, fl. iv.1948, *Rodin* 4418 (K; PRE; SRGH). W: Matobo, Quaringa, fr. iv.1957, *Miller* 4303 (EA; SRGH). C: Marandellas, fl. 5.iv.1950, *Wild* 3312 (SRGH). E: Engwa, fl. 3.ii.1955, *E.M. & W.* 167 (BM; SRGH). S: Ndanga, fl. 28.viii.1922, *Eyles* 3657 (SRGH). **Nyasaland.** N: 48 km. SW. of Mzuzu, st. 16.vi.1953, *Langdale Brown* 62 (EA). C: Dedza, fl. 19.xii.1957, *Adlard* 265 (SRGH). S: Zomba, fr. 8.vi.1946, *Brass* 16324 (BM; K; SRGH). **Mozambique.** N: Maniamba, fl. 29.v.1948, *Pedro & Pedrógão* 4072 (EA; LMJ). Z: Morrumbala Mts., fr. iv.1943, *Torre* 5250 (LISC). T: between Furancungo and Angónia, fr. 25.viii.1941, *Torre* 3327 (LISC). MS: Gorongosa, Mandirgue, fl. 11.vii.1947, *Simão* 1446 (LM; SRGH). LM: Maputo, fr. 21.xi.1944, *Mendonça* 2998 (LISC).

Widespread throughout tropical Africa and in the Cape, Natal, Transvaal and Swaziland. Also in the Yemen. A species of thicket, forest edges or woodland tolerating a wide range of climatic conditions with annual rainfall from 38–150 cm. and a wide range of altitude from 0–2000 m. (at least).

An extremely variable species, especially in S. Africa. The shape of the leaflets, their size, the degree and depth of toothing and the length of the petioles all show extreme variability. The form represented by *R. erythrodes*, with its rounded rather than cuneate leaflet-bases and generally longer petioles, at first sight seems worth separation as a subspecies mainly distributed from the Transvaal northwards but the number of intermediates is far too large to make this worth while. The type of *R. tridentata* has smaller leaflets than the tropical material and there are only 3–4 dentations per leaflet. This form is connected by intermediates, however, with the other common S. African form represented by the type of *R. cuneifolia*. Gilg and Brandt (loc. cit.) were mistaken in giving the name *Rhoicissus cirrhiflora* to this species. The type of *Rhus cirrhiflora* L. f. is, in fact, a form of *Rhoicissus digitata* (L. f.) Gilg & Brandt.

3. **Rhoicissus rhomboidea** (E. Mey. ex Harv.) Planch. in A. & C. DC., Mon. Phan. **5**, 2: 467 (1887).—Gilg & Brandt in Engl., Bot. Jahrb. **46**: 439 (1911).—Bak. f. in Journ. Linn. Soc., Bot. **40**: 47 (1911).—Eyles in Trans. Roy. Soc. S. Afr. **5**: 408 (1916).—Burtt Davy, F.P.F.T. **2**: 473 (1932).—Suesseng. in Engl. & Prantl, Nat. Pflanzenfam., ed. 2, **20d**: 332 (1953). Syntypes from Natal.

Cissus rhomboidea E. Mey. ex Harv. in Harv. & Sond., F.C. **1**: 252 (1860). Syntypes as above.

Scrambling vine or tall liane; young parts ferruginous-villous with long weak hairs, soon glabrescent; tendrils glabrescent. Leaves 3-foliolate; petiole up to 2·3 cm. long; stipules not seen; leaflet-lamina up to 6 × 4 cm., terminal obovate, laterals asymmetric, rhomboid, apex acuminate, margins with c. 6 acuminate dentations, base cuneate; petiolules up to 1–4 cm. long. Inflorescences of moderately dense cymes, branches of inflorescence ± persistently ferruginous-villous; peduncle up to 1 cm. long; pedicels c. 2 mm. long; bracts c. 8 × 6 mm., rotund, ferruginous-villous. Flowers glabrous. Fruit c. 1 cm. diam., subglobose, glabrous. Seeds 1–2, c. 8 × 7 mm., ellipsoid-globose, one side flattened when seeds in pairs; testa pale brown, slightly rugose, not furrowed.

S. Rhodesia. E: Chirinda Forest, fl. x.1905, *Swynnerton* 92 (BM; K; SRGH).
Also in Cape Province, Natal and the Transvaal. A tall climber in closed forest or at forest edges.

4. **Rhoicissus revoilii** Planch. in A. & C. DC., Mon. Phan. **5**, 2: 469 (1887).—Gilg & Brandt in Engl., Bot. Jahrb. **46**: 440 (1911).—Brenan, T.T.C.L.: 31 (1949).—Suesseng. in Engl. & Prantl, Nat. Pflanzenfam., ed. 2, **20d**: 332 (1953).—Keay, F.W.T.A., ed. 2, **1**, 2: 681 (1958).—Willems, F.C.B. **9**: 566 t. 56 (1960). Type from Somaliland.

Rhoicissus sansibarensis Gilg in Engl., Pflanzenw. Ost-Afr. **C**: 257 (1895).—Bak. f. in Journ. Linn. Soc., Bot. **40**: 47 (1911).—Eyles in Trans. Roy. Soc. S. Afr. **5**: 408 (1916). Type from Zanzibar.

Rhoicissus schlechteri Gilg & Brandt, tom. cit.: 438 (1911).—Burtt Davy, F.P.F.T. **2**: 473 (1932).—Suesseng., tom. cit.: 330 (1953). Syntypes: Mozambique, Delagoa Bay, *Schlechter* 11990 (B†; BM); *Junod* 304 (Z).

Rhoicissus erythrodes var. *ferruginea* sensu Eyles, loc. cit.

Rhoicissus spp.—Eyles, loc. cit.

Shrub or small tree to 3·3 m. tall with drooping branches or sometimes with sarmentose branches or more rarely a climber; young branches appressed-ferruginous-pubescent or densely fulvous- or ferruginous-pubescent or -tomentose, later glabrescent; tendrils tomentose, pubescent or glabrescent, often absent in herbarium specimens. Leaves 3-foliolate; petioles up to c. 4 cm. long, indumentum as in young stems; stipules not seen; leaflet-lamina up to 12 × 6 cm., terminal leaflets narrowly lanceolate, lanceolate, oblanceolate or ovate, laterals lanceolate, subfalcate-oblong or rhomboid, apex obtuse or acute, margins entire or slightly undulate-lobed on the lateral leaflets, sometimes revolute, base from narrowly to broadly cuneate, from thinly pubescent to glabrous above, from ferruginous- or fulvous-tomentose to glabrous below; axils of main nerves sometimes provided with domatia; petiolules up to 2 cm. long. Inflorescences of rather lax or moderately dense cymes, branches from appressed-ferruginous-pubescent to fulvous- or ferruginous-tomentose; peduncle up to c. 4 cm. long; pedicels c. 1 mm. long; bracts up to 2 mm. long, lanceolate to ovate, very caducous. Calyx densely ferruginous-pubescent to fulvous- or ferruginous-tomentose. Petals glabrous or with hairs near the apex. Fruit ± black when ripe, 1–1·5 cm. in diam., globose, glabrous. Seeds 1–2, c. 5–7 mm. in diam., subglobose with a flattened side when seeds paired, rugose, with a shallow but distinct longitudinal furrow.

N. Rhodesia. E: Fort Jameson, fr. 3.vi.1958, *Fanshawe* 4514 (FHO; K; NDO). **S. Rhodesia.** N: Umvukwes, Ruorka Ranch, fl. 16.xii.1952, *Wild* 3924 (PRE; SRGH). W: Shangani Reserve, fr. iii.1951, *Davies* 516 (SRGH). C: Serui, fl. 4.ii.1950, *Hornby* 3126 (SRGH). E: Inyanga, fl. xi.1931, *Brain* 7277 (K; LISC; SRGH). S: Lundi R., fr. 30.vi.1930, *Hutchinson & Gillett* 3303 (BM; K). **Nyasaland.** S: Zomba, fl., *Clements* 733 (FHO). **Mozambique.** N: Mogincual, st. 26.vii.1948, *Pedro & Pedrógão* 4675 (EA; LMJ). Z: between Mocuba and Milange, fr. 9.vi.1949, *Barbosa & Carvalho* 2994 (LM; LMJ; SRGH). T: between Fíngoè and Chicoa, fr. 29.vi.1949, *Barbosa & Carvalho* 3389 (LM; LMJ). MS: Vila de Manica, fl. & fr. 3.i.1948, *Mendonça* 3575 (LISC). SS: Macia, Muianga, fr. 2.vii.1947, *Pedro & Pedrógão* 1442 (COI; K; LMJ; PRE; SRGH). LM: 20 km. from Lourenço Marques on Umbeluzi road, fl. 18.ii.1955, *E.M. & W.* 488 (BM; LISC; SRGH).
Also in S. Arabia, Comoros, Somaliland, E. Africa, Ghana, the Congo and the Transvaal. A woodland species occurring from 0–2000 m. At the higher altitudes in *Brachystegia* woodland often recorded from rocky hills.

As with *R. tridentata* this species shows a tendency towards differentiation into two subspecies. In the Transvaal, in the Lourenço Marques and Sul do Save divisions of Mozambique and in the SE. part of S. Rhodesia there is a form represented by the type of *R. schlechteri* with narrowly oblong obtuse leaflets. To the north of this area we have the form represented by the type of *R. revoilii sens. strict.* with much broader acute leaflets the lateral ones being much more asymmetric. There are many intermediates, however, particularly in the Zimbabwe region of S. Rhodesia. Widely distributed in S. Rhodesia there is another form with densely ferruginous-tomentose leaves and stems. Once more, however, intermediates are common and this form, as far as indumentum is concerned, can be matched with specimens from Zanzibar I.

5. **Rhoicissus digitata** (L. f.) Gilg & Brandt in Engl., Bot. Jahrb. **46**: 439 (1911).—Suesseng. in Engl. & Prantl, Nat. Pflanzenfam., ed. 2, **20d**: 330 (1953). Type from Cape Province.
Rhus digitata L. f., Suppl. Pl.: 184 (1781). Type as above.
Rhus cirrhiflora L. f., loc. cit. Type from Cape Province.
Cissus thunbergii Eckl. & Zeyh., Enum. Pl. Afr. Austr.: 56 (1835).— Harv. in Harv. & Sond., F.C. **1**: 250 (1860) *nom. illegit.* Type as above.
Rhoicissus thunbergii Planch. in A. & C. DC., Mon. Phan. **5**, 2: 469 (1887) *nom. illegit.* Type as above.
Rhoicissus cirrhiflora (L. f.) Gilg & Brandt, tom. cit.: 438 (1911) quoad syn. *Rhus cirrhiflorum* et *Cissus cirrhiflora* excl. syn. alt. et specim. cit.

Woody climber or scandent shrub with young parts ferruginous-villous with rather long hairs, later glabrescent and with pale lenticels; tendrils glabrescent. Leaves 3–5-foliolate (specimens almost invariably with some 5-foliolate leaves); petiole up to 2·8 cm. long; stipules c. 5 mm. long, ± oblong, ferruginous-villous, very caducous; leaflet-lamina up to 9 × 2·8 cm., narrowly oblong to oblong-oblanceolate, the terminal leaflet the longest, apex acute or obtuse, margins entire or with a few dentations, ± revolute, base cuneate, glabrous above, appressed-ferruginous-pubescent below at least on the main nerves and midrib; main nerves ± prominent below and usually having domatia in their axils; petiolules up to 0·8 mm. long. Inflorescences of rather lax cymes; branches of inflorescence ferruginous-hairy; peduncle up to 5 mm. long; pedicels 1–2 mm. long; bracts not seen. Calyx and corolla appressed-ferruginous-pubescent. Ripe fruits not seen.

Mozambique. LM: Namaacha, st. 9.i.1947, *Barbosa* 79 (LM).
Also in the Cape Province, Natal and the Transvaal. A climber usually found in riverine fringing vegetation.

3. CISSUS L.

Cissus L., Sp. Pl. **1**: 117 (1753); Gen. Pl. ed. 5: 53 (1754).—Descoings in Not. Syst. **16**: 113–125 (1960).
Cissus Sect. *Eucissus* Planch. in A. & C. DC., Mon. Phan. **5**, 2: 471 (1887).

Erect or climbing perennial herbs or shrubs; tendrils leaf-opposed or absent. Leaves simple or more rarely lobed or digitately 3–7-foliolate, margins variously toothed, rarely entire; stipules present. Inflorescences in leaf-opposed or terminal or rarely axillary cymes with the flowers in umbels on the ultimate branchlets. Flowers 4-merous. Flower-bud conical, not constricted at the middle. Calyx entire or 4-lobed. Petals cucullate at the apex, becoming deflexed after anthesis, caducous. Anthers on short filaments. Disk annular, ± adnate to ovary, entire or lobed. Style simple, subulate; stigma subulate or subcapitate. Fruit usually 1-seeded. Seed oblong, ovoid or subspherical, often abruptly narrowed at one end and sometimes with a dorsal crest.

Leaves simple:
  Stems and branches green, succulent, quadrangular, winged but wings never corky, frequently leafless - - - - - - - - - 16. *quadrangularis*
  Stems and branches not as above:
    Leaves lobed:
      Plant a climber with tendrils - - - - - - - - 7. *cucumerifolia*
      Plant erect without tendrils - - - - - - - 17. *bathyrhakodes*
    Leaves not lobed (occasionally with some lobed leaves in *C. rotundifolia* and *C. trothae*):
      Plant erect, without tendrils:
        Leaves very narrowly elliptic to elliptic:

Plant not above 30 cm. tall, stems often zigzag; petiole up to 1 mm. long 13. *guerkeana*
Plant 1 m. or more tall; petiole 2–7 mm. long - - - 14. *cornifolia*
Leaves circular to broadly ovate or very broadly oblong:
Plant without shining black medifixed hairs - - 8. *trothae*
Plant provided with shining black medifixed hairs - - 10. *quarrei*
Plant a tendrilled climber:
Flowering before leaves:
Plant entirely glabrous - - - - 14. *cornifolia*
Plant pubescent - - - - - 15. *fanshawei*
Flowering with leaves:
Flower-bud and ovary pubescent or pilose:
Plant woolly-pilose with reddish-brown hairs - - - 12 *rubiginosa*
Plant without reddish-brown hairs:
Leaves, stem and inflorescence with shining medifixed hairs; fruit glabrescent - - - - - 9. *schmitzii*
Shining black medifixed hairs absent; fruit with long thick soft hairs 11. *grisea*
Flower-bud and ovary glabrous:
Leaves thick, succulent, circular to ovate - - - 5. *rotundifolia*
Leaves not succulent:
Stems quadrangular:
Leaves ovate-oblong, truncate or rounded at base, cuspidate at apex; older stems not developing corky wings; flower-bud acute or acuminate - - - - - 6. *producta*
Leaves circular (to broadly ovate), cordate, acuminate at apex; older stems developing corky wings; flower-bud obtuse at apex 3. *petiolata*
Stems never quadrangular or developing corky wings; leaves cuspidate:
Tertiary veins raised on lower surface of leaf; fruit up to 6·5 mm. long, pyriform - - - - - 2. *aristolochiifolia*
Tertiary veins not raised on lower surface of leaf; fruit ellipsoid, 10–20 mm. long:
Leaf margin entire or obscurely dentate - - 1. *integrifolia*
Leaf margin serrate with teeth up to 1·5 mm. long - 4. *welwitschii*
Leaves digitate:
Plant erect; without tendrils:
Plant glabrous - - - - - - 20. *crusei*
Plant pubescent - - - - - - 21. *nigropilosa*
Plant climbing; with tendrils:
Plant a liane with succulent stems; leaves opposite the cymes, digitate; flower-bud 4–5 mm. long; pedicels pubescent - - - - - 18. *aralioides*
Plant a slender climber; leaves opposite the cymes mostly simple; flower-bud 2 mm. long; pedicels glabrous - - - - - - 19. *faucicola*

1. **Cissus integrifolia** (Bak.) Planch. in A. & C. DC., Mon. Phan. **5**, 2: 483 (1887).—Gilg & Brandt in Engl., Bot. Jahrb. **46**: 462 (1911).—Bak. f. in Journ. Linn. Soc., Bot. **40**: 46 (1911).—Brenan, T.T.C.L.: 27 (1949).—Suesseng. in Engl. & Prantl, Nat. Pflanzenfam. ed. 2, **20d**: 256 (1953).—Dewit, F.C.B. **9**: 513, fig. 13F (1960).—White, F.F.N.R.: 231 (1962). Syntypes: Mozambique, between Sena and Lupata, *Kirk* (K); Chupanga, *Kirk* (K).
Vitis integrifolia Bak. in Oliv., F.T.A. **1**: 391 (1868). Syntypes as above.

Climber to tops of trees; older stems woody, not winged, producing gumlike exudate when cut; plant with scattered caducous hairs when young, otherwise ± glabrous; tendrils 2-fid, becoming woody on older stems. Leaves simple; petiole up to 5 cm. long; leaf-lamina up to 10·3 × 9·3 cm. but usually smaller, broadly ovate, acuminate at the apex, margin entire or obscurely dentate, usually truncate at the base or sometimes slightly cordate or rounded, tertiary veins on lower surface not raised; stipules 2 × 2 mm., very broadly ovate, ciliate, caducous. Cymes lax; peduncle up to 6 cm. long; pedicels 4–5 mm. long; bracts and bracteoles up to 1 × 1 mm., broadly ovate. Flower-bud 2·5 × 2 mm., very broadly oblong-cylindric. Calyx c. 0·5 mm. long, ± entire. Petals yellowish-green. Ovary glabrous; style 1·5 mm. long. Ripe fruit red with a bloom, up to 20 × 10 mm., ellipsoid, glabrous. Seed 1, 16 × 8 mm., with a strong dorsal crest and a few lateral ridges.

Caprivi Strip. Katima Mulilo, 5.i.1959, *Killick & Leistner* 3303 (PRE; SRGH). **N. Rhodesia.** B: Sesheke Distr., 37 km. SW. of Machili, fl. 21.xii.1952, *Angus* 996 (EA;

FHO). N: Lake Mweru, fl. 12.xi.1957, *Fanshawe* 3924 (K; NDO). E. Fort Jameson Distr., E. Chitandika area, fr. 13.iii.1956, *Grout* 153 (NDO). S: Mumbwa, fl. 1912, *Macaulay* 1046 (K). **S. Rhodesia.** N: Urungwe Distr., Msukwe R., fl. 17.xi.1953, *Wild* 4164 (K; SRGH). E: 21 km. S. of Umtali, fl. & fr. 26.xii.1955, *Chase* 5927 (BM; SRGH). S: Buhera Distr., Sabi R., fr. iii.1954, *Davies* 718 (SRGH). **Nyasaland.** C: Lilongwe Distr., fl. 1943, *Barker* 536 (EA). S: Upper Shire and Blantyre, fl. xii.1893, *Scott Elliot* 8431 (K). **Mozambique.** N: Mandimba, fr. 10.i.1942, *Hornby* 4801 (PRE). Z: Mocuba, Namagoa, fl. 1.xii.1948, *Faulkner* 375 (COI; K; SRGH). T: between Tete and Quebrabasa, fl. xi.1858, *Kirk* (K). MS: Chimoio, between Vila Pery and Amatongas, fl. 10.xii.1943, *Torre* 6278 (LISC). SS: between Inharrime and Zavala, fl. 10.xii.1944, *Mendonça* 3363 (LISC).

Also in Kenya, Tanganyika and the Congo. Riverine vegetation, 0–c. 1100 m.

2. **Cissus aristolochiifolia** Planch. in A. & C. DC., Mon. Phan. **5,** 2: 488 (1887).—Gilg & Brandt in Engl., Bot. Jahrb. **46**: 462 (1911).—Suesseng. in Engl. & Prantl, Nat. Pflanzenfam. ed. 2, **20d**: 256 (1953). Type: Nyasaland, Shire Highlands, *Buchanan* 276(K).

Climber with scattered caducous hairs when young; tendrils 2-fid. Leaves simple; petiole up to 4·8 cm. long; leaf-lamina up to 8·5 × 9 cm., but usually smaller, broadly to very broadly ovate, obtuse or acute at the apex, margin minutely serrate with teeth up to 1 mm. in length, cordate to truncate at base, some of tertiary veins on under surface frequently slightly raised; stipules 1·5 × 1·5 mm., very broadly ovate, caducous. Cymes lax; peduncle up to 6 cm. long; pedicels up to 6 mm. long in fruit; bracts and bracteoles up to 1 × 1 mm. long, broadly ovate. Flower-bud 1·5 × 1·5 mm., very broadly cylindric. Calyx c. 0·5 mm. long, entire. Petals greenish. Ovary glabrous; style 0·5–1 mm. Ripe fruits black, 6·5 × 5·5 mm. when dry, pyriform, glabrous. Seed 1, 6 × 5 mm., dorsal crest weak and without lateral ridges.

Nyasaland. S: Zomba Mt., fr. 12.iii.1955, *E.M. & W.* 814 (BM; SRGH). **Mozambique.** Z: Gúruè, fl. & fr. 10.v.1943, *Torre* 5180 (LISC).

Apparently confined to southern Nyasaland and the Zambezia Province of Mozambique; in riverine vegetation.

3. **Cissus petiolata** Hook f. in Hook., Niger Fl.: 262 (1849).—Planch. in A. & C. DC. Mon. Phan. **5,** 2: 492 (1887).—Gilg & Brandt in Engl., Bot. Jahrb. **46**: 464 (1911) pro parte excl. syn. *Vitis welwitschii* et *Cissus welwitschii*.—Brenan, T.T.C.L.: 27 (1949).—Suesseng. in Engl. & Prantl, Nat. Pflanzenfam. ed. 2, **20d**: 257 (1953) pro parte excl. syn. *Vitis welwitschii* et *Cissus welwitschii*.—Keay, F.W.T.A. ed. 2, **1,** 2: 680 (1958).—Dewit, F.C.B. **9**: 515 (1960).—White, F.F.N.R.: 231 (1962). Type from Nigeria.

Vitis suberosa Welw. ex Bak. in Oliv., F.T.A. **1**: 392 (1868). Type from Angola.

Cissus suberosa (Welw. ex Bak.) Planch in A. & C. DC., tom cit.: 481 (1887).—Exell & Mendonça, C.F.A. **2,** 1: 46 (1954). Type as above.

Cissus petiolata var. *pubescens* Dewit, tom. cit.: 516, t. 51 (1960). Type from the Congo.

Cissus quadrangularis sensu White, F.F.N.R.: 231 (1962) pro parte quoad specim. Miller.

Vigorous climber; stems ± quadrangular, glabrous, becoming woody when old and developing longitudinal corky wings up to 1·2 cm. broad on the angles; tendrils simple. Leaves simple, glabrous or with a few hairs on the petioles and lower surface of the leaf when young; petiole up to 10·8 cm. long but usually shorter; leaf-lamina up to 13 × 13 cm. but usually somewhat smaller, circular to broadly ovate, acuminate at the apex, margin repand-denticulate with teeth usually c. 1 mm. long, cordate at the base, 3–5-nerved at the base with 4 pairs of nerves above; stipules subcircular, glabrous, up to 1·5 × 1·5 mm. Cymes terminal and axillary, much-branched; peduncle up to 4·5 cm. long, glabrous; primary branches of inflorescence sparingly pubescent; pedicels up to 8 mm. long, pubescent, lengthening to 13 mm. in fruit and becoming glabrous. Flower-bud conical, up to 3 mm. long. Calyx 1 mm. long, pubescent when young. Petals glabrous. Ovary glabrous; style 1–2 mm. long. Fruit c. 8 mm. in diam., glabrous. Seeds up to 4 per fruit, only 1 usually becoming fully developed, smooth, c. 6 mm. long with one well-marked crest and 2 subsidiary ones.

N. Rhodesia. N: Chiengi, fr. 7.x.1938, *Miller* 226 (FHO; NDO). W: Mwinilunga Distr., Matonchi R., fl. 9.xi.1938, *Milne-Redhead* 4511 (K; PRE). C: Broken Hill, fr.

vi.1920, *Rogers* 26089 (K). **Mozambique.** Z: Milange, fl. 23.ii.1943, *Torre* 4804 (LISC). T: Tete, outskirts of Zóbuè, fr. 21.viii.1943, *Torre* 5776 (LISC). MS: Gorongosa, between Vila Paiva and R. Vandúzi, fr. 21.vii.1941, *Torre* 3135 (LISC).

From Guinée Republic to Ethiopia and southwards through the Congo, Uganda and Tanganyika and reaching its southern limit in Angola, N. Rhodesia and Mozambique. In fringing forest.

4. **Cissus welwitschii** (Bak.) Planch. in A. & C. DC., Mon. Phan. **5,** 2: 489 (1887).—Exell & Mendonça, C.F.A. **2,** 1: 46 (1954). Type from Angola.

Vitis welwitschii Bak. in Oliv., F.T.A. **1**: 393 (1868). Type as above.

Cissus petiolata sensu Gilg & Brandt in Engl., Bot. Jahrb. **46**: 464 (1911) pro parte quoad syn. *Vitis welwitschii* et *Cissus welwitschii.*

Vigorous climber; stems cylindric, never quadrangular and not developing corky wings. Tendrils simple. Leaves simple, glabrous or with a few hairs when young; petiole up to 4·5 cm. long; leaf-lamina up to 9 × 8·5 cm., ovate to broadly ovate, acuminate at the apex, margin serrate with teeth up to 1·5 mm. long, truncate to slightly cordate at the base, 3–5-nerved at the base with 3 or more pairs of secondary nerves; stipules up to 3 mm. long, broadly oblong, glabrous. Cymes axillary, few-flowered, sparsely branched; peduncle up to 2 cm. long, glabrous or puberulous; pedicels up to 6 mm. long lengthening to up to 15 mm. in fruit, puberulous to glabrous. Flower-bud up to 3·5 mm. long, conical, glabrous. Calyx 1 mm. long, entire, puberulous to glabrous. Ovary glabrous; style 1·5 mm. long, ± capitate. Fruit 1·5 × 1 cm., ellipsoid. Seed 1, 10 × 6 × 4 mm. flattened, ellipsoid, smooth.

N. Rhodesia. E: between Hofmeyr turn-off and Kachalalo, fl. 12.xii.1958, *Robson* 914 (K; SRGH). S: Mapanza, fl. 28.xi.1957, *Robinson* 2513 (K; SRGH). **S. Rhodesia.** N: Urungwe, fl. 6.i.1958, *Goodier* 539 (PRE; SRGH). W: Wankie Distr., Lutope–Gwai Junction, fl. 26.ii.1963, *Wild* 6013 (K; SRGH). E: Lusitu R., fr. 3.iv.1960, *Pole Evans* (SRGH).

Also in Angola. In riverine fringes and on termite mounds in *Brachystegia* woodland. This species is very closely related to *C. fragilis* E. Mey., a native of Natal and the Transvaal, but from the material available appears distinct.

5. **Cissus rotundifolia** (Forsk.) Vahl, Symb. Bot. **3**: 19 (1790).—Planch. in A. & C. DC., Mon. Phan. **5,** 2: 512 (1887).—Bak. f. in Journ. Linn. Soc., Bot. **40**: 46 (1911).—Gilg & Brandt in Engl., Bot. Jahrb. **46**: 465 (1912).—Eyles in Trans. Roy. Soc. S. Afr. **5**: 409 (1916).—Burtt Davy, F.P.F.T., **1,** 475 (1926).—Brenan, T.T.C.L.: 27 (1949).—Suesseng. in Engl. & Prantl, Nat. Pflanzenfam. ed. 2, **20d**: 257 (1953).—Mogg in Macnae & Kalk, Nat. Hist. Inhaca I., Moçamb.: 149 (1958).—Dewit, F.C.B. **9**: 513 (1960). Type from Arabia.

Saelanthus rotundifolius Forsk., Fl. Aegypt.-Arab.: CV, 35 (1775); Icones, t. 4 (1776). Type as above.

Vitis crassifolia Bak. in Oliv., F.T.A. **1**: 391 (1868). Type: Mozambique, between Sena and Lupata, *Kirk* (K).

Cissus crassifolia (Bak.) Planch., tom. cit.: 508 (1887). Type as above.

Vigorous climber; stems often 4–5-angled, pubescent or glabrescent; tendrils 2-fid. Leaves simple or occasionally lobed; petiole pubescent, up to 1 cm. long; leaf-lamina up to 8 × 8 cm., circular to ovate, obtuse at the apex, margin crenate, cordate at the base, pubescent or glabrescent on both sides, thick and fleshy; stipules up to 4 mm. long, ± semicircular, glabrous, caducous. Cymes leaf-opposed and terminal, lax; peduncle 3 cm. long; pedicels 4–5 mm. long lengthening to 10 mm. in fruit, appressed-pubescent. Flower-bud 3·5 × 1·5 mm. Calyx 1 mm. long, entire. Petals green. Ovary glabrous; style 0·5 mm. long. Fruit 1·5 × 1·3 cm., red when ripe. Seeds 1–2 per fruit, 9 mm. long, smooth with single crest.

S. Rhodesia. E: Hotsprings, fl. 22.x.1948, *Chase* 1549 (BM; SRGH). S: Bikita, Devuli Ranch, fr. 24.vii.1958, *Chase* 6974 (SRGH). **Nyasaland.** S: Mpatamanga Gorge, 16.viii.1960, *Leach* 10456 (SRGH). **Mozambique.** T: between Lupata and Tete, fl.ii.1859, *Kirk* (K). MS: Boka, Lower R. Buzi, fl. 10.xii.1906, *Swynnerton* 2072 (BM). SS: between Inharrime and Zavala, fr. 20.xi.1941, *Torre* 3871 (LISC). LM: between Quinta do Umbeluzi and Boane, fl. 20.xi.1946, *Gomes e Sousa* 3472 (COI; K).

Extending from Arabia through eastern Africa southwards to Mozambique and the Transvaal; also recorded from the Congo. Distribution mainly coastal but extending inland in our area up the Sabi and Zambesi R. systems. In drier types of woodland and bush.

6. **Cissus producta** Afz. in Remed. Guin.: 63 (1815).—Planch. in A. & C. DC., Mon. Phan. **5**, 2: 493 (1887) pro parte excl. specim. Barter.—Gilg & Brandt in Engl., Bot. Jahrb. **46**: 477 (1912) excl. syn. *C. arguta*.—Brenan, T.T.C.L.: 27 (1949).—Suesseng. in Engl. & Prantl, Nat. Pflanzenfam. ed. 2, **20d**: 259 (1953) excl. syn. *C. arguta*.—Exell & Mendonça, C.F.A. **2**, 1: 51 (1954).—Keay, F.W.T.A., ed. 2, **1**, 2: 678 (1958).—Dewit, F.C.B. **9**: 523, fig. 13B (1960).—White, F.F.N.R.: 231 (1962). Type from Sierra Leone.
Vitis producta (Afz.) Bak. in Oliv., F.T.A. **1**: 389 (1868). Type as above.

Climber; stems quadrangular, very sparingly pubescent or glabrescent; tendrils present. Leaves simple; petiole 1–6·5 cm., leaf-lamina up to 10·4 × 6·2 cm., ovate-oblong, cuspidate at the apex, margin remotely serrate, slightly cordate, truncate or rounded at the base, glabrous except sometimes pubescent at the base underneath and along principal nerves; stipules 2 mm. long, triangular, auriculate. Cymes 4–10 cm. long, leaf-opposed; peduncle ± 1 cm. long lengthening to 2 cm. in fruit, glabrescent; pedicels ± 5 mm. long lengthening to 10 mm. or more in fruit, appressed-pubescent or glabrescent; bracts and bracteoles 0·5 mm. long, glabrous, ciliate. Flower-bud ± 4 mm. × 2 mm., conical, acute or acuminate at the apex, glabrous. Calyx 1 mm. long, subentire, glabrous. Ovary glabrous; style 1·5 mm. long; stigma subcapitate. Fruit black or purplish-red, 1·1–1·8 × 0·8–1 cm. Seed 1, up to 13 × 8 mm., smooth, with dorsal crest.

N. Rhodesia. N: Samfya Mission, fr. 22.viii.1952, *White* 3125 (FHO). **S. Rhodesia.** E: Pungwe R. valley below gorge, fr. 17.vii.1948, *Chase* 848 (BM; SRGH).

From Senegal to Uganda and Tanganyika and southwards to Angola and Rhodesia. Fringing forest.

7. **Cissus cucumerifolia** Planch. in A. & C. DC., Mon. Phan. **5**, 2: 474 (1887).—Gilg & Brandt in Engl., Bot. Jahrb. **46**: 467 (1912).—Suesseng. in Engl. & Prantl, Nat. Pflanzenfam. ed. 2, **20d**: 257 (1953). Type: Nyasaland, Chibisa, Shire R., *Kirk* (K).
Cissus macrantha Werderm. in Notizbl. Bot. Gart. Berl. **13**: 279 (1936).—Brenan T.T.C.L.: 27 (1949).—Suesseng. in Engl. & Prantl, Nat. Pflanzenfam. ed. 2, **20d**: 257 (1953). Type from Tanganyika (Lindi).

Liane; stems 4–5-angled with lax white crisped pubescence or glabrescent; tendrils 2-fid. Leaves simple; petiole up to 15 cm. long, pubescence as on stems; leaf-lamina up to 15 × 16 cm., broadly transversely elliptic to very broadly ovate, usually 3-lobed up to halfway down, apiculate at the apex, margin subsinuate-dentate, cordate at the base, pubescence as on stems but sparse on upper surface; stipules 4 mm. long, broadly oblong to ± circular. Cymes leaf-opposed and terminal; peduncle up to 3 cm. long with pubescence as on stems; pedicels up to 10 mm. long, becoming thickened and up to 13 mm. long in fruit, densely pubescent, becoming glabrous in fruit. Bracts and bracteoles up to 1·5 mm. long, pubescent. Flower-bud 4·5 × 2·5 mm., glabrous. Calyx c. 1·5 mm. long, densely pubescent when young. Stamens with filaments 0·5 mm. long; anthers 1 mm. long. Ovary glabrous; style 1 mm. long; stigma subcapitate. Fruit 17 × 10 mm. when dry, glabrous; seed 1, 15 mm. long, smooth with a single crest.

Nyasaland. C: Dedza Distr., Mua Livulezi Forest, fr. 16.v.1960, *Adlard* 362 (SRGH). S: Chibisa, Shire R., fl. iii.1859, *Kirk* (K). **Mozambique.** N: Porto Amélia, fl. 13.iii.1959, *Myre & Macedo* 3516 (LM; SRGH). MS: Vila Machado, st. 17.iv.1948, *Mendonça* 4005 (LISC).

Known only from the coastal strip of southern Tanganyika and northern Mozambique and extending inland into Nyasaland. Riverine forest.

8. **Cissus trothae** Gilg & Brandt in Engl., Bot. Jahrb. **46**: 474 (1912).—Brenan, T.T.C.L.: 28 (1949).—Suesseng. in Engl. & Prantl, Nat. Pflanzenfam. ed. 2, **20d**: 259 (1953). Syntypes from Tanganyika.
Ampelocissus cinnamochroa sensu Brenan, T.T.C.L.: 26 (1949).

Erect shrublet from perennial rootstock, flowering before adult leaves are produced; whole plant except petals and ovary densely ferruginous-tomentose; stem cylindric. Tendrils usually absent. Leaves simple; petiole 4–18 cm. long; leaf-lamina up to 23 × 22 cm., circular to broadly ovate, apex obtuse, sometimes obscurely 3-lobed, margin dentate or sinuate-dentate, base deeply cordate; stipules up to 15 × 8 mm., densely pubescent outside, glabrous within. Cymes leaf-opposed; peduncle 2–4 cm. long; pedicels congested, 4 mm. long, lengthening to 10 mm. in fruit; bracts and bracteoles up to 4 mm. long. Flower-bud 3 mm. long, puberulous towards apex. Calyx 1 mm. long. Petals purple. Ovary glabrous; style 1 mm. long; stigma subcapitate. Fruit 9 × 6 mm., glabrous.

N. Rhodesia. E: between Hofmeyr turn-off and Kachalolo, fl. 12.xii.1958, *Robson* 920 (K; SRGH). **S. Rhodesia.** N: Urungwe Distr., Msukwe R., fl. 18.xi.1953, *Wild* 4203 (K; PRE; SRGH). **Nyasaland.** fr. 1937, *Clements* 740 (K).

Also in Tanganyika. Apparently rare or of very scattered occurrence. Rocky slopes in *Brachystegia* woodland and edges of swamps.

9. **Cissus schmitzii** Dewit in Bull. Jard. Bot. Brux. **29**: 292 (1959); F.C.B. **9**: 541 (1960). TAB. **95**, fig. B. Type from the Congo (Katanga).

Cissus sciaphila sensu White, F.F.N.R.: 231 (1962).

Woody climber; stems cylindric, pubescent when young with scattered shining black medifixed hairs, glabrescent when older; tendrils 2-fid. Leaves simple; petiole up to 13 cm. long, tomentellous, with scattered black medifixed hairs; leaf-lamina up to 20 × 16 cm., broadly ovate to subcircular, abruptly acuminate at the apex, margin serrate, cordate at the base, sparsely pubescent above, tomentellous below with scattered shining black medifixed hairs on both surfaces; basal nerves 3–5; secondary nerves in 5–8 pairs; stipules up to 5 mm. long, oblong, pubescent. Cymes lax; whole inflorescence from peduncle to buds with a crisped white indumentum densely interspersed with shining black medifixed hairs; peduncle up to 10 cm. long; pedicels up to 7 mm. long, lengthening to 12 mm. in fruit. Flower-bud 2 mm. long. Calyx up to 1 mm. long. Ovary with black medifixed hairs; style 1–1·5 mm. long. Fruit 10–15 × 8 mm., ovoid, glabrescent. Seed 1, 10 × 5 mm., with a strong dorsal crest and a lateral ridge on each side branching into smaller ridges.

N. Rhodesia. W: Ndola, fl. 10.i.1954, *Fanshawe* 669 (K; NDO; SRGH).

Recorded only from Katanga and the Western division of N. Rhodesia. Termite mounds in woodland.

10. **Cissus quarrei** Dewit in Bull. Jard. Bot. Brux. **29**: 291 (1959); F.C.B. **9**: 536 (1960). Type from the Congo (Katanga).

Erect herb; stem cylindrical, striate, tomentellous with short crisped hairs and scattered shining black medifixed hairs. Leaves simple; petiole up to 3 cm. long, with pubescence as on stems; leaf-lamina up to 8·5 × 8·5 cm., very broadly oblong to circular, apiculate at the apex, margin serrate, subcordate at the base, pubescent above and below, with the hairs beneath ± confined to the veins, with scattered shining black medifixed hairs on both surfaces but especially beneath; basal nerves 3–5; secondary nerves in 4–5 pairs; stipules triangular with pubescence as on stems. Cymes trichotomously branched; peduncle 4·5 cm. long, indumentum as for stems; pedicels up to 5 mm. long, thickening and lengthening to 10 mm. in fruit, densely covered with shining black caducous medifixed hairs; bracts and bracteoles up to 1·5 mm. long, pubescent. Flower-bud up to 2 mm. long, pubescent with white hairs in lower $\frac{1}{3}$ and very densely covered with shining black medifixed hairs in upper $\frac{2}{3}$. Calyx up to 1 mm. long, entire, densely pubescent with white hairs and with scattered black medifixed hairs. Ovary with black medifixed hairs; style 0·5–1 mm. long. Fruit glabrescent; (ripe fruits unknown).

N. Rhodesia. N: Kawambwa, Mbereshi R., fl. 2.xii.1961, *Richards* 15482 (K; SRGH).

Also in Katanga and SW. Tanganyika. Riverine vegetation.

11. **Cissus grisea** (Bak.) Planch. in A. & C. DC., Mon. Phan. **5**, 2: 622 (1887).—Gilg & Brandt in Engl., Bot. Jahrb. **46**: 472 (1912).—Suesseng. in Engl. & Prantl, Nat. Pflanzenfam. ed. 2, **20d**: 259 (1953). TAB. **95** fig. A. Type: Nyasaland, Shire R., *Kirk* (K).

Vitis grisea Bak. in Oliv., F.T.A. **1**: 395 (1868). Type as above.

Climber, ± woody; stems cylindric with short grey pubescence; tendrils 2-fid. Leaves simple; petiole 2–8 cm. long, densely pubescent; leaf-lamina up to 15 × 15 cm., rotund or ± circular, base cordate with a wide sinus, apex apiculate, margin denticulate, both surfaces pubescent with short spreading hairs especially on

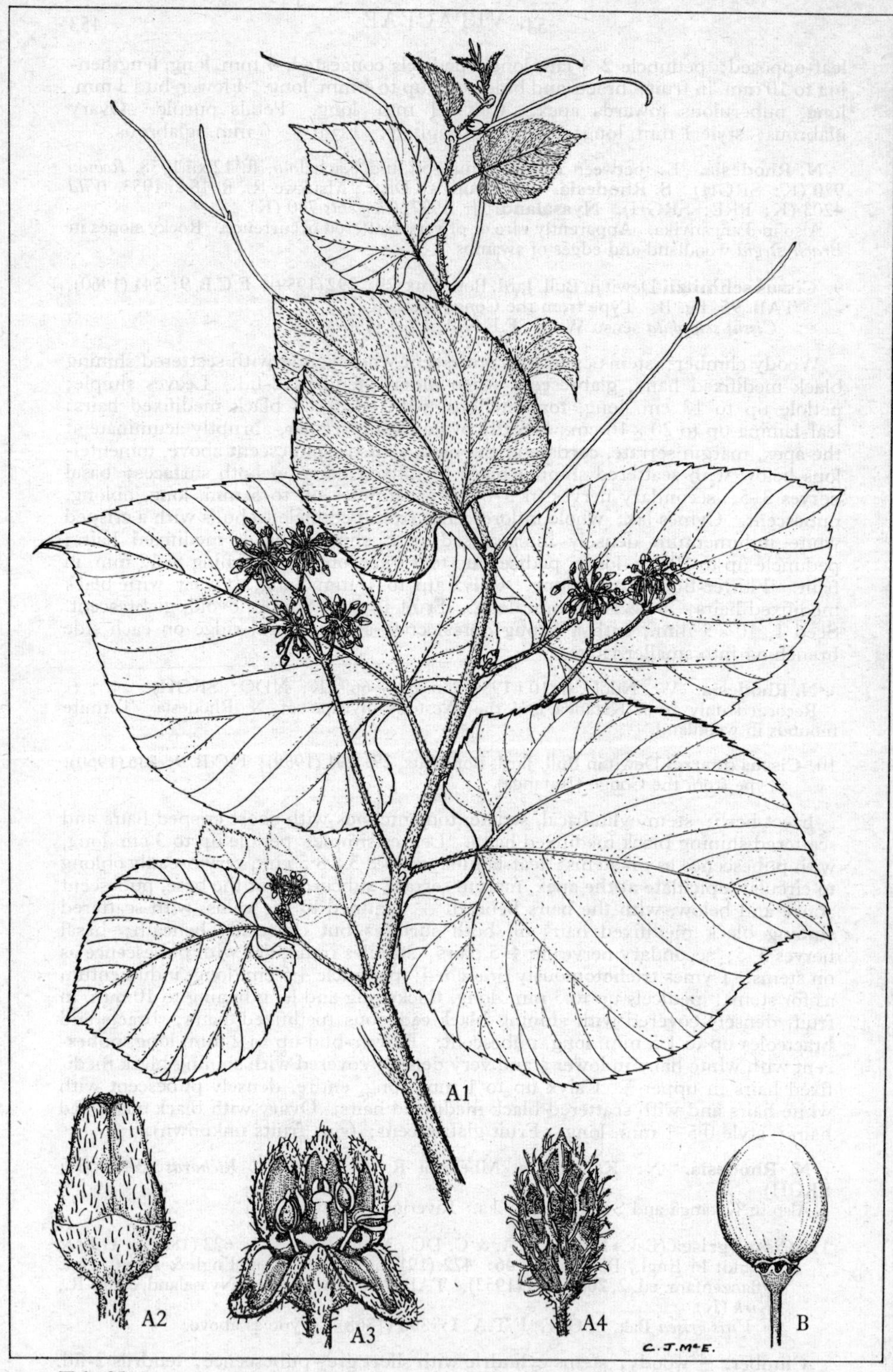

Tab. 95. A.—CISSUS GRISEA. A1, flowering branch (× ⅔) *Drummond* 5399; A2, flower-bud (× 8) *Drummond* 5399; A3, flower (× 8) *Bullock* 3510; A4, fruit (× 2) *Lovemore* 389. B.—CISSUS SCHMITZII, fruit (× 2) *Fanshawe* 948.

the nerves; stipules 3 × 3 mm., apiculate at the apex. Cymes leaf-opposed; peduncle up to 9·5 cm. long, pubescent; pedicels 5 mm. long, lengthening to 13 mm. in fruit, densely pubescent. Bracts and bracteoles 1·5 mm. long, pubescent. Flower-bud 2–2·5 × 15 mm., densely pubescent. Calyx 1 mm. long, densely pubescent. Ovary pubescent; style 1 mm. long; stigma subcapitate. Fruit 9 × 6 mm. when dry, covered with long thick soft hairs or processes which are themselves pubescent; seed 1, 8 × 5 mm.

N. Rhodesia. S: Kariba Hills, fl. 2.ii.1958, *Drummond* 5442 (K; PRE; SRGH). **S. Rhodesia.** N: Chirundu, fr. 21.ii.1954, *Lovemore* 389 (SRGH). **Nyasaland.** N: Likoma I., fr. 28.vi.1900, *Johnson* (K). S: Shire R., fr. 1863, *Kirk* (K).

Recorded also from the Rukwa Valley in SW. Tanganyika. Riverine vegetation.

12. **Cissus rubiginosa** (Welw. ex Bak.) Planch. in A. & C. DC., Mon. Phan. **5**, 2: 485 (1887).—Gilg & Brandt in Engl., Bot. Jahrb. **46**: 475, fig. 5 E–G (1912).—Suesseng. in Engl. & Prantl, Nat. Pflanzenfam., ed. 2, **20d**: 259 (1953).—Exell & Mendonça, C.F.A. **2,** 1: 51 (1954).—Keay, F.W.T.A., ed. 2, **1,** 2: 677 (1958).—Dewit, F.C.B. **9**: 538, t. 53 (1960). Type from Angola.

Vitis rubiginosa Welw. ex Bak. in Oliv., F.T.A. **1**: 394 (1868). Type as above.

Cissus bussei Gilg & Brandt, tom. cit.: 475 (1912).—Brenan T.T.C.L.: 27 (1949). —Suesseng. in Engl. & Prantl, Nat. Pflanzenfam. ed. 2, **20d**: 259 (1953). Syntypes from Tanganyika and Nyasaland, Missale, Upper Luangwa R, *Nicholson* (K).

Climber; all parts woolly-pilose with reddish-brown hairs; stems cylindric; tendrils 2-fid. Leaves simple; petiole up to 8 cm. long; leaf-lamina up to 13·5 × 15 cm., circular to broadly elliptic or ovate, acute, obtuse or apiculate at the apex, margin dentate with projecting subulate teeth up to 2 mm. long, deeply cordate at the base, with woolly indumentum, less dense on upper surface; stipules 7 × 3 mm., oblong-falcate. Cymes leaf-opposed; peduncle 2·6–6·5 cm. long; pedicels 4 mm. long, lengthening to 11 mm. in fruit; bracts and bracteoles up to 4 × 1·5 mm., lanceolate. Flower-bud 2–2·5 mm. long, woolly-pilose. Calyx 1 mm. long, subentire, woolly-pilose. Ovary pilose; style 1 mm. long; stigma subcapitate. Fruit 7 × 5 mm. when dry, often with persistent style, glabrous. Seed 1, 6 × 4 mm., with dorsal crest.

N. Rhodesia. N: Lake Mweru, 900 m., fr. 22.iv.1957, *Richards* 9402 (K; SRGH). E: 48 km. from Fort Jameson on Lundazi road, fr. 26.iv.1952, *White* 2469 (FHO). **Nyasaland.** N: Misuku Hills, fr. vi.1896, *Whyte* (K). C: Kota Kota, fl. 6.xi.1944, *Benson* 707 (PRE). S: Dedza–Golomoti road, fl. 19.iii.1955, *E.M. & W.* 1036a (BM; SRGH). **Mozambique.** N: Mutuáli, fl. & fr. 25.ii.1954, *Gomes e Sousa* 4220 (COI; K; LMJ; PRE).

From Sierra Leone to the Sudan and reaching its southern limit in Angola and our area. In rocky places in *Brachystegia* woodland and fringing forest.

13. **Cissus guerkeana** (Büttn.) Dur. & Schinz in Acad. Roy. Belg. Mém. **53,** 4: 94 (1896).—Gilg & Brandt in Engl., Bot. Jahrb. **46**: 479 (1912).—Suesseng. in Engl. & Prantl, Nat. Pflanzenfam. ed. 2, **20d**: 261 (1953).—Exell & Mendonça, C.F.A. **2,** 1: 48 (1954).—Dewit, F.C.B. **9**: 520 (1960). Type from the Congo.

Vitis guerkeana Büttn. in Verh. Bot. Verein. Brand. **31**: 89 (1889). Type as above.

Erect herb to 30 cm. tall; stems annual from a tuberous rootstock; tubers up to 9 × 3 cm., yellowish; stems and branches glabrous or with a few scattered hairs, cylindric, longitudinally ribbed, internodes 1 to 3·5 cm. long, often zig-zag; tendrils absent. Leaves simple; petiole up to 1 mm. long, glabrous; leaf-lamina 3·5–6 × 0·5–1·5 cm., very narrowly elliptic to elliptic, margin distantly dentate, glabrous or sometimes with sparse hairs along midrib beneath; stipules 3 × 2 mm., triangular. Cymes leaf-opposed, of few-flowered congested umbels; peduncle 1 cm. long, glabrous; pedicels 2–3 mm. long, glabrous; bracts and bracteoles c. 1 mm. long, triangular. Flower-bud 1·25 mm. long, glabrous. Calyx 0·75 mm. long. Stigma subcapitate. Fruit red, 7 × 4 mm. Seed 1, 5 × 3·5 mm., smooth.

N. Rhodesia. B: 88 km. W. of Mongu on Mankoya road, fl. 19.xi.1959, *Drummond & Cookson* 6611 (K; PRE; SRGH).

Also from the Congo and Angola. *Cryptosepalum* woodland on Kalahari Sand.

14. **Cissus cornifolia** (Bak.) Planch. in A. & C. DC., Mon. Phan. **5,** 2: 492 (1887). Gilg & Brandt in Engl., Bot. Jahrb. **46**: 480 (1912).—Gilg & Fr. in R.E.Fr., Wiss.

Ergebn. Schwed. Rhod.-Kongo-Exped. **1**: 134 (1914).—Brenan, T.T.C.L.: 27 (1949).—Suesseng. in Engl. & Prantl, Nat. Pflanzenfam. ed. 2, **20d**: 261 (1953).—Wild, Guide Fl. Vict. Falls: 151 (1953).—Williamson, Useful Pl. Nyasal.: 35 (1955).—Keay, F.W.T.A. ed. 2, **1**, 2: 679 (1958).—Dewit, F.C.B. **9**: 518 (1960).—White, F.F.N.R.: 230 (1962). Syntypes from Nigeria, Sudan and Uganda.

Vitis cornifolia Bak. in Oliv., F.T.A. **1**: 390 (1868). Types as above.

Cissus marlothii sensu Eyles in Trans. Roy. Soc. S. Afr. **5**: 409 (1916).

Cissus lonicerifolia C.A.Sm. in Burtt Davy, F.P.F.T. **2**: xx, 475 (1932).— O. B. Mill. in Journ. S. Afr. Bot. **18**: 52 (1952). Type from the Transvaal.

Erect or sometimes semiscandent shrub; stem and branches cylindric; young shoots often ferruginous-tomentose becoming glabrescent when adult, but sometimes almost glabrous; tendrils occasionally present. Leaves simple, appearing after the flowers; petiole 2–7 mm. long; leaf-lamina up to 7·8 × 5 cm., narrowly elliptic to elliptic, acute to obtuse at the apex, dentate at the margin, cuneate, rounded or occasionally cordate at the base; stipules 2–5 × 2–3 mm., oblong to triangular, caducous. Cymes leaf-opposed; peduncle 4–9 cm. long, pubescent to glabrescent; pedicels 2 mm. long, lengthening to 5 mm. in fruit, glabrous; bracts and bracteoles up to 2 mm. long, pubescent to glabrous. Flower-bud 2 mm. long, glabrous. Calyx 1 mm. long, subentire, glabrous. Ovary glabrous; style 0·5 mm. long; stigma subcapitate. Fruit purple-black, 10 × 8 mm., ovoid. Seed 1, 9 × 5 mm., smooth with dorsal crest.

Bechuanaland Prot. N: Matetsi, fl. ix.1949, *Miller* 927 (PRE). SE: from Mahalapye to 64 km. north, fl. & fr. 14.x.1954, *Story* 4836 (PRE; SRGH). **N. Rhodesia.** N: Lake Mweru, fr. 12.xi.1957, *Fanshawe* 4009 (FHO; K; NDO). W: Chilanga, fl. viii.1929, *Sandwith* 31 (K). E: Lundazi R., fr. 19.xi.1958, *Robson* 674 (K; SRGH). S: Mazabuka, fl. & fr. 11.x.1930, *Milne-Redhead* 1251 (K; PRE). **S. Rhodesia.** N: Mtoko Distr., fl. 14.ix.1953, *Phelps* 60 (SRGH). W: Matopos Research Station, fr. 12.xii.1951, *Plowes* 1373 (SRGH). C: Poole Farm, Hartley, fr. 12.x.1947, *Hornby* 2934 (SRGH). E: Umtali, Commonage, fl. 5.x.1948, *Chase* 1618 (BM; SRGH). S: Nuanetsi Distr., Tswiza, fr. 1.xi.1955, *Wild* 4700 (SRGH). **Nyasaland.** S: Mlanje, Tuchila plateau, fl. & fr. xi.1933, *Topham* 937 (FHO). **Mozambique.** N: Mutuáli, fl. 16.ix.1953, *Gomes e Sousa* 4116 (COI; K; LMJ; PRE). Z: Namagoa, fl. & fr. ix.1945, *Faulkner* P.83 (COI; SRGH). T: Marávia, between Fíngoè and Chicoa, fl. & fr. 25.ix.1942, *Mendonça* 395 (LISC). MS: Chironda Búzi, fl. 19.ix.1946, *Simão* 927 (LM; SRGH). SS: Mapai, fr. 3.xi.1944, *Mendonça* 2729 (LISC). LM: Magude, between Maéle and Mapulanguene, fr. 30.xi.1944, *Mendonça* 3163 (LISC).

Throughout tropical Africa. Found in all types of woodland from coast to 1800 m.

5. **Cissus fanshawei** Wild & Drummond in Kirkia, **2**: 141 (1961). Type: N. Rhodesia, Ndola, *Fanshawe* 1578 (EA; FHO; K, holotype; NDO; SRGH).

Woody liane, completely glabrous; stems cylindric; tendrils 2-fid. Leaves simple; petiole up to 3 cm. long; leaf-lamina up to 7 × 6 cm., broadly ovate, apiculate at the apex, margin serrate, cordate at the base; stipules 2·5 × 1·5 mm., oblong. Cymes up to 17 × 8 cm., lax, much-branched; peduncle 0·5–3·5 cm. long; pedicels 2·5 to 4 mm. long; bracts and bracteoles up to 2·5 × 2·5 mm., circular, auriculate at the base, minutely ciliate. Flower-bud 1·5–1·75 × 1·5 mm. Calyx 0·25–0·5 mm. long, subentire. Petals yellow-green. Ovary glabrous; style 1 mm. long; stigma subcapitate. Young fruit glabrous.

N. Rhodesia. N: Kawambwa, fl. 25.viii.1957, *Fanshawe* 3634 (K; NDO). W: Ndola, fl. 29.ix.1954, *Fanshawe* 1578 (EA; FHO; K; NDO; SRGH); Kitwe, st. 15.iii.1957, *Fanshawe* 3048 (NDO).

Endemic in N. Rhodesia. On termite mounds in *Brachystegia* woodland.

16. **Cissus quadrangularis** L., Syst. Nat. ed. 12, **2**: 124 et Mant. Pl.: 39 (1767).—Planch. in A. & C. DC., Mon. Phan. **5**, 2: 509 (1887).—Gilg & Brandt in Engl., Bot. Jahrb. **46**: 481 (1912).—Engl., Pflanzenw. Afr. **1**, 1: 70, fig. 56 (1915).—Burtt Davy, F.P.F.T. **2**: 475 (1932).—Brenan, T.T.C.L.: 28 (1949).—Suesseng. in Engl. & Prantl, Nat. Pflanzenfam. ed. 2, **20d**: 262 (1953).—Exell & Mendonça, C.F.A. **2**, 1: 48, t. 9B (1954).—Keay, F.W.T.A. ed. 2, **1**, 2: 676 (1958).—Mogg in Macnae & Kalk, Nat. Hist. Inhaca I., Moçamb.: 149 (1958).—Dewit, F.C.B. **9**: 512 (1960).—White, F.F.N.R.: 231 (1962) pro parte quoad specim. Martin. Type from India.

Vitis quadrangularis (L.) Wall. ex Wight & Arn., Prodr. Fl. Penins. Ind. Or. **1**: 125 (1834).—Bak. in Oliv., F.T.A. **1**: 399 (1868).—Type as above.

Cissus tetragona Harv. in Fl. Cap. **1**: 249 (1860).—Planch., loc. cit. Type from S. Africa.

Cissus cactiformis Gilg in Engl., Pflanzenw. Ost-Afr. **C**: 258 (1895).—Gilg & Brandt, tom cit.: 482, fig. 5A–D (1912).—Brenan, loc. cit.—Suesseng., loc. cit.—White, op. cit.: 230 (1962). Type from Tanganyika.

Vitis succulenta Galpin in Kew Bull. **1895**: 144 (1895). Type from the Transvaal.

Cissus succulenta (Galpin) Burtt Davy, loc. cit. Type as above.

Cissus quadrangularis var. pubescens Dewit in Bull. Jard. Bot. Brux. **29**: 297 (1959); F.C.B. **9**: 513 (1960). Type from the Congo.

Climbing herb; stem 1–5 cm. in diam. excluding wings, fleshy, quadrangular with wings at the angles 2–15 mm. broad, glabrous or pubescent or pubescent only at the angles, the angles with a reddish brown marginal line; tendrils present; frequently leafless. Leaves simple, fleshy, glabrous or sparingly pubescent; petiole 0·5–3 cm. long; leaf-lamina up to 4 × 3 cm., very broadly ovate, sometimes 3-lobed or deeply dissected, dentate at the margin, apex obtuse, base truncate to cordate. Cymes axillary, up to 10 cm. in length, few-flowered, sparingly branched; peduncle up to 2 cm. long, glabrous or pubescent. Pedicels 3 mm. long lengthening to 9 mm. in fruit. Flower-bud 2–3 mm. long, glabrous. Calyx 1 mm. long, entire, glabrous. Ovary glabrous; style up to 1·5 mm. long. Fruit red when ripe, up to 8 × 8 mm., ovoid to globular, glabrous. Seeds 5 × 4 mm., smooth with single dorsal crest.

N. Rhodesia. W: Ndola, fr. 10.v.1954, *Fanshawe* 1176 (K; NDO). S: Gwembe Valley, st. 29.iii.1952, *White* 2358 (FHO). **S. Rhodesia.** N: Kariba, fl. 16.x.1957, *Phipps* 796 (K; SRGH). W: 24 km. NE of Sebungwe-Zambezi R. confluence, fr. v.1956, *Plowes* 1985 (K; SRGH). E: Hot Springs, fl. 22.x.1948, *Chase* 1453 (BM; K; SRGH). S: Chibi Distr., 13 km. N. of Triangle turn-off, fr. 28.xi.1959, *Leach* 9525 (K; LISC; PRE; SRGH). **Nyasaland.** N: Livingstonia-Karonga road, fl. & fr. 9.i.1959, *Richards* 10574 (K; SRGH). **Mozambique.** Z: Morrumbala, st. xii.1858, *Kirk* (K). T: between Tete and Chioco, fl. 27.ix.1942, *Mendonça* 465 (LISC). MS: Vila Machado, fr. 8.ix.1942, *Mendonça* 169 (LISC). SS: between Chissane and Licilo, Bilene, fl. 9.x.1957, *Barbosa & Lemos* 7976 (COI; K). LM: Marracuene, between Umbeluzi and Matola, fl. 2.xii.1948, *Barbosa* 627 (LISC).

In the drier parts of tropical Africa and Madagascar, extending through Arabia, India, Ceylon, Malaya and as far east as the Philippines. Widespread in a variety of habitats but always in areas of low rainfall.

The specimen *Fries* 1192 (UPS) from the Lufu R., N. Rhodesia, referred to in R.E. Fr., Wiss. Ergebn. Schwed. Rhod.-Kongo-Exped. **1**: 134 (1914) and by White, F.F.N.R.: 231 (1962) under the name *Cissus fischeri* Gilg consists of leaves and stems only. These are certainly much more hairy than any of our available material of *Cissus quadrangularis* but as the type of *C. fischeri* Gilg is probably destroyed and as the Fries material is so poor we prefer to defer a decision as to whether it represents a form of *C. quadrangularis* or a distinct species.

17. **Cissus bathyrhakodes** Werderm. in Notizbl. Bot. Gart. Berl. **13**: 279 (1936).—Suesseng. in Engl. & Prantl, Nat. Pflanzenfam. ed. 2, **20d**: 262 (1953). Type from Tanganyika (Lindi).

Erect perennial herb up to 1 m. tall; stems glabrous with a powdery white bloom; tendrils absent. Leaves shallowly to deeply (3) 5 (7)-lobed; petiole up to 13 cm. long, glabrous; leaf-lamina up to 16 × 16 cm., subcircular in outline, acute at apex, margin serrate, truncate to shallowly cordate at base, varying from shallowly 3-lobed to deeply 5–7-lobed to within 1 cm. of base of lamina, lobes ± elliptic, glabrous; stipules 6 mm. long, broadly ovate, somewhat falcate. Cymes on glabrous peduncles 2–3·5 cm. long; pedicels 3 mm. long lengthening to 6 cm. in fruit, sparsely pubescent; bracts and bracteoles up to 1·5 mm. long. Flower-bud 1·5 mm. long, glabrous. Calyx 0·5 mm. long, 4-lobed, glabrous. Ovary glabrous; style 0·5 mm. long. Fruit red when ripe, 10 × 6 mm. Seed 1, 8 × 5 mm., smooth.

Mozambique. Z: Moebede road, Lugela, fl. 26.xi.1948, *Faulkner* K372 (COI; K). MS: Chimoio, fr. 16.ii.1948, *Garcia* 250 (LISC).

Coastal strip from S. Tanganyika to Manica e Sofala. Woodland.

18. **Cissus aralioides** (Welw. ex Bak.) Planch. in A. & C. DC., Mon. Phan. **5,** 2: 513 (1887).—Gilg & Brandt in Engl., Bot. Jahrb. **46**: 485 (1912).—Brenan, T.T.C.L.: 28 (1949).—Suesseng. in Engl. & Prantl, Nat. Pflanzenfam. ed. 2, **20d**: 262 (1953)

—Exell & Mendonça, C.F.A. **2**, 1: 54 (1954).—Keay, F.W.T.A. ed. 2, **1**, 2: 679 (1958).—Dewit, F.C.B. **9**: 547, fig. 13C (1960). Type from Angola.
Vitis aralioides Welw. ex Bak. in Oliv., F.T.A. **1**: 411 (1868). Type as above.

Liane; stems cylindric, somewhat succulent, constricted at the nodes, glabrous; tendrils 2-fid, glabrous. Leaves digitate, glabrous; petiole 4–15 cm. long, glabrous; leaflet-lamina up to 18 × 5·8 cm., elliptic to oblanceolate, shortly caudate to acute at the apex, margin serrate, attenuate at the base into a petiolule, glabrous. Stipules up to 8 × 4 mm., falcate. Cymes up to 10 cm. long; peduncle 2–3 cm. long, pubescent to glabrescent; pedicels 4 mm. long lengthening to 15 mm. in fruit, pubescent; bracts and bracteoles 1·5 mm. long. Flower-bud 4–5 mm. long, glabrous. Calyx 1 mm. long, subentire, glabrous. Ovary glabrous; style 1 mm. long. Fruit up to 3 × 2 cm., ovoid. Seed 1, 1·5 × 8 cm., smooth with a dorsal crest.

Mozambique. N: R. Rovuma, fl. 28.iii.1861, *Kirk* (K). Z. between Mocuba and Maganja da Costa, fl. 19.iv.1943, *Torre* 5190 (LISC).
From Senegal to Sudan and south to Angola and Mozambique. Riverine forest.

19. **Cissus faucicola** Wild & Drummond in Kirkia, **2**: 142 (1961). Type: N. Rhodesia, Abercorn, *Richards* 7343 (K, holotype; SRGH).

Herbaceous climber; tendrils 2-fid. Leaves digitately 3–5-foliolate; petiole up to 7·5 cm. long, glabrous or sometimes with scabrid hairs; leaflets 9 × 3 cm., subsessile or with petiolules up to 8 mm. long, elliptic, acute to acuminate at the apex, margin serrate with setiform teeth up to 1 mm. long, narrowly cuneate at the base, glabrous but sometimes with scabrid hairs on midrib and nerves below; leaves on flowering branches reduced, simple, petiolate, sometimes 3-lobed, narrowly to broadly ovate, acute at the apex, cuneate to truncate at the base; stipules up to 5 × 4 mm., narrowly ovate to ovate. Cymes few-flowered, congested, leaf-opposed; peduncle 1–15 mm. long; pedicels 2·5 mm. long, lengthening to 5 mm. in young fruit; bracts and bracteoles 0·5–2 mm. long. Flower-bud 2 × 1·5 mm. Calyx up to 1 mm. long, subentire or obscurely 4-lobed. Petals yellow-green. Ovary glabrous; style 1 mm. long; stigma subcapitate. Young fruit glabrous.

N. Rhodesia. N: Lake Mweru, fl. 14.x.1957, *Fanshawe* 3968 (K; NDO). **Nyasaland.** N: Misuku Hills, fl. 12.i.1959, *Richards* 10613 (K; SRGH). S: above Neno, Kirk Range, fl. 8.xii.1960, *Chapman* 1075 (BM; PRE; SRGH).
Also in Tanganyika. In evergreen forest, 1400–2100 m.

20. **Cissus crusei** Wild & Drummond in Kirkia, **2**: 141 (1961). Type: N. Rhodesia, Mufulira, *Cruse* 464 (K, holotype).

Erect completely glabrous herb up to 40 cm. tall; stem often unbranched, zig-zag, deeply sulcate or winged; tendrils absent. Leaves digitately 5–7-foliolate; petiole 3–9·5 cm. long; leaflet-lamina up to 13·5 × 3·5 cm., subsessile, narrowly obovate, apiculate at the apex, margin serrate to irregularly pinnatifid, narrowly cuneate at the base; stipules up to 6 cm. long, ovate. Cymes relatively few-flowered, axillary, up to 3 cm. in diam.; peduncle up to 3 cm. long; pedicels 3–4 mm. long; bracts and bracteoles 0·5–2 mm. long, ovate. Flower-bud 2·5 × 1·5 mm., conical. Calyx up to 1 mm. long, subentire. Petals reddish-green. Ovary 0·5 mm. long; style 1 mm. long; stigma slightly capitate. Fruit 9 × 8 mm., globose. Seed 1, 5 × 4 mm., ± smooth.

N. Rhodesia. W: Mwinilunga, S. of Matonchi Farm, fl. & fr. 13.xii.1937, *Milne-Redhead* 365 (K). C: Lusaka, fl. 16.xii.1955, *King* 248 (K). S: Mazabuka, Dundwa Agricultural Station, fr. 23.ii.1960, *White* 7316 (FHO).
Known only from N. Rhodesia. In *Brachystegia* woodland on termite mounds.

21. **Cissus nigropilosa** Dewit in Bull. Jard. Bot. Brux. **29**: 296 (1959); F.C.B. **9**: 546 (1960). Type from the Congo (Katanga).

Erect herb to c. 1 m. tall; stem cylindric, striate, pubescent, with short curved hairs interspersed with black medifixed hairs; tendrils absent. Leaves digitately 5-foliolate; petiole up to 6·5 cm. long, with indumentum as on stems; leaflet-lamina 5–11 × 0·5–1 cm., linear to occasionally oblanceolate, apex acute to slightly acuminate, margin ciliate-dentate especially towards apex, base narrowly cuneate, upper surface pubescent with white hairs and a few scattered caducous black medifixed hairs, under surface densely covered with black medifixed hairs especially when young and principal nerves with stiff curved white hairs; stipules 2–4 × 2–4 mm., triangular, ciliate, upper surface glabrous, lower surface with simple white hairs and black medifixed hairs. Cymes leaf-opposed; peduncle up to 5·5 cm. long, with indumentum as on stems; pedicels 1·5–2·5 mm. long lengthening to 6 mm. in fruit, densely covered with black medifixed hairs; bracts and bracteoles c. 1 mm. long, triangular, ciliate and with a few simple white hairs and black medifixed hairs. Flower-bud up to 1·5 mm. long, pubescent and with black medifixed hairs. Calyx 0·25 mm. long, pubescent and with black medifixed hairs. Ovary glabrous; style 1 mm. long; stigma slightly notched. Fruit dark purple-black, 8 × 6 mm., ovoid. Seed c. 6 × 4 mm., ovoid.

N. Rhodesia. N: Chilongowelo, fl. 2.xi.1954, *Richards* 2197 (K). **Nyasaland.** N: 67 km. S. of Karonga, fl. 9.i.1959, *Robinson* 3127 (EA; K; SRGH).

Among rocks in *Brachystegia* woodland. Known elsewhere only from the Congo (Katanga) and Tanganyika (Ufipa Distr.).

4. CYPHOSTEMMA (Planch.) Alston

Cyphostemma (Planch.) Alston in Trimen, Handb. Fl. Ceyl. **6,** Suppl.: 53 (1931).—Descoings in Notul. Syst **16**: 113–125 (1960).

Cissus Sect. *Cyphostemma* Planch. in A. & C. DC., Mon. Phan. **5,** 2: 472 (1887) ("Cyphostomma").

Erect, prostrate or climbing perennial herbs or shrubs; tendrils opposite the leaves or absent. Leaves digitately 3–9-foliolate or rarely simple and entire or lobed, very rarely pedate, margins variously toothed; stipules present. Inflorescences of leaf-opposed or axillary corymbose cymes. Flowers 4-merous. Flower-bud ± cylindric or lageniform, apex rounded and often ± inflated, ± constricted at or near the middle. Calyx entire or ± 4-toothed. Petals cucullate at the apex, becoming deflexed after anthesis, very caducous. Anthers usually with ± straight filaments not bending noticeably over the gynoecium. Disk of 4 rather fleshy truncate or conical glands, ± free from each other but ± adnate to the ovary. Style simple, subulate; stigma minutely 2-fid, or subentire and minutely subcapitate. Seed usually 1, usually with a distinct dorsal crest or crests and often ± rugose.

We agree with Alston (loc. cit.) and Descoings (loc. cit.) that *Cyphostemma* is a good genus and should be separated from *Cissus* and *Cayratia*. However, in order to save space we have not considered it necessary to cite the many new combinations in *Cyphostemma* made by Descoings when they refer to species that we consider synonymous with other species.

Tendrils absent:
  Leaflets with scattered sessile discoid glands:
    Plant usually scrambling or climbing; stem glands sessile; leaflets elliptic to narrowly obovate-oblong - - - - - - - - 29. *gigantophyllum*
    Plant usually erect; stem with glandular hairs up to 2 cm. long; leaflets linear-lanceolate to narrowly elliptic - - - - - - - 13. *vanmeelii*
  Leaflets without sessile discoid glands:
    Plants prostrate or rarely climbing:
      Leaves simple (occasionally 3-foliolate or 3-lobed); leaf-lamina from oblong-elliptic to almost circular:
        Leaves fleshy; pedicels glandular, otherwise glabrous - - 1. *humile*
        Leaves not fleshy; pedicels pubescent - - - - - 2. *wittei*
      Leaves always 3-, 4-, 5- or 6-foliolate:
        Peduncles and/or pedicels glandular:
          Leaves ± fleshy; leaflets oblanceolate-lorate to very narrowly elliptic 4. *hereroense*
          Leaves not fleshy, narrowly elliptic, oblanceolate or obovate:
            Corolla glabrous - - - - - 34. *elisabethvilleanum*
            Corolla pubescent or with glandular hairs:
              Pedicels with glandular hairs - - - - 5. *richardsiae*
              Pedicels without glandular hairs - - - - 40. *mildbraedii*
        Peduncles and pedicels eglandular:
          Leaves subsessile (petiole up to 2 mm. long) - - - 6. *hermannioides*
          Leaves distinctly petiolate:

Leaves shortly pubescent on the nerves and main veins below
7. *obovato-oblongum*
Leaves whitish- or brownish-villous below (or pilose when old)
8. *abercornense*
Plants erect or if ± prostrate then with long red glandular hairs up to 1·4 cm. long on the stems:
Plant quite glabrous except for the margins of the stipules - - 9. *junceum*
Plants ± hairy:
Pedicels glandular:
Stems glandular:
Leaves all sessile, 3-foliolate - - - - - 10. *crotalarioides*
Leaves petiolate, or if some sessile then stems with red glandular hairs up to 2 cm. long:
Leaves ferruginous- or brownish- or whitish-tomentose below:
Glands on vegetative parts sparse and short (± hidden by the ferruginous tomentum) - - 11. *chloroleucum*
Glands on vegetative parts very dense:
Leaflets viscid above, whitish-tomentellous beneath 18. *viscosum*
Leaflets not viscid above, brownish-tomentose or -lanate beneath
12. *manikense*
Leaves not ferruginous- or brownish- or whitish-tomentose below:
Corolla glabrous, eglandular - - 34. *elisabethvilleanum*
Corolla puberulous or pubescent and glandular:
Stems with long red glandular hairs up to 1·4 cm. long 30. *barbosae*
Stems with dark or purplish glandular hairs up to 0·5 cm. long; leaflets with whitish eglandular hairs and short blackish capitate glandular hairs - - - - - 14. *heterotrichum*
Stems not glandular:
Leaves glabrous below; flower-bud glabrous; ovary glabrous
15. *fugosioides*
Leaves with midrib pubescent below; flower-bud pubescent; ovary puberulous - - - - - - 16. *nanellum*
Pedicels not glandular:
Stems glandular:
Leaves puberulous or pubescent only on the nerves below - 17. *zombense*
Leaves ferruginous-tomentose or ferruginous-pubescent especially on the main nerves:
Flower-bud pubescent and with a few capitate glands near the apex
19. *stenolobum*
Flower-bud ferruginous-pilose - - - - - - 20. *princeae*
Stems not glandular:
Leaves ferruginous- or whitish-tomentose below:
Leaves 7–9 foliolate; leaflets linear-elliptic to very narrowly elliptic, tapering gradually to the base; flower-bud glabrous
21. *septemfoliolatum*
Leaves 3–5(7)-foliolate; leaflet narrowly elliptic to elliptic; flower-bud pubescent (rarely glabrescent) - - - - 22. *kerkvoordei*
Leaves sparsely pilose or glabrescent below - - - 23. *rhodesiae*
Tendrils present:
Leaves sessile or subsessile:
Plant glabrous and glaucous - - - - - 3. *schlechteri*
Plant pubescent and/or with some glandular hairs:
Ovary pubescent; fruit densely pubescent - - - 25. *congestum*
Ovary glabrous; fruit glabrous or with one or two glandular hairs:
Pedicels puberulous or glabrous, eglandular; leaves ± glaucous 26. *subciliatum*
Pedicels pubescent and densely glandular; leaves not glaucous 27. *setosum*
Leaves petiolate (some petioles at least 1·5 cm. long):
Leaves pedate with 5 leaflets - - - - - 24. *adenocaule*
Leaves digitate:
Stipules circular to broadly transversely elliptic, up to 2·5 cm. in diam.
55. *rotundistipulatum*
Stipules never circular or broadly transversely elliptic:
Pedicels glandular (if eglandular then leaflets with sessile discoid glands scattered near the apex of the leaves):
Leaflets with sessile discoid glands usually near the margins or towards the apex on one or both sides:
Glandular hairs on stems elongated and reddish - - 28. *hildebrandtii*
Glandular hairs on stems absent or sessile - - 29. *gigantophyllum*
Leaflets without sessile glands:
Corollas (at least some in each inflorescence) with glandular hairs (or if

absent then plant drying black and with some glandular hairs on fruits, inflorescence branches and vegetative parts):
Plant suberect or erect; stems with glandular hairs up to 1·4 cm. long 30. *barbosae*
Plant climbing or prostrate; stems with glands not more than 5 mm. long:
Pedicels glabrous except for a few glands near the middle 50. *amplexum*
Pedicels pubescent and glandular:
Leaves always 3-foliolate; calyx usually glandular:
Plant very densely covered with dark purplish or blackish short glandular hairs, otherwise only sparsely pubescent or glabrescent 31. *lynesii*
Plant not densely purplish- or blackish-glandular; leaves tomentellous or densely pubescent below; flower-bud distinctly pubescent as well as glandular - - - 37. *simulans*
Leaves normally 5-foliolate:
Stems glabrescent or glabrous except for glands; leaves membranous:
Fruit glabrous; stem-glands sparse, short and inconspicuous 38. *milleri*
Fruit pubescent and glandular; stem-glands dense, up to 5 mm. long - - - - 33. *glandulosissimum*
Stems pubescent or tomentose:
Leaves whitish- or greyish-tomentose below:
Fruit densely pubescent - - - - - 43. *lanigerum*
Fruit pubescent and glandular - - - 32. *bambuseti*
Leaves not tomentose below; fruit pubescent and glandular:
Plant drying black, climbing - - - 36. *buchananii*
Plant not drying black, prostrate or climbing 37. *simulans*
Corollas without glandular hairs:
Leaflets quite glabrous or sometimes with a few glands on the midrib below:
Leaflets membranous, elliptic to broadly elliptic - 48. *bororense*
Leaflets linear-lanceolate - - - - - 52. *graniticum*
Leaflets variously hairy:
Leaflet scabrous-pubescent on both sides 51. *trachyphyllum*
Leaflets not scabrous:
Corolla glabrous; leaves 3-foliolate - - 34. *elisabethvilleanum*
Corolla puberulous or pubescent; leaves 3–5-foliolate:
Flower-bud 3–5 mm. long, densely yellowish-pubescent 61. *cyphopetalum*
Flower-bud 2–3 mm. long, grey-pubescent:
Leaflets very narrowly elliptic to oblanceolate, very sparsely puberulous on both sides; pedicels with subsessile glands 54. *saxicola*
Leaflets narrowly elliptic to obovate, densely pubescent below; pedicel glands not subsessile - - - 44. *puberulum*
Pedicels eglandular:
Leaflets tomentose below, or, if only densely pubescent below then flower-buds with a few capitate glands near or at the apex;
Leaflets whitish-tomentose below:
Flower-bud densely pubescent; calyx eglandular - 45. *hypoleucum*
Flower-bud thinly pubescent at the apex and on the petal-margins; calyx often glandular - - - 46. *vandenbrandeanum*
Leaflets sometimes tomentose but not whitish below; flower-bud often with a few capitate glands - - - - - 40. *mildbraedii*
Leaflets never tomentose below:
Petals with a single apical gland - - - - - 49. *robsonii*
Petals without apical glands:
Leaves never more than 3-foliolate:
Ovary and fruit glabrous - - - 57. *kilimandscharicum*
Ovary and fruit pubescent - - - - - 58. *jaegeri*
Leaves 3–5-foliolate (some 5-foliolate leaves on each plant):
Leaflets quite glabrous above (sometimes with a few hairs on the midrib):
Corolla and calyx quite glabrous:
Petiolules when mature c. 3 cm. long - - 59. *lovemorei*
Petiolules 0–3 mm. long - - - - - - 60. *quinatum*
Corolla and calyx ± puberulous or pubescent:
Stems with glandular hairs:
Leaves quite glabrous beneath except for glands on midrib, membranous; stipules ovate-lanceolate - 35. *kirkianum*

Leaves pubescent at least on the nerves beneath, chartaceous; stipules lanceolate-acuminate - - - 41. *montanum*
Stems without glandular hairs:
Bracts and bracteoles up to 6 mm. long, ovate-lanceolate 56. *paucidentatum*
Bracts and bracteoles c. 1 mm. long, linear - 42. *kaessneri*
Leaflets puberulous or pubescent above:
Leaflets linear-lanceolate, puberulous near margins above, otherwise glabrous - - - - - - - 53. *tenuissimum*
Leaflets oblanceolate, obovate or lanceolate to ovate, variously hairy:
Ovary and fruit pubescent and glandular; leaves without glands 39. *masukuense*
Ovary and fruit pubescent (if a few glands on fruit then leaves also with some glands or midrib):
Stems glandular - - - 47. *cirrhosum* subsp. *rhodesicum*
Stems eglandular - 47. *cirrhosum* subsp. *transvaalense*

1. **Cyphostemma humile** (N.E. Br.) Descoings in Notul. Syst. **16**: 122 (1960). Type from Natal.
Vitis humilis N.E. Br. in Hook., Ic. Pl. **16**: t. 1565 (1887). Type as above.
Cissus humilis (N.E. Br.) Planch. in A. & C. DC., Mon. Phan. **5**, 2: 629 (1887).—Gilg & Brandt in Engl., Bot. Jahrb. **46**: 488 (1912).—Suesseng. in Engl. & Prantl, Nat. Pflanzenfam. ed. 2, **20d**: 240 (1953). Type as above.

Prostrate herb or occasionally climbing; stems up to c. 0·6 m. long, often zig-zag, with scattered stalked glandular hairs and often with long eglandular whitish hairs in addition; tendrils absent. Leaves simple, or occasionally deeply 3-lobed or 3-foliolate; leaf-lamina rather thick and fleshy when fresh, up to 20 × 12 cm., from oblong-elliptic to almost circular, apex rounded or acute, margin coarsely serrate, base from broadly cuneate to truncate, sometimes asymmetric, glabrous above or puberulous with longer hairs on the main nerves, densely pubescent or puberulous below or with a few hairs restricted to the main nerves, rarely quite glabrous; petioles 0·2–2·5 cm. long; stipules up to 2 × 0·8 cm., lanceolate-oblong, acuminate, margins ciliate. Cymes rather lax; peduncle up to 12 cm. long, at the apex of axillary branches, with a few scattered glandular hairs c. 1 mm. long and often also with whitish eglandular hairs; pedicels with denser and often shorter glandular hairs, otherwise glabrous; bracts and bracteoles 1–2 mm. long, lanceolate-acuminate, ciliate on the margins. Flower-bud c. 3 mm. long, oblong-cylindric, only slightly constricted in the middle, glabrous. Calyx c. 0·7 mm. long, entire. Petals green, red-tipped. Ovary glabrous; style c. 1·2 mm. long, slightly capitate. Fruit c. 2·2 × 1·7 cm. when fresh, ellipsoid, glabrous. Seed usually 1, c. 1·1 × 0·6 cm., dorsal crest very faint, with 2 pairs of faint lateral furrows, faintly rugose and foveolate (markings often almost invisible in old seeds).

Subsp. **dolichopus** (C.A. Sm.) Wild & Drummond in Kirkia, **3**: 20 (1962). Type from the Transvaal (Soutpansberg).
Cissus dolichopus C.A. Sm. in Burtt Davy, F.P.F.T. **2**: xx, 476 (1932). Type as above.

Differs from subsp. *humile* in the following characters: dries a darker colour, has scattered often puberulous hairs between the nerves on the under side of the leaves which are often also puberulous above, the leaves tend to be broader, the stipules smaller and the hairs on the pedicels tend to be sparser and longer.

S. Rhodesia. N: Darwin Distr., Umvukwes Range, Mvurkwe, fl. 21.xii.1952, *Wild* 3976 (K; PRE; SRGH). W: Bulawayo, fl. xii.1897, *Rand* 54 (BM). S: Belingwe, fl. and fr. 26.xii.1959, *Leach* 9717 (K; PRE; SRGH).
Also in the Transvaal and Natal. Often on stony ground and appearing to have a ready tolerance for serpentine or chrome-bearing soils.
Subsp. *humile* is apparently confined to Natal and subsp. *dolichopus* is mainly in the Transvaal and S. Rhodesia although an occasional specimen of the latter subspecies (e.g. *Codd* 2499 (PRE; SRGH) from Utrecht) has been recorded in Natal.

2. **Cyphostemma wittei** (Staner) Wild & Drummond in Kirkia, 2: 141 (1961). Type from the Congo (Katanga).
Cissus wittei Staner in De Wild. & Staner, Contr. Fl. Katanga, Suppl. **4**: 49 (1932).—Dewit, F.C.B. **9**: 470, fig. 12A (1960). Type as above.

Prostrate herb from a thick perennial rootstock; stems c. 1 m. long, brownish- or whitish-tomentose or -pubescent, sometimes glandular; tendrils absent. Leaves simple, sessile or subsessile; leaf-lamina up to 15 × 13 cm., broadly oblong-elliptic, in large older leaves sometimes shallowly 3-lobed and very broadly ovate, apex rounded, base shallowly cordate, upper surface glabrous or with a few hairs on the nerves, lower surface whitish- or brownish-tomentose or hairy only on the nerves and veins; stipules up to c. 1 cm. long, lanceolate, densely pubescent. Cymes axillary, trichotomous; peduncle up to 4 (7) cm., pubescent or tomentose; branches of inflorescence sometimes with stalked subcapitate glands; pedicels 2–3 mm. long, pubescent; bracts and bracteoles c. 2 mm. long, lanceolate, pubescent. Flower-bud 2–4 mm. long, constricted near the middle, puberulous or glabrous. Calyx 0·5 mm. long, entire, pubescent at the margin. Petals red. Ovary glabrous; style c. 1·5 mm. long, apex minutely 2-fid. Fruit purple-black, 1 × 0·8 cm., ellipsoid-globose, glabrous. Seed 1, c. 8 × 6 mm., ellipsoid, deeply foveolate.

N. Rhodesia. N: Abercorn Distr., Kambole escarpment, fl. & fr. 2.i.1955, *Richards* 3852 (K). W: Luanshya, fl. 2.xii.1953, *Fanshawe* 527 (K; NDO).

Also in the Congo (Katanga). Woodland.

At first sight it might appear that the form with glandular stems and inflorescence branches would warrant at least varietal rank but one gathering (*Milne-Redhead* 2839 (K)) from Mwinilunga, mounted on two sheets, is patently not a mixed gathering but on one sheet the plant is quite eglandular, the other has glands.

3. **Cyphostemma schlechteri** (Gilg & Brandt) Descoings in Notul. Syst. **16**: 124 (1960). Type: Mozambique, Ressano Garcia, *Schlechter* 11893 (B, holotype †).

Cissus schlechteri Gilg & Brandt in Engl., Bot. Jahrb. **46**: 489 (1912).—Suesseng. in Engl. & Prantl, Nat. Pflanzenfam. ed. 2, **20d**: 240 (1953). Type as above.

Cissus unguiformifolius C.A. Sm. in Burtt Davy, F.P.F.T. **2**: xx, 476 (1932).—I.C. Verdoorn in Fl. Pl. Afr. **25**: t. 972 (1946). Type from the Transvaal (Messina).

Prostrate or procumbent herb, glabrous and ± glaucous in all parts; stems green or pinkish, up to 1 m. long, with tendrils as a rule along the whole length. Leaves digitately 3–5-foliolate, sessile or subsessile; leaflet-lamina subfleshy, up to c. 12 × 3·5 cm., lateral leaflets smaller, linear-lanceolate to narrowly oblong-obovate, acute and often apiculate at the apex, margins sparsely but coarsely serrate, narrowly cuneate to the subsessile base. Cymes lax at the apex of axillary branches; peduncles up to c. 6 cm. long; pedicels slender, up to c. 6 mm. long; bracts and bracteoles c. 2mm. long, subulate. Flower-bud 2·5–3 mm. long, oblong-cylindric, slightly constricted at the middle. Calyx 0·7 mm. long, entire or minutely 4-toothed. Petals greenish-yellow with purplish tips. Ovary glabrous; style c. 1·75 mm. long, scarcely capitate. Fruit peach-red, 1·4 × 0·8 cm., ellipsoid-oblong. Seed 1, 7 × 4 mm., almost smooth but with a fairly strong dorsal furrow.

Bechuanaland Prot. SE: Derdepoort, fl. 29.xi.1954, *Codd* 8852 (PRE). **S. Rhodesia.** S: Bubye-Limpopo junction, Chikwarekware, fl. 23.ii.1961, *Wild* 5349 (K; PRE; SRGH). **Mozambique.** LM: Goba-Fronteira, fl. 13.xii.1947, *Barbosa* 720 (LISC; LM).

Also in the Transvaal. Usually on rocky or sandy soils.

The fruit is reputed by Bechuanaland Africans to be poisonous.

4. **Cyphostemma hereroense** (Schinz) Descoings in Notul. Syst. **16**: 121 (1960). Syntypes from SW. Africa.

Cissus hereroensis Schinz in Bull. Herb. Boiss., Sér. 2, **8**: 640 (1908).—Gilg & Brandt in Engl., Bot. Jahrb. **46**: 496, fig. 8 (1912).—Suesseng. in Engl. & Prantl, Nat. Pflanzenfam. ed. 2, **20d**: 241 (1955). Syntypes as above.

Procumbent herb with branches up to c. 60 cm. long from an elongated tuberous rootstock; stems rather thick, with elongate striations, densely pubescent; tendrils absent. Leaves somewhat fleshy, digitately 3–6-foliolate, sessile; leaflet-lamina 10 × 1·7 cm., subsessile, from oblanceolate-lorate to very narrowly elliptic, apex acute, margin coarsely serrate, base narrowly cuneate, glabrous or glabrescent above, densely pubescent below with scattered stalked capitate glands at the margin; stipules up to 1 × 0·4 cm., oblong-lanceolate, pubescent. Cymes at the apex of short axillary branches; peduncle c. 6 cm. long, pubescent, sometimes with a few glandular hairs; pedicels up to 6 mm. long, puberulous and with dense short glandular hairs; bracteoles c. 1 mm. long, linear, ciliolate. Flower-bud c. 2·7 mm. long, oblong-cylindric, slightly constricted at the middle, puberulous. Calyx

0·7 mm. long, puberulous or glabrescent. Ovary puberulous, glabrous; style 1·7 mm. long, apex subcapitate or shortly bifid. Fruit c. 1·4 × 1·2 cm., subglobose, puberulous. Seed 1, c. 1·1 × 1 cm., not prominently crested, horizontally rugose.

Bechuanaland Prot. SE: 1·25 km. S. of Lobatsi, fr. 17.i.1960, *Leach & Noel* 244 (CAH; K; SRGH).

Also in SW. Africa, Northern Cape Province and Orange Free State. Low rainfall areas.

5. **Cyphostemma richardsiae** Wild & Drummond in Kirkia, **2**: 136 (1961). Type: N. Rhodesia, Abercorn Distr., Kasama road, *Richards* 3984 (K, holotype; SRGH).

Prostrate herb; stems pubescent and with glandular-capitate hairs; tendrils absent. Leaves digitately 3–5-foliolate; petiole up to 3·5 cm. long, indumentum as for stems; leaflet-lamina narrowly obovate to obovate, apex obtuse, margin serrate-crenate, base cuneate, very sparsely pilose above, sparsely pilose below but nerves and veins pubescent and principal nerves with capitate-glandular hairs; median leaflet up to 16 × 8 cm.; lateral leaflets up to 9 × 4·5 cm.; petiolules 0–1 cm. (2 cm. in median leaflet) long; stipules up to 1 cm. long, ovate-lanceolate or lanceolate, acuminate, pubescent and with margin sparsely glandular. Cymes on axillary branches or subterminal; peduncle up to 7·5 cm. long, pubescent and with capitate glandular hairs; pedicels up to 3·5 mm. long, pubescent and with capitate glandular hairs; bracts and bracteoles c. 1·5 mm. long, subulate, pubescent. Flower-bud up to 3·5 mm. long, oblong-cylindric, slightly constricted near the middle, glabrous but with a single capitate glandular hair near the apex of each petal. Calyx c. 0·3 mm. long, subentire, pubescent. Petals red and yellow. Ovary glabrous; style 1 mm. long, apex scarcely subcapitate. Fruit (immature) glabrous.

N. Rhodesia. N: Abercorn, fl. 9.i.1955, *Bock* 217 (PRE; SRGH). W: 32 km. N. of Solwezi, 19.iii.1961, *Drummond & Rutherford-Smith* 7054 (SRGH).

Not known outside N. Rhodesia. *Brachystegia* woodland.

6. **Cyphostemma hermannioides** Wild & Drummond in Kirkia, **2**: 133 (1961). Type: N. Rhodesia, Lunzua Valley, *Richards* 3594 (K, holotype; SRGH).

Prostrate herb; stems tomentose to pubescent; tendrils absent. Leaves digitately 3-foliolate; petiole 0–2 mm. long, tomentose to densely pubescent; leaflet-lamina with an apiculate-crenate margin, glabrous above but puberulous at the margin and on the midrib, densely pubescent on the nerves below; median leaflet up to 9·5 × 6 cm., broadly elliptic to narrowly obovate, apex obtuse, base cuneate; lateral leaflets up to 3·5 × 3 cm., broadly ovate, obovate or broadly obovate, apex acute or rounded, base asymmetrically rounded; petiolules up to 2 mm. (6 mm. in median leaflet) long; stipules up to 6 mm. long, oblong-lanceolate, acute or acuminate, ± falcate, pubescent. Cymes on densely pubescent peduncles up to 2·5 cm. long; pedicels up to 1·7 mm. long, pubescent or very sparsely pubescent; bracts and bracteoles c. 1 mm. long, subulate, pubescent. Flower-bud c. 1·7 cm. long, oblong-cylindric, slightly constricted near the middle, glabrous. Calyx c. 0·25 mm. long, subentire, puberulous. Petals yellow with red tips. Ovary glabrous; style c. 1 mm. long, apex minutely 2-fid. Fruit red, c. 8 mm. in diam., subglobose, glabrous. Seed 1, c. 6 × 4 mm., ellipsoid, rugose.

N. Rhodesia. B: Kabompo, 23.iii.1961, *Drummond & Rutherford-Smith* 7228 (SRGH). N: Lake Tanganyika, Mpulungu, fr. 29.xii.1951, *Richards* 181 (K). W: Kabompo Gorge, fl. 23.xi.1962, *Richards* 17480 (SRGH).

Not known outside N. Rhodesia. Grassland or woodland at about 1,000 m. above sea level.

7. **Cyphostemma obovato-oblongum** (De Wild.) Descoings in Notul. Syst. **16**: 123 (1960). Type from the Congo (Katanga).

Cissus obovato-oblonga De Wild. in Fedde, Repert. **13**: 202 (1914); Ann. Soc. Sci. Brux. **38**, 2: 389 (1914).—Dewit, F.C.B. **9**: 481, t. 48 (1960). Type as above.

Creeping herb from tuberous rootstock; stems striate, pubescent; tendrils absent. Leaves digitately 3–5-foliolate; petiole up to 5 (6) cm. long, pubescent; leaflet-lamina, up to 16 × 7 cm., narrowly oblong-elliptic to narrowly obovate, apex acute to obtuse, margin serrate or serrate-crenate, base narrowly cuneate, ±

glabrous above, shortly pubescent on the nerves and main veins below; petiolules up to 2 cm. long, longer in the median leaflet; stipules up to 1·5 cm. long, triangular-lanceolate, apex acute to acuminate, pubescent at least at the margins. Cymes terminal on short axillary branches, trichotomous; peduncle up to 14 (20) cm. long, pubescent; pedicels up to 3·5 mm. long, pubescent; bracts and bracteoles c. 1–2 mm. long, lanceolate, pubescent at the margins. Flower-bud c. 2·8 mm. long, oblong-cylindric, ± deeply constricted in the upper half, very sparsely pubescent or glabrous. Calyx 0·5 mm. long, entire, pubescent. Petals yellowish-crimson. Ovary glabrous; style 1·7 mm. long, minutely 2-fid at the apex. Fruit c. 7 × 5 mm. (not quite ripe). Seed 1, c. 5 × 4 mm., with a dorsal and 2 lateral crests, rugose.

N. Rhodesia. W: Ndola, fl. 9.i.1954, *Fanshawe* 642 (K; NDO).
Also in the Congo. *Brachystegia* woodland, sometimes on termite mounds.

Two other specimens from N. Rhodesia (*White* 3640 (FHO) from Kawambwa (N) and *Cruse* 125 (K) from Mufulira (W)) almost certainly belong here but have one or two glands on the stems.

8. **Cyphostemma abercornense** Wild & Drummond in Kirkia, **2**: 132 (1961). Type: N. Rhodesia, Abercorn Distr., Kawimbe, *Richards* 7294 (K, holotype).

Prostrate herb; stems brownish-tomentose to pubescent; tendrils absent. Leaves digitately 3–5-foliolate; petiole up to 4 cm. long, tomentose or densely pubescent; leaflet-lamina oblanceolate, apex acute or subobtuse, margin serrate, base cuneate, glabrous above but midrib puberulous, whitish- or brownish-villous or sometimes pilose below, veins reticulate; median leaflets up to 15 × 5 cm.; lateral leaflets up to 9 × 2·5 cm.; petiolules 0–1 cm. long; stipules up to 1·5 × 0·8 cm., oblong-lanceolate, acuminate, tomentose or pubescent. Cymes leaf-opposed; peduncle up to 4·5 cm. long, tomentose or pubescent; pedicels up to 3 mm. long, pubescent or sparsely pubescent; bracts and bracteoles up to 3 mm. long, pubescent or sparsely pubescent. Flower-bud up to 2·3 mm. long, oblong-cylindric, slightly constricted near the middle, glabrous or almost so. Calyx 0·5 mm. long, subentire, pubescent. Petals yellowish or reddish. Ovary glabrous; style 1·5 mm. long; stigma scarcely capitate. Fruit (immature) glabrous.

N. Rhodesia. N: Abercorn Distr., Mpulungu road, fl., *Richards* 2250 (K; SRGH).
Not known outside N. Rhodesia. Rocky hills.

9. **Cyphostemma junceum** (Webb) Descoings in Notul. Syst. **16**: 122 (1960). Type from Sudan (Nubia).
Cissus juncea Webb, Fragm. Fl. Aethiop.-Aegypt.: 57 (1854).—Planch. in A. & C. DC., Mon. Phan. **5**, 2: 578 (1887).—Gilg & Brandt in Engl., Bot. Jahrb. **46**: 490, fig. 5 M (1912).—Suesseng. in Engl. & Prantl, Nat. Pflanzenfam. ed. 2, **20d**: 240 (1953).—Dewit, F.C.B. **9**: 472 (1960). Type as above.
Vitis juncea (Webb) Bak. in Oliv., F.T.A. **1**: 401 (1868). Type as above.
Vitis jatrophoides Welw. ex Bak., tom. cit.: 400 (1868). Type from Angola.
Cissus jatrophoides (Welw. ex Bak.) Planch. in A. & C. DC., Mon. Phan. **5**, 2: 579 (1887) pro parte.—Gilg & Brandt, tom. cit.: 491, fig. 5 M (1912).—R.E. Fr., Schwed. Rhod.-Kongo-Exped. **1**: 134 (1914).—Eyles in Trans. Roy. Soc. S. Afr. **5**: 409 (1916).—Suesseng., loc. cit.—Exell & Mendonça, C.F.A. **2**, 1: 57 (1954).—Dewit, loc. cit. Type as above.

Erect herb to 1·3 m. tall, glabrous except for a few ± caducous cilia on the stipules; ± glaucous especially when young; tendrils absent; rootstock tuberous. Leaves digitately 3- or very rarely 5-foliolate, sessile or petiole up to 10 cm. long and as a rule increasing in length down the stem; leaflet-lamina up to 30 × 5 (8) cm., lateral leaflets smaller, narrowly oblong-lanceolate, often folded along the midrib and falcate when young, apex acuminate, margin sharply serrate to serrate-crenate, base cuneate; stipules up to 2·5 × 1 cm., lanceolate-acuminate, ciliate. Cymes lax; peduncle terminal, up to c. 25 cm. long; pedicels c. 3 mm. long, slender, lengthening in fruit; bracts and bracteoles c. 2 × 0·5 mm., subulate, caducous. Flower-bud c. 3 mm. long, oblong-cylindric, constricted somewhat above the middle. Calyx c. 0·5 mm. long, entire or shallowly lobed. Petals greenish-red. Ovary glabrous; style c. 2 mm. long, capitate. Fruit reddish-purple turning violet with a bloom, c. 1 × 0·6 cm., ellipsoid. Seed usually 1, c. 7 × 4 mm., with 3 strong dorsal ridges and a few horizontal ridges.

N. Rhodesia. B: Balovale, fl. 15.x.1953, *Gilges* 43 (PRE; SRGH). N: Abercorn, fl. 23.x.1952, *Robertson* 166 (EA; PRE; SRGH). W: Ndola, fl. 26.xi.1947, *Brenan* 8387 (FHO; K). C: Broken Hill, fl. x.1909, *Rogers* 8429 (SRGH). E: Fort Jameson, st., *Bush* 22 (K). S: Mumbwa, fl., *Macaulay* 940 (K). **S. Rhodesia.** N: Urungwe, Mgunje, fl. & fr. 16.xi.1953, *Wild* 4152 (SRGH). C: Hartley, Poole Farm, fl. 27.ii.1948, *Hornby* 2936 (K; SRGH). S: Fort Victoria, st., *Monro* 331 (SRGH). **Nyasaland.** C: Lilongwe, fl. 9.x.1951, *Jackson* 641 (K). S: Mt. Mlanje, fl., *Whyte* 120 (BM; K). **Mozambique.** N: Mandimba, fl. 23.xi.1941, *Hornby* 3463 (K). Z: Milange, fl. 13.xi.1942, *Mendonça* 1431 B (LISC). T: Furancungo, fl. 29.ix.1942, *Mendonça* 515 (LISC).

Also in the Sudan, Guineé Republic, Ivory Coast, Ghana, Togo Republic, Dahomey, Nigeria, Cameroun, Congo, Kenya, Uganda, Tanganyika and Angola. Usually in *Brachystegia* woodlands.

10. **Cyphostemma crotalarioides** (Planch.) Descoings in Notul. Syst. **16**: 121 (1960). Syntypes from Sudan, Nigeria and Nyasaland: Shire Highlands, *Buchanan* 587 (B; BM; K), 748 (B; K).

Cissus crotalarioides Planch. in A. & C. DC., Mon. Phan. **5**, 2: 577 (1887).—Gilg & Brandt in Engl., Bot. Jahrb. **46**: 498 (1912).—R.E. Fr., Schwed. Rhod.-Kongo-Exped. **1**: 134 (1914).—Eyles in Trans. Roy. Soc. S. Afr. **5**: 408 (1916).—Suesseng. in Engl. & Prantl, Nat. Pflanzenfam. ed. 2, **20d**: 241 (1953).—Dewit, F.C.B. **9**: 512 (1960). Syntypes as above.

Vitis variifolia Bak. in Kew Bull. **1897**: 248 (1897). Type: Nyasaland, Zomba, *Whyte* (K, holotype).

Cissus variifolia (Bak.) Gilg & Brandt, tom. cit.: 500 (1912).—Suesseng., tom. cit.: 242 (1953). Type as above.

Cissus hypargyrea sensu Eyles, tom. cit.: 409 (1916).

Erect herb c. 1 m. tall; rootstock probably thickened and tuberous; stems rather stout, sulcate, tomentellous or densely pubescent and with short glandular-capitate hairs; tendrils absent. Leaves digitately 3-foliolate, sessile; leaflet-lamina up to 14 × 6·5 cm., narrowly oblong-obovate to obovate, apex acute, margin serrate, base cuneate, puberulous and with scattered short glandular-capitate hairs above, densely pubescent to tomentose below and with scattered glands; petiolules 0–3 cm. long. Cymes in the upper axils; peduncle up to c. 4 cm. long, tomentose and with capitate glands; pedicels c. 2·5 mm. long, tomentose and with short capitate glands; bracts and bracteoles c. 2 mm. long, linear, tomentose. Flower-bud c. 3 mm. long, oblong-cylindric, very slightly constricted in the upper half, densely pubescent and with short capitate glands denser at the apex. Calyx c. 0·5 mm. long, entire, densely pubescent. Ovary glandular-puberulous; style 1·7 mm. long, shortly 2-fid at the apex. Fruit c. 1 × 0·8 cm., ellipsoid-globose, densely pubescent with short glandular-capitate hairs also. Seed 1, c. 0·9 × 0·6 cm., somewhat reniform, without a prominent crest but prominently rugose.

N. Rhodesia. B: Mankoya, fl. 8.xi.1959, *Drummond & Cookson* 6224 (K; PRE; SRGH). N: Abercorn, Msisi, fl. 25.xi.1911, *Fries* 1326 (S). W: Bwana Mkubwa, fl. 10.xi.1953, *Fanshawe* 488 (EA; K; NDO). C: Mkushi, fl. 1.xii.1952, *White* 3816 (FHO). S: Bombwe, fl. 10.xi.1932, *Martin* 370/32, (NDO). **S. Rhodesia.** N. Urungwe, Msukwe, fl. 19.xi.1953, *Wild* 4206 (SRGH). C: Salisbury Kopje, fl. xi.1908, *Rand* 1340 (BM). **Nyasaland.** C: Dedza, fl. 5.i.1961, *Chapman* 1109 (SRGH). S: Shire Highlands, *Buchanan* 587 (BM; K). **Mozambique.** N: Vila Cabral, fl. xi.1933, *Gomes e Sousa* 1580 (COI).

Also in Dahomey, Nigeria, Cameroun, Ubangi-Shari, Sudan and the Congo. *Brachystegia* woodlands.

There is a good deal of variation in the density of the indumentum on the underside of the leaves in this species. It is probable that *C. michelii* Dewit from the Congo represents a more glabrescent form.

11. **Cyphostemma chloroleucum** (Welw. ex Bak.) Descoings in Notul. Syst. **16**: 120 (1960). Type from Angola (Huila).

Vitis chloroleuca Welw. ex Bak. in Oliv., F.T.A. **1**: 406 (1868). Type as above.

Cissus chloroleuca (Welw. ex Bak.) Planch. in A. & C. DC., Mon. Phan. **5**, 2: 592 (1887).—Gilg & Brandt in Engl., Bot. Jahrb. **46**: 505 (1912).—Suesseng. in Engl. & Prantl, Nat. Pflanzenfam. ed. 2, **20d**: 245 (1953).—Exell & Mendonça, C.F.A. **2**, 1: 60 (1954). Type as above.

Erect herb c. 0·6 m. tall; stems rather stout, sulcate, ferruginous-tomentose with scattered long red glandular-capitate hairs; tendrils absent. Leaves digitately 3–5-foliolate; petiole with indumentum like the stems, up to 11 cm. long; leaflet-lamina up to c. 24 × 10 cm., elliptic to obovate-elliptic, apex acute, margin serrate, base cuneate, upper surface puberulous with scattered short glandular-capitate hairs, lower surface ferruginous-tomentose with a few glandular hairs on the midrib. Cymes spreading, trichotomous towards the ends of the branches; peduncle 4·5 (8) cm. long, ± tomentose and with dense red capitate-glandular hairs; pedicels c. 2 mm. long, pubescent and with dense short capitate-glandular hairs; bracts and bracteoles c. 1–2 mm. long, linear, glandular-pubescent. Flower-bud up to 2·5 mm. long, constricted in the upper half, pubescent and with glandular-capitate hairs. Calyx 0·5 mm. long, densely pubescent and with short capitate glands. Ovary pubescent and with short capitate glands; style 2 mm. long, minutely 2-fid or capitate at the apex. Fruit purple-black, c. 1 × 0·7 cm., subglobose, tomentose or pubescent with short capitate glands. Seed 1, c. 6 × 4·5 mm., ovoid, without an obvious dorsal crest, minutely tuberculate.

N. Rhodesia. W: Mwinilunga, Kalenda Ridge, fl. & fr. xi.1937, *Milne-Redhead* 3056 (K).
Also in the Congo and Angola. *Brachystegia* woodland and termite mounds.

12. **Cyphostemma manikense** (De Wild.) Descoings in Notul. Syst. **16**: 122 (1960) (" mainkensis "). Type from the Congo.
Cissus manikensis De Wild. in Fedde, Repert. **13**: 201 (1914) (" mainkensis "); Ann. Soc. Sci. Brux. **38,** 2: 387 (1914).—Dewit, F.C.B. **9**: 510 (1960). Type as above.
Cissus pseudomanikensis Dewit in Bull. Jard. Bot. Brux. **29**: 263 (1959); F.C.B. **9**: 511 (1960). Type from the Congo (Katanga).

Erect herb c. 0·6 m. tall with stout rootstock; stems stout, shortly and densely pubescent or tomentose and with very dense purplish glandular-capitate hairs. Leaves digitately 5-foliolate; petiole up to 10 cm. long, indumentum as for stems; leaflet-lamina 10 (12) × 4 (5) cm., narrowly obovate to elliptic, apex acute, margin serrate, base cuneate, upper surface pubescent and with ± dense capitate-glandular hairs, lower surface pale-brownish-tomentose or -lanate, veins and nerves pubescent but with dense capitate-glandular hairs; petiolules up to 1 cm. long; stipules up to 2·2 cm. long, lanceolate-acuminate, pubescent and with glandular hairs. Cymes ± terminal; peduncle c. 4 cm. long, pubescent and with dense capitate-glandular hairs; pedicels c. 1·5 mm. long, indumentum as for peduncles; bracts and bracteoles c. 1·5 mm. long, linear-lanceolate, pubescent and glandular at the margins. Flower-bud c. 2·5 mm. long, constricted just above the middle, puberulous and with very dense capitate-glandular hairs at the apex. Calyx 0·5 mm. long, ± 4-toothed, pubescent and with capitate-glandular hairs. Petals red at base, greenish-red above. Ovary pubescent; style 0·75 mm. long, minutely 2-fid at the apex. Ripe fruits not known.

N. Rhodesia. N: Abercorn, fl. 15.xi.1952, *White* 3666 (FHO). W: Ndola, fl. 26.xi.1947, *Brenan* 8379 (FHO; K).
Also in the Congo (Katanga). *Brachystegia* woodland.

13. **Cyphostemma vanmeelii** (Lawalrée) Wild & Drummond in Kirkia, **2**: 140 (1961). TAB. **96,** fig. A. Type from the Congo (Katanga).
Cissus vanmeelii Lawalrée in Bull. Jard. Bot. Brux. **19**: 219 (1949).—Dewit, F.C.B. **9**: 493 (1960). Type as above.
Cissus stipulaceoides Dewit in Bull. Jard. Bot. Brux. **29**: 494 (1959); F.C.B. **9**: 494 (1960). Type from the Congo (Katanga).

Erect herb c. 0·6 m. tall or occasionally stems prostrate; stems striate, tomentose with long whitish weak hairs or glabrescent later and also with long slender red patent glandular hairs up to 2 cm. long; tendrils absent. Leaves digitately 3–5-foliolate, sessile or with petioles up to 4 cm. long, indumentum like that of stems; leaflet-lamina up to 28 × 8·5 cm., linear-lanceolate to narrowly elliptic, apex acute to acuminate, margin serrate, base narrowly cuneate, ± sparsely pubescent above with short whitish hairs, more densely pubescent below, especially on the nerves, with long whitish hairs and also with sessile scattered discoid glands; stipules up to c. 1·5 cm. long, ovate-acuminate, densely pilose with long hairs especially at the margins. Cymes terminal or in the upper axils, usually trichotomous but not very wide-spreading; peduncle up to c. 9 cm. long, indumentum as for stems; pedicels up to 2 mm. long but often shorter, pubescent and with short capitate-glandular hairs; bracts and bracteoles linear-subulate, with long hairs on

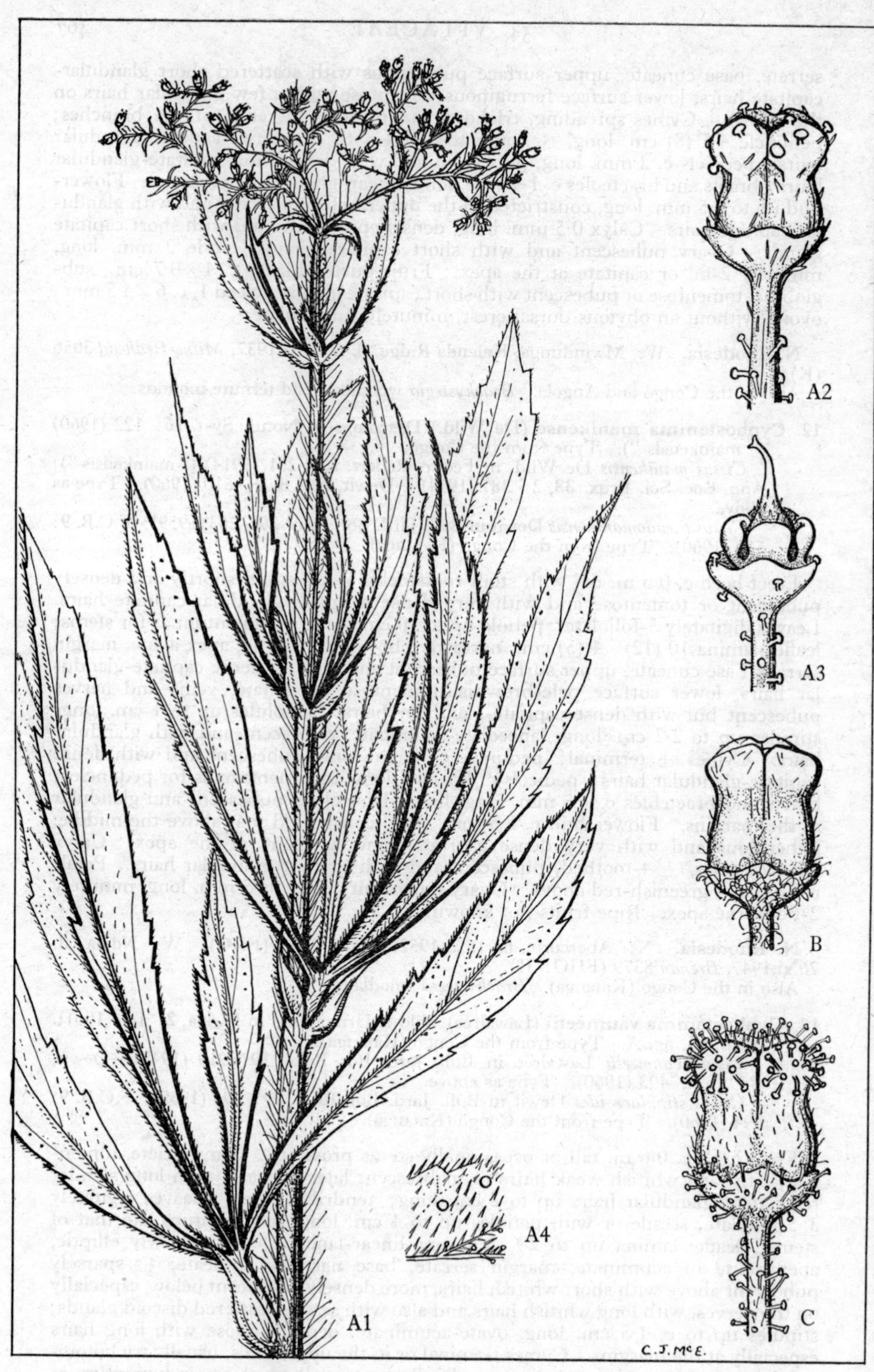

Tab. 96. A.—CYPHOSTEMMA VANMEELII. A1, flowering stem (×4/9) *Richards* 7310; A2, bud (×8) *Milne-Redhead* 3643; A3, flower-bud with petals removed (×8) *Macaulay* 1022; A4, portion of leaf with glands (×12) *Richards* 7310. B.—CYPHOSTEMMA KERKVOORDEI, flower-bud (×8) *Corby* 558. C.—CYPHOSTEMMA HETEROTRICHUM, flower-bud (×8) *Wild* 4658.

the outside and on the margins. Flower-bud cream with reddish tip, 2–3 mm. long, constricted just above the middle, pubescent and with short capitate-glandular hairs especially towards the apex. Calyx 0·5 mm. long, pubescent. Ovary pubescent; style up to 2·5 mm. long, shortly 2-fid at the apex. Fruit crimson, c. 1 cm. in diam., globose, pubescent and with long red gland-tipped hairs. Seed 1, c. 5 × 3·5 mm., ellipsoid with a dorsal crest and 2 faintly marked lateral crests, remainder minutely tuberculate.

N. Rhodesia. N: Abercorn Distr., Chilongowelo, fl. 19.xii.1956, *Richards* 7310 (K). W: Mwinilunga, fl. 13.xii.1937, *Milne-Redhead* 3643 (K). S: Mumbwa, fl., *Macaulay* 1073 (K).

Also in the Congo and Tanganyika. *Brachystegia* woodland, sometimes on termite mounds.

14. **Cyphostemma heterotrichum** (Gilg & Fr.) Descoings in Notul. Syst. **16**: 122 (1960). Tab. **96** fig. C. Type: from Rwanda-Burundi.
Cissus heterotricha Gilg & Fr. in R.E. Fr., Schwed. Rhod.-Kongo-Exped. **1**: 134 (1914).—Suesseng. in Engl. & Prantl, Nat. Pflanzenfam. ed. 2, **20d**: 243 (1953). Type as above.
Cissus reedii Dewit in Bull. Jard. Bot. Brux. **29**: 279 (1959); F.C.B. **9**: 505 (1960). Type from Rwanda-Burundi.

Erect branching herb c. 1 m. tall; stems rather stout, sulcate, with short whitish hairs, slightly longer and darker (often purplish or blackish) glandular hairs and usually a few long (up to 5 mm.) glandular whitish hairs; tendrils absent. Leaves digitately 3–5-foliolate, petiole up to 14 cm. long, indumentum like the stems but the longer eglandular hairs often denser; leaflet-lamina sessile, up to c. 12 (16) × 5 (9) cm., elliptic to obovate, apex acute to acuminate, margin serrate to crenate, base cuneate, upper surface pubescent with short white eglandular hairs and short usually blackish glandular-capitate hairs, lower surface similar but hairs somewhat denser, especially on the nerves; stipules up to c. 1·7 × 1 cm., indumentum as for leaves. Cymes ± terminal, trichotomous; peduncle up to 7 cm. long; pedicels up to 3 mm. long; all branches of inflorescence densely pubescent with short pale hairs and short dark capitate-glandular hairs; bracts and bracteoles up to 3 mm. long, lanceolate, with short eglandular and glandular hairs. Flower-bud up to 2·7 mm. long, oblong-cylindric, constricted just above the middle, pubescent with short eglandular and glandular-capitate hairs especially at the apex. Calyx 0·7 mm. long, entire, puberulous and sometimes glandular. Ovary minutely puberulous; style c. 1·8 mm. long, minutely 2-fid at the apex. Fruit up to 8 × 5 mm., ellipsoid, with short glandular and eglandular hairs. Seed 1, up to 7 × 4 mm., with a dorsal crest and 2 faint lateral crests, rugose.

N. Rhodesia. N: Mpanda, fl. 11.ii.1955, *Richards* 4468 (K). **S. Rhodesia.** N: Urungwe, Zambezi escarpment, fl. 31.i.1958, *Drummond* 5391 (K; PRE; SRGH). E: Umtali Distr., Sabi Drift, Odzi Rd., fl. 1.xii.1954, *Wild* 4658 (K; PRE; SRGH).

Also in Kenya, Uganda, Rwanda Burundi and the Congo. Woodlands and river banks.

15. **Cyphostemma fugosioides** (Gilg) Descoings in Notul. Syst. **16**: 121 (1960). Type from Angola.
Cissus fugosioides Gilg in Warb., Kunene-Samb.-Exped. Baum: 294 (1903).—Gilg & Brandt in Engl., Bot. Jahrb. **46**: 491 (1912).—Suesseng. in Engl. & Prantl, Nat. Pflanzenfam. ed. 2, **20d**: 240 (1953).—Exell & Mendonça, C.F.A. **2**, 1: 57 (1954). Type from Angola.
Cissus gracillimoides Dewit in Bull. Jard. Bot. Brux. **29**: 265 (1959); F.C.B. **9**: 485 (1960). Type from the Congo.

Erect herb 12–50 cm. tall from a tuberous rootstock; stems slender, pubescent; tendrils absent. Leaves digitately 3–5-foliolate, sessile or with petioles up to 3·3 (4·5) cm. long, sparsely pubescent or glabrescent; leaflet-lamina up to 8·5 × 1 cm., subsessile, linear to lorate-lanceolate, apex acute or obtuse, mucronate, margin sparsely and acutely serrate, base narrowly cuneate, glabrous; stipules c. 7 mm. long, lanceolate-acuminate, ciliate. Cymes lax, terminal; peduncle 2 (6) cm. long, pubescent; pedicels 1–2 mm. long, with short capitate glands; bracts and bracteoles 1–2 mm. long, lanceolate-triangular, margins ciliate. Flower-bud c. 2 mm. long, slightly constricted at the middle, glabrous. Calyx c. 0·3 mm. long, glabrous. Petals yellow. Ovary glabrous; style 1·5 mm. long, scarcely capitate. Fruit scarlet, c. 8 × 7 mm., ellipsoid, glabrous. Seed 1, c. 5 × 3 mm., with a strong dorsal crest and a wrinkled pit on each side of the crest.

N. Rhodesia. W: Mwinilunga, Cha Mwana Plain, fl. & fr. 14.x.1937, *Milne-Redhead* 2777 (K).

Also in Angola and the Congo. Woodland or open grassland.

16. **Cyphostemma nanellum** (Gilg & Fr.) Descoings in Notul. Syst. **16**: 123 (1960). Type: N. Rhodesia, Kalungwishi R., *Fries* 1132 (UPS, holotype).

Cissus nanella Gilg & Fr. in R.E. Fr., Schwed. Rhod.-Kongo-Exped. **1**: 135, t. 10 fig. 4 (1914).—Suesseng. in Engl. & Prantl, Nat. Pflanzenfam. ed. 2, **20d**: 244 (1953). Type as above.

Erect herb to 25 cm. tall from a tuberous root; stems slender, pubescent; tendrils absent. Leaves digitately 3-foliolate; petiole up to 3·2 cm. long, pubescent, with a few short glands near the apex; leaflet-lamina up to 4·3 × 0·6 cm. (young leaves), linear to narrowly oblanceolate-lorate, margin serrate near the apex, glabrous above, pubescent on the midrib below; stipules up to 7 × 3 mm., lanceolate-acuminate, pubescent. Cymes ± terminal, trichotomous; peduncle up to 1·3 cm. long, pubescent; pedicels up to 2·3 mm. long, pubescent with short purple-black capitate glands. Flower-bud c. 2 mm. long, oblong-cylindric, very slightly constricted near the middle, pubescent. Calyx purplish at the margins, 0·5 mm. long, puberulous. Ovary puberulous; style 1 mm. long, slender, minutely capitate. Fruits not seen.

N. Rhodesia. N: between Luvingo and Mporokoso, fl. 28.x.1911, *Fries* 1132 (UPS).

Apparently confined to N. Rhodesia. Woodland on burnt ground.

17. **Cyphostemma zombense** (Bak.) Descoings in Notul. Syst. **16**: 125 (1960). Type: Nyasaland, Mt. Zomba, *Whyte* (K, holotype).

Vitis apodophylla Bak. in Kew Bull. **1897**: 248 (1897) non *V. apodophylla* Bak. (1894) ex Arabia. Type as above.

Vitis zombensis Bak., op. cit. **1898**: 302 (1898). Type as above.

Cissus zombensis (Bak.) Gilg & Brandt in Engl., Bot. Jahrb. **46**: 499 (1912).—Suesseng. in Engl. & Prantl, Nat. Pflanzenfam. ed. 2, **20d**: Type as above.

Cissus upembaënsis Dewit in Bull. Jard. Bot. Brux. **29**: 273 (1959); F.C.B. **9**: 474 (1960). Type from the Congo (Katanga).

Erect herb c. 1 m. tall with a swollen root; stems striate, pubescent and with capitate-glandular hairs; tendrils absent. Leaves digitately 3–5-foliolate, sessile; leaflet-lamina sessile, up to 24 × 6 cm., narrowly oblong-oblanceolate, apex acuminate, margin serrate, puberulous or glabrous above, glabrous except for the pubescent nerves below or sometimes the interspaces minutely puberulous, sometimes with some glandular hairs; stipules up to 1·5 cm. long, oblong-triangular, pubescent. Cymes subterminal or in the axils of the upper leaves; peduncle up to 12 cm. long, indumentum as for stems; pedicels up to 3 mm. long, pubescent; bracts and bracteoles c. 1·5 mm. long, lanceolate-triangular, ciliate. Flower-bud 2–2·5 mm. long, oblong-cylindric or lageniform, scarcely constricted, pubescent. Ovary glabrous; style c. 1 mm. long, apex minutely 2-fid or subcapitate. Ripe fruits so far not known.

N. Rhodesia. N: Kawambwa, Kafulwe Mission, fl. 4.xi.1952, *White* 3595 (FHO). **Nyasaland.** S: Shire Highlands, fl., *Buchanan* 277 (K).

Also in the Congo (Katanga). Savanna woodland or open woodland.

18. **Cyphostemma viscosum** (Gilg & Fr.) Descoings in Notul. Syst. **16**: 125 (1960). Type: N. Rhodesia, Lake Bangweulu, *Fries* 1081 (UPS, holotype).

Cissus viscosa Gilg & Fr. in R.E. Fr., Schwed. Rhod.-Kongo-Exped. **1**: 136 (1914).—Suesseng. in Engl. & Prantl, Nat. Pflanzenfam. ed. 2, **20d**: 245 (1953). Type as above.

Erect herb up to 1 m. tall from a large tuberous rootstock; stems stout, very viscid with subsessile capitate glands and stalked glands; tendrils absent. Leaves digitately 3-foliolate; petiole up to 5 cm. long, glandular like the stems; leaflet-lamina up to 38 × 16 cm., narrowly obovate, obovate, or elliptic, apex acute or obtuse, margin crenate or undulate-crenate, base cuneate, glabrous above except for sessile glands which give a varnished appearance, whitish-tomentellous beneath; petiolules up to 3 cm. long; stipules up to 3 cm. long, ovate-acuminate, viscid with sessile glands. Cymes terminal, trichotomous; peduncle up to 8 cm. long, indumentum as for stems; pedicels up to 2 mm. long, puberulous and glandular; bracts and bracteoles 1–2 mm. long, subulate, puberulous. Flower-bud c. 3 mm. long, oblong-cylindric, constricted just above the middle, puberulous and with subsessile glands. Calyx 0·5 mm. long, puberulous and glandular. Petals green. Ovary glandular-puberulous; style 1·5 mm. long, apex subcapitate. Fruit plum-coloured, 1·8 × 1·2 cm. (fresh), ellipsoid, tapering slightly to the apex, puberulous and glandular with subsessile and stalked glands, stalk of the latter puberulous. Seed 1, c. 1 × 0·7 cm., ellipsoid, with a dorsal crest, rugose and surface rough.

N. Rhodesia. B: 80 km. E. of Mankoya, fr. 21.xi.1959, *Drummond & Cookson* 6721 (K; LISC; PRE; SRGH). N: Kasama, fr. 26.xi.1952, *Angus* 853 (FHO). W: Mwinilunga, fr. 2.xi.1937, *Milne-Redhead* 3055 (K). C: 8 km. E. of Lusaka, fl. 26.x.1955, *King* 183 (K). S: Monze to Magoye, st. 20.ii.1960, *White* 7272 (FHO). **S. Rhodesia.** N: 3 km. NE. of Trelawney, fr. 2.xii.1963, *Drummond* 8433 (K; SRGH).

Not known outside Rhodesia. *Brachystegia* woodland or in *Pericopsis-Pterocarpus* savanna (Chipya).

19. **Cyphostemma stenolobum** (Welw. ex Planch.) Descoings in Notul. Syst. **16**: 124 (1960). Type from Angola (Huila).

Vitis stenoloba Welw. ex Bak. in Oliv., F.T.A. **1**: 408 (1868) pro parte excl. *Welwitsch* 1451. Type as above.

Cissus stenoloba (Welw. ex Bak.) Planch. in A. & C. DC., Mon. Phan. **5**, 2: 578 (1887).—Gilg & Brandt in Engl., Bot. Jahrb. **46**: 496 (1912).—Suesseng. in Engl. & Prantl. Nat. Pflanzenfam. ed. 2, **20d**: 241 (1953).—Exell & Mendonça, C.F.A. **2**, 1: 58 (1954). Type as above.

Erect herb up to 0·6 (1·0) m. tall from a narrowly elongated tuberous root; stems rather slender, ferruginous-pilose or -tomentose with a few capitate glands especially at the nodes; tendrils absent. Leaves digitately 3–5-foliolate, sessile or the lower leaves with ferruginous-pilose petioles up to 1·4 cm. long; leaflet-lamina up to 10 × 1 cm., linear-oblanceolate to very narrowly elliptic, apex acute, margin serrate but entire towards the base, base narrowly cuneate, glabrous or glabrescent above, ferruginous-pubescent or -tomentose below, especially on the main nerves; stipules up to 1 × 0·6 cm., lanceolate-acuminate, ferruginous-pilose. Cymes ± terminal, trichotomous; peduncle up to 2·5 (5) cm. long, ferruginous-pilose with glandular-capitate hairs also; pedicels c. 2·5 mm. long, pubescent; bracts and bracteoles c. 1 mm. long, triangular, membranous. Flower-bud up to 2 mm. long, constricted near the middle, pubescent and with a few capitate-glandular hairs near the apex. Calyx 0·3 mm. long, entire, pubescent. Petals green with brownish-red tips. Ovary glabrous; style 1·3 mm. long, minutely 2-fid at the apex. Fruit 1 × 0·7 cm., ellipsoid, glabrous. Seed 1, up to 0·7 × 0·4 cm., ellipsoid, with a dorsal crest and two faint lateral crests, minutely tuberculate and horizontally rugose.

N. Rhodesia. W: Mwinilunga, Matonchi, fl. & fr. 16.xi.1937, *Milne-Redhead* 3260 (K).

Also in Angola. *Cryptosepalum* woodland on Kalahari Sand.

20. **Cyphostemma princeae** (Gilg & Brandt) Descoings in Notul. Syst. **16**: 123 (1960). Type from Tanganyika (Uhehe).

Cissus princeae Gilg & Brandt in Engl., Bot. Jahrb. **46**: 495 (1912).—Suesseng. in Engl. & Prantl, Nat. Pflanzenfam. ed. 2, **20d**: 241 (1953). Type as above.

Erect herb c. 1 m. tall; stems rather stout, ferruginous-tomentose with scattered red capitate-glandular hairs; tendrils absent. Leaves digitately 3–5-foliolate, sessile near the inflorescence, lower leaves with petioles up to 7 cm. long, indumentum as for stems; leaflet-lamina subsessile, up to c. 16·5 × 7·5 cm., lateral leaflets shorter, oblanceolate (when young) to obovate, apex acute, margin serrate, base cuneate, upper surface glabrous or with a few very sparse hairs when young, whitish- or ferruginous-tomentose below and with long glandular-capitate hairs on the midrib, venation reticulate and less tomentose than the interspaces; stipules up to 3 cm. long, oblong-acuminate, ferruginous-tomentose. Cymes spreading, corymbose; peduncle terminal, up to 14 cm. long, di- or trichotomous, ferruginous-tomentose and with long capitate-glandular hairs; pedicels up to 2·5 mm. long, ferruginous-pilose; bracts and bracteoles c. 2 mm. long, ferruginous-pilose. Flower-bud c. 2 mm. long, oblong-cylindric, slightly constricted at the middle, ferruginous-pilose. Calyx c. 0·7 mm. long, entire, ferruginous-pilose. Ovary sparsely puberulous; style c. 1·8 mm. long, ± sparsely puberulous towards the base, apex often shortly 2-fid. Fruit glabrescent (not seen when mature).

N. Rhodesia. N: Abercorn, fl. 14.xi.1954, *Richards* 2242 (K).
Also in Tanganyika.

21. **Cyphostemma septemfoliolatum** Wild & Drummond in Kirkia, **2**: 138 (1961). Type: N. Rhodesia, Abercorn Distr., Chilongowelo, *Richards* 2295 (K, holotype).

Erect herb c. 0·8 m. tall with a swollen root; stems pubescent; tendrils absent. Leaves digitately 7–9-foliolate; petiole up to 4 cm. long, pubescent; leaflet-lamina up to 26 × 4 cm., linear-elliptic to very narrowly elliptic, apex acuminate, margin serrate, base very narrowly cuneate, glabrous above but midrib puberulous, whitish-villous or -tomentose below but nerves and veins brown and glabrescent, venation reticulate; petiolules 0–2 cm. long; stipules up to 1 × 0·4 cm., oblong-lanceolate, apex acuminate, densely pubescent. Cymes subterminal; peduncle up to 8 cm. long, pubescent; pedicels up to 3 mm. long, pubescent; bracts and bracteoles 1–2 mm. long, subulate, pubescent. Flower-bud c. 1·75 mm. long, oblong-cylindric, slightly constricted near the middle, glabrous. Calyx c. 0·25 mm. long, subentire, pubescent. Petals yellowish or red. Ovary glabrous; style c. 1 mm. long, apex subcapitate. Fruit (immature) glabrous.

N. Rhodesia. N: Kalambo Falls, fl. 6.i.1955, *Richards* 3925 (K; SRGH). C: Serenje Distr., Kamona Rest House, fl. 30.xi.1952, *White* 3802 (FHO).
Also in Tanganyika. *Brachystegia* woodland.

22. **Cyphostemma kerkvoordei** (Dewit) Descoings in Notul. Syst. **16**: 122 (1960) (" kervoordei "). TAB. **96** fig. B. Type from the Congo (Katanga).
Cissus kerkvoordei Dewit in Bull. Jard. Bot. Brux. **29**: 275 (1959); F.C.B. **9**: 479 (1960). Type as above.

Erect herb to 1·2 m. tall with a tuberous root; stems stout, striate, densely ferruginous-puberulous or glabrescent; tendrils absent. Leaves digitately 3–5(7)-foliolate; petiole 5 (10) cm. long, indumentum as for stems; leaflet-lamina up to 26 × 10 cm., narrowly elliptic to elliptic, apex acute, margin irregularly serrate, base cuneate, upper surface sparsely pubescent or glabrous, somewhat bullate when young, lower surface brownish- or greyish-tomentose; nerves reticulate; petiolules up to 1·5 cm. long or leaflets subsessile; stipules up to 2 (2·5) cm. long, oblong-lanceolate, densely brown-puberulous or glabrescent. Cymes in the upper axils or apparently terminal; peduncle up to 10 cm. long, densely pubescent or glabrescent; pedicels up to 2·5 mm. long, pubescent, rarely glabrescent; bracts and bracteoles 1–2 mm. long, pubescent. Flower-bud c. 3 mm. long, constricted above the middle, pubescent or sometimes glabrescent. Calyx 0·5 mm. long, ± 4-toothed, pubescent. Petals yellowish. Ovary glabrous; style up to 2 mm. long, apex minutely 2-fid. Fruit c. 7 mm. in diam., subglobose, glabrous. Seed 1, c. 6 × 4 mm., ellipsoid, strongly tuberculate-rugose.

N. Rhodesia. N: Abercorn, Kasama road, fr. 9.i.1955, *Richards* 3996 (K). W: Ndola, fr. 29.i.1954, *Fanshawe* 726 (FHO; K; NDO). **S. Rhodesia.** N: 3 km. NE. of Trelawney, fl. 2.xii.1963, *Grosvenor* 49 (K; LISC; SRGH). C: Salisbury, Lochinvar, fl. 17.xi.1945, *Wild* 385 (K; SRGH). E: Inyanga, fl. x.1956, *Miller* 3856 (SRGH). **Nyasaland.** N: Mzimba Distr., Mt. Hora, fl. & fr. 11.i.1950, *Krippner* 17 (BM). C: Chintembwe Distr., fl., *Herklots* 123 (EA). S: Neno Hills, fl. 9.ii.1938, *Lawrence* 628 (K).
Also in Tanganyika and the Congo. *Brachystegia* woodland. *Krippner* 17 represents a very glabrescent form of this species.

23. **Cyphostemma rhodesiae** (Gilg & Brandt) Descoings in Notul. Syst. **16**: 124 (1960). Type: S. Rhodesia, near Salisbury, *Engler* 3053 (B, holotype †). Neotype: Salisbury, between Avondale and Mabelreign, fl., *Drummond* 4932 (K; SRGH, neotype).
Cissus rhodesiae Gilg & Brandt in Engl., Bot. Jahrb. **46**: 500 (1912).—Eyles in Trans. Roy. Soc. S. Afr. **5**: 409 (1916).—Suesseng. in Engl. & Prantl, Nat. Pflanzenfam. ed. 2, **20d**: 243 (1953). Type as above.
Cissus zombensis sensu Bak. f. in Journ. Linn. Soc., Bot. **40**: 46 (1911).—Eyles, loc. cit.—Hopkins, Bacon & Gyde, Comm. Veld Fl.: 69 (1940).

Erect herb to c. 1·3 m. tall; stems rather stout, sulcate, with a loose ferruginous indumentum or glabrescent; tendrils absent. Leaves digitately 3–5-foliolate, sessile or subsessile; leaflet-lamina up to c. 25 × 14 cm., oblong-elliptic to obovate, apex rounded or acute, margin serrate, base cuneate, glabrous above, sparsely pilose below, densely so on the nerves when young, later glabrescent; veins reticulate; petiolules up to 3 cm. long; stipules up to c. 2 × 0·8 cm., lanceolate-acuminate, densely pubescent or glabrescent with hairs on the margins. Cymes terminal or in the upper axils, di- or trichotomous; peduncle up to c. 10 cm. long, from densely pubescent to glabrescent; pedicels c. 2 mm. long, from densely ferruginous-pilose to pubescent; bracts and bracteoles 1–2 mm. long, linear, ferruginously hairy. Flower-bud up to 2·5 mm. long, oblong-cylindric, constricted in the upper half, pilose. Calyx 0·5 mm. long, pilose. Ovary sparsely puberulous or glabrous; style 1·5 mm. long, slightly capitate. Fruit purple-black, c. 1·2 × 0·8 cm., subglobose, glabrous. Seed 1, c. 0·8 × 0·7 cm., obovoid, without a strong dorsal crest but rugose.

N. Rhodesia. B: 80 km. E. of Mankoya, fl. 21.xi.1959, *Drummond & Cookson* 6871 (SRGH). W: Solwezi, fl. 16.x.1953, *Fanshawe* 424 (K; NDO). C: Chilanga, fl. 10.x.1929, *Sandwith* 72 (K). S: Mazabuka, fl. 10.ix.1931, *Trapnell* 446 (K; PRE). **S. Rhodesia.** N: Sinoia, fl. 2.xi.1931, *Rattray* 385 (SRGH). C: Salisbury, fl. & fr. 28.x.1922, *Eyles* 3692 (SRGH). E: Himalayas, Engwa, fr. 1.ii.1955, *E.M. & W.* 13 (BM; SRGH). S: Zimbabwe, fl. & fr. 6.xii.1960, *Leach* 10565 (K; PRE; SRGH). **Nyasaland.** S: Zomba, fl. ix.1891 *Whyte* (BM). **Mozambique.** T: Furancungo, fl., *Torre* 3340 (LISC). MS: Chimoio, fr., *Mendonça* 3624 (LISC).

Not recorded outside our area. *Brachystegia* woodland.

24. **Cyphostemma adenocaule** (Steud. ex A. Rich.) Descoings in Notul. Syst. **16**: 120 (1960). Syntypes from Ethiopia.

Cissus adenocaulis Steud. ex. A. Rich., Tent. Fl. Abyss. **1**: 111 (1847).—Gilg & Brandt in Engl., Bot. Jahrb.: **46**: 516 (1912).—Suesseng. in Engl. & Prantl, Nat. Pflanzenfam. ed. 2, **20d**: 250 (1953).—Exell & Mendonça, C.F.A. **2,** 1: 66 (1954).—Dewit, F.C.B. **9**: 468, t. 47 (1960). Syntypes as above.

Climbing herb; rootstock tuberous; stems glabrous or glabrescent but usually with a few scattered short glandular hairs; tendrils present, long and slender. Leaves pedate, 5-foliolate; petioles up to 4·5 cm. long, indumentum as for stems; leaflet-lamina up to 10 × 3 (5) cm., elliptic to broadly ovate, acuminate at the apex, margin serrate, subcordate or rounded at the base, glabrous or sparsely puberulous on both sides; petiolules up to 2 cm. long but usually less, indumentum as for stems; stipules up to 8 × 5 mm., ovate- or lanceolate-acuminate, sometimes falcate, glabrous with puberulous margins. Cymes lax; peduncle up to 7 cm. long, puberulous; pedicels up to 3 mm. long, densely puberulous; bracts and bracteoles 1–2 mm. long, subulate, pubescent. Flower-bud up to 4·5 mm. long, cylindric, swollen at the apex, broadening to the base, puberulous. Calyx 0·5 mm. long, entire, puberulous. Ovary densely pubescent; style up to 3 mm. long; stigma minutely capitate. Fruit red, c. 6 × 5 mm., obovoid, puberulous or glabrescent. Seed 1, c. 4 × 3·5 mm., ovoid, faintly tuberculate.

Nyasaland. S: Fort Johnston, fr. 14.iii.1955, *E.M. & W.* 867 (BM; SRGH). **Mozambique.** N: Nametil, R. Mluli, fl. 11.iii.1937, *Torre* 1320 (COI; LISC). MS: Beira, fl. 1894, *Kuntze* (K).

Widespread also in East and West Tropical Africa. Forest climber.

Several varieties of this species have been described by Dewit (loc. cit.) from the Congo varying in the amount of indumentum and the presence or absence of glandular hairs. All our material seen so far would fall into the type variety, var. *adenocaule*, of which the vegetative parts are glabrous or sparsely pubescent and the stems have a few glandular hairs.

25. **Cyphostemma congestum** (Bak.) Descoings in Notul. Syst. **16**: 120 (1960). Type: Nyasaland, Chibiza, *Meller* (K, holotype).

Vitis congesta Bak. in Oliv., F.T.A. **1**: 412 (1868). Type as above.

Cissus congesta (Bak.) Planch. in A. & C. DC., Mon. Phan. **5,** 2: 590 (1887).—Gilg & Brandt in Engl., Bot. Jahrb. **46**: 520 (1912).—Eyles in Trans. Roy. Soc. S. Afr. **5**: 408 (1916).—Suesseng. in Engl. & Prantl, Nat. Pflanzenfam. ed. 2, **20d**: 250 (1953). Type as above.

Cissus fleckii Schinz in Bull. Herb. Boiss., Sér. 2, **8**: 640 (1908).—Gilg & Brandt, loc. cit.—Suesseng., loc. cit. Type from SW. Africa (Hereroland).

Prostrate or climbing herb with a tuberous root; stems striate, greyish-pubescent or -puberulous with scattered ± long capitate-glandular hairs, sometimes glands very few, rarely quite absent; tendrils present. Leaves ± subfleshy, 3–5-foliolate, sessile or subsessile; leaflet-lamina up to 12 × 5 cm., from narrowly elliptic to obovate when young and ± folded along the midrib, falcate, apex acute, base cuneate, from very sparsely to fairly densely puberulous or pubescent on both surfaces; petiolule up to 1·5 cm. long, grey-pubescent or -puberulous, sometimes with a few scattered capitate glands; stipules c. 5 mm. long, ovate-acuminate, ± falcate, pubescent. Cymes lax, on short axillary branches, trichotomous; peduncle up to c. 5 cm. long, pubescent, sometimes with scattered glandular-capitate hairs; pedicels c. 2 mm. long, densely whitish-pubescent; bracts and bracteoles c. 2 mm. long, linear, pubescent. Flower-bud c. 2·7 mm. long, oblong-cylindric, constricted just above the middle, puberulous. Calyx c. 0·3 mm. long, densely pubescent. Petals greenish or pinkish. Ovary pubescent; style c. 1·7 mm. long, minutely subcapitate. Fruit red, c. 8 × 5 mm., ellipsoid, apiculate, ± densely pubescent. Seed 1, c. 6 × 4 mm., with a dorsal crest, 2 fainter lateral crests, faintly rugose and minutely tuberculate.

Caprivi Strip, Linyanti, fl. & fr. 27.xii.1958, *Killick & Leistner* 3154 (PRE; SRGH). **Bechuanaland Prot.** N: Lake Ngami, fr. 5.v.1930, *van Son* in Herb. Mus. Transv. 29036 (PRE; SRGH). SE: Kgatla, st. 28.xi.1955, *Reyneke* 442 (PRE; SRGH). **N. Rhodesia.** B: Sesheke, fl., *Macaulay* 34 (K). N: Samfya, Lake Bangweulu, fl. & fr. 30.i.1959, *Watmough* 168 (K; LISC; PRE; SRGH). S: Mazabuka, Kafue Flats, fl. & fr. 7.iv.1955, *E.M. & W.* 1420 (BM; SRGH). **S. Rhodesia.** N: 8 km. N. of Banket, fr. 29.iv.1948, *Rodin* 4394 (K; SRGH). W: Victoria Falls, fl. & fr. 7.vii.1930, *Hutchinson & Gillett* 3464 (BM; K; SRGH). C: Hartley, Umfuli R., fr. 7.iv.1954, *Wild* 4550 (SRGH). **Nyasaland.** S: Lake Nyasa, Boadzulu I., fl. 14.iii.1955, *E.M. & W.* 876 (BM; SRGH). **Mozambique.** MS: Mungári, fr. 6.vi.1941, *Torre* 2817 (LISC); Nova Lusitania, near Muda, fl. 23.iii.1960, *Wild & Leach* 5226 (SRGH).

Also in the Transvaal and SW. Africa. Woodland or near river banks.

Specimens with more densely glandular stems are more frequent at the eastern end of the range of the species but moderately glandular specimens can be found even as far west as SW. Africa.

26. **Cyphostemma subciliatum** (Bak.) Descoings in Notul. Syst. **16**: 125 (1960). TAB. **97** fig. A. Syntypes: Mozambique, near Tete, *Kirk* (K); banks of R. Luabo, *Kirk* (K).

Vitis subciliata Bak. in Oliv., F.T.A. **1**: 409 (1868). Type as above.

Cissus subciliata (Bak.) Planch. in A. & C. DC., Mon. Phan. **5,** 2: 594 (1887).—Gilg & Brandt in Engl., Bot. Jahrb. **46**: 518 (1912).—Suesseng. in Engl. & Prantl, Nat. Pflanzenfam. ed. 2, **20d**: 250 (1953). Syntypes as above.

Cissus subglaucescens Planch., tom. cit.: 591 (1887). Type: Nyasaland, Shire Highlands, *Buchanan* 281 (E; K, holotype).

Cissus subciliata var. *stuartii* Planch., tom. cit.: 595 (1887). Syntypes: Mozambique, between Tete and Lupata, *Kirk* (K), Kirk drawing 36, partly (K); between R. Chire and Chupanga, *Stewart* (K).

Cissus subciliata var. *kirkii* Planch., loc. cit. Type: Mozambique, Kirk drawing No. 35, partly.

Cissus marionae Exell & Mendonça in Bol. Soc. Brot., Sér. 2, **26**: 230, t. 9 (1952); C.F.A. **2,** 1: 62, t. 13 (1954). Type: N. Rhodesia, Mwinilunga, *Milne-Redhead* 3527 (BM; K, holotype).

Climbing herb with a tuberous napiform root, stems with scattered glandular-capitate hairs, otherwise glabrous or sparsely pubescent; tendrils present. Leaves subfleshy, digitately 3–5-foliolate, sessile or subsessile; leaflet-lamina up to c. 10 × 4 cm., narrowly obovate, narrowly elliptic, ovate-elliptic or elliptic, apex acute or obtuse, margins crenate-serrate or serrate, base cuneate or narrowly truncate, sometimes asymmetric, sparsely puberulous when young but soon glabrous except that sometimes the midrib remains crisped-puberulous above and with a few glandular-capitate hairs below; petiolules longest on the median leaflet, up to c. 1·5 cm. long, lateral leaflets sometimes subsessile, crisped-puberulous, sometimes with a few capitate glands; stipules up to 1 × 0·5 cm., obliquely ovate to ovate-lanceolate, margin glandular. Cymes terminal on short side branches, lax; peduncle up to c. 4 cm. long, puberulous or glabrous, often with a few capitate glands; pedicels up to 4·5 mm. long, puberulous or glabrous; bracts and bracteoles c. 2 mm. long, lanceolate-acuminate, pubescent but with a few glands at the margins. Flower-bud

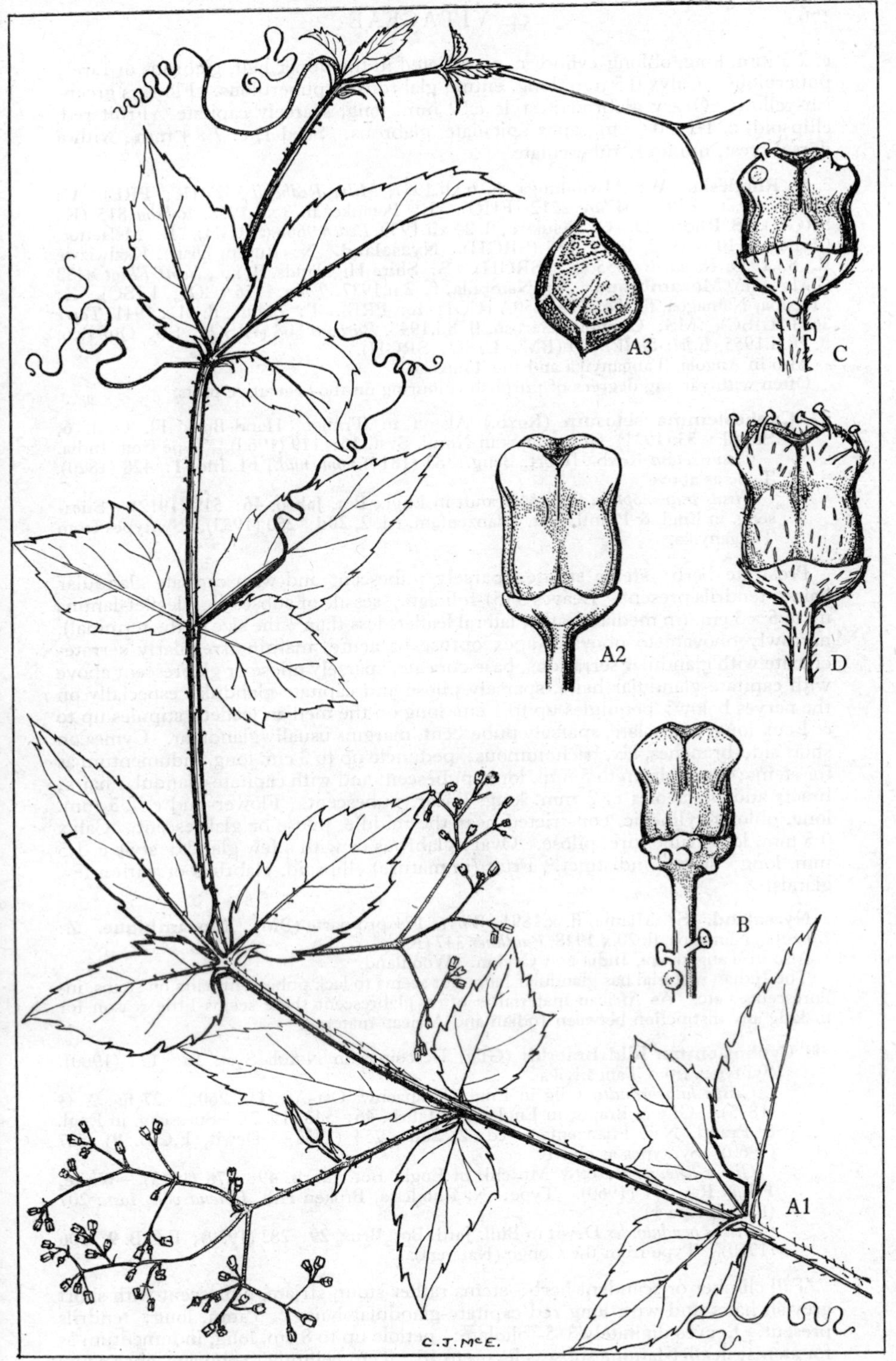

Tab. 97. A.—CYPHOSTEMMA SUBCILIATUM. A1, flowering branch ($\times\frac{2}{3}$) *Wild* 3883; A2, flower-bud ($\times 10$) *Wild* 3883; A3, seed ($\times 3$) *Torre* 1486. B.—CYPHOSTEMMA BORORENSE, flower-bud ($\times 8$) *Drummond* 5371. C.—CYPHOSTEMMA GIGANTOPHYLLUM, flower-bud ($\times 8$) *Wild* 3913. D.—CYPHOSTEMMA BUCHANANII, flower-bud ($\times 8$) *Eyles* 4620.

c. 2·5 mm. long, oblong-cylindric, constricted in the upper half, glabrous or rarely puberulous. Calyx 0·5 mm. long, entire, glabrous or puberulous. Flowers greenish-yellow. Ovary glabrous; style c. 2 mm. long, scarcely capitate. Fruit red, ellipsoid, c. 1·2 × 0·6 cm., apex apiculate, glabrous. Seed 1, c. 7 × 4 mm., with a dorsal crest, minutely tuberculate.

N. Rhodesia. W: Mwinilunga, fl. 6.xii.1937, *Milne-Redhead* 3527 (K; PRE). C: Lusaka, fr. 6.iii.1952, *White* 2212 (FHO). E: Petauke, fl. 3.xii.1958, *Robson* 815 (K; SRGH). **S. Rhodesia.** C: Selukwe, fl. 24.xii.1959, *Leach* 9666 (SRGH). E: Melsetter, Cashel, fl. 10.xi.1952, *Wild* 3883 (SRGH). **Nyasaland.** N: Rumpi Distr., Kaziwizwe R., 8.i.1959, *Richards* 1055 (K; SRGH). S: Shire Highlands, fl. xii., *Scott Elliot* 8482 (BM; K). **Mozambique.** N: Nampula, fl. 2.ii.1937, *Torre* 1176 (COI; LISC). Z: Mocuba, Namagoa, fl., *Faulkner* 250A (COI; K; PRE). T: Zóbuè, fl. 21.x.1941, *Torre* 3693 (LISC). MS: Chimoio, Garuso, fl. 8.i.1948, *Barbosa* 808 (LISC). SS: Quissico, fl. 28.ii.1955, *E.M. & W.* 699 (BM; LISC; SRGH).

Also in Angola, Tanganyika and the Transvaal.

Often with varying degrees of purplish colouring on the vegetative parts.

27. **Cyphostemma setosum** (Roxb.) Alston in Trimen, Hand-Book Fl. Ceyl. **6**, Suppl.: 53 (1931).—Descoings in Notul. Syst. **16**: 119 (1960). Type from India.

Cissus setosa Roxb., [Hort. Beng.: 83 (1814) *nom. nud.*] Fl. Ind. **1**: 428 (1820). Type as above.

Cissus psammophila Gilg & Brandt in Engl., Bot. Jahrb. **46**: 519 (1912).—Suesseng. in Engl. & Prantl, Nat. Pflanzenfam. ed. 2, **20d**: 250 (1953). Syntypes from Tanganyika.

Prostrate herb; stems striate, sparsely pubescent and with capitate glandular hairs; tendrils present. Leaves 3 (5)-foliolate; sessile or subsessile; leaflet-lamina up to 5 × 3 cm. (in median leaflet, lateral leaflets less than ½ the size of the terminal), narrowly obovate to obovate, apex obtuse or acute, margin irregularly serrate-crenate with glandular serrations, base cuneate, sparsely pilose or glabrescent above with capitate-glandular hairs, sparsely pilose and capitate-glandular especially on the nerves below; petiolules up to 1 cm. long on the median leaflet; stipules up to c. 1 cm. long, lanceolate, sparsely pubescent, margins usually glandular. Cymes on short side-branches, lax, trichotomous; peduncle up to 3 cm. long, indumentum as for stems; pedicels up to 5 mm. long, pubescent and with capitate-glandular hairs; bracts and bracteoles c. 2 mm. long, linear, pubescent. Flower-bud c. 2·5 mm. long, oblong-cylindric, constricted near the middle, pilose or glabrescent. Calyx 0·5 mm. long, subentire, pilose. Ovary glabrous or with a few glands; style c. 1·5 mm. long; stigma indistinct. Fruit (immature) ellipsoid, glabrous or with a few glands.

Nyasaland. S: Mlanje, fl. x.1891, *Whyte* 174 pro parte (BM). **Mozambique.** Z: Mocuba, Namagoa, fl. 20.x.1948, *Faulkner* 347 (K).

Also in Tanganyika, India and Ceylon. Woodland.

The Indian material has glandular hairs but seems to lack pubescent hairs on stems, inflorescence, etc. As African material is often glabrescent there seems little reason for making any distinction between Indian and African material.

28. **Cyphostemma hildebrandtii** (Gilg) Descoings in Notul. Syst. **16**: 122 (1960). Syntypes from Tanganyika.

Cissus hildebrandtii Gilg in Engl., Pflanzenw. Ost-Afr. **C**: 260, t. 27 fig. A–G (1895).—Gilg & Brandt in Engl., Bot. Jahrb. **46**: 543 (1912).—Suesseng. in Engl. & Prantl, Nat. Pflanzenfam. ed. 2, **20d**: 254 (1953).—Dewit, F.C.B. **9**: 507 (1960). Syntypes as above.

Cissus helenae Busc. & Muschl. in Engl., Bot. Jahrb. **49**: 478 (1913).—White, F.F.N.R.: 231 (1960). Type: N. Rhodesia, Broken Hill, *Helena von Aosta* 207 (B, holotype †).

Cissus penduloides Dewit in Bull. Jard. Bot. Brux. **29**: 283 (1959); F.C.B. **9**: 506 (1960). Type from the Congo (Katanga).

Tall climber or scandent herb; stems rather stout, striate, pubescent with short greyish hairs and with long red capitate-glandular hairs c. 5 mm. long; tendrils present. Leaves digitately 3–5-foliolate; petiole up to 8 cm. long, indumentum as for stems; leaflet-lamina subsessile, up to 15 × 8 cm., elliptic to obovate, apex acute or acuminate, margins crenate-serrate, base cuneate, pubescent or puberulous above, sometimes with stalked glands on the midrib, densely tomentose or densely pubescent below, sometimes with stalked glands near the base of the midrib, with sessile capitate glands concentrated towards the margins on both sides of the leaflet; stipules up to 2 cm. long, lanceolate-acuminate, pubescent and with some glands. Cymes on axillary branches, lax; peduncle c. 4 cm. long, indumentum as for stems; pedicels c. 3 mm. long, pubescent and with stalked capitate glands; bracts and bracteoles up to 1·3 cm. long, linear-lanceolate to filiform, pubescent. Flower-bud c. 3 mm. long, oblong-cylindric, constricted just above the middle, puberulous or glabrescent but with stalked capitate-glandular hairs at the apex. Calyx 0·5 mm. long, entire, pubescent. Ovary glandular-pubescent; style 1·5 mm. long. Fruit red, c. 10 × 7 mm., ellipsoid, pubescent and with long red capitate-glandular hairs. Seed 1, c. 1 × 0·6 cm., ellipsoid with a deep marginal furrow, rugose-tuberculate.

N. Rhodesia. N: Lake Bangweulu, Mwamfuli Point, fl. 14.ii.1959, *Watmough* 256 (K; SRGH). W: Chingola, fr. 17.iv.1954, *Fanshawe* 1096 (K; NDO). C: 48 km. E. of Lusaka, fl. 19.i.1961, *Best* 275 (SRGH). **Mozambique.** N: Mutuáli, fl. 12.xi.1954, *Gomes e Sousa* 4185 (COI; K; LMJ; PRE).
Also in Kenya, Tanganyika and the Congo. Forest edges or woodland.

Allied to *C. gigantophyllum* but distinguished by its long glandular hairs on the stems stalked glands on the corollas and longer bracteoles.

29. **Cyphostemma gigantophyllum** (Gilg & Brandt) Descoings in Notul. Syst. **16**: 121 (1960). TAB. **97** fig. C. Type from S. Tanganyika.
Cissus gigantophylla Gilg & Brandt in Engl., Bot. Jahrb. **46**: 453 (1912).—Suesseng. in Engl. & Prantl, Nat. Pflanzenfam. ed. 2, **20d**: 252 (1953). Type from S. Tanganyika.
Cissus leucadenia Suesseng., Mitt. Bot. Staatssamml. Münch. **6**: 184 (1953). Type: S. Rhodesia, Rusape, *Dehn* (M, holotype).

Scrambling or climbing herb, rarely erect; stems striate, densely and shortly pubescent, often with sessile capitate glands; tendrils present. Leaves digitately 3–5-foliolate, rarely 7-foliolate; petiole up to 20 cm. long but mostly shorter, indumentum as for stems; leaflet-lamina up to 20 × 8 cm. but usually smaller, elliptic to narrowly obovate-oblong, apex acute, margin from serrate to crenate, base cuneate, pubescent or glabrescent above, ± densely pubescent below or rarely subtomentose, with scattered sessile capitate glands usually concentrated near the apex of the leaflet, subsessile or shortly petiolulate; stipules up to c. 1 (2) cm. long, ovate-lanceolate or lanceolate, pubescent. Cymes lax, on short axillary branches; peduncle c. 5 cm. long, indumentum as for stems; pedicels up to 8 mm. long, densely pubescent, usually with sessile capitate glands and occasionally with shortly stalked capitate glands also; bracts and bracteoles up to 5 mm. long, linear-lanceolate, apex acuminate, pubescent. Flower-bud c. 2·5 mm. long, constricted above the middle, pubescent, sometimes with a few subsessile capitate glands. Calyx 0·5 mm. long, entire, pubescent. Petals cream with pink tips. Ovary glandular-pubescent; style c. 1·3 mm. long, apex scarcely capitate. Fruit scarlet, c. 1·2 × 0·8 cm., ellipsoid, pubescent with short hairs and with long red capitate glands up to c. 6 mm. long. Seed 1, c. 7·5 × 5 mm., ellipsoid, with a faint dorsal crest, otherwise almost smooth.

N. Rhodesia. C: Lusaka, fr. 6.iii.1952, *White* 2211 (FHO). S: Mazabuka, fl. 6.ii.1958, *Drummond* 5510 (K; PRE; SRGH). **S. Rhodesia.** N: Darwin Distr., Chiswiti Reserve, 22.i.1960, *Phipps* 2368 (SRGH). W: Shangani, fr. 1.iv.1954, *Davies* in GHS 46270 (SRGH). C: Prince Edward Dam, fl. 8.i.1952, *Wild* 3743 (SRGH). **Nyasaland.** S: Lower Kasupe, fr. 13.iii.1955, *E.M. & W.* 831 (BM; SRGH). **Mozambique.** Z: Mocuba-Régulo Namabida, fr. 2.vi.1949, *Barbosa & Carvalho* 2958 (LM; SRGH). MS: Chimoio, Garuso, fl. 5.i.1948, *Mendonça* 3587 (LISC).
Also in Tanganyika. Woodland.

The Zambezia specimen (*Barbosa & Carvalho* 2958) differs from all the other specimens in having 7-foliolate leaves.

30. **Cyphostemma barbosae** Wild & Drummond in Kirkia, **2**: 132 (1961). Type: Mozambique, Maputo, *Barbosa* 728 (LISC, holotype; LM).

Erect or suberect herb with a tuberous root; stems pubescent or glabrescent and with dense elongate red glandular hairs up to 1·4 cm. long; tendrils present or sometimes absent. Leaves digitately 3–7-foliolate; petiole up to 1·6 cm. long, indumentum as for stems; leaflet-lamina subsessile, up to 19 × 8 cm., broadly elliptic to circular, apex acuminate, base cuneate, margin crenate with apiculate teeth, sparsely and shortly capitate-glandular above, main nerves with long glandular capitate hairs below; stipules up to 2 cm. long, lanceolate-acuminate, ± falcate, pubescent and sparsely glandular at the margins. Cymes on axillary branches; peduncle up to 12 cm. long, indumentum as for stems; pedicels up to 1·2 cm. long, puberulous and with short capitate-glandular hairs; bracts and bracteoles up to 1·2 cm. long, linear-lanceolate, acuminate, margin pubescent and sparsely glandular. Flower-bud c. 3·5 mm. long, oblong-cylindric, constricted near the middle, pubescent and with capitate glandular hairs at the apex. Calyx c. 0·7 mm. long, subentire, densely pubescent. Ovary glandular; style 2 mm. long, apex scarcely capitate. Fruit red, c. 1 × 0·7 cm., ellipsoid, with long glandular hairs. Seed 1, c. 8 × 5 mm., ± smooth.

Mozambique. LM: Namaacha, fr. 22.xii.1944, *Torre* 6930 (LISC).
Not known outside the Lourenço Marques District. Open woodlands and dry stony places.

This species is near *C. woodii* Gilg & Brandt, the most obvious difference being that the latter species does not possess the elongate glandular stem hairs up to 1·4 cm. long of *C. barbosae*. *C. woodii* was recorded by Gilg & Brandt (Engl., Bot. Jahrb. **46**: 493 (1912)) as coming from Delagoa Bay in Mozambique but the specimen he cites, *Junod* 1313 (Z), is actually from the Transvaal (Forêt de Marovougne which is near Shilouvane, south of Leydsdorp).

31. **Cyphostemma lynesii** (Dewit) Descoings in Notul. Syst. **16**: 122 (1960). Type from the Congo (Katanga).
Cissus lynesii Dewit in Bull. Jard. Bot. Brux. **29**: 285 (1959); F.C.B. **9**: 496 (1960). Type as above.

Climbing or scandent herb; stems striate, puberulous and with dense rather short capitate glands (drying rather dark); tendrils present. Leaves digitately 3-foliolate, very rarely 5-foliolate; petiole up to 10 cm. long, indumentum as for stems; leaflet-lamina sessile, up to 12 × 7·5 cm., elliptic to obovate, apex acute to acuminate, base cuneate, lateral leaflets very asymmetric at the base, margin crenate-dentate, sparsely pubescent on both sides and with numerous capitate-glandular hairs; stipules up to c. 1 cm. long, ovate-lanceolate, acuminate, densely pubescent and with short capitate-glandular hairs. Cymes lax, on axillary branches, trichotomous; peduncles up to 5 cm. long, indumentum as for stems; pedicels c. 3 mm. long, pubescent and with dense short capitate glands; bracts and bracteoles c. 2 mm. long, linear, pubescent and glandular. Flower-bud up to 3 mm. long, oblong-cylindric, constricted near the middle, glabrescent as far as eglandular hairs are concerned but densely covered with short capitate-glandular hairs. Calyx 0·5 mm. long, subentire, puberulous and with dense capitate glands. Ovary glandular; style c. 1·3 mm. long, apex subcapitate. Fruit black, up to 7 × 4 mm., ellipsoid, pubescent and with dense capitate-glandular hairs. Seed 1, c. 4 × 3 mm., obovoid, with dorsal crest and 2 lateral lines of tubercles, strongly rugose.

N. Rhodesia. B: Sesheke, fl. 22.i.1952, *White* 1948 (FHO). W: Ndola, fl. 13.ii.1954, *Fanshawe* 804 (K; NDO; SRGH). E. Luangwa R., fr. 25.iii.1955, *E.M. & W.* 1185 (BM; SRGH). **S. Rhodesia.** E. Stapleford Forestry Reserve, Nyumquarara Valley, fl. ii.1935, *Gilliland* 1530 (BM; K). **Nyasaland.** S. Likabula Forest Station, fl. & fr. 19.ii.1957, *Chapman* 441 (BM; FHO; SRGH). **Mozambique.** Z. Milange, fl. 24.ii.1943, *Torre* 4823 (LISC). MS: between R. Revuè & R. Munhinga, fr. 27.iv.1948, *Barbosa* 1605 (LISC).
Also in Tanganyika and the Congo. Termite mounds, rocky hills and river sides.

32. **Cyphostemma bambuseti** (Gilg & Brandt) Descoings in Notul. Syst. **16**: 120 (1960). Type from the Congo.
Cissus bambuseti Gilg & Brandt in Engl., Bot. Jahrb. **46**: 544 (1912); in Mildbr., Deutsch. Zentr.-Afr.-Exped. 1907–8, **2**: 493 (1912).—Suesseng. in Engl. & Prantl, Nat. Pflanzenfam. ed. 2, **20d**: 254 (1953).—Dewit, F.C.B. **9**: 508 (1960). Type as above.

Climber up to c. 6 m. tall; stems pubescent and with rather short-stalked capitate glandular hairs; tendrils present. Leaves digitately 3–5-foliolate; petiole 4 (6) cm. long, indumentum as for stems; leaflet-lamina up to 8 (10) ×

4(5) cm., obovate-elliptic or elliptic, apex acuminate, margin serrate, base cuneate or ± rounded on lateral leaflets, sparsely pubescent or glabrous above, tomentellous or densely pubescent below; petiolules up to 7 mm. long; stipules up to 1 cm. long, lanceolate-acuminate, pubescent. Cymes lax, on axillary branches; peduncle 2 (10) cm. long, densely pubescent and with short capitate glands; pedicels c. 2 mm. long, densely pubescent and with short capitate glands; bracts and bracteoles c. 1 mm. long, lanceolate, tomentose. Flower-bud c. 4·5 mm. long, cylindric, swollen at the apex, pubescent and with short capitate-glandular hairs on the sides and near the apex. Calyx 0·5 mm. long, entire, pubescent and with some stalked capitate-glandular hairs. Ovary glandular-pubescent; style 2·5 mm. long; apex scarcely capitate. Fruit c. 1 × 0·6 cm., ellipsoid, densely pubescent and with capitate-glandular hairs. Seed 1, c. 6 × 4 mm., ellipsoid, with a dorsal crest at the bottom of a furrow with two lateral crests, pitted.

N. Rhodesia. N: Kasuto, fl. 2.iii.1955, *Richards* 4745 (K).
Also in Kenya, Uganda and the Congo. Forest edges and thickets.

33. **Cyphostemma glandulosissimum** (Gilg & Brandt) Descoings in Notul. Syst. **16**: 121 (1960). Type: Nyasaland, Mt. Malosa, *Whyte* (K).
Vitis cirrhosa sensu Bak. in Oliv., F.T.A. **1**: 410 (1868) pro parte excl. specim. Kirk.
Cissus kirkiana var. *livingstonei* Planch. in A. & C. DC., Mon. Phan. **5,** 2: 601 (1887) (" livingstonii "). Syntypes from Uganda and Nyasaland, Zomba Mt., *Buchanan* 278 (K, holotype).
Cissus glandulosissima Gilg & Brandt in Engl., Bot. Jahrb. **46**: 525 (1912).—Suesseng. in Engl. & Prantl, Nat. Pflanzenfam. ed. 2, **20d**: 251 (1953). Type as above.

Climbing or procumbent herb; stems striate, densely clothed with ± long capitate glandular hairs; tendrils present. Leaves digitately 5-foliolate; petiole up to 7 cm. long, indumentum as for the stems; leaflet-lamina up to 10 × 6 cm., elliptic, apex acute, margin crenate, base cuneate, glabrous above, pubescent on the nerves below and with scattered capitate glands, sessile or subsessile; stipules up to 3 × 2 cm., ovate- or lanceolate-acuminate, margins pubescent and glandular. Cymes on axillary branches, trichotomous; peduncle c. 3 cm. long, pubescent and with glandular hairs; pedicels up to c. 4 mm. long, pubescent and with numerous rather short capitate-glandular hairs; bracts and bracteoles membranous, up to 2 × 0·8 cm., oblong-lanceolate, margins pubescent and often glandular. Flower-bud c. 2·7 mm. long, oblong-cylindric, constricted near the middle, with a single capitate stalked gland near the apex of each petal. Calyx 0·5 mm. long, entire, pubescent. Ovary pubescent and with capitate glands; style 2 mm. long, apex minutely subcapitate. Fruit 1 × 0·8 cm., ellipsoid, pubescent and with shortly stalked capitate glands. Seed 1, c. 7 × 4·5 mm., obovoid-ellipsoid, with a faint dorsal crest and two strong lateral crests, otherwise almost smooth.

N. Rhodesia. W: Mwinilunga, Kaoomba R., fl. 22.xii.1937, *Milne-Redhead* 3777 (K). **Nyasaland.** S: Shire Highlands, *Buchanan* 278 (K). **Mozambique.** Z: Ile, fr. 26.vi.1943, *Torre* 5585 (LISC).
Also in Uganda. On rocky hills (more precise ecology not known).
Robinson 4252 (SRGH) from Kasama (NR: N) is very near this species but has shorter bracteoles and more glands on the corolla. More material is needed.

34. **Cyphostemma elisabethvilleanum** (Dewit) Descoings in Notul. Syst. **16**: 121 (1960). Type from the Congo (Katanga).
Cissus elisabethvilleana Dewit in Bull. Jard. Bot. Brux. **29**: 268 (1959); F.C.B. **9**: 486 (1960). Type as above.

Suberect, prostrate or scrambling herb; stems up to c. 0·6 m. long, pubescent and with slightly longer capitate-glandular hairs; tendrils present or often absent. Leaves digitately 3-foliolate; petiole up to 1·5 cm. long, indumentum as for stems; leaflet-lamina up to 7·5 × 1·7 cm., narrowly oblong to elliptic, apex acute or subacute, margin serrate, base cuneate or ± asymmetrically rounded in the lateral leaflets, strigose-pilose on both sides, some of the hairs gland-tipped; petiolule up to 0·5 cm. long but leaflets often subsessile; stipules c. 5 mm. long, lanceolate-acuminate, puberulous, glandular-ciliate. Cymes on axillary branches; peduncle up to 5 cm. long, indumentum as for stems; pedicels up to 3 mm. long, pubescent or glabrescent and with some capitate-glandular hairs; bracts and bracteoles c. 2 mm. long, linear, pubescent. Flower-bud c. 2 mm. long, cylindric, constricted near the middle, glabrous. Calyx 0·25 mm. long, entire, puberulous or glabrous. Ovary glabrous; style c. 0·75 mm. long, apex scarcely subcapitate. Fruit red, up to 7 × 5 mm., ellipsoid-globose, with a few scattered capitate glands. Seed 1, c. 0·6 mm. in diam., subglobose, with two parallel crests and a few tuberculate ridges.

N. Rhodesia. B: Mankoya, fl. & fr. 22.ii.1952, *White* 2122 (FHO). N: Abercorn Distr., Kawimbe, fr. 24.i.1957, *Richards* 7965 (K; SRGH). W: Ndola, fl. & fr. 10.i.1954, *Fanshawe* 663 (K; NDO).

Also in Tanganyika and the Congo. *Brachystegia* and other woodlands.

35. **Cyphostemma kirkianum** (Planch.) Descoings in Notul. Syst. **16**: 122 (1960). Syntypes: Mozambique, near Tete, *Kirk* (K); between Sena and Lupata, *Kirk* (K).

Vitis cirrhosa sensu Bak. in Oliv., F.T.A. **1**: 410 (1868) pro parte quoad specim. Kirk.

Cissus kirkiana Planch. in A. & C. DC., Mon. Phan. **5**, 2: 601 (1887).—Gilg & Brandt in Engl., Bot. Jahrb. **46**: 529 (1912).—Suesseng. in Engl. & Prantl, Nat. Pflanzenfam. ed. 2, **20d**: 252 (1953). Syntypes as above.

Vitis paucidentata sensu Kuntze, Rev. Gen. **3**, 2: 41 (1898).

Climbing herb, stems sulcate-striate, with ± sparse long glandular hairs, sometimes sparsely puberulous also; tendrils present. Leaves 3–5-foliolate; petioles up to 5 cm. long, indumentum as for stems or occasionally quite glabrous; leaflet-lamina membranous, up to 12 × 6·3 cm., elliptic, apex acuminate, margin crenate-serrate, base cuneate, glabrous on both sides but often with a few stalked glands on the midrib; petiolule up to 1 cm. long in the median leaflet, puberulous; stipules up to 2 cm. long, ovate-lanceolate, glabrescent. Cymes lax, axillary; peduncle up to 5 cm. long, puberulous and with long scattered glandular hairs; pedicels up to 1 cm. long, densely pubescent; bracts and bracteoles up to 7 mm. long, lanceolate or linear-lanceolate, puberulous at least on the margins. Flower-bud c. 2 mm. long, oblong-cylindric, puberulous. Calyx 0·5 mm. long, entire, puberulous. Petals greenish, orange-tipped. Ovary puberulous; style 2 mm. long, apex capitate. Fruit c. 1 × 0·7 cm., puberulous. Seed 1, c. 7 × 4 mm., ellipsoid, with a dorsal crest, rugose.

N. Rhodesia. S: Mazabuka, fl. 1.ii.1958, *Drummond* 5428 (K; PRE; SRGH). **S. Rhodesia.** N: Urungwe, Zambezi Valley, Rifa R., fl. 25.2.1953, *Wild* 4079 (K; SRGH). Mozambique. Z: Mopeia, Chamo, fl. iii.1859, *Kirk* (K). T: near Tete, fl. ii.1860, *Kirk* (K). MS: Chimoio, Serra de Chindaza, fr. 27.iii.1948, *Garcia* 759 (LISC).

Also in Tanganyika. Thickets, woodland and riverine fringes.

36. **Cyphostemma buchananii** (Planch.) Descoings in Notul. Syst. **16**: 120 (1960). TAB. **97** fig. D. Type: Nyasaland, Shire Highlands, *Buchanan* 279 (K, holotype).

Cissus buchananii Planch. in A. & C. DC., Mon. Phan. **5**, 2: 601 (1887) pro parte excl. Welw. 1469 & 1492.—Bak. f. in Journ. Linn. Soc., Bot. **40**: 46 (1911) pro parte.—Gilg & Brandt in Engl., Bot. Jahrb. **46**: 538 (1912).—Eyles in Trans. Roy. Soc. S. Afr. **5**: 408 (1916).—Suesseng. in Engl. & Prantl, Nat. Pflanzenfam. ed. 2, **20d**: 253 (1953).—Dewit, F.C.B. **9**: 475 (1960). Type as above.

Climbing herb; drying black; stems with a short pubescence with slightly longer capitate-glandular hairs; tendrils present. Leaves digitately 3–5-foliolate; petioles up to c. 5 cm. long, indumentum as for stems; leaflet-lamina up to 11 × 7 cm., elliptic to narrowly obovate, apex acute, margin crenate-serrate, base usually somewhat rounded on the lateral leaflets, pubescent or glabrescent above, tomentose or pubescent below, often with a few capitate-glandular hairs towards the base; petiolules up to 1·5 cm. long, indumentum as for stems; stipules up to c. 7 mm. long, lanceolate-acuminate, ± falcate, pubescent and with capitate-glandular hairs. Cymes lax on axillary side-branches; peduncle up to c. 7 cm. long, indumentum as for stems; pedicels c. 2 mm. long, pubescent and with some capitate-glandular hairs; bracts and bracteoles c. 2 mm. long, linear-lanceolate, pubescent and with some capitate-glandular hairs. Flower-bud up to 3·5 mm. long, oblong-cylindric, constricted above the middle, pubescent and sometimes with a few capitate-glandular hairs. Calyx 0·5 mm. long, densely pubescent. Petals reddish. Ovary glandular-pubescent; style 1·5 mm. long, apex scarcely subcapitate. Fruit red, ellipsoid, apiculate, pubescent and with short capitate-glandular hairs. Seed c. 7 × 5 mm., ellipsoid, with a faint dorsal crest, minutely tuberculate.

Bechuanaland Prot. N: Francistown, fl. i.1926, *Rand* (BM). **N. Rhodesia.** B: Kabompo, fr. 24.iii.1961, *Drummond & Rutherford-Smith* 7263 (K; PRE; SRGH). N: Abercorn, Mpulungu, fl. 12.ii.1957, *Richards* 8179 (K; SRGH). E: Fort Jameson, fr. iii.1950, *Gilges* 33 (PRE; SRGH). **S. Rhodesia.** N: Urungwe, Zwipani, fl. 5.ii.1958, *Goodier* 523 (SRGH). W: Bulawayo, fl. xii.1897, *Rand* (BM). C: Rusape, fl. & fr. 23.viii.1939, *Hopkins* in GHS 7039 (SRGH). E: Chirinda, fl. xii.1937, *Obermeyer* 2216 (PRE; SRGH). S: S. of Lundi R., Beitbridge road, fr. 15.ii.1955, *E.M. & W.* 379 (BM; SRGH). **Nyasaland.** C: Kota Kota, fr. ii.1944, *Benson* 720 (PRE). S: Mlanje, fl. 1919, *Shinn* (BM). **Mozambique.** N: Nampula, fl. & fr. 10.iv.1961, *Balsinhas & Marrime* 366 (LMJ; SRGH). Z: Mocuba, Namagoa, fl. 26.i.1948, *Faulkner* 250 (K). T: Boroma, fl. iii.1892, *Menyhart* 934 (K). MS: Chimoio, Garuso Mt., fr. 3.iii.1948, *Barbosa* 1093 (LISC). SS: Inhachengo, fl. 26.ii.1955, *E.M. & W.* 632 (BM; LISC; SRGH).

Also in Tanganyika, Zanzibar, Congo and the Transvaal. Forest edges or woodland.

The material from the midlands of S. Rhodesia often has a few capitate-glandular hairs on the petals but they are often only on a small proportion of the flowers and so this form does not seem to constitute a distinct taxon. Very occasional glands can be found on specimens from other localities (e.g. *Schlieben* 1607 (BM) from Tabora in Tanganyika). *Drummond & Rutherford-Smith* 6936 (K; SRGH) from Chingola (NR: W) is somewhat anomalous in that it does not dry very dark and is somewhat glabrescent for this species but it probably belongs here.

37. **Cyphostemma simulans** (C.A. Sm.) Wild & Drummond in Kirkia, 2: 140 (1961). Type from the Transvaal (White River).
Cissus simulans C.A. Sm. in Burtt Davy, F.P.F.T. 2: xx, 476 (1932). Type as above.

Prostrate or climbing herb; stems shortly pubescent and with capitate-glandular hairs; tendrils present. Leaves digitately 3–5-foliolate; petiole up to 4·5 cm. long; indumentum as for stems; leaflet-lamina up to 10 × 7 cm., narrowly elliptic to broadly obovate, apex acute, margins serrate, base cuneate, very sparsely pubescent above, pubescent and glandular at least on the nerves below; petiolule up to 1·5 cm. long; stipules c. 5 mm. long, falcate, lanceolate, pubescent and with a few glands. Cymes opposite the leaves, trichotomous, lax; peduncle up to 10 cm. long, indumentum as for stems; pedicels 2–3 mm. long, pubescent and with short capitate-glandular hairs; bracts and bracteoles 1–2 mm. long, lanceolate-subulate, pubescent. Flower-bud c. 3 mm. long, oblong-cylindric, constricted near the middle, pubescent and with capitate-glandular hairs near the apex. Calyx 0·5 mm. long, ± entire, densely pubescent. Ovary glandular-pubescent; style 1·5 mm. long, apex minutely 2-fid. Fruit c. 9 × 7 mm., ellipsoid, pubescent and with capitate-glandular hairs. Seed 1.

S. Rhodesia. S: Ndanga, Triangle, fl. & fr. 22.xii.1951, *Wild* 3714 (SRGH). **Nyasaland.** S: Shire Highlands, fl. & fr. 6.vii.1879, *Buchanan* 163 (K). **Mozambique.** Z: Mocuba, Namagoa, fl. vii.1943, *Faulkner* 50 (BM; EA; PRE).

Also in the Transvaal. Open woodland. Near *C. buchananii* but does not dry black and the glandular hairs, especially on the fruits, are much shorter. The Namagoa specimens have very well-developed petiolules whilst the type has sessile or subsessile leaflets. There are intermediates, however.

38. **Cyphostemma milleri** Wild & Drummond in Kirkia, 2: 135 (1961). Type: S. Rhodesia, Matopos, *Miller* 5778 (K; SRGH, holotype).

Climbing herb; stems sparsely pubescent or glabrescent; tendrils present. Leaves digitately 3–5-foliolate; petiole up to 6 cm. long, with a few capitate-glandular hairs; leaflet-lamina membranous, up to 12 × 8 cm., broadly elliptic to rotund, apex acute or obtuse, margins coarsely serrate, base cuneate, glabrous above, glabrous below except for a few glandular hairs on the nerves; petiolules up to 2 cm. long, very sparsely glandular; stipules up to 8 mm. long, narrowly lanceolate, falcate, puberulous. Cymes on short side-branches, lax; peduncle up to 5 cm. long, sparsely pubescent; pedicels up to 6 mm. long, pubescent and with capitate-glandular hairs; bracts and bracteoles c. 1 mm. long, subulate, pubescent. Flower-bud c. 2·5 mm. long, oblong-cylindric, constricted near the middle, pubescent and with capitate-glandular hairs near the apex. Calyx 0·5 mm. long, subentire, pubescent. Ovary glabrous; style c. 0·8 mm. long; stigma indistinct. Fruit yellow, c. 9 × 7 mm., ellipsoid, glabrous. Seed 1, c. 7 × 5 mm., obovoid, slightly rugose.

S. Rhodesia. W: Matobo Distr., Besna Kobila, fr. v.1955, *Miller* 2851 (SRGH). Not known elsewhere. Woodland.

39. **Cyphostemma masukuense** (Bak.) Descoings in Notul. Syst. **16**: 123 (1960). Type: Nyasaland, Misuku plateau near Karonga, *Whyte* (K, holotype).
Vitis masukuensis Bak. in Kew Bull. **1897**: 249 (1897). Type as above.
Cissus buchananii sensu Bak. f. in Journ. Linn. Soc., Bot. **40**: 46 (1911) pro parte quoad *Swynnerton* 231 pro parte.
Cissus masukuensis (Bak.) Gilg & Brandt in Engl., Bot. Jahrb. **46**: 528 (1912).—Suesseng, in Engl. & Prantl, Nat. Pflanzenfam. ed. 2, **20d**: 252 (1953). Type as above.

Climbing herb; stems pubescent and with a few capitate-glandular hairs; tendrils present. Leaves digitately 3–5-foliolate; petiole up to 6 cm. long, indumentum as for stems; leaflet-lamina up to 12 × 6 cm., broadly elliptic, apex acute or acuminate, margin serrate-crenate, base ± rounded, sparsely pilose above, densely pubescent below; petiolules up to 7 mm. (19 mm. in median leaflet) long; stipules c. 8 mm. long, caducous, lanceolate-acuminate, ± falcate, pubescent. Cymes on lateral branches; peduncle up to 6 cm. long, pubescent; pedicels c. 2 mm. long, pubescent; bracts and bracteoles c. 0·7 mm. long, deltate, pubescent. Flower-bud c. 3·5 mm. long, oblong-cylindric, constricted at the middle, pubescent. Calyx 0·5 mm. long, subentire, pubescent. Ovary glandular-pubescent; style c. 2·2 mm. long, apex minutely 2-fid. Fruits (immature) ellipsoid, pubescent with short capitate glands. Seed 1.

S. Rhodesia. E: Chirinda Forest, fl. 29.iii.1950, *Hack* 177/50 (SRGH). **Nyasaland.** N: Misuku Plateau, fl. vii.1896, *Whyte* (K).
Also in southern Tanganyika. Climbing in forest or at forest edges.

40. **Cyphostemma mildbraedii** (Gilg & Brandt) Descoings in Notul. Syst. **16**: 123 (1960). Type from Rwanda-Burundi.
Cissus mildbraedii Gilg & Brandt in Engl., Bot. Jahrb. **46**: 501 (1912); in Mildbr., Deutsch. Zentr.-Afr.-Exped. 1907–8, **2**: 492 (1914).—Suesseng. in Engl. & Prantl, Nat. Pflanzenfam. ed. 2, **20d**: 244 (1953).—Dewit, F.C.B. **9**: 498, t. 50 (1960). Type as above.
Cissus centrali-africana Gilg & Fr. in R.E.Fr., Schwed. Rhod.-Kongo-Exped. **1**: 136 (1914).—Suesseng., tom. cit.: 245 (1953). Type: N. Rhodesia, Kalambo Falls, *Fries* 1352 (UPS, holotype).
Cissus termetophyla De Wild. in Fedde, Repert. **13**: 202 (1914); in Ann. Soc. Sci. Brux. **38**, 2: 390 (1914).—Dewit, tom. cit.: 477 (1960). Type from the Congo (Katanga).

Climbing or prostrate herb, stems striate, pubescent and with capitate-glandular hairs; tendrils present or sometimes absent. Leaves digitately 3–5 (7)-foliolate; petiole up to 6·5 (9) cm. long, indumentum as for stems; leaflet-lamina up to 12 × 6 cm., narrowly elliptic, oblanceolate or obovate, apex obtuse or acute, margin serrate-crenate, base cuneate, sparsely pubescent above, somewhat more densely pubescent or tomentose below especially on the nerves, with some capitate glands near the base and on the midrib; petiolules up to 1·5 cm. long; stipules up to 1·6 (2) cm. long, lanceolate-acuminate, ± falcate, pubescent. Cymes lax, on axillary branches; peduncle up to 20 cm. long, indumentum as for stems; pedicels 2–4 mm. long, pubescent; bracts and bracteoles up to 3 mm. long, filiform, pubescent. Flower-bud c. 2·3 mm. long, oblong-cylindric, constricted near the middle, puberulous to densely pubescent, often with capitate glands at the apex of the petals. Calyx 0·5 mm. long, entire, densely pubescent. Ovary pubescent; style 1·5 mm. long, apex scarcely capitate. Fruit reddish-purple, c. 8 × 6 mm., ellipsoid, pubescent and sometimes with a few short glands. Seed 1, c. 7 × 5 mm., ellipsoid, with a faint dorsal crest and two lateral crests, ± rugose.

N. Rhodesia. B: 32 km. NE. of Mongu, fl. 10.xi.1959, *Drummond & Cookson* 6295 (K; PRE; SRGH). N: Kalambo, fl. & fr. 27.xi.1911, *Fries* 1352 (UPS). W: Ndola, fr. 18.ii.1956, *Fanshawe* 2780 (K; NDO; SRGH). S: Mapanza Mission, fl. 16.i.1953, *Robinson* 52 (K).

Also in Rwanda-Burundi and the Congo. Grassland, Chipya savannas, termitaria or *Brachystegia* woodland.

C. mildbraedii is unusual for a *Cyphostemma* species in that it appears to show considerable variation in two characters that are normally fairly constant in this genus, i.e., the presence or absence of tendrils and the presence or absence of glands on the corolla.

41. **Cyphostemma montanum** Wild & Drummond in Kirkia, **2**: 135 (1961). Type: S. Rhodesia, Himalayas, Engwa, *E.M. & W.* 77 (BM; SRGH, holotype).

Scandent or climbing herb; stems sparsely pubescent and with a few capitate-glandular hairs; tendrils present. Leaves digitately 3–5-foliolate; petiole up to 5 cm. long, indumentum as for stems; leaflet-lamina up to 7 × 4·5 cm., sessile or subsessile, elliptic to obovate, apex acuminate, margin crenate-serrate, base cuneate, glabrous above, sparsely pubescent and with capitate-glandular hairs on the nerves below; stipules c. 7 mm. long, lanceolate-acuminate, ± falcate, pubescent. Cymes on axillary branches; peduncle up to 8 cm. long, pubescent and often with some capitate-glandular hairs; pedicels up to 2 mm. long, pubescent or puberulous; bracts and bracteoles 1–2 mm. long, filamentous, pubescent. Flower-bud up to 2 mm. long, oblong-cylindric, constricted near the middle, slightly pubescent at the petal margins. Calyx 0·5 mm. long, subentire, puberulous. Ovary glabrous; style c. 1·2 mm. long, apex minutely 2-fid. Fruit c. 7 × 5 mm., ellipsoid, glabrous. Seed 1, c. 6 × 5 mm., ellipsoid, with a dorsal and two lateral crests, rugose.

S. Rhodesia. E: Vumba, Norseland, fl. 8.iii.1946, *Chase* 991 (BM; SRGH).
Also in S. Tanganyika. Common at about 2000 m. on rocky crags from Inyanga southwards to the Chimanimani Mts.

42. **Cyphostemma kaessneri** (Gilg & Brandt) Descoings in Notul. Syst. **16**: 122 (1960). Type from the Congo (Katanga).
Cissus kaessneri Gilg & Brandt in Engl., Bot. Jahrb. **46**: 528 (1912).—Suesseng. in Engl. & Prantl, Nat. Pflanzenfam. ed. 2, **20d**: 252 (1953).—Dewit, F.C.B. **9**: 478 (1960). Type as above.

Climbing herb; stems striate, puberulous; tendrils present. Leaves digitately 5-foliolate; petiole up to 5 cm. long, puberulous; leaflet-lamina up to 8 × 4 cm., apex acute or subacute, margin serrate-crenate, base cuneate, glabrous except sometimes on the midrib above, puberulous below, more densely so on the nerves; petiolules up to 10 mm. long, puberulous; stipules up to 7 × 3 mm., lanceolate-falciform, puberulous. Cymes lax, on axillary branches, trichotomous; peduncle up to 7 cm. long, puberulous; pedicels c. 3 mm. long, densely pubescent; bracts and bracteoles c. 1 mm. long, linear, pubescent. Flower-bud 2–3·5 mm. long, oblong-cylindric, constricted above the middle, densely puberulous. Calyx 0·5 mm. long, entire, puberulous. Petals cream with reddish tips. Ovary minutely puberulous; style up to 2·5 mm. long, apex scarcely capitate. Ripe fruits not known.

N. Rhodesia. B: Sesheke, fl. 23.xii.1952, *Angus* 1006 (BM; FHO). W: Kitwe, fl. 14.ii.1954, *Fanshawe* 816 (FHO; K; NDO). S: Mazabuka, fl. 12.i.1952, *White* 1901 (FHO).
Also in the Congo. Riverine fringes or on termitaria in *Brachystegia* woodland.

43. **Cyphostemma lanigerum** (Harv.) Descoings in Notul. Syst. **16**: 122 (1960). Syntypes from the Transvaal and Natal.
Cissus lanigera Harv., Thes. Cap. **1**: 41, t. 65 (1859); in Harv. & Sond., F.C. **1**: 252 (1860).—Gilg & Brandt in Engl., Bot. Jahrb. **46**: 542 (1912).—Suesseng. in Engl. & Prantl, Nat. Pflanzenfam. ed. 2, **20d**: 254 (1953). Syntypes as above.
Cissus lanigera var. *holubii* Planch. in A. & C. DC., Mon. Phan. **5**, 2: 600 (1887). —Burtt Davy, F.P.F.T. **2**: 476 (1932). Syntypes from S. Africa.

Climbing herb; stems densely pubescent or tomentose and usually with some scattered capitate glands; tendrils present. Leaves 3–5-foliolate; petiole up to 2 cm. long, indumentum as for stems; leaflet-lamina subsessile, up to 5·5 × 3 cm., narrowly obovate to obovate, apex obtuse or acute, margin serrate or crenate-serrate, base cuneate, sparsely pubescent above, greyish-tomentose below; stipules up to 1 cm. long, ovate-falcate, densely pubescent. Cymes opposite the leaves, lax, di- or trichotomous; peduncle up to 7 cm. long, indumentum as for stems; pedicels pubescent and with capitate glandular hairs; bracts and bracteoles up to 2 mm.

long, subulate, pubescent. Flower-bud up to 3·5 mm. long, oblong-cylindric, ± constricted near the middle, densely pubescent and with a few glandular-capitate hairs at the apex (not all the flowers in each inflorescence possess glands). Calyx tomentose. Stigma minutely subcapitate. Ripe fruit not seen.

S. Rhodesia. C: Selukwe, fl. 25.xii.1959, *Leach* 9669 (K; PRE; SRGH). E: Odzani R., fl. 1915, *Teague* 515 (K).
Also in the Transvaal and Natal. Recorded in Rhodesia from granite kopjes.

44. **Cyphostemma puberulum** (C.A. Sm.) Wild & Drummond in Kirkia, **2**: 140 (1961). Type: Bechuanaland Prot., Bamangwato Country, *Holub* (K, holotype).
Cissus lanigera var. *sosong* Planch. in A. & C. DC., Mon. Phan. **5,** 2: 600 (1887) (" sosang ").—Gilg & Brandt in Engl., Bot. Jahrb. **46**: 542 (1912). Type as above.
Cissus puberula C.A. Sm. in Burtt Davy, F.P.F.T. **2**: xx, 476 (1932). Type as above.

Climbing herb; stems shortly grey-pubescent; tendrils present. Leaves 3–5-foliolate; petiole up to 1 (4) cm. long, grey-pubescent; leaflet-lamina slightly fleshy, up to 8 × 2·3 (4·5) cm., narrowly elliptic to obovate, apex obtuse or acute, margin serrate-crenate, base cuneate, puberulous or glabrescent above, densely pubescent below, petiolules 0–2 mm. long (up to 1 cm. on central leaflet); stipules c. 5 mm. long, lanceolate- or ovate-falcate, densely pubescent outside. Inflorescence on short side-branches or leaf-opposed, trichotomous, lax; peduncle up to 7 cm. long, densely pubescent; pedicels up to c. 2 mm. long, pubescent and with short capitate-glandular hairs; bracts and bracteoles 1–2 mm. long, subulate, pubescent. Flower-bud up to 3 mm. long, oblong-cylindric, ± constricted near the middle, densely pubescent. Calyx 0·5 mm. long, subentire, pubescent. Ovary pubescent; style up to 2 mm. long, apex subcapitate. Fruit c. 8 × 6 mm., subglobose, pubescent. Seed 1, c. 7 × 5 mm., subglobose, with a dorsal keel, tuberculate-rugose.

Bechuanaland Prot. SE: Bamangwato Reserve, *Holub* (K). **S. Rhodesia.** E: Sabi R., E. bank near Chibuwe, fr. 11.v.1959, *Savory* 451 (SRGH). S: 24 km. N. of Sabi-Lundi junction, fr. 9.vi.1950, *Chase* 2452 (BM; SRGH).
Also in the Transvaal. Often along river banks.

Nearly related to *C. lanigera,* but has no glands on the corolla, has subfleshy leaves and is confined to hotter and drier areas.

45. **Cyphostemma hypoleucum** (Harv.) Descoings in Notul. Syst. **16**: 122 (1960). Syntypes from Natal.
Cissus hypoleuca Harv. in Harv. & Sond., F.C. **1**: 252 (1860).—Planch. in A. & C. DC., Mon. Phan. **5,** 2: 598 (1887).—Gilg & Brandt in Engl., Bot. Jahrb. **46**: 543 (1912).—Suesseng. in Engl. & Prantl, Nat. Pflanzenfam. ed. 2, **20d**: 254 (1953). Syntypes as above.

Climbing herb; stems pubescent; tendrils present. Leaves 3–5-foliolate; petioles 3·5(5) cm. long, pubescent; leaflet-lamina up to 10 × 6 cm., elliptic to obovate or rarely broadly obovate, apex obtuse or acute, margin crenate or serrate-crenate, base cuneate, sparsely pubescent or glabrescent above, grey-tomentose below; petiolule of central leaflet up to 1 cm. long, petiolules of lateral leaflets 0–2·5 mm. long; stipules up to c. 7 mm. long, ovate-falcate, densely pubescent outside. Cymes on short side-branches, lax, trichotomous; peduncle up to 7 cm. long, densely grey-pubescent; pedicels up to 3·5 mm. long, grey-tomentose or densely pubescent; bracts and bracteoles c. 2 mm. long, subulate, pubescent. Flower-bud c. 3 mm. long, oblong-cylindric, ± constricted near the middle, densely pubescent. Calyx 0·5 mm. long, subentire, densely pubescent. Petals yellowish. Ovary pubescent; style up to 3 mm. long, apex minutely 2-fid. Fruit (immature) c. 6 × 5 mm., densely pubescent.

S. Rhodesia. E: Umtali, Zimunya Reserve, fl. 13.i.1952, *Chase* 4327 (BM; SRGH).
Also in Natal and the Transvaal. Recorded from granite hillsides.

46. **Cyphostemma vandenbrandeanum** (Dewit) Descoings in Notul. Syst. **16**: 125 (1960). Type from the Congo (Katanga).
Cissus vandenbrandeana Dewit in Bull. Jard. Bot. Brux. **29**: 264 (1959); F.C.B. **9**: 475 (1960). Type as above.

Climbing or prostrate herb; stems pubescent and sometimes with capitate-glandular hairs near the nodes; tendrils present. Leaves digitately 3–5-foliolate; petiole up to 4·5 cm. long but often shorter, pubescent and sometimes with a few scattered capitate glands; leaflet-lamina sessile or subsessile, up to 10 × 5 cm., elliptic, obovate or ovate, apex acute, margin crenate, base cuneate, glabrous above but with a few hairs on the midrib, whitish-tomentose below, veins reticulate; stipules up to 1·2 cm. long, lanceolate-acuminate, ± falcate, pubescent and sometimes with a few capitate glands at the margins. Cymes on side-branches, ± compact, trichotomous; peduncle up to 7 cm. long, densely pubescent, bracts and bracteoles 1–2 mm. long, pubescent. Flower-bud up to 3 mm. long, constricted near the middle, thinly pubescent along the petal margins and near the apex. Calyx 0·5 mm. long, ± 4-toothed, pubescent and often with a few subsessile glands. Petals yellowish with red tips. Ovary pubescent; style c. 2 mm. long, apex minutely 2-fid. Ripe fruits not known.

N. Rhodesia. E: Nyika Plateau, fl. 2.i.1959, *Richards* 10412 (K; SRGH).
Also in Tanganyika and the Congo. Edge of forest patches or riverine forest.

47. **Cyphostemma cirrhosum** (Thunb.) Descoings in Notul. Syst. **16**: 120 (1960). Type from Cape Province.
Vitis cirrhosa Thunb., Prodr. Pl. Cap. **1**: 44 (1794). Type as above.
Cissus cirrhosa (Thunb.) Willd. in L., Sp. Pl. ed. 4, **1**: 657 (1798).—Harv. in Harv. & Sond., F.C. **1**: 252 (1860) excl. var. *β glabra*.—Planch. in A. & C. DC., Mon. Phan. **5,** 2: 603 (1887).—Gilg & Brandt in Engl., Bot. Jahrb. **46**: 524 (1912).—Suesseng. in Engl. & Prantl, Nat. Pflanzenfam. ed. 2, **20d**: 251 (1953). Type as above.

Climbing herb; stems pubescent, or in one subspecies with capitate-glandular hairs also; tendrils present. Leaves 3–5-foliolate; petiole up to 5 (8) cm. long, indumentum as for stems; leaflet-lamina up to c. 13 × 7·5 cm., ovate, broadly elliptic or elliptic, apex rounded or acute, margin serrate or crenate-serrate, base cuneate or rounded, ± pubescent on both sides; petiolules 0–3 cm. long; stipules up to c. 1 cm. long, lanceolate or ovate, falcate, pubescent. Cymes on short side-branches, lax, di- or trichotomous; peduncle up to c. 9 cm. long, pubescent; pedicels c. 3 mm. long, pubescent. Flower-bud 2–3 mm. long, oblong-cylindric, constricted near the middle, ± densely pubescent. Calyx 0·5 mm. long, pubescent. Ovary pubescent; style 1·5–2 mm. long, apex minutely bifurcate. Fruit up to c. 9 × 7 mm., ellipsoid, pubescent. Seed 1, c. 8 × 6 mm., obovoid, slightly rugose.

Cape Province, Natal, Transvaal, SW. Africa, Bechuanaland Prot., S. Rhodesia and Nyasaland.

Subsp. **transvaalense** (Szyszyl.) Wild & Drummond in Kirkia, **2**: 139 (1961). Type from the Transvaal.
Vitis cirrhosa var. *transvaalensis* Szyszyl., Polypet. Thalam. Rehm.: 45 (1888). Type as above.
Cissus cirrhosa var. *transvaalensis* (Szyszyl.) Burtt Davy & Pott-Leendertz in Ann. Transv. Mus. **3**: 121, 151 (1912). Type as above.
Cissus sandersonii var. *transvaalensis* (Szyszyl.) C.A. Sm. in Burtt Davy, F.P.F.T., **2**: xx, 476 (1932). Type as above.

Stems eglandular. Leaflets petiolate, slightly fleshy. Pubescence rather short.

Bechuanaland Prot. N: Kwebe Hills, fl. ii.1897, *Lugard* 187 (K). **N. Rhodesia.** B: Machili, fl. 10.i.1961, *Fanshawe* 6130 (SRGH). **S. Rhodesia.** W: Wankie, fl. iv.1954, *Levy* 1137 (SRGH).
Also in Cape Province, Natal, Transvaal and SW. Africa. Drier types of woodland.

Subsp. **rhodesicum** Wild & Drummond in Kirkia, **2**: 139 (1961). Type: S. Rhodesia Salisbury, *Rogers* 5775 (K; SRGH, holotype).

Stems with capitate-glandular hairs (sometimes very sparse). Petiolule length variable. Leaflets usually elliptic with ± acute apex. Pubescence coarser than in subsp. *transvaalense*.

S. Rhodesia. N: between Banket and Umvukwes, fr. 23.iv.1948, *Rodin* 4394 (K; PRE). C: Salisbury, Twentydales, fl. 18.xii.1951, *Wild* 3706 (SRGH). E: Inyanga, fl.

22.i.1942, *Hopkins* in GHS 8628A (SRGH). **Nyasaland.** C: Dedza, fl. 7.i.1961, *Chapman* 1112 (SRGH).

Not recorded outside S. Rhodesia and Nyasaland. Common in *Brachystegia* woodland.

48. **Cyphostemma bororense** (Klotzsch) Descoings in Notul. Syst. **16**: 120 (1960). TAB. **97** fig. B. Type: Mozambique, Boror, *Peters* (B, holotype †).

Cissus bororensis Klotzsch in Peters, Reise Mossamb. Bot. **1**: 179 (1861).—Planch. in A. & C. DC., Mon. Phan. **5**, 2: 608 (1887).—Gilg & Brandt in Engl., Bot. Jahrb. **46**: 523 (1912).—Suesseng. in Engl. & Prantl, Nat. Pflanzenfam. ed. 2, **20d**: 251 (1953). Type: Mozambique, Boror, *Peters* (B, holotype †).

Vitis bororensis (Klotzsch) Bak. in Oliv., F.T.A. **1**: 411 (1868). Type as above.

Cissus agnus-castus Planch., tom. cit.: 598 (1887). Type: Mozambique, Tete, *Kirk* (K, holotype).

Cissus zambesica Wild & Drummond in Kirkia, **1**: 12 (1960). Type: S. Rhodesia, Chirundu, *Drummond* 5371 (K; PRE; SRGH, holotype).

Climbing or creeping herb; stems striate, glabrous or with a few capitate-glandular hairs especially near the nodes. Leaves digitately 3–5-foliolate; petiole up to c. 3 cm. long, puberulous on the upper side and sometimes with stalked capitate glands near the apex; leaflet-lamina up to 7·5 × 5 cm., membranous, margin serrate-crenate with the teeth glandular-apiculate, elliptic to broadly elliptic, apex mucronate, base cuneate or in lateral leaflets rounded, puberulous on the nerves above, glabrous or with a few capitate-glandular hairs on the midrib below; petiolule of median leaflet up to 5 mm. long, lateral leaflets sessile or subsessile; stipules up to c. 7 mm. long, lanceolate- to ovate-acuminate, often ± falcate, margin glandular. Cymes on short axillary branches; peduncle up to 4·5 cm. long, glabrous or with a few capitate-glandular hairs towards the apex; pedicels up to c. 6 mm. long, glabrous but with shortly stalked capitate-glandular hairs in the middle portion; bracts and bracteoles c. 1·5 mm. long, lanceolate, glabrous. Flower-bud c. 2·5 mm. long, oblong-cylindric, constricted near the middle, glabrous. Petals red-tipped. Calyx 0·5 mm. long, subentire, often with a few subsessile glands. Ovary glabrous; style c. 1·5 mm. long, apex scarcely thickened. Fruit red, c. 8 × 5 mm., ellipsoid-globose, glabrous. Seed 1, 5 × 4 mm., obovoid, with a dorsal crest and two fainter lateral crests, rugose-tuberculate.

Caprivi Strip. Mpilila I., fl. 15.i.1959, *Killick & Leistner* 3393 (PRE; SRGH). **N. Rhodesia.** B: Sesheke, 3 km. W. of Masese, fr. 25.i.1952, *White* 1959 (FHO). W: Ndola, fl. 8.i.1955, *Fanshawe* 1781 (K; NDO; SRGH). S: Ngwesi R., fr. 15.i.1956, *Gilges* 553 (PRE; SRGH). **S. Rhodesia.** N: Binga Distr., Dett R., fl. 8.xii.1956, *Lovemore* 519 (SRGH). **Mozambique.** N: R. Rovuma, *Kirk* (K). Z: Boror, *Peters* (B†). T: Tete, *Kirk* (K). MS: Mavita, fl. 22.i.1948, *Barbosa* 866 (LISC; LM).

Also in Tanganyika. Woodland, forest edges and river sides.

The Rhodesian material differs from the type in having conspicuous glands on the apex of the petioles and on the underside of the leaflet midribs. This form was described as *Cissus zambesica*. However, several Mozambique specimens have a few short glands at the apices of the petioles and since glands on the vegetative parts of *Cyphostemma* spp., as opposed to glands on the inflorescences, do not seem to be significant as a rule at the specific level, *C. zambesica* is not now considered a good species.

49. **Cyphostemma robsonii** Wild & Drummond in Kirkia, **2**: 137 (1961). Type: Nyasaland, N. of Chitala on Kasache road, *Robson* 1576 (K; SRGH, holotype).

Climbing herb with tuberous root, stems pubescent or glabrescent; tendrils present. Leaves digitately 3–5-foliolate; petiole up to 5 cm. long, pubescent; leaflets up to 6 × 2 cm., oblanceolate to narrowly obovate, apex acute, margin serrate-crenate with gland-tipped teeth, base cuneate, scabrid-puberulous above, sparsely puberulous especially on the nerves below; lateral leaflets sessile or subsessile, median leaflet with petiole up to 1 cm. long; stipules up to 1 cm. long, lanceolate, acuminate, minutely ciliate. Cymes on short side-branches, lax; peduncle up to 4 cm. long, glabrous; pedicels up to 4 mm. long, glabrous; bracts and bracteoles up to 2 mm. long, lanceolate, glabrous. Flower-bud up to 2·5 mm. long, oblong-cylindric, constricted near the middle, glabrous except for a single stipitate gland near the apex of each petal. Calyx up to 0·4 mm. long, subentire, glabrous. Petals red. Ovary glabrous; style c. 1·5 mm. long; stigma subcapitate. Ripe fruit not known.

Nyasaland. C: N. of Chitala on Kasache road, fl. 12.ii.1959, *Robson* 1576 (K; SRGH). Also in Tanganyika. *Brachystegia* woodland or grassland.

50. **Cyphostemma amplexum** (Bak.) Descoings in Notul. Syst. **16**: 120 (1960). Type: Mozambique (? or Tanganyika), R. Rovuma, *Kirk* (K, holotype).
Vitis amplexa Bak. in Oliv., F.T.A. **1**: 403 (1868). Type as above.
Cissus amplexa (Bak.) Planch. in A. & C. DC., Mon. Phan. **5**, 2: 593 (1887). Type as above.
Cissus bororensis sensu Gilg & Brandt in Engl., Bot. Jahrb. **46**: 523 (1912) pro parte quoad specim. Kirk. ex R. Rovuma.

Climbing herb with a napiform root; stems glabrous except for subsessile capitate glands near the nodes; tendrils present. Leaves digitately 3–5-foliolate; petiole up to 4 cm. long, sparsely pilose and usually with a few glandular hairs; leaflets up to 5 × 2·5 cm., narrowly elliptic to broadly elliptic, apex acute, margin serrate-crenate with gland-tipped teeth, base ± cuneate, sparsely puberulous on both sides, petiolule 0–1 cm. long; stipules up to 1 cm. long, ovate-lanceolate or lanceolate, acuminate, margin sometimes glandular. Cymes on short side-branches, lax, trichotomous; peduncle up to 4 cm. long, glabrous; pedicels up to 5 mm. long, glabrous except for stipitate glands near the middle; bracts and bracteoles up to 2 mm. long, lanceolate, glabrous. Flower-bud up to 5 mm. long, oblong-cylindric, constricted near the middle, glabrous except for a single stipitate gland near the apex of each petal. Calyx c. 0·5 mm. long, subentire, sometimes glandular. Petals wine-coloured. Ovary glabrous; style c. 2 mm. long; stigma indistinct. Fruit 8 × 6 mm., ellipsoid, glabrous. Seed 1, 7 × 5 mm., ellipsoid, slightly rugose, scarcely crested.

Mozambique. N: Nampula, fl. 21.i.1937, *Torre* 1181 (COI; LISC).
Not known elsewhere. Forest climber.

51. **Cyphostemma trachyphyllum** (Werderm.) Descoings in Notul. Syst. **16**. 125 (1960). Type from Tanganyika (Lindi Distr.).
Cissus trachyphylla Werderm. in Notizbl. Bot. Gart. Berl. **13**: 281 (1936). Type as above.

Climbing herb; stems rather slender, puberulous; tendrils present. Leaves digitately 3–5-foliolate; petiole up to 4 cm. long, densely puberulous; leaflet-lamina 5 (6) × 2 (3·2) cm., elliptic to narrowly obovate, apex acute, margin serrate-crenate, base cuneate, scabrous-pubescent on both surfaces; petiolules up to 5 mm. long; stipules up to 8 mm. long, lanceolate-acuminate, sparsely pubescent. Cymes lax, on axillary branches; peduncle up to 3 cm. long, glabrous; pedicels up to 3 (5) mm. long, glabrous except for a few very shortly stalked capitate glands. Flower-bud 1·5 mm. long, swollen at the apex, glabrous. Calyx entire, glabrous. Ovary glabrous. Fruit c. 7 × 5 mm., ellipsoid, glabrous. Seed 1, c. 6 × 4 mm., ellipsoid, with a dorsal crest, ± rugose.

Mozambique. N: Cabo Delgado, Montepuez, fr. 2.iii.1959, *Myre & Macedo* 3502 (LM; SRGH).
Also in S. Tanganyika. Sandy soils.

52. **Cyphostemma graniticum** (Wild & Drummond) Wild & Drummond in Kirkia, 2: 140 (1961). Type: S. Rhodesia, Inyanga, Inyangombe Falls, *Wild* 4883 (BRLU: K; LISC; PRE; SRGH, holotype).
Cissus granitica Wild & Drummond in Kirkia, **1**: 11 (1960). Type as above.

Scandent herb; stems up to 1·5 m. long, striate, glabrous; tendrils present. Leaves digitately 3–9-foliolate; petioles up to 2·5 cm. long, glabrous; leaflet-lamina subsessile, up to 11 × 0·7 cm., linear-lanceolate, apex acuminate, margin sparsely serrate, base narrowly cuneate, teeth glandular, glabrous; stipules up to 1 cm. long, lanceolate to ovate-acuminate, margin glandular-toothed. Cymes on axillary branches, lax, trichotomous; peduncle up to 6 cm. long, glabrous; pedicels up to 5 mm. long, with sparse capitate glands; bracts and bracteoles up to 2 mm. long, lanceolate, margins glandular-toothed. Flower-bud 2–3·5 mm. long, oblong-cylindric, constricted near the middle, glabrous. Calyx pink, up to 1 mm. long, subentire, glabrous. Petals pale yellow. Ovary glabrous; style c. 1 mm. long apex scarcely capitate. Fruit red, c. 1·2 × 0·8 cm., ellipsoid, glabrous. Seed 1, c. 7 × 5 mm., ellipsoid, with dorsal crest and prominently horizontally rugose.

S. Rhodesia. N: Umvukwes, Ruorka Ranch, fl. 16.xii.1952, *Wild* 3919 (SRGH). C: Salisbury Distr., Domboshawa, fl. 29.xi.1946, *Greatrex* in GHS 15555 (SRGH). E: Inyanga Distr., Inyangombe Falls, fl. 3.xii.1959, *Wild* 4883 (BRLU; K; LISC; PRE; SRGH).

Not known outside S. Rhodesia. Usually on granite or serpentine rocks.

53. **Cyphostemma tenuissimum** (Gilg & Fr.) Descoings in Notul. Syst. **16**: 125 (1960). Type: N. Rhodesia, Kalungwishi R., *Fries* 1152 (UPS, holotype).

Cissus tenuissima Gilg. & Fr. in R.E. Fr., Schwed. Rhod.-Kongo-Exped. **1**: 137, t. 10, fig. 5 (1914).—Suesseng. in Engl. & Prantl, Nat. Pflanzenfam. ed. 2, **20d**: 251 (1953). Type as above.

Scandent herb with a tuberous root; stems slender, striate, glabrous; tendrils present. Leaves digitately 3–5-foliolate; petiole up to 2 cm. long, puberulous above otherwise glabrous; leaflet-lamina up to 4 × 0·5 cm., linear-lanceolate to narrowly lanceolate, apex acuminate, margin sparsely serrate, base cuneate, puberulous near the margins above, otherwise glabrous; petiolule c. 1 mm. long or leaflets subsessile; stipules up to 3·5 mm. long, lanceolate, margin puberulous. Cymes lax, on axillary branches; peduncle up to 4 cm. long, glabrescent or glabrous; pedicels up to 3 mm. long, glabrous or minutely puberulous; bracts and bracteoles c. 1 mm. long, subulate, puberulous. Flower-bud 2 mm. long, oblong-cylindric, constricted near the middle, glabrous. Calyx purplish, 0·5 mm. long, entire, glabrous. Petals greenish with deep purple tips. Ovary glabrous; style 0·7 mm. long, apex scarcely capitate. Fruit not seen.

N. Rhodesia. N: Kalungwishi R., fl. 29.x.1911, *Fries* 1152 (UPS).

Known so far only from the type collection. Rocky ground.

54. **Cyphostemma saxicola** (Gilg & Fr.) Descoings in Notul. Syst. **16**: 125 (1960). Type: N. Rhodesia, Kalambo Falls, *Fries* 1394 (UPS, holotype).

Cissus saxicola Gilg & Fr. in R.E. Fr., Schwed. Rhod.-Kongo-Exped. **1**: 138, fig. 10 (1914).—Suesseng. in Engl. & Prantl, Nat. Pflanzenfam. ed. 2, **20d**: 251 (1953). Type as above.

Climbing or scandent herb; stems to 0·6 m. long, slender, shortly pubescent and with a very few short capitate glands; tendrils present. Leaves digitately 3–5-foliolate; petiole up to 3–4 cm. long, indumentum as for stems; leaflet-lamina up to 6 × 1·6 cm., very narrowly elliptic to oblanceolate, apex acute, margin with a few serrations towards the apex, base cuneate, very sparsely puberulous on both surfaces; petiolules up to 1 cm. long; stipules c. 3 mm. long, lanceolate, puberulous. Cymes on short axillary branches; peduncle up to 1·3 cm. long, indumentum as for stems; pedicels c. 2 mm. long, puberulous with scattered subsessile capitate glands; bracts and braceoles c. 1 mm. long, subulate, puberulous. Flower-bud 2 mm. long, oblong-cylindric, scarcely constricted above the middle, puberulous. Calyx 0·5 mm. long, subentire, puberulous. Ovary puberulous; style 0·4 mm. long, apex not capitate. Fruit scarlet, c. 8 × 6 mm., ellipsoid, puberulous. Seed 1, c. 7 × 5 mm., obovoid, with a dorsal crest, rugose-tuberculate.

N. Rhodesia. N: Abercorn Distr., Chilongowelo, fl., *Richards* 2292 (K).

Not known so far outside the Northern Province of N. Rhodesia. Dense woodland.

55. **Cyphostemma rotundistipulatum** Wild & Drummond in Kirkia, **2**: 137 (1961). Type: N. Rhodesia, Abercorn Distr., old road to Stembewe, *Richards* 4068 (K, holotype; SRGH).

Scandent herb; stems sparsely pubescent or glabrescent; tendrils present. Leaves digitately 3-foliolate; petiole up to 4 cm. long, indumentum as for stems; leaflet-lamina membranous, apex acute to acuminate, margin crenate, glabrous above but midrib and larger nerves puberulous, glabrous below; median leaflet up to 9·5 × 3·5 cm., narrowly ovate, base cuneate; lateral leaflets up to 7 × 3 cm., ovate, base asymmetrically obtuse; petiolules 1–7 mm. long; stipules up to 2·5 cm. in diam., persistent, circular to broadly transversely elliptic, apex obtuse, margin sparsely crenate, base slightly cordate, glabrous. Cymes on sparsely pubescent peduncles up to 4 cm. long; pedicels c. 4 mm. long, pubescent and usually with subsessile glands; bracts and bracteoles c. 1 mm. long, subulate, pubescent. Flower-bud up to 2·5 mm. long, oblong-cylindric, constricted near the middle, glabrous. Calyx 0·5 mm. long, subentire, pubescent and densely glandular.

Petals red. Ovary glabrous; style 1·5 mm. long, apex scarcely capitate. Fruit probably red, c. 8 × 6 mm., subglobose, glabrous. Seed 1, c. 5 × 4 mm., obovoid, ± smooth.

N. Rhodesia. N: Abercorn Distr., Sunzu, fl. 8.i.1955, *Bock* 175 (PRE; SRGH).
Not known outside N. Rhodesia. *Brachystegia* woodland.

56. **Cyphostemma paucidentatum** (Klotzsch) Descoings in Notul. Syst. **16**: 123 (1960). Type: Mozambique, Inhambane, *Peters* (B, holotype †).
Cissus paucidentata Klotzsch in Peters, Reise Mossamb. Bot. **1**: 178 (1861).—Planch. in A. & C. DC., Mon. Phan. **5**, 2: 602 (1887).—Gilg & Brandt in Engl., Bot. Jahrb. **46**: 522 (1912).—Suesseng. in Engl. & Prantl, Nat. Pflanzenfam. ed. 2, **20d**: 251 (1953). Type as above.
Vitis paucidentata (Klotzsch) Bak. in Oliv., F.T.A. **1**: 410 (1868). Type as above.

Climbing herb, stems striate, glabrous or rarely sparsely puberulous; tendrils present. Leaves digitately 3–5-foliolate; petiole up to c. 3·5 cm. long, glabrous; leaflet-lamina membranous, up to 11 × 3·5 cm., elliptic, apex acute to acuminate, margin crenate but often entire towards the base, base cuneate, glabrous on both sides; petiolule up to 0·5 cm. long on median leaflet but remaining leaflets sessile or subsessile; stipules up to 1·3 cm. long, oblong-lanceolate, glabrous. Cymes lax, on axillary or leaf-opposed side-branches; peduncle up to 7 cm. long, glabrous; pedicels up to 1 cm. long, pubescent or puberulous; bracts and bracteoles up to 6 mm. long, ovate-lanceolate, glabrous but with puberulous margins. Flower-bud c. 3·5 mm. long, oblong-cylindric, constricted above the middle, puberulous. Calyx 0·5 mm. long, entire, densely puberulous. Petals greenish. Ovary puberulous; style 2 mm. long, minutely 2-fid at the apex. Fruit c. 0·8 × 6 mm. (not mature), ellipsoid, glabrescent.

S. Rhodesia. S: S. of Lundi R., Beitbridge road, fl. 15.ii.1955, *E.M. & W.* 370 (BM; SRGH). **Mozambique.** MS: Inhamitanga, fl. 5.vii.1946, *Simão* 735 (LM; LMJ; SRGH). SS: 19 km. from Mabote, towards Vilanculos-Mambone crossroads, fl. & fr. 28.iii.1952, *Myre & Balsinhas* 5054 (LM; SRGH).
Also in the Transvaal and Tanganyika. Forest or rather dense woodland.

57. **Cyphostemma kilimandscharicum** (Gilg) Descoings in Notul. Syst. **16**: 122 (1960). Type from Tanganyika (Kilimanjaro).
Cissus kilimandscharica Gilg in Engl., Bot. Jahrb. **19**, Beibl. **47**: 39 (1894).—Gilg & Brandt in Engl., Bot. Jahrb. **46**: 526 (1912).—Suesseng. in Engl. & Prantl, Nat. Pflanzenfam. ed. 2, **20d**: 251 (1953). Type as above.

Climber or liane; stems glabrous (pubescent in Congo material); tendrils present. Leaves digitately 3-foliolate; petiole up to 3·5 (7) cm. long, glabrous except for a few hairs near the apex; leaflet-lamina up to 7 × 5 cm., ovate or more rarely lanceolate, apex acute to acuminate, margins serrate to serrate-crenate, terminal leaflet cuneate at base, laterals rounded or subcordate, sparsely pubescent on the nerves on both sides or glabrescent; petiolules up to 1·8 cm. long, sparsely pubescent; stipules caducous, c. 1 cm. long, falcate-ovate, glabrous. Cymes lax, trichotomous, on axillary branches; peduncle 3 (10) cm. long, sparsely pubescent or glabrescent; pedicels c. 5 mm. long, ± densely puberulous; bracts and bracteoles c. 2 mm. long, lanceolate, puberulous, sometimes with a few short capitate-glandular hairs also. Flower-bud up to 4·5 mm. long, oblong-cylindric, swollen at base and apex, puberulous and sometimes with a few longer hairs. Calyx 0·5 mm. long, entire, puberulous. Petals yellow-green. Ovary glabrous; style up to 3·5 mm. long, apex scarcely capitate. Fruit red, c. 1 × 0·6 cm., ellipsoid, apiculate, glabrous. Seed c. 8 × 4·5 mm., ellipsoid, with a dorsal crest and 2 lateral crests, rugose.

S. Rhodesia. E: Vumba, fl. 14.i.1961, *Drummond* 6882 (K; LISC; PRE; SRGH). **Nyasaland.** N: Misuku Hills, Mugesse Forest, fr. x.1953, *Chapman* 169 (FHO).
Also in Kenya, Tanganyika and the Congo. Forest at 1000 m. or above.

58. **Cyphostemma jaegeri** (Gilg & Brandt) Descoings in Notul. Syst. **16**: 122 (1960). Type from Tanganyika.
Cissus jaegeri Gilg & Brandt in Engl., Bot. Jahrb. **46**: 526 (1912).—T.C.E. Fr. in Notizbl. Bot. Gart. Berl. **8**: 560 (1923).—Brenan in T.T.C.L.: 29 (1949). Type as above.

Climber or liane; stems glabrous or pubescent; tendrils present. Leaves digitately 3-foliolate; petiole up to c. 7 cm. long, glabrous or pubescent; leaflet-lamina up to 12 × 7 cm., ovate, apex acuminate, margin coarsely sinuate-dentate or crenate, base rounded, subcordate or broadly cuneate, very sparsely pubescent or glabrescent on both sides; petiolules up to c. 1·3 cm. long, pubescent or glabrescent; stipules caducous, c. 1 cm. long, broadly falcate, apex acute or subacute, pubescent or glabrescent. Cymes lax, trichotomous, on axillary branches; peduncle up to c. 10 cm. long, pubescent; pedicels up to 8 mm. long, pubescent; bracts and bracteoles 1–2 mm. long, lanceolate. Flower-bud up to 4·5 mm. long, oblong-cylindric, swollen at base and apex, densely pubescent. Calyx 0·5 mm. long, entire, pubescent. Ovary pubescent; style up to 4 mm. long, apex subcapitate. Fruit c. 1 × 0·7 cm., pubescent. Mature seed not seen.

N. Rhodesia. N: Ndundu Machito, fl. & fr. 10.iii.1955, *Richards* 4864 (K; SRGH). Also in Kenya and Tanganyika. Liane of forest patches.

Very near *C. kilimandscharicum* but has a pubescent ovary and fruit. The type has probably been destroyed and if this is so *R.E. & T. C.E. Fries* 783 (UPS) from the Aberdare Mts. in Kenya should be chosen as neotype.

59. **Cyphostemma lovemorei** Wild & Drummond in Kirkia, **2**: 134 (1961). Type: S. Rhodesia, Sebungwe Distr., Lubu R., *Lovemore* 545 (K; SRGH, holotype).

Climbing herb; stems glabrous or with a few capitate-glandular hairs near the nodes; tendrils present. Leaves digitately 5-foliolate; petiole up to 2 cm. long, glabrous; leaflet-lamina semi-succulent, glaucous, up to 9 × 4 cm., oblong-elliptic to narrowly ovate, apex acute to subacute, margin ± sparsely serrate with acuminate teeth, base cuneate to obtuse, glabrous on both sides; petiolules up to 3 cm. long; stipules c. 7 mm. long, lanceolate-acuminate, ± falcate, glabrous. Cymes on side-branches; peduncle up to 4 cm. long, glabrous; pedicels up to 7 mm. long, glabrous; bracts and bracteoles up to 1 mm. long, subulate, glabrous. Flower-bud 2–3 mm. long, oblong-cylindric, apex inflated, glabrous. Calyx 0·5 mm. long, subentire, glabrous. Petals yellowish-green. Ovary glabrous; style c. 2 mm. long, apex scarcely capitate. Fruit c. 8 × 7 mm., ellipsoid-globose, glabrous. Seed 1, c. 7 × 6 mm., ellipsoid-globose, slightly rugose.

N. Rhodesia. S: 3 km. N. of Chirundu Bridge, fl. 1.ii.1958, *Drummond* 5412 (K; PRE; SRGH). **S. Rhodesia.** N: Urungwe Distr. Rifa R., fl. 24.ii.1953, *Wild* 4099 (SRGH).

Not known outside the Zambezi valley area. *Colophospermum mopane* woodland or riverine thicket.

Close to *C. quinatum* but distinguished by the elongated petiolules. With further collecting it may prove to be a subspecies of *C. quinatum.*

60. **Cyphostemma quinatum** (Dryand.) Descoings in Notul. Syst. **16**: 124 (1960). Type: a specimen cultivated at Kew from the Cape of Good Hope.

Cissus quinata Dryand. in Ait., Hort. Kew. ed. 2, **1**: 260, t. 1 (1810).—Planch. in A. & C. DC., Mon. Phan. **5**, 2: 603 (1887).—Gilg & Brandt in Engl., Bot. Jahrb. **46**: 521 (1912).—Suesseng. in Engl. & Prantl, Nat. Pflanzenfam. ed. 2, **20d**: 250 (1953). Type as above.

Cissus cirrhosa var. *β*, *glabra* Harv. in Harv. & Sond., F.C. **1**: 252 (1860). Type from Cape Province.

Climbing herb; stems glabrous or with a few scattered hairs; tendrils present. Leaves 3–5-foliolate; petiole 2 (4) cm. long, sparsely pubescent or glabrous; leaflet-lamina subfleshy, up to 5 (8) × 2·4 (5) cm., broadly elliptic to obovate, apex acute or obtuse, margin coarsely serrate, base cuneate, glabrous on both sides or very sparsely puberulous on the midrib; petiolules c. 2 (5) mm. long, glabrous or pubescent and often with one or two glandular hairs; stipules c. 7 mm. long, ovate-falcate, glabrous but with glandular margins. Cymes on short side-branches, lax, trichotomous; peduncle up to 5 (6) cm. long, glabrous; pedicels up to 4 cm. long, glabrous; bracts and bracteoles c. 1 mm. long, subulate, glabrous. Flower-bud c. 3 mm. long, oblong-cylindric, ± constricted near the middle, glabrous. Calyx 0·5 mm. long, entire, glabrous. Ovary glabrous; style 1·5 mm. long, apex minutely 2-fid. Fruit c. 10 × 7·5 mm., ellipsoid, glabrous. Seed 1, c. 8 × 7 mm., with a dorsal keel and slightly rugose.

Mozambique. SS: S. Martinho do Bilene, fl. 11.iii.1952, *Barbosa & Balsinhas* 4876 (LMJ; SRGH).

Also in the Eastern Cape Province. A coastal species.

61. **Cyphostemma cyphopetalum** (Fresen.) Descoings in Notul. Syst. **16**: 121 (1960). Syntypes from Ethiopia.

Cissus cyphopetala Fresen. in Mus. Senckenb. **2**: 282 (1837).—Planch. in A. & C. DC., Mon. Phan. **5**, 2: 609 (1887).—Gilg & Brandt in Engl., Bot. Jahrb. **46**: 540 (1912).—Suesseng. in Engl. & Prantl, Nat. Pflanzenfam. ed. 2, **20d**: 254 (1953).—Dewit, F.C.B. **9**: 484 (1960). Syntypes as above.

Vitis cyphopetala (Fresen.) Bak. in Oliv., F.T.A. **1**: 407 (1868). Syntypes as above.

Cissus tenuipes Gilg & Fr. in R.E. Fr., Schwed. Rhod.-Kongo-Exped. **1**: 139 (1914). Type from Rwanda-Burundi.

Climbing herb with a tuberous root; stems striate, densely pubescent; tendrils present. Leaves digitately 3–5-foliolate; petiole c. 4 cm. long, pubescent; leaflet-lamina 8 (9) × 5·5 cm., obovate, broadly obovate, ovate or elliptic-oblong, apex acute or subacute, margin coarsely serrate-dentate, base broadly cuneate or rounded, puberulous on both sides, more densely so below; petiolules up to 1·2 cm. long; stipules c. 1 cm. long, ovate-falcate, apex acute, pubescent. Cymes lax, on axillary branches; peduncle 4 (8) cm. long, densely pubescent and sometimes with a few capitate-glandular hairs; pedicels up to 0·8 cm. long, densely yellowish-pubescent and with short capitate glands; bracts and bracteoles c. 1 mm. long, lanceolate, pubescent. Flower-bud up to 5 mm. long, swollen at the apex, densely yellowish-pubescent. Calyx 0·5 mm. long, entire, densely pubescent. Petals creamy-green. Ovary pubescent; style c. 3·3 mm. long, apex scarcely capitate. Fruit 1 × 0·5 cm., ellipsoid, densely pubescent. Seed 1, c. 7 × 5 mm., ellipsoid, 3-crested, rugose.

N. Rhodesia. N: Abercorn Distr., Kalambo R., Sansia Falls, fl. 1.i.1957, *Richards* 7431 (K; SRGH).

Also in Ethiopia, Eritrea, Sudan, Kenya, Tanganyika, Uganda, Rwanda-Burundi and the Congo. River-sides.

5. CAYRATIA Juss.

Cayratia Juss., Dict. Sci. Nat. **10**: 103 (1818) in obs.; Dict. Class. Hist. Nat. **4**: 346 (1823) *nom. conserv.*

Climbing perennial herbs; tendrils leaf-opposed. Leaves 3-foliolate or pedate, margins toothed; stipules present. Inflorescence a divaricately-branched cyme, usually axillary. Flowers 4-merous; bud depressed-globose, not constricted at the middle. Calyx entire. Petals cucullate at the apex, caducous. Anthers on short filaments. Disk annular, thin, undulate. Style simple, subulate; stigma capitate. Fruit 2–4 seeded.

Leaves pedate - - - - - - - - - - - - 1. *gracilis*
Leaves 3-foliolate - - - - - - - - - - - 2. *ibuensis*

1. **Cayratia gracilis** (Guill. & Perr.) Suesseng. in Engl. & Prantl, Nat. Pflanzenfam. ed. 2, **20d**: 278 (1953). TAB. **98** fig. B. Type from Senegal.

Cissus gracilis Guill. & Perr. in Guill., Perr. & Rich., Fl. Senegamb. Tent. **1**: 134 (1832).—Planch. in A. & C. DC., Mon. Phan. **5**, 2: 565 (1887).—Gilg & Brandt in Engl., Bot. Jahrb. **46**: 487, fig. 6 A–C (1912).—Eyles in Trans. Roy. Soc. S. Afr. **5**: 409 (1916).—Burtt Davy, F.P.F.T. **2**: 476 (1932).—Brenan, T.T.C.L.: 28 (1949).—Wild, Guide Fl. Vict. Falls: 151 (1953).—Exell & Mendonça, C.F.A. **2**, 1: 55 (1954).—Keay, F.W.T.A. ed. 2, **1**, 2: 679 (1958).—Willems, F.C.B. **9**: 552 (1960). Type as above.

Vitis gracilis (Guill. & Perr.) Bak. in Oliv., F.T.A. **1**: 404 (1868) pro parte; non *Vitis gracilis* Wall. (1824). Type as above.

Climbing herb; stems and branches glabrous or glabrescent; tendrils present. Leaves pedate, 5–8-foliolate; petioles up to 6 cm. long, glabrous or glabrescent; leaflet-lamina up to 7·5 × 3·5 cm., narrowly ovate to ovate, acuminate at the apex, cordate to obtuse at the base, margin coarsely serrate, puberulous especially on nerves beneath to glabrescent; median leaflet with a petiolule 2·5 cm. long; stipules ± 1 mm. long, triangular, caducous. Cymes axillary, c. 10 cm. long: peduncle up to 4 cm. long; pedicels 1·5 mm. long; bracts and bracteoles minute.

Flower-bud 2 mm. long, globose. Calyx entire. Ovary glabrous; style 0·5 mm. long; stigma capitate. Fruit up to 6·5 × 6·5 mm. when dry, globose or depressed globose. Seeds (2) 4 to a fruit, c. 5·5 mm. long, with the two inner faces flat and the outer concave.

N. Rhodesia. N: Lunzua Valley, fl. & fr. 5.iii.1955, *Richards* 4810 (K). W: Mwinilunga Distr., Matonchi Farm, fl. & fr. 20.xii.1937, *Milne-Redhead* 3754 (BM; K). C: Mumbwa, fl. 1912, *Macaulay* 1140 (K). E: Fort Jameson, Lunkwakwa, fr. 23.iii.1955, *E.M. & W.* 1132 (BM; SRGH). S: Mapanza, fl. & fr. 7.iii.1954, *Robinson* 598 (K; SRGH). **S. Rhodesia.** N: Chiswiti Reserve, Darwin Distr., fl. 27.i.1960, *Phipps* 2430 (K; PRE; SRGH). C: Salisbury, Epworth, fl. 3.ii.1946, *Wild* 1831 (SRGH). E: Imbeza Valley, fl. 2.iii.1957, *Chase* 6345 (EA; SRGH). S: Victoria Distr., fr. 1909, *Monro* 1043 (BM; SRGH). **Nyasaland.** N: fr., *Whyte* (K). S: Mlanje, Swazi Estate, fl. & fr. 29.iii.1949, *Faulkner* K391 (K; SRGH). **Mozambique.** N: Massangulo, fr. iv.1933, *Gomes e Sousa* 1403 (COI; K). Z: Milange, fl. & fr. 23.ii.1943, *Torre* 4815 (LISC). T: between Lupata and Tete, fl. ii.1859, *Kirk* (K). MS: Chimoio, fr. 14.iii.1948, *Garcia* 603 (LISC). LM: Libombo Marsh, fl. & fr., *Junod* 525 (PRE).

Widespread in tropical Africa. From sea-level to 1500 m. in various types of woodland.

2. **Cayratia ibuensis** (Hook. f.) Suesseng. in Engl. & Prantl, Nat. Pflanzenfam. ed. 2, **20d**: 278 (1953). TAB. 98, fig. A. Syntypes from Nigeria.

Cissus ibuensis Hook. f. in Hook., Niger Fl.: 265 (1849).—Gilg & Brandt in Engl., Bot. Jahrb. **46**: 486 (1912).—Brenan, T.T.C.L.: 28 (1949).—Exell & Mendonça, C.F.A. **2**, 1: 55 (1954).—Keay, F.W.T.A. ed. 2, **1**, 2: 679 (1958).—Willems, F.C.B. **9**: 547 (1960).—Syntypes as above.

Herbaceous climbing perennial; tendrils present; stem, leaves and inflorescences crisped-puberulous to glabrescent. Leaves digitately 3-foliolate; petioles 1–5 cm. long; leaflet-lamina 8 × 4 cm., ovate to elliptic, acute at the apex, margin serrate-denticulate, sometimes lobed, rounded at the base; petiolule of median leaflet up to 1·7 cm. long; stipules up to 4 mm. long, very broadly ovate. Cymes axillary; peduncle up to 9 cm. long; pedicels 1·5 mm. long; bracts and bracteoles 1 mm. long, ciliate or glabrous. Flower-bud 1·5 mm. long, depressed-globose, puberulous to glabrous. Calyx entire, puberulous. Ovary glabrous; style cylindric; stigma capitate. Fruit 1 cm. in diam. when fresh, subglobose. Seeds 2–4 to a fruit.

N. Rhodesia. W: Mwinilunga Distr., W. of Matonchi Farm, fl. & fr. 29.xii.1937, *Milne-Redhead* 3860 (K). **Nyasaland.** S: Shire Highlands, fl., *Buchanan* 284 (K). **Mozambique.** MS: between Sena and Lupata, fl. & fr. 26.i.1859, *Kirk* (K).

Also in Egypt, Sudan, Nigeria, Cameroun, Uganda, Tanganyika, Congo and Angola. Riverine vegetation, alluvial soils and, rarely, in *Brachystegia* woodland.

55. LEEACEAE

By H. Wild

Trees, shrubs or shrublets, rarely somewhat sarmentose. Leaves alternate, pinnate, bipinnate or occasionally tripinnate; stipules petiolar, often caducous; tendrils absent. Inflorescences leaf-opposed, of spreading much-branched cymes. Flowers actinomorphic, bisexual. Calyx cupular, 5-lobed. Corolla 5-lobed, lobes joined at base, becoming reflexed. Stamens 5, opposite the petals, inflexed in bud but exserted and erect in the open flower, borne between the truncate-emarginate lobes of a tube adnate to the base of the corolla-tube and possessing a downward deflexed sleeve of tissue arising at about its middle (a complicated structure considered to be derived from the fusion of the filaments and distinguishing the family from the Vitaceae, which have free filaments); anthers 2-locular, dorsifixed, with ± longitudinal dehiscence. Disk apparently absent. Ovary superior, 3–8-locular with 1 anatrapous ovule in each loculus; style cylindric; stigma subcapitate. Fruit depressed-subglobose, 3–8-lobed. Seeds 3–8, laterally compressed and triangular in transverse section; endosperm ruminate.

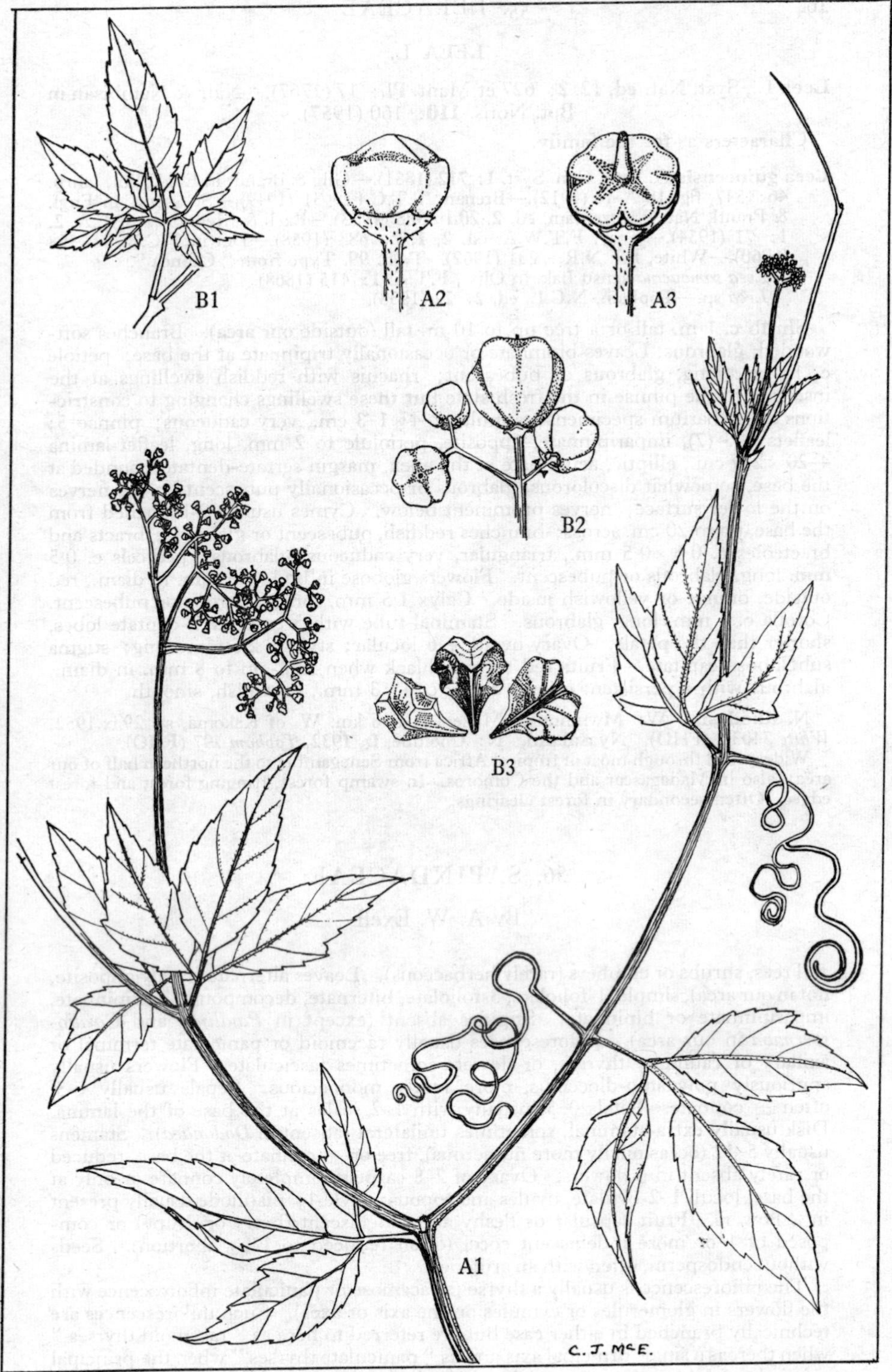

Tab. 98. A.—CAYRATIA IBUENSIS. A1, flowering branch (×⅔) *Kirk* s.n.; A2, flower-bud (×8) *Kirk* s.n.; A3, flower-bud with petals removed (×8) *Kirk* s.n. B.—CAYRATIA GRACILIS. B1, leaf (×⅔) *Richards* 727; B2, portion of infructescence (×2) *White* 2365; B3, seeds (×2) *White* 2365.

LEEA L.

Leea L., Syst. Nat. ed. 12, **2**: 627 et Mant. Pl.: 17 (1767).—Nair & Nambisan in Bot. Notis. **110**: 160 (1957).

Characters as for the family.

Leea guineensis G. Don, Gen. Syst. **1**: 712 (1831).—Gilg & Brandt in Engl., Bot. Jahrb. **46**: 547, fig. 18 A–K (1912).—Brenan, T.T.C.L.: 31 (1949).—Suesseng. in Engl. & Prantl, Nat. Pflanzenfam. ed. 2, **20d**: 388 (1953).—Exell & Mendonça, C.F.A. **2**, 1: 71 (1954).—Keay, F.T.W.A. ed. 2, **1**, 2: 683 (1958).—Dewit, F.C.B. **9**: 568 (1960).—White, F.F.N.R.: 231 (1962). TAB. **99**. Type from " Guinea ".
Leea sambucina sensu Bak. in Oliv., F.T.A. **1**: 415 (1868).
Leea sp.—Topham, N.C.L. ed. 2: 22 (1958).

Shrub c. 1 m. tall or a tree up to 10 m. tall (outside our area). Branches soft-wooded, glabrous. Leaves bipinnate or occasionally tripinnate at the base; petiole c. 12 cm. long, glabrous or pubescent; rhachis with reddish swellings at the insertion of the pinnae in the fresh state but these swellings changing to constrictions in herbarium specimens; stipules 2–4 × 1–3 cm., very caducous; pinnae 5; leaflets 3–5 (7), imparipinnate, opposite, petiolule to 2 mm. long, leaflet-lamina 4–20 × 2–7 cm., elliptic, acuminate at the apex, margin serrate-dentate, rounded at the base, somewhat discolorous, glabrous or occasionally pubescent on the nerves on the lower surface; nerves prominent below. Cymes usually 2-branched from the base, up to 20 cm. across; branches reddish, pubescent or glabrous; bracts and bracteoles c. 0·5 × 0·5 mm., triangular, very caducous, glabrous; pedicels c. 0·5 mm. long, glabrous or pubescent. Flowers globose in bud, 2–5 mm. in diam., red outside, orange or yellowish inside. Calyx 1·5 mm. long, glabrous or pubescent. Corolla c. 5 mm. long, glabrous. Staminal tube with 5 truncate 2-dentate lobes, shorter than the petals. Ovary ovoid, 4–6-locular; style c. 2·5 mm. long; stigma subglobose-capitate. Fruits red turning black when ripe, up to 8 mm. in diam., glabrous with a persistent calyx. Seeds c. 4 × 3 mm., brownish, smooth.

N. Rhodesia. W: Mwinilunga, Muzera R., 16 km. W. of Kakoma, st. 29.ix.1952, *White* 3403A (FHO). **Nyasaland.** N: Chombe, fr. 1932, *Topham* 797 (FHO).
Widespread through most of tropical Africa from Senegambia to the northern half of our area; also in Madagascar and the Comoros. In swamp forest, fringing forest and forest edges. Often secondary in forest clearings.

56. SAPINDACEAE

By A. W. Exell

Trees, shrubs or climbers (rarely herbaceous). Leaves alternate (rarely opposite, not in our area), simple, 1-foliolate, 3-foliolate, biternate, decompound, paripinnate, imparipinnate or bipinnate. Stipules absent (except in *Paullinia* and *Cardiospermum* in our area). Inflorescences usually racemoid or paniculate terminal or axillary or caulinary thyrses, or flowers sometimes fasciculate. Flowers usually spuriously polygamo-dioecious, more rarely monoecious. Sepals usually 4–5, often ± connate. Petals 0–5, usually with 1–2 scales at the base of the lamina. Disk usually extra-staminal, sometimes unilateral (absent in *Dodonaea*). Stamens usually 5–12 (occasionally more numerous), free or ± connate at the base, reduced or rarely absent in ♀ flowers. Ovary of 2–8 carpels completely connate or only at the base, loculi 1–2-ovulate, ovules anatropous; style 1; pistillode usually present in ♂ flowers. Fruit capsular or fleshy and indehiscent (berry or drupe) or composed of 1 or more indehiscent cocci (often reduced to 1 by abortion). Seeds without endosperm, often with an arillode.

The inflorescence is usually a thyrse (a racemose or paniculate inflorescence with the flowers in glomerules or cymules on the axis or axes). Such inflorescences are technically branched in either case but are referred to here as " racemoid thyrses " when there is a single principal axis and as " paniculate thyrses " when the principal axis is branched to form a panicle. The distinction is not always clear (see note under *Allophylus*, p. 497).

According to Van der Pijl (in Act. Bot. Néerl. **6**: 608 (1957)) the structures

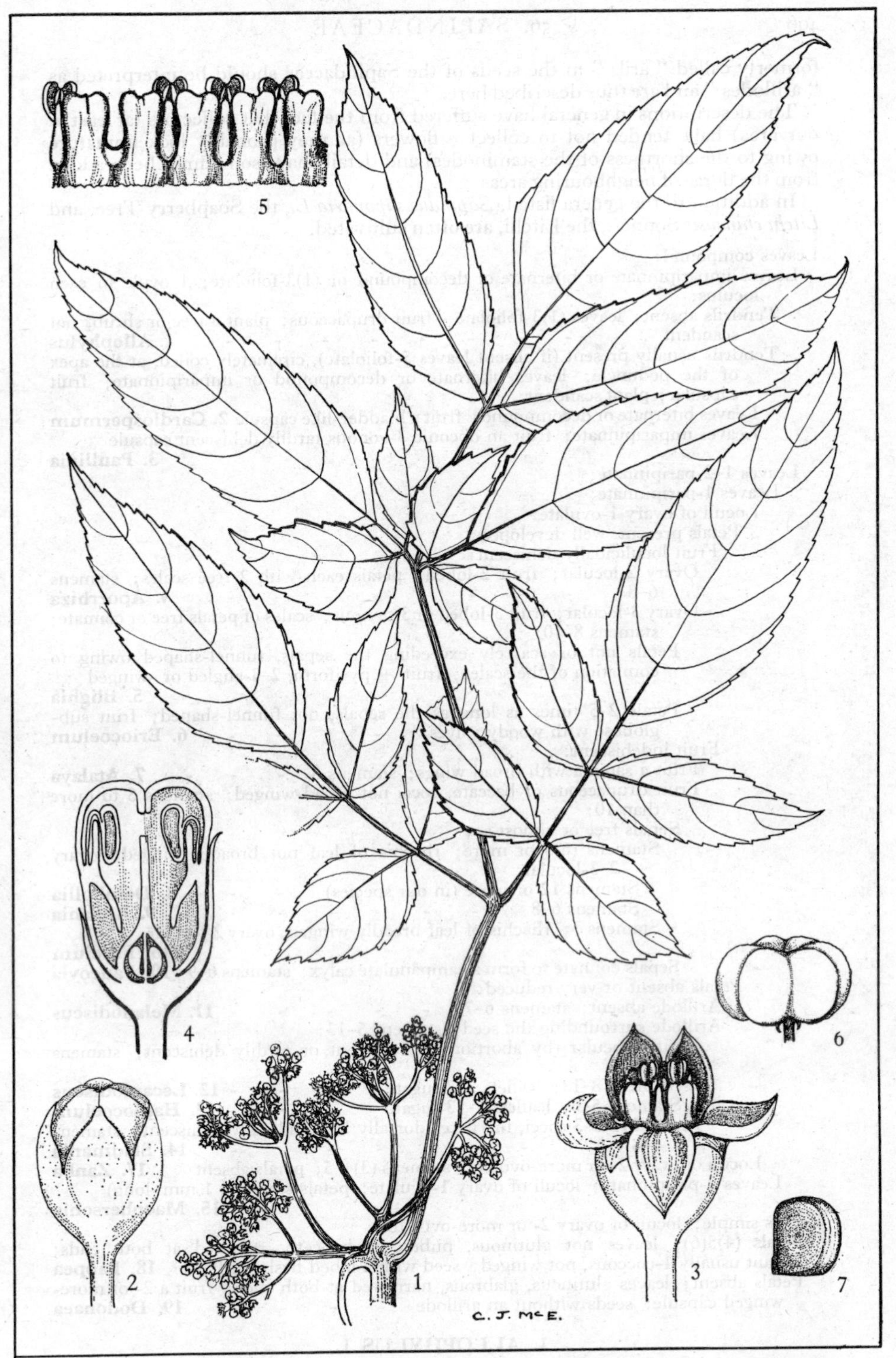

Tab. 99. LEEA GUINEENSIS. 1, flowering branch ($\times\frac{4}{9}$); 2, flower-bud ($\times 8$); 3, flower ($\times 8$); 4, longitudinal section of flower ($\times 8$); 5, staminal tube and stamens spread out (with one stamen removed) ($\times 8$); 6, fruit ($\times 3$); 7, seed ($\times 3$). 1–5 from *Battiscombe* 499, 6 & 7 from *Verdcourt* 1667.

formerly called "arils" in the seeds of the Sapindaceae should be interpreted as "arillodes" and are thus described here.

The descriptions in general have suffered from the fact that collectors (at least in our area) have tended not to collect ♀ flowers (as they probably look immature owing to the shortness of the staminodes) and details have sometimes been added from the floras of neighbouring areas.

In addition to the genera listed, *Sapindus saponaria L.*, the Soapberry Tree, and *Litchi chinensis* Sonner., the Litchi, are often cultivated.

Leaves compound:
  Leaves imparipinnate or biternate or decompound or (1)3-foliolate; 1 ovule in each loculus:
    Tendrils absent; leaves (1)3-foliolate; fruit drupaceous; plant a tree or shrub, not scandent - - - - - - - - - - **1. Allophylus**
    Tendrils usually present (if absent leaves 5-foliolate), circinately coiled, at the apex of the peduncle; leaves biternate or decompound or imparipinnate; fruit capsular; plant scandent:
      Leaves biternate or decompound; fruit a bladder-like capsule **2. Cardiospermum**
      Leaves imparipinnate; fruit an obconic 3-gonous tardily dehiscent capsule **3. Paullinia**
  Leaves 1–2-paripinnate:
    Leaves 1-paripinnate:
      Loculi of ovary 1-ovulate:
        Petals present, well developed:
          Fruit loculicidally dehiscent:
            Ovary 2-locular; fruit 2-lobed; petals each with 2 free scales; stamens 6–8 - - - - - - - - - - **4. Aporrhiza**
            Ovary 3-locular; fruit 3-lobed or 3-gonous; scales of petals free or connate; stamens 8–10:
              Petals not or scarcely exceeding the sepals, funnel-shaped owing to connation of the scales; fruit ± pyriform, 2–3-angled or -winged **5. Blighia**
              Petals 2–5 times as long as the sepals, not funnel-shaped; fruit subglobose with woody valves - - - - **6. Eriocoelum**
          Fruit indehiscent:
            Fruit a samara with broad wings; stamens 8 - - - **7. Atalaya**
            Fruit drupaceous or baccate, cocci not broad-winged; stamens 5 to more than 10:
              Sepals free or almost so:
                Stamens (6)8 or more; rhachis of leaf not broadly winged; ovary 2–3-locular:
                  Stamens 12 or more (in our species) - - **8. Deinbollia**
                  Stamens 6–8 - - - - - - - - - **9. Aphania**
                Stamens 5; rhachis of leaf broadly winged; ovary 2-locular **16. Filicium**
              Sepals connate to form a campanulate calyx; stamens 6–9 **10. Pancovia**
        Petals absent or very reduced:
          Arillode absent; stamens 6–7 - - - - - - **11. Melanodiscus**
          Arillode surrounding the seed; stamens 5–13:
            Fruit 1-locular (by abortion), indehiscent or tardily dehiscent; stamens 5–13:
              Stamens 8–13; leaflets 3–7-jugate - - - **12. Lecaniodiscus**
              Stamens 5–6; leaflets 5–13-jugate - - - **13. Haplocoelum**
            Fruit of (1)2–3 cocci, the latter dorsally longitudinally dehiscent; stamens 6–8 - - - - - - - - - - - **14. Stadmania**
      Loculi of ovary 2- or more-ovulate; stamens (3)4–5; petals absent **17. Zanha**
    Leaves 2-paripinnate; loculi of ovary 1-ovulate; petals small (c. 1 mm. long) **15. Macphersonia**
Leaves simple; loculi of ovary 2- or more-ovulate:
  Petals (4)5(6); leaves not glutinous, pubescent beneath, rounded at both ends; fruit usually 1-coccous, not winged; seed with a lobed fleshy arillode **18. Pappea**
  Petals absent; leaves glutinous, glabrous, narrowed at both ends; fruit a 2- or more-winged capsule; seeds without an arillode - - - - **19. Dodonaea**

1. ALLOPHYLUS L.

Allophylus L., Sp. Pl. **1**: 348 (1753); Gen. Pl. ed. 5: 164 (1754).

Trees or shrubs. Leaves (1) 3-foliolate, usually minutely glandular beneath. Inflorescence a racemoid or paniculate thyrse. Flowers monoecious (? rarely polygamo-dioecious) usually ♂ and ♀ in the same inflorescence (the ♀ usually apparently ⚥ but the anthers sterile), small, greenish, white or yellowish. Sepals 4, the 2 outer ones larger, imbricate. Petals 4, clavate or spathulate with an emarginate apex, each with a usually (always in our species) hairy scale. Disk unilateral, usually 4-lobed. Stamens 8 (rarely fewer), slightly connate at the base, somewhat shorter in the ♀ flowers (staminodes). Ovary deeply 2–3-lobed; loculi 1-ovulate; style 2–3-fid. Fruit drupaceous, of 2–3 cocci or frequently of 1 coccus by abortion. Seeds obovoid, without arillodes, usually (always in our species) glabrous.

As far as my dissections show, the functionally ♀ flowers usually develop first and for this reason the length given for the filaments in the ♂ flowers may sometimes be too short owing to their immaturity. The glomerule or cymule usually (or at least often) consists of one ♀ flower and several ♂ ones.

The taxonomy of the species of this genus is unusually difficult as nearly all the species appear to hybridize (though I know of no experimental evidence of this), the leaflets are very variable in shape and indumentum and the flowers provide few characters of value in classification. It is difficult to avoid using the character of " branched " or " unbranched " inflorescences in the key and in good specimens this is usually not difficult to determine; but poor or depauperated specimens, probably normally " branched ", may not show the character, while species with normally " unbranched " inflorescences may occasionally show a small branch. The inflorescences are never truly simple as the flowers are grouped in small subsessile glomerules or shortly pedunculate cymules along the axis or axes and this is technically branching.

The classification here proposed is still far from satisfactory but only experimental work can disentangle the taxonomy. The fruits may well provide useful characters but so relatively few fruiting specimens have been collected that there is little point in using fruit-characters in the key.

Branchlets and inflorescences minutely tomentellous; style usually 3-fid 1. *abyssinicus*
Branchlets and inflorescences pubescent to tomentose or more rarely glabrous; style usually 2-fid (sometimes 3-fid in *A. didymadenius*):
  Terminal leaflet usually at least 4–5 times as long as the lateral ones which are sometimes completely reduced (rarely only twice as long in a few leaves) 2. *congolanus*
  Terminal leaflet $1\frac{1}{2}$–2 times as long as the lateral ones (the 3 leaflets occasionally subequal):
    Flowers rather large for the genus with outer sepals 2–2·5 mm. long; rhachis of cymule up to 3–6 mm. long: - - - - - - 3. *didymadenius*
    Flowers small to average for the genus with outer sepals 1·3–2 mm. long; rhachis of cymule usually not more than 3 mm. long or glomerules subsessile:
      Terminal leaflet 3–4 times as long as broad, narrowly elliptic - 4. *natalensis*
      Terminal leaflet up to $1\frac{1}{2}$–3 times as long as broad, usually elliptic or obovate-elliptic:
        Terminal leaflet usually lobed (in our area) to half the distance to the midrib, 2·5–5 cm. long; inflorescence 1·5–6 cm. long, unbranched - 5. *decipiens*
        Terminal leaflet with margin subentire to crenate or serrate, usually more than 5 cm. long (except in *A. alnifolius*); inflorescence usually more than 5 cm. long (except sometimes in *A. alnifolius*):
          Inflorescence unbranched (racemoid):
            Terminal leaflet typically ± obovate (usually broadest above the middle) and narrowing abruptly from the middle towards the narrowly cuneate base so that the margin (at least in most leaves) is somewhat concave between the broadest part of the lamina and the base:
              Branchlets and petioles almost glabrous or sparsely appressed- or crisped-pubescent or sparsely pilose - - - - - - 6. *alnifolius*
              Branchlets and petioles densely ± patent-pubescent or tomentose 7. *rubifolius*
            Terminal leaflet usually elliptic and broadest near the middle and not conspicuously narrowing in the lower half (or only occasionally so towards the very base):
              Under surface of leaf densely pubescent or pilose to subtomentose; margin of leaflets conspicuously serrate - - - 8. *richardsiae*
              Under surface of leaf almost glabrous to pubescent:
                Under surface of leaf almost glabrous (except for occasional tufts of hairs on the lower surface in the axils of the lateral nerves) 9. *chaunostachys*

Under surface of leaf pubescent:
Terminal leaflet 7–15 cm. long, usually toothed or crenate for at least $\frac{2}{3}$ of the length of the margin:
Terminal leaflet 2–3 times as long as broad, margin usually rather sharply dentate or serrate - - - - 10. *mossambicensis*
Terminal leaflet $1\frac{1}{2}$–2 times as long as broad, margin crenate to shallowly dentate or subentire - - - - 11. *whitei*
Terminal leaflet 5–8 cm. long, usually toothed only in the upper half of the margin - - - - - - 12. *rhodesicus*
Inflorescence branched (paniculate) when well developed:
Terminal leaflet 3–9 cm. long; inflorescence 2·5–10 cm. long:
Branchlets and petioles almost glabrous or sparsely appressed- or crisped-pubescent or sparsely pilose - - - - 6. *alnifolius*
Branchlets and petioles densely ± patent-pubescent or tomentose 7. *rubifolius*
Terminal leaflet up to 15–17 cm. long; inflorescence 9–27 cm. long:
Leaves membranous to thinly papyraceous; inflorescence sparsely pubescent to glabrous - - - - - - 13. *chirindensis*
Leaves papyraceous to subcoriaceous; inflorescence densely pubescent 14. *africanus*

1. **Allophylus abyssinicus** (Hochst.) Radlk. in Engl. & Prantl, Nat. Pflanzenfam. **3,** 5: 313 (1896); in Sitz.-Ber. Bayer. Akad. **38,** 2: 223 (1909); in Engl., Pflanzenr. IV, 165, **1,** 2: 534 (1932).—Bak. f. in Journ. of Bot. **57**: 186 (1919).—Brenan, T.T.C.L.: 554 (1949).—Hauman, F.C.B. **9**: 303 (1960).—Dale & Greenway, Kenya Trees and Shrubs: 504 (1961).—White, F.F.N.R: 223 (1962). Type from Ethiopia.
Schmidelia abyssinica Hochst. in Flora, **26**: 80 (1843). Type as above.
Schmidelia africana sensu Bak. in Oliv., F.T.A. **1**: 421 (1868) pro parte quoad syn. *S. abyssinica.*

Tree up to 20 m. tall; wood hard, white, brittle; bark smooth, greyish-green; branchlets minutely tomentellous, eventually glabrescent. Leaves 3-foliolate; petiole 2–9 cm. long, minutely tomentellous; leaflets subequal or the terminal one up to nearly twice as long as the lateral ones; petiolules 2–9 mm. long; lamina of terminal leaflet 4·5–14 × 2–6 cm., elliptic, chartaceous, minutely tomentellous (sometimes subsericeous) on both surfaces when young and soon glabrescent, minutely glandular beneath, apex acute and slightly acuminate, margin rather shallowly and often irregularly (sometimes doubly) serrate or dentate, base cuneate and sometimes slightly narrowed; lateral nerves 8–10 pairs, often with tufts of hairs in their axils beneath. Inflorescence up to 8 cm. long (in our area), 1–2-branched, minutely tomentellous. Flowers in few-flowered subsessile glomerules, pedicels up to 1 mm. long, minutely tomentellous. Outer sepals 1·5 × 1·5 mm., subcircular, minutely pubescent, inner 1·5 × 1 mm., elliptic, minutely pubescent. Petals 1 × 0·7 mm., spathulate. Stamens with filaments 1 mm. long in ♂ flowers; staminodes with filaments 0·5 mm. long in ♀ flowers. Style 1·5 mm. long, usually 3–fid. Fruit (not seen from our area) 6–8 × 6 mm., obovoid-globose, sparsely pubescent, with 1 or 2 cocci developing.

S. Rhodesia. E: Banti Forest Reserve, fl. 16.ix.1947, *English* 25/47 (FHO; SRGH). **Nyasaland.** C: Dedza Mt., 1980m., *Chapman* 1237 (FHO; SRGH).

From Ethiopia and the Congo southwards to S. Rhodesia and Nyasaland. Submontane forest.

Distinguished from all other species of the genus in our area by having a usually 3-fid style and by its minute tomentellum.

2. **Allophylus congolanus** Gilg in Engl., Bot. Jahrb. **24**: 294 (1897).—Radlk. in Engl., Pflanzenr. IV, 165, **1,** 2: 527 (1932).—Burtt Davy & Hoyle, N.C.L.: 69 (1936).—Brenan, T.T.C.L.: 554 (1949).—Hauman, F.C.B. **9**: 314 (1960). Type from the Congo.
Allophylus appendiculato-serratus Gilg, op. cit. **30**: 348 (1901).—Bak. f. in Journ. of Bot. **57**: 184 (1919). Type from Tanganyika.
Allophylus congolanus var. *monophyllus* Bak. f., loc. cit. Type probably from Tanganyika (" E. coast of Lake Nyasa ").

Small tree or shrub; branchlets fulvous-tomentose. Leaves 1-foliolate or 3-foliolate with the lateral leaflets much reduced; petiole 2–4 mm. long, fulvous-tomentose; terminal leaflet usually at least 4–5 times as long as the lateral ones (rarely only twice as long in a few leaves), the latter sometimes completely reduced;

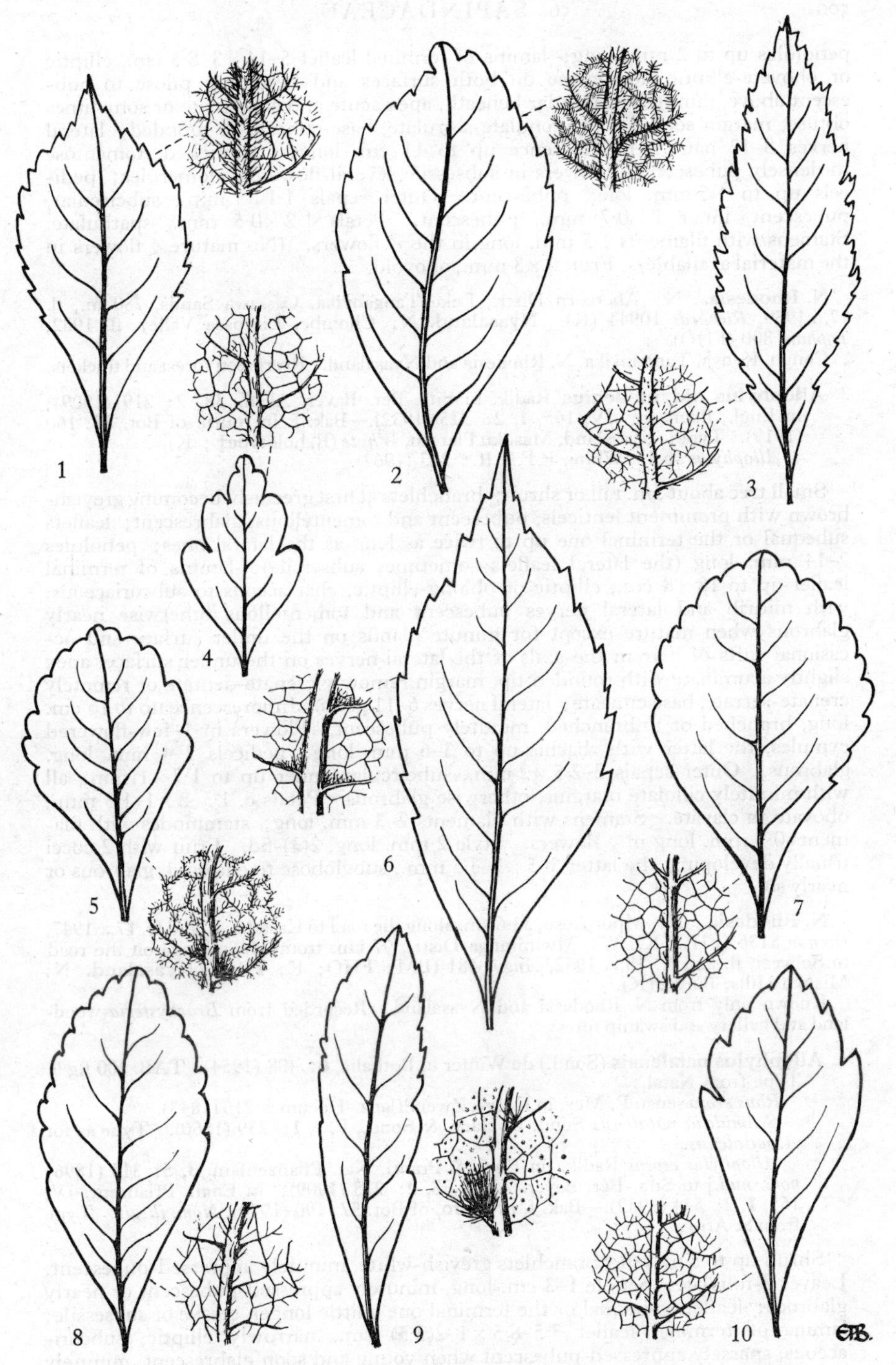

Tab. 100. Leaflets of ALLOPHYLUS (all × ⅔; reticulation × 4). 1, A. RICHARDSIAE; 2, A. WHITEI; 3, A. MOSSAMBICENSIS; 4, A. DECIPIENS; 5, A. RUBIFOLIUS; 6, A. CHAUNOSTACHYS; 7, A. ALNIFOLIUS; 8, A. AFRICANUS (CATARACTARUM form); 9, A. NATALENSIS; 10, A. RHODESICUS.

petiolules up to 2 mm. long; lamina of terminal leaflet 5–14 × 3–8·5 cm., elliptic or obovate-elliptic, tomentose on both surfaces and becoming pilose to pubescent above, minutely glandular beneath, apex acute and acuminate or sometimes obtuse, margin serrulate or crenulate-serrulate, base cuneate to rounded; lateral nerves 8–12 pairs. Inflorescence up to 11 cm. long, unbranched, tomentose to densely pubescent. Flowers in subsessile several-flowered glomerules; pedicels up to 1·5 mm. long, pubescent. Outer sepals 1·1 × 1 mm., subcircular, pubescent, inner 1 × 0·7 mm., pubescent. Petals 1·2 × 0·5 mm., spathulate. Stamens with filaments 1·5 mm. long in the ♂ flowers. (No mature ♀ flowers in the material available). Fruit 6 × 3 mm., obovoid.

N. Rhodesia. N: Abercorn Distr., Lake Tanganyika, Cassawa Sands, 780 m., fl. 17.ii.1959, *Richards* 10944 (K). **Nyasaland.** N: Chombe, Limpasa Valley, fl. 1932, *Topham* 800 (FHO).

Congo, Kenya, Tanganyika, N. Rhodesia and Nyasaland. Evergreen forest and thickets.

3. **Allophylus didymadenius** Radlk. in Sitz.-Ber. Bayer. Akad. **38,** 2: 219 (1909); in Engl., Pflanzenr. IV, 165, **1,** 2: 525 (1932).—Bak. f. in Journ. of Bot. **57**: 160 (1919). Type: Nyasaland, Masuku Plateau, *Whyte* (B, holotype† ; K).
Allophylus sp.1—White, F.F.N.R.: 223 (1962).

Small tree about 5m. tall or shrub; branchlets at first greenish, becoming greyish-brown with prominent lenticels, pubescent and tomentellous, glabrescent; leaflets subequal or the terminal one up to twice as long as the lateral ones; petiolules 2–14 mm. long (the lateral leaflets sometimes subsessile); lamina of terminal leaflet up to 18 × 8 cm., elliptic or oblong-elliptic, chartaceous to subcoriaceous, with midrib and lateral nerves pubescent and tomentellous otherwise nearly glabrous when mature except for minute glands on the under surface and occasional tufts of hair in the axils of the lateral nerves on the under surface, apex slightly acuminate with rounded tip, margin remotely crenate-dentate or remotely crenate-serrate, base cuneate; lateral nerves 6–11 pairs. Inflorescence up to 16 cm. long, branched or unbranched, minutely pubescent. Flowers in 1–few-flowered cymules, the latter with rhachis up to 3–6 mm. long; pedicels 2–4 mm. long, glabrous. Outer sepals 2–2·5 × 2 mm., subcircular, inner up to 1·7 × 1 mm., all with minutely ciliolate margins, otherwise glabrous. Petals c. 1·5–2 × 1–1·3 mm., obovate to clavate. Stamens with filaments 2–3 mm. long; staminodes with filaments 0·5 mm. long in ♀ flowers. Style 2 mm. long, 2(3)-fid. Fruit with 2 cocci usually developing, the latter 3–5 × 2–3·5 mm., subglobose to ellipsoid, glabrous or nearly so.

N. Rhodesia. N: Mporokoso, 33·6 km. along the road to Kasama, fl. and fr. 17.x.1947, *Brenan* 8136 (FHO; K). W: Mwinilunga Distr., 97 km. from Mwinilunga on the road to Solwezi, fl. and fr. 18.ix.1952, *Angus* 481 (BM; FHO; K; PRE). **Nyasaland.** N: Misuku Hills, *Whyte* (K).

Known only from N. Rhodesia and Nyasaland. Recorded from *Brachystegia*-woodland and gallery and swamp forest.

4. **Allophylus natalensis** (Sond.) de Winter in Bothalia, **6**: 408 (1954). TAB. **100** fig 9. Type from Natal.
Rhus erosa sensu E. Mey. in Drège, Zwei Pflanz.-Docum.: 216 (1843).
Schmidelia natalensis Sond. in Harv. & Sond., F.C. **1**: 239 (1860). Type as for *A. natalensis.*
Allophylus erosus Radlk. [in Engl. & Prantl, Nat. Pflanzenfam. **3,** 5: 313 (1896) *nom. nud.*] in Sitz.-Ber. Bayer. Akad. **38,** 2: 225 (1909); in Engl., Pflanzenr. IV, 165, **1,** 2: 544 (1932).—Bak. f. in Journ. of Bot. **57**: 190 (1919). *Nom. illegit.* Type from S. Africa.

Shrub up to 5 m. tall; branchlets greyish-white, minutely appressed-pubescent. Leaves 3-foliolate; petiole 1–3 cm. long, minutely appressed-pubescent or nearly glabrous; leaflets subequal or the terminal one a little longer, sessile or subsessile; lamina of terminal leaflet 3·5–8·5 × 1–2(2·5) cm., narrowly elliptic, subcoriaceous, sparsely appressed-pubescent when young and soon glabrescent, minutely glandular and punctate on the under surface, apex rounded, margin rather remotely and shallowly serrulate, base cuneate; lateral nerves 5–8 pairs. Inflorescence 3–7 cm. long, usually branched, minutely appressed-pubescent to nearly glabrous. Flowers in few-flowered subsessile glomerules; pedicels up to 2·5 mm. long, glabrous. Outer sepals 2 × 2 mm., subcircular, minutely pubescent, inner

1·8 × 1·2 mm., broadly elliptic, minutely pubescent. Petals 1·5 × 1 mm., spathulate. Staminodes (no ♂ flowers seen) with filaments 0·5 mm. long in ♀ flowers. Ovary 2-lobed, minutely pubescent; style up to 1·5 mm. long, 2-fid. Fruit bright red, 6 mm. in diam., subglobose, sparsely pubescent when mature, 1 coccus usually developing.

Mozambique. LM: Inhaca I., fl. 14.vii.1957, *Barbosa* 7715 (LMJ; PRE; SRGH); Marracuene, fr. i.vi.1959, *Barbosa & Lemos* 8546 (COI; LISC; LMJ; PRE; SRGH).
Also in S. Africa. Coastal dunes (in our area).

5. **Allophylus decipiens** (Sond.) Radlk. in Engl. & Prantl, Nat. Pflanzenfam. **3,** 5: 313 (1896); in Engl., Pflanzenr. IV, 165, **1,** 2: 522 (1932).—Bak. f. in Journ. of Bot. **57**: 160 (1919).—de Winter in Bothalia, **6**: 407 (1954). TAB. **100** fig. 4. Type from S. Africa.
Rhus spicata Thunb., Fl. Cap. **2**: 217 (1818). Type from S. Africa.
Schmidelia decipiens [Presl, Bot. Bemerk.: 41 (1844) *nom. nud.*] Sond. in Harv. & Sond., F.C. **1**: 239 (1860). Type as for *A. decipiens.*
Allophylus spicatus (Thunb.) Fourcade in Trans. Roy. Soc. S. Afr. **21**: 100 (1932) non *A. spicatus* (Poir.) Radlk. (1896).

Many-stemmed shrub; branchlets silver-grey, appressed-pubescent when young, soon glabrescent. Leaves 3-foliolate; petiole 0·5–3·5 cm. long, appressed- or crisped-pubescent; terminal leaflet usually c. 1½ times as long as the lateral ones; petiolules up to 1 mm. long or leaflets subsessile; lamina of terminal leaflet 1–4·5 × 0·5–2·5 cm., papyraceous, shortly and sparsely appressed- or crisped-pubescent on the midrib and lateral nerves and with occasional tufts of hairs on the under surface of the axils of the latter, otherwise glabrous, lobed usually to about halfway to the midrib (at least in some leaves), lobes generally rounded but apex occasionally acute, base cuneate and often narrowed; lateral nerves 3–6 pairs. Inflorescence 1·5–6 cm. long, unbranched, appressed-pubescent. Flowers whitish, in few-flowered very shortly pedunculate cymules; pedicels up to 2 mm. long. Outer sepals 1·3 × 1·3 mm., subcircular, inner 1·3 × 1 mm., elliptic, all glabrous or nearly so. Petals 1·2 × 1 mm., obovate. Stamens with pilose filaments 2 mm. long; staminodes with filaments 0·7 mm. long in ♀ flowers. Style 1 mm. long, 2-fid. Fruit (not known from our area) 3·5 mm. in diam., globose.

Mozambique. LM: Umbeluzi, fl. 18.ii.1955, *Pedro* 5005 (LMJ).
Also in S. Africa. Thickets and thicket margins of riverine forests at low altitudes.

The above description (apart from the fruit) is of Mozambique specimens. In S. Africa some specimens have much less deeply lobed leaflets.

6. **Allophylus alnifolius** (Bak.) Radlk. in Engl. & Prantl, Nat. Pflanzenfam. **3,** 5: 313 (1896); in Engl., Pflanzenr. IV, 165, **1,** 2: 521 (1932).—Bak. f. in Journ. of Bot. **57**: 159 (1919).—Burtt Davy & Hoyle, N.C.L.: 69 (1936).—Brenan, T.T.C.L.: 553 (1949).—Williamson, Useful Pl. Nyasal.: 16 (1956).—Hauman, F.C.B. **9**: 304 (1960). —Dale & Greenway, Kenya Trees and Shrubs: 504 (1961). TAB. **100** fig 7. Type: Mozambique, locality uncertain, *Forbes* (BM; K, holotype).*
Schmidelia alnifolia Bak. in Oliv., F.T.A. **1**: 422 (1868).—Sim, For. Fl. Port. E. Afr.: 31 (1909). Type as above.
Schmidelia repanda Bak., loc. cit.—Sim, loc. cit. Syntypes: Mozambique, R. Rovuma, *Kirk* (K); Lower Chire Valley, *Kirk* (K); Chire, Elephant Marsh, *Meller* (K).
Allophylus repandus (Bak.) Engl., Bot. Jahrb. **17**: 160 (1893).—Bak. f., tom. cit.: 185 (1919). Type as for *S. repanda.*

Shrub up to 6 m. tall (once recorded as a tree); bark grey; branchlets usually greyish-brown or white, appressed-pubescent when young (rarely sparsely pilose), soon glabrescent. Leaves 3-foliolate; petiole up to 5·5 cm. long, appressed- or crisped-pubescent or glabrous; leaflets subequal or terminal one 1½–2 times as long as the lateral ones; petiolules up to 8 mm. long in the terminal leaflet but more usually all 3 leaflets subsessile; lamina of terminal leaflet 3–8·5 × 2–4·5 cm., obovate to obovate-elliptic, usually broadest well above the middle, papyraceous, sparsely pilose or appressed-pubescent mainly on the midrib and nerves beneath

* There are two unnumbered and not precisely localised *Forbes* specimens of *A. alnifolius* in the Kew Herbarium. The one stamped "Herbarium Hookerianum" with "*S. alnifolia* Baker!" written on it in Baker's hand is obviously the holotype. The other specimen stamped "Herbarium Benthamianum" was written up later in another hand.

and often with tufts of hairs on the under surface in the axils of the lateral nerves, minutely glandular beneath, otherwise nearly glabrous, apex rounded to obtuse (rarely acute), margin subentire or crenate or crenate-serrate or slightly lobed, base usually very narrowly cuneate and distinctly narrowed from the middle downwards; lateral nerves 4–6 pairs. Inflorescence 2·5–12 cm. long, branched or unbranched, appressed- or crisped-pubescent. Flowers yellowish-green or cream or white, in few-flowered subsessile glomerules; pedicels up to 1·3 mm. long. Outer sepals 1·5 × 1·5 mm., inner 1 × 1 mm., all subcircular, glabrous or sparsely minutely pubescent. Petals 1·5 × 1 mm., obcuneate. Disk 4-lobed, unilateral. Stamens with glabrous filaments 1·2 mm. long; staminodes 1 mm. long in the ♀ flowers. Style 1·5 mm. long, 2-fid. Fruits red or orange (one record) or black (one record), 6 × 4 mm., ellipsoid-globose, sparsely pubescent when young.

S. Rhodesia. W: Matobo Distr., Farm Shimbashawa, 1370 m., fl.i.1954, *Miller* 2134 (K; LISC; SRGH). E: Chipinga Distr., Sabi Valley, fl.i.1957, *Davies* 2402 (K; LISC; SRGH). S: Belingwe Distr., Ngobi Dip, fl. 11.xii.1953, *Wild* 4347 (K; LISC; SRGH). **Mozambique.** N: Ilha Tecomaze, fl. 29.iii.1961, *Gomes e Sousa* 4674 (K; PRE; SRGH). Z or T: Lower Chire Valley, fl. i.1862, *Kirk* (K). MS: Mossurize, fl. 22.ii.1907, *Johnson* 150 (K) SS: Guijá, fl. 10.vi.1947, *Pedrógão* 285 (COI; K; LISC; LMJ; PRE).

Congo (*fide* Hauman), S. Rhodesia and Mozambique. *Brachystegia*- and *Colophospermum mopane*-woodlands, thickets, wooded grassland and riverine fringes up to c. 1400 m.

7. **Allophylus rubifolius** (Hochst. ex A. Rich.) Engl., Hochgebirgsfl. Trop. Afr.: 292 (1892) pro parte excl. specim. *Meyer* 158 fide Radlk.—Schinz in Denkschr. Math. Naturw. Akad. Wiss. Wien, **78**: 427 (1905).—Eyles in Trans. Roy. Soc. S. Afr. **5**, 4: 405 (1916) (" rubrifolius ").—Bak. f. in Journ. of Bot. **57**: 159 (1919)—Radlk. in Engl., Pflanzenr. IV, 165, **1**, 2: 520 (1932).—Hauman, F.C.B. **9**: 305 (1960).—Dale & Greenway, Kenya Trees and Shrubs: 504 (1961). TAB. **100** fig. 5. Syntypes from Ethiopia.

Schmidelia rubifolia Hochst. ex A. Rich., Tent. Fl. Abyss. **1**: 103 (1847).—Bak. in Oliv., F.T.A. **1**: 423 (1868).—Sim, For. Fl. Port E. Afr.: 32 (1909). Syntypes as above.

Allophylus tristis Radlk. in Sitz.-Ber. Bayer. Akad. **38**, 2: 225, 237 (1909); in Engl., Pflanzenr. IV, 165, **1**, 2: 542 (1932).—Bak. f., tom. cit: 188 (1919). Syntypes: Mozambique, Sena, Chiramba and Chupanga, *Kirk* (K); Quelimane, *Stuhlmann* 668(B†); Puguruni, *Stuhlmann* 670 (B†).

Shrub (rarely shrublet) up to 5m. tall or small tree; branchlets usually pale coloured, tomentose or densely pubescent and only tardily glabrescent. Leaves 3-foliolate; petiole up to 5 cm. long, tomentose or densely pubescent; leaflets subequal or more often the terminal one up to twice as long as the lateral ones; petiolules up to 4 mm. long but leaflets usually subsessile; lamina of terminal leaflet 5–9 × 2·5–5 cm., obovate-elliptic or elliptic or rarely narrowly elliptic, usually broadest near the middle, papyraceous to chartaceous, densely pubescent to tomentose on both surfaces, minutely glandular beneath, usually with conspicuous tufts of hairs on the under surface in the axils of the lateral nerves and sometimes also in the axils of the secondary nerves, apex rounded to acute, margin irregularly crenate-serrate or dentate or remotely serrulate, base narrowly cuneate and distinctly narrowed from the middle downwards; lateral nerves 4–8 pairs. Inflorescence 4–10 cm. long, branched or unbranched. Flowers white, greenish-white or yellow, in few-flowered subsessile glomerules; pedicels up to 1·5 mm. long, sparsely pubescent or glabrous. Outer sepals 1·8 × 1·9 mm., subcircular, inner 1 × 1 mm., subcircular, all minutely pubescent or nearly glabrous, with minutely ciliate margins. Petals 1·2–1·8 × 0·8–1 mm., clavate or spathulate. Stamens with filaments 1·6 mm. long., staminodes with filaments 0·6 mm. long in the ♀ flowers. Ovary 2-lobed, densely pubescent; style 2 mm. long, 2-fid. Fruit red or orange, 4–6 mm. in diam., subglobose, sparsely pubescent.

N. Rhodesia. E: Petauke-Sasare road, 800 m., fl. 25.1.1959; *Robson* 831 (K). **S. Rhodesia.** Umtali Commonage, fl. 17.1.1950. *Chase* 1923 (BM; K; SRGH). **Nyasaland.** S: Port Herald Distr., near Namyala R., fr. 25.iii.1961, *Phipps* 2740 (SRGH). **Mozambique.** N: Nampula, fl. 21.xi.1937, *Torre* 1230 (LISC). Z: Quelimane Distr., Lugela-Mocuba, Namagoa, fl. 17.iii.1949, *Faulkner* 439 (COI; K; SRGH). MS: Chimoio, Serra de Chibata, fl. 11.ii.1948, *Garcia* 165 (LISC).

From Ethiopia and the Congo southwards to S. Rhodesia, Nyasaland, Mozambique and Natal. Dense or open *Brachystegia* woodland, bush and thickets.

If this species should eventually prove to be conspecific with *A. melanocarpus* (E. Mey.) Radlk. (said to have black or white fruits and not recorded by Radlkofer outside S. Africa) that name would have priority.

Several specimens from N. and S. Rhodesia (*Davies* 2697 (SRGH); *Goodier* 22 (LISC; SRGH); *Lovemore* 299 (LISC; SRGH); *Phelps* 20 (SRGH); *Phipps* 2279 (K; PRE; SRGH), 2359 (SRGH); *Robinson* 1782 (SRGH)) seem to be predominantly *A. rubifolus* but apparently transitional to *A. africanus* (Group B. *griseo-tomentosus*). There are also many specimens with branched inflorescences which seem nevertheless to belong to *A. rubifolius* although Radlkofer (loc. cit.) and Hauman (loc. cit.) restrict the species to plants with unbranched inflorescences.

8. **Allophylus richardsiae** Exell in Bol. Soc. Brot., Sér. 2, **38**: 109 (1964). TAB. **100** fig. 1. Type: N. Rhodesia, Abercorn Distr., Itembe Gorge, *Richards* 12070 (K, holotype; SRGH).

Shrub c. 2·5 m. tall or small tree; branchlets fulvous-tomentose. Leaves 3-foliolate; petiole up to 7 cm. long, fulvous-tomentose; leaflets subequal or the terminal one c. 1½ times as long as the lateral ones; petiolules up to 4 mm. long but the lateral leaflets often subsessile; lamina of terminal leaflet 6–10 × 3–4·5 cm., elliptic, drying dark brown above and greyish-green beneath, papyraceous, pubescent to densely pubescent (especially on the nerves) or pilose above and densely pubescent or pilose to subtomentose and minutely glandular and often with tufts of hairs in the axils of the lateral nerves on the under surface, apex acuminate, margin serrate, base cuneate; lateral nerves 8–10 pairs. Inflorescence 5–13 cm. long, unbranched, densely pubescent. Flowers in few-flowered subsessile glomerules; pedicels up to 3 mm. long, nearly glabrous. Outer sepals 1·5 × 2–2·5 mm., subreniform, with minutely ciliate margins but otherwise glabrous, inner 1·3 × 1·1 mm., subcircular. Petals 1·2 × 0·8 mm., spathulate. Stamens with filaments 1–1·5 mm. long; staminodes with filaments 0·8 mm. long in ♀ flowers. Style 1·2 mm. long, 2-fid. Fruit 5 × 4 mm., broadly ellipsoid, minutely pubescent, 1 or 2 cocci developing.

N. Rhodesia. N: Abercorn Distr., Itembe Gorge, 1525 m., fl. 3.i.1960, *Richards* 12070 (K; SRGH). **Nyasaland.** S: Zomba Distr., fl. 1937, *Clements* 716 (FHO). Without precise locality, *Buchanan* 427 pro parte (BM).

Known only from N. Rhodesia and Nyasaland. Ecology insufficiently known (recorded from cliff-faces).

Buchanan 427 (B†) was one of the numerous syntypes of *A. griseo-tomentosus* Gilg. *Buchanan* 427 (BM) consists of two specimens which are clearly different gatherings. The upper one, with unbranched inflorescences, is *A. richardsiae*; the lower one, with very immature branched inflorescences appears to be *A. griseo-tomentosus* (placed in this work under *A. africanus*).

9. **Allophylus chaunostachys** Gilg in Engl., Bot. Jahrb. **30**: 349 (1901).—Radlk. in Engl., Pflanzenr. IV, 165, **1**, 2: 524 (1932).—Brenan, T.T.C.L.: 553 (1949).—Meikle in Mem. N.Y. Bot. Gard. **8**, 3: 240 (1953). TAB. **100** fig. 6. Type from Tanganyika.

Allophylus buchananii Gilg ex Radlk. in Sitz.-Ber. Bayer. Akad. **38**, 2: 219 (1909).—Bak. f. in Journ. of Bot. **57**: 182 (1919).—Radlk. in Engl., Pflanzenr. IV, 165, **1**, 2: 524 (1933).—Burtt Davy & Hoyle, N.C.L.: 69 (1936).—Meikle, loc. cit.—Hauman, F.C.B. **9**: 308 (1960).—White, F.F.N.R.: 223 (1962).—Chapman, Veg. Mlanje Mount.: 49 (1962). Type: Nyasaland, *Buchanan* 294 (B, holotype†; BM).

Allophylus tenuifolius Radlk. in Sitz.-Ber. Bayer. Akad. **38**, 2: 221 (1909); in Engl. Pflanzenr IV, 165, **1**, 2: 529 (1932).—Bak. f., tom. cit.: 184 (1919).—Burtt Davy & Hoyle, loc. cit. Type: Nyasaland, *Buchanan* 363 (B, holotype†; BM; K).

Allophylus gazensis Bak. f., tom. cit.: 182 (1919). Type: S. Rhodesia, Chimanimani, *Swynnerton* 321 (BM, holotype; K; SRGH).

Allophylus sp. nr. *buchananii*.—Meikle, loc. cit.

Large shrub or small tree up to 3–7 m. tall (but see note below); branchlets appressed-pubescent when very young, soon glabrescent, usually pale with rather prominent whitish lenticels. Leaves 3-foliolate; petiole up to 7 cm. long, pubescent to nearly glabrous; terminal leaflet usually 1½ times as long as the lateral ones; petiolules up to 15 mm. long but usually much shorter in the often subsessile lateral leaflets; lamina of terminal leaflet up to 15·5 × 6 cm., elliptic to narrowly elliptic, papyraceous (when young) to subcoriaceous, nearly glabrous except for occasional tufts of hairs in the axils of the primary and sometimes of the secondary nerves on the under surface, apex usually acuminate, margin rather remotely serrate or serrulate or crenate-serrulate, base cuneate to broadly cuneate (rarely slightly narrowed, perhaps due to hybridization with *A. alnifolius*); lateral nerves 8–10 pairs. Inflorescence up to 24 cm. long, unbranched, sparsely pubescent to glabrous. Flowers white or greenish, in few-flowered subsessile glomerules or short cymules with rhachis 0·5–1 mm. long; pedicels up to 3 mm. long. Outer sepals 1·5–1·8 × 1·5 mm., subcircular, inner 1·5 × 1 mm., elliptic, all glabrous or sparsely pubescent. Petals 1·5 × 1 mm., spathulate. Stamens with filaments 1·8 mm. long; staminodes 0·5 mm. long in ♀ flowers. Ovary 2-lobed, pubescent; style 2 mm. long, 2-fid. Fruit 6 mm. in diam., subglobose, minutely pubescent, 1 or 2 cocci developing.

N. Rhodesia. N: Luwingu, fl. 7.v.1958, *Fanshawe* 4425 (FHO). W: Chingola Distr., fl. 18.i.1956, *Fanshawe* 2746 (K; SRGH). E: Lundazi Distr., Nyika Plateau, Kangampande Mt., 2130 m., *White* 2730 (BM; FHO; K). **S. Rhodesia.** E: Melsetter Distr., Tarka Forest Reserve, fl. iii.1957, *Barrett* 136/57 (K; LISC; SRGH). **Nyasaland.** N: Rumpi Distr., Nyika Plateau, above Nchena-nchena, fl. 4.v.1952, *White* 2585 (FHO; K). S: Mt. Mlanje, Luchenya Plateau, fl. & fr. (immat.) 12.vii.1946, *Brass* 16802 (BM; K; SRGH). **Mozambique.** Z. Gúruè, fr. 20.ix.1944, *Torre* 2191 (LISC).

Also in Tanganyika, eastern Congo and the Transvaal. Submontane forest.

The stature of this species is somewhat puzzling. In general it is described by collectors as a large shrub or small tree up to 3 m. (rarely 6 m.) tall. *Brass* 16802 from Mt. Mlanje was described as a " tree up to more or less 30 m. (actually 30 cm. on the ticket!) high . . . common in primary forest as a canopy tree ". Meikle (in Mem. N.Y. Bot. Gard. **8**, 3: 240 (1953)) separated this specimen from *A. buchananii* (which I consider to be a synonym of *A. chaunostachys*) saying " stature and the coarse, subcoriaceous texture of leaves distinguish this species from *A. buchananii*; the inflorescences are also longer and more robust, and the twigs more prominently lenticellate than in that species ". Subcoriaceous leaves, robust inflorescences and almost equally prominent lenticels occur in *White* 2730 (probably not seen by Meikle) from the Nyika Plateau, described as a shrub 4 m. tall. It may be suspected that the height (30 m.) given for *Brass* 16802 and the note " common in primary forest as a canopy tree " is erroneous. *Newman & Whitmore* 213 (BM), from *Widdringtonia*-forest on Mt. Mlanje, is described as a " small tree to 3 m." and is clearly the same species.

The leaves of this species become thicker with age and are very young in the type of *A. tenuifolius* Radlk.

10. **Allophylus mossambicensis** Exell in Bol. Soc. Brot., Sér. 2, **38**: 107 (1964). TAB. **100** fig. 3. Type: Mozambique, Sul do Save, Vila João Belo, *Lemos & Balsinhas* 45 (K; LISC, holotype; LMJ; SRGH).

Shrub up to 2·5 m. tall; branchlets pale-coloured, pubescent when young and usually rather tardily glabrescent. Leaves 3-foliolate; petiole up to 7 cm. long, densely pubescent; leaflets subequal or the terminal one up to twice as long as the lateral ones; petiolules up to 12 mm. long in the terminal leaflet but usually much shorter in the sometimes subsessile lateral ones; lamina of terminal leaflet up to 15 × 7 cm., elliptic, thinly papyraceous, rather shortly pubescent on both surfaces, minutely glandular beneath and with rather inconspicuous tufts of hairs in the axils of the lateral nerves, apex acute to rather blunt, margin usually rather sharply dentate or serrate and occasionally slightly lobed, base cuneate and sometimes narrowed towards its extremity; lateral nerves 6–9 pairs. Inflorescence 3–8 cm. long, usually unbranched, pubescent. Flowers in few-flowered subsessile glomerules or in cymules with rhachis up to 3 mm. long; pedicels up to 2 mm. long, pubescent. Outer sepals 1·8 × 2 mm., very broadly transversely elliptic, inner 1·2 × 1·1 mm., subcircular, concave, all minutely pubescent with minutely ciliate margins. Petals 1·2 × 1 mm., spathulate. Sexual organs immature.

Mozambique. SS: Inharrime, Ponta de Závora, fl. 4.iv.1959, *Barbosa & Lemos* 8498 (COI; K; LISC; LMJ; PRE; SRGH).

Known only from the Sul do Save Province of Mozambique. Coastal dunes, mixed forests and forest margins.

11. **Allophylus whitei** Exell in Bol. Soc. Brot., Sér. 2, **38**: 110 (1964). TAB. **100** fig. 2. Type: N. Rhodesia, Livingstone Distr., Malanda Forest Reserve, *White* 7677 (FHO, holotype).

Shrub 1·5–2 m. tall; branchlets densely fulvous-pubescent to fulvous-tomentose, tardily glabrescent. Leaves 3-foliolate; petiole 2·5–5 cm. long, densely fulvous-pubescent; leaflets subequal or the terminal one up to 1½ times as long as the lateral ones; petiolules up to 10 mm. long in the terminal leaflet and up to 4 mm. long in the lateral ones, fulvous-pubescent; lamina of terminal leaflet up to 10 × 6·5 cm., elliptic, papyraceous, pubescent on both surfaces (densely so beneath), minutely glandular beneath and with occasional rather inconspicuous tufts of hairs in the axils of the lateral nerves, apex acute to rounded, margin rather remotely and shallowly serrate or crenate-serrate, base cuneate; lateral nerves 6–8 pairs. Inflorescence up to 12 cm. long, unbranched, densely fulvous-pubescent. Flowers (immature) in rather dense subsessile glomerules; pedicels up to 2·5 mm. long at time of fruiting, sparsely pubescent or glabrous. Outer sepals 1·5 × 1·2 mm., elliptic to broadly elliptic, inner 1·5 × 1 mm., elliptic, all sparsely pubescent and with margins minutely ciliate. Fruit up to 6 × 5 mm., broadly obovoid-ellipsoid to subglobose, sparsely pubescent, 1 coccus usually developing.

N. Rhodesia. S: Mazabuka Distr., 16 km. from Choma on the road to Pemba, fl. 30.i.1960, *White* 6619 (FHO).

Known only from the Southern Prov. of N. Rhodesia. In dense thickets on Kalahari Sand with *Baikiaea plurijuga* as emergent and in dense thickets elsewhere with other emergents.

This species is near to *A. welwitschii* Gilg, from which it differs in the densely pubescent under surface of the leaves, and to *A. andongensis* Bak. f., which has much more coarsely serrate to dentate margins of the leaflets and the latter are also more sharply acuminate. Both these Angolan taxa are forest species though *A. welwitschii* has also been recorded from savanna.

12. **Allophylus rhodesicus** Exell in Bol. Soc. Brot., Sér. 2, **38**: 108 (1964). TAB. **100** fig. 10. Type: S. Rhodesia, Chilimanzi Distr., Shashe R., *Seward* 21/51 (K; SRGH, holotype).

Shrub or small tree; branchlets densely fulvous-pubescent, fairly soon glabrescent and then greyish-brown. Leaves 3-foliolate; petiole up to 5·5 cm. long, pubescent; leaflets subequal or the terminal one up to twice as long as the lateral ones; petiolules up to 4(7) mm. long, usually shorter in the often subsessile lateral leaflets, pubescent; lamina of terminal leaflet up to 8 × 4 cm., elliptic, chartaceous, pubescent on both surfaces (mainly on the nerves) or nearly glabrous above, minutely glandular beneath, apex acute to rounded, margin irregularly and often remotely crenate-serrate or crenate-dentate, base cuneate to narrowly cuneate (rarely appreciably narrowed); lateral nerves 6–9 pairs. Inflorescence 3·5–13 cm. long, unbranched or branched (? in hybrids), pubescent. Flowers yellowish, in several-flowered subsessile glomerules; pedicels up to 2 mm. long, glabrous or pubescent. Outer sepals 1·6 × 2 mm., broadly transversely elliptic, glabrous, inner 1·6 × 0·7 mm., elliptic to narrowly elliptic, glabrous. Petals 1·2 × 0·8 mm., spathulate. Stamens with filaments 1 mm. long; staminodes 0·5 mm. long in ♀ flowers. Ovary 2-lobed, glabrous or sparsely pubescent; style 1 mm. long, 2-fid. Fruit 6 × 4 mm., subglobose-ellipsoid, sparsely pubescent or glabrous when mature.

S. Rhodesia. W: Victoria Falls, fl. & fr. iii.1960, *Armitage* 93/60 (SRGH). C: Chilimanzi Distr., 6·4 km. W. of Umvuma, fl. 7.ii.1951, *McGregor* 21/51 (K; SRGH). E: Chipinga Distr., Sabi Valley Experimental Station, fl. iii.1961, *Soane* 294 (K; SRGH).

S. Rhodesia and probably in Southern Prov. of N. Rhodesia. Savanna woodland.

Very similar in appearance to *A. africanus* (Group D *cataractarum*) but with unbranched inflorescences. Specimens with branched inflorescences otherwise scarcely distinguishable from *A. rhodesicus* (*Drysdale* 2/51 (SRGH); *Goodier* 340 (SRGH); *Lovemore* 225 (K; LISC; SRGH); *White* 2248 (FHO), 3875 (FHO), the two latter from N. Rhodesia) may be hybrids with *A. africanus* and perhaps also *Pardy* 4586 (SRGH) with unbranched inflorescences but very hairy leaves.

13. **Allophylus chirindensis** Bak. f. in Journ. Linn. Soc., Bot. **40**: 48 (1911); in Journ. of Bot. **57**: 188 (1919). Type: S. Rhodesia, Chirinda Forest, *Swynnerton* 112 (BM, holotype).

Small to medium-sized tree up to 15 m. tall; bark silver-grey; branchlets greyish-white with prominent lenticels, glabrous. Leaves 3-foliolate; petiole 3–11 cm. long, glabrous; leaflets subequal or the terminal one up to 1½ times as long as the lateral ones; petiolules up to 12 mm. long in the terminal leaflet, usually much shorter in the lateral ones; lamina of terminal leaflet up to 17 × 9 cm., elliptic, membranous to thinly papyraceous, sometimes minutely pubescent on the nerves and with occasional tufts of hairs in the axils of the lateral nerves on the under surface, otherwise glabrous, apex acute to rounded and sometimes slightly acuminate, margin crenate or crenate-serrate or serrate, base cuneate to narrowly cuneate; lateral nerves 10–13 pairs. Inflorescence up to 27 cm. long, branched, sparsely pubescent to glabrous. Flowers white or yellow in few- to several-flowered subsessile glomerules; pedicels up to 4 mm. long, glabrous. Outer sepals 1·8 × 1·8 mm., subcircular, glabrous, inner 1·8 × 1 mm., elliptic, concave, glabrous. Petals 1·5 × 1 mm., spathulate. Stamens with filaments 2 mm. long; staminodes 0·6 mm. long in the ♀ flowers. Ovary 2-lobed, minutely pubescent; style 1·5 mm. long, 2-fid. Fruit red, 7 mm. in diam., subglobose, minutely pubescent when young, glabrescent, 1 coccus usually developing.

S. Rhodesia. E: Mt. Selinda Forest, fl. 4.i.1948, *McGregor* 1/48 (BM; FHO; SRGH).

Known only from the Eastern Distr. of S. Rhodesia. Medium-altitude evergreen forest.

14. **Allophylus africanus** Beauv., Fl. Owar. & Benin, **2**: 75, t. 107 (1819 or 1820).—Sim, For. Fl. Port. E. Afr.: 31 (1909).—R.E.Fr. in Wiss. Ergebn. Schwed. Rhod.-Kongo-Exped. **1**: 131 (1914).—Bak. f. in Journ. of Bot. **57**: 188 (1919).—Radlk. in Engl., Pflanzenr. IV, 165, **1**, 2: 536 (1932).—Burtt Davy & Hoyle, N.C.L.: 69 (1936).—Exell & Mendonça, C.F.A., **2**, 1: 77 (1954).—Williamson, Useful Pl. Nyasal.: 16 (1956).—Keay, F.W.T.A. ed. 2, **1**, 2: 713 (1958).—Hauman, F.C.B. **9**: 291 (1960).—Dale & Greenway, Kenya Trees and Shrubs: 504 (1961).—White, F.F.N.R.: 223 (1962). Type from Nigeria.

Schmidelia africana (Beauv.) DC., Prodr. **1**: 610 (1824) pro parte.—Bak. in Oliv. F.T.A. **1**: 421 (1869) pro parte. Type as above.

Allophylus griseo-tomentosus Gilg in Engl., Bot. Jahrb. **24**: 290 (1897).—Bak. f., tom. cit: 189 (1919).—Radlk., tom. cit.: 540 (1932).—Burtt Davy & Hoyle, loc. cit.—Dale & Greenway, op. cit.: 504 (1961). Syntypes from Tanganyika and Nyasaland: *Buchanan* 377 (B†), 427 (B†), 563 (B†), 573 (B†), 800 (B†).

Allophylus cataractarum Bak. f., tom. cit.: 189 (1919).—Wild, Guide Fl. Vict. Falls: 152 (1952).—White, loc. cit. Type: Victoria Falls, *Rogers* 5538 (BM, holotype; SRGH).

Allophylus holubii Bak. f., loc. cit.—O.B. Mill., B.C.L.: 36 (1948). Type: Bechuanaland Prot., Leshuma Valley, *Holub* (K, holotype).

Allophylus spragueanus Burtt Davy in Kew Bull. **1921**: 280 (1921). Type: S. Rhodesia, Victoria Falls, *Allen* 143 (K, holotype).

Allophylus melanocarpus sensu Steedman, Trees etc. S. Rhod.: 43, t. 42 (1933). The apparently 5-merous flowers in the plate are wrongly drawn.

Small tree up to 10 m. tall or shrub; branchlets fulvous-tomentose to nearly glabrous. Leaves 3-foliolate; petiole up to 7 cm. long, nearly glabrous to densely pubescent; leaflets subequal or with the terminal one up to $1\frac{1}{2}$ times as long as the lateral ones; petiolules up to 12 mm. long in the terminal leaflet but rarely exceeding 2 mm. in the sometimes subsessile lateral ones; lamina of terminal leaflet up to 15 × 8 cm. (sometimes larger outside our area) but often much smaller (especially in Groups C and D, q.v.), obovate to elliptic, chartaceous to subcoriaceous, almost glabrous to pubescent above, almost glabrous to tomentose beneath, minutely glandular beneath and sometimes with tufts of hairs in the axils of the lateral nerves, apex acute or rounded and sometimes slightly acuminate, margin shallowly crenate-serrate or shallowly serrate or sometimes subentire, base cuneate to narrowly cuneate (rarely appreciably narrowed except in some possible hybrids); lateral nerves 6–9 pairs. Inflorescence 9–26 cm. long, branched, densely pubescent. Flowers white, cream, yellow or green, in few-flowered subsessile glomerules; pedicels up to 2 mm. long. Outer sepals 1·8 × 1·8 mm., subcircular, inner 1·5 × 1 mm., elliptic, all glabrous to pubescent. Petals 1·2 × 0·9 mm., spathulate. Stamens with filaments 1·5 mm. long; staminodes 0·5 mm. long in ♀ flowers. Ovary 2-lobed; style 1·8 mm. long, 2-fid. Fruit red or orange turning red, 4–6 mm. in diam., subglobose in our area (sometimes ellipsoid elsewhere), sparsely pubescent to pubescent when young, glabrous to pubescent when mature, 1 coccus (rarely 2) developing.

Widespread in our area and throughout tropical Africa. Occurring in a wide range of habitats.

It seems advisable to avoid creating new names for intraspecific taxa in this complex series as it is difficult to make progress with the classification until the plants have been studied genetically. In our area the species can be divided into 4 moderately distinct groups with some geographical and ecological separation but with considerable intergrading and overlapping. There are also intermediates with *A. rubifolius* and *A. rhodesicus* presumably due to past or present hybridization. As some populations are fairly homogeneous and ecologically distinct, I have indicated the epithets which can be applied to them, if desired, within our area.

Group A (*africanus*).

Leaves nearly glabrous on the under surface (except for occasional tufts of hairs in the axils of the lateral nerves) to sparsely pubescent, usually drying brown. Lamina of the terminal leaflet up to 15 × 8 cm. (usually c. 8–9 × 4–5 cm.), obovate to elliptic.

N. Rhodesia. W: Chingola, fl. 5.ii.1957, *Fanshawe* 2993 (K; SRGH). **S. Rhodesia.** N: Urungwe Reserve, fl. 4.i.1958, *Goodier* 515 (LISC; PRE; SRGH). **Nyasaland.** N: Nyika Plateau, fl. ii-iii, 1903, *McClounie* 29 (K). C: Chulu, Kasungu-Lundazi Road, 1050 m., fl. 15.i.1959, *Robson* 1185 (K). S: Limbe, 1220 m., fl. 1.ii.1948, *Goodwin* 53 (BM). **Mozambique.** N: Massangulo, fl. iii.1933, *Gomes e Sousa* 1309 (COI). Z: between Mocuba and Milange, fl. 18.iii.1943, *Torre* 4944 (LISC). T: Angónia, banks of R. Mauè, fr. 12.v.1948, *Mendonça* 4199 (LISC).

Mainly in the wetter parts of our area, in riverine forest, in woodland (often on termite mounds) and in thickets.

Group A appears to be the typical form of *A. africanus* though in our area it usually (but not always) has smaller leaves than in West Africa.

Torre 5220 (LISC), from Massingire in the Sul do Save Prov. of Mozambique, seems to be no more than a luxuriant specimen with large leaves and large 3-times-branched inflorescences. *Garcia* 186 (LISC), from Chimoio in the Manica e Sofala Prov. of Mozambique, with terminal leaflet rather abruptly narrowed at the base and under surface nearly glabrous, might be: (1) *africanus* (*A*) × *rubifolius;* (2) *africanus* (*A*) × *alnifolius; or* (3) *africanus* (*B*) × *alnifolius.* Any of these combinations might give a similar plant but the comparitive rarity of *A. alnifolius* in Mozambique makes the first suggestion the most probable. There are transitions to Group B (*griseo-tomentosus*) in Mozambique and a few transitions to Group D (*cataractarum*) in N. Rhodesia.

Group B (*griseo-tomentosus*).

Leaves pubescent to tomentose on the under surface, usually drying brown or dark greyish-green. Terminal leaflet mainly as in Group A in size and shape but not attaining as large a size as is sometimes reached in the latter. Small trees.

N. Rhodesia. N: Abercorn, fl. iv.1931, *Miller* 230/31 (FHO). S: Katambora, fl. 15.ii.1956, *Gilges* 578 (K; PRE; SRGH). **S. Rhodesia.** N: Lomagundi Distr., 67 km. from Sanyati-Sinoia, Copper Queen road, fl. 30.i.1960, *Lovemore* 579 (SRGH). **Nyasaland.** N: Vipya, Lusangazi, Mzuzu, fl. iii.1954, *Chapman* 187 (FHO; K). S? without precise locality, fl. 1934, *Clements* 401 (FHO). **Mozambique.** N: between Mutuáli and R. Lúrio, fr. 9.iv.1953, *Gomes e Sousa* 4101 (COI; LISC; LMJ; PRE; SRGH).

A. griseo-tomentosus does not appear to be more than a hairy form of *A. africanus.* It has much the same distribution and *A. griseo-tomentosus* forma *glabrior* Radlk. (in Engl., Pflanzenr. IV, 165, **1**, 2: 540 (1932)) is an intermediate. Plants transitional to Group C (*holubii*) tend to have smaller leaves and there are also specimens transitional to *A. rubifolius* with leaflet-shape characteristic of the latter but with branched inflorescences. One of these (*Lemos & Macuácua* 114 (COI; LMJ; PRE; SRGH) from the Manica e Sofala Prov. of Mozambique) is described as a shrub. *Fanshawe* 4495 (FHO), from Fort Jameson, with leaves yellowish-tomentose beneath with rather sharply dentate margins and with very short inflorescences, may belong to Group B or may prove to be new.

Group C (*holubii*).

Leaves with indumentum much as in Group B but with leaflets smaller in size and relatively narrower, usually drying brown. Terminal leaflet usually up to 5–7 × 2–3 cm. Shrubs or more rarely small trees.

Caprivi Strip. Kakumba I., fl. 17.i.1959, *Killick & Leistner* 3413 (K; SRGH). **Bechuanaland Prot.** N: Leshuma Valley, *Holub* (K). **S. Rhodesia.** N: Mtoko Distr., Chazanini, fl. ii-iii.1958, *Brayne* 16 (SRGH). W: Insiza, fl. ii.1941, *Hopkins* in GHS 7726 (SRGH). C: Featherstone, fl. 26.iii.1947, *Goldsmith* 6/47 (SRGH).

From the Caprivi Strip and northern Bechuanaland Prot. across S. Rhodesia. Usually on termite mounds and also recorded from *Brachystegia* woodland.

This is the common *Allophylus* in the Central Division of S. Rhodesia: it has rarely been collected elsewhere. Luxuriant specimens show transition to Group B (*griseotomentosus*) and some specimens are transitional to Group D (*cataractarum*). *Brayne* 16 (cited above) may be transitional to Group B and *Lovemore* 579 (cited under Group B) may be transitional to Group C. *Wild* 4780 (LISC; SRGH), from Wankie Game Reserve, S. Rhodesia (W), tentatively identified on the sheet as *A. rubifolius*, a shrub but with leaves unusually large for Group C, may be transitional to Groups A or B or may indeed have characters derived from *A. rubifolius*. The placing of these numerous transitional forms is guess-work without genetical study. I have cited a few of them to show why an orthodox intraspecific classification has not been attempted.

Group D (*cataractarum*). TAB. **100** fig.8.

Leaves rather densely pubescent to shortly pilose beneath, especially on the midrib, nerves and reticulation, usually drying a rather bright green. Terminal leaflet usually c. 6–8(10) × 3–4 cm. Small trees or shrubs up to 3 m. tall.

N. Rhodesia. B: Machili, fr. 22.ii.1961, *Fanshawe* 6306 (FHO; K; SRGH). C: Mt. Makulu, fl. 8.i.1960, *White* 6159 (FHO). S: Livingstone Distr., Victoria Falls, fl. 7.i.1953, *Angus* 1117 (BM; K). **S. Rhodesia.** W: common around Victoria Falls but the specimens do not indicate which ones were collected S. of the Zambezi.

Common in the Victoria Falls Region and extending northwards and westwards to the adjacent districts. Thickets, riverine fringes and termite mounds.

Forming a fairly homogeneous population around Victoria Falls but with transitional forms towards the limits of its area. Intermediates with Groups A (*africanus*) and C (*holubii*) make it difficult to maintain *A. cataractarum* even as a subspecies or variety.

Undescribed and unindentified species

Allophylus bussei Gilg in Engl., Pflanzenw. Afr. **3,** 2: 270 (1921) *nom. nud.*

Said to occur in " Mossambikküstenland, Muera Plateau " but Muera Plateau is in Tanganyika. The specimen has presumably been destroyed.

Doubtful record

Allophylus monophyllus (E. Mey.) Radlk. in Engl. & Prantl, Nat. Pflanzenfam. **3,** 5: 312 (1896).—Sim, For. Fl. Port. E. Afr.: 31 (1909) (as *Schmidelia monophylla* Presl). Sim (loc. cit.) states that this species is " scarce but present throughout the Province ".

I have seen no specimen from Mozambique.

2. CARDIOSPERMUM L.

By Lorna F. Bowden

Cardiospermum L., Sp. Pl. **1**: 366 (1753); Gen. Pl. ed. 5: 171 (1754).

Annual or rarely perennial climbing herbs (sometimes slightly woody); stems longitudinally ridged. Leaves usually stipulate, petiolate, biternately compound to apparently pinnate (in our area). Inflorescence an axillary corymbose paniculate thyrse with a pair of coiled tendrils at the apex of the peduncle. Flowers spuriously polygamo-dioecious, zygomorphic; bracts scale-like, minute. Sepals 4, lateral ones small, anterior and posterior ones larger. Petals 4, with scale-like appendages, those of the 2 posterior ones complex in form and sometimes joined to those of the anterior ones. Disk unilateral, bearing posteriorly 2 glands and elongated anteriorly into a short androgynophore. Stamens 8, free or slightly connate at the base; staminodes shorter in the ♀ flowers. Ovary 3–locular, loculi 1-ovulate; style 3-fid; pistillode vestigial in ♂ flowers. Fruit an inflated membranous capsule. Seeds black, globose with a white ovate to reniform hilum.

A genus of 12–20 species, mainly in the New World, in tropical and subtropical regions.

Flowers 2–6 mm. long; leaves biternate or some pinnate:
- Leaves biternate; leaflets incised-serrate; fruit obversely tetrahedroid 1. *halicacabum*
- Leaves some at least pinnate; leaflets crenate to lobed; fruit ovoid to obversely tetrahedroid - - - - - - - - - - - 2. *corindum*

Flowers 7–10 mm. long; leaves biternate; fruits broadly ellipsoid 3. *grandiflorum*

1. **Cardiospermum halicacabum** L., Sp. Pl. **1**: 366 (1753) pro parte.—Sond. in Harv. & Sond., F.C. **1**: 237 (1860).—Klotzsch in Peters, Reise Mossamb. Bot. **1**: 118 (1861).—Bak. in Oliv., F.T.A. **1**: 417 (1868).—Eyles in Trans. Roy. Soc. S. Afr. **5**: 405 (1916).—Burtt Davy, F.P.F.T. **2**: 488 (1932).—Radlk. in Engl., Pflanzenr. IV, 165, **1**, 2: 379 (1932).—Exell & Mendonça, C.F.A. **2**, 1: 75 (1954).—Williamson, Useful Pl. Nyasal.: 31 (1955).—Keay, F.W.T.A. ed. 2, **1**, 2: 711 (1958).—Hauman, F.C.B. **9**: 286 (1960).—Mitchell in Puku, **1**: 117 (1963). TAB. **101** fig. A. Type from India.

Annual straggling minutely puberulous herbaceous or slightly woody climber, sometimes flowering when only 25 cm. tall; stems with internodes 4–7 cm. long. Leaves biternate, triangular; petiole (0·5)0·9–3·5 cm. long; rhachis 0·4–2 cm. long; terminal leaflet 2·5–4·5(6) × 0·7–2 cm; lateral leaflets 2·5–4·5(5) × 0·8–2(3·5) cm.; lamina bright green with a paler under surface, only the main veins raised, apex acuminate, margin incised-serrate, base attenuate; stipules 2 minute caducous scales. Inflorescence a reduced complex axillary corymbose thyrse, abortively 3-flowered; peduncle 4–9 cm. long, very sparsely puberulous, multi-bracteate. Flowers white to yellowish, 2–3 mm. long. Sepals with lateral ones c. 1 mm. long, anterior and posterior ones c. 3 mm. long. Petals 4–5 mm. long, posterior ones bearing recurved clavate appendages provided with hairs which cause the two free ends to cohere. Stamens with filaments 3 mm. long, free or slightly connate at the base; anthers 0·5 mm. long. Ovary hirsute; style-branches clavate, shortly plumose. Seed 5 mm. in diam., reniform.

Throughout the tropics and subtropics in the wetter regions.

Fruit more than 2 cm. long - - - - - - - - - var. *halicacabum*
Fruit less than 2 cm. long - - - - - - - - - var. *microcarpum*

Var. **halicacabum**

Bechuanaland Prot. N: Kapoko, Okovango Valley, fl. & fr. 29.viii.1933, *Schoenfelder* S. 151 (PRE). SE: Kgatla, 1150 m., fl. & fr. iii.1955, *Reyneke* 240 (LISC; PRE). **N. Rhodesia.** B: Kaunga, Mashi R., fl. 13.ix.1961, *Mubita* in CAH 10501 (SRGH). C: Zambezi-Kafue confluence, fl. & fr. 12.viii.1957, *Angus* 1647 (SRGH). S: Mazabuka, fl. & fr. 26.ix.1931, *Veterinary Officer* C.R.S. 437 (PRE). **S. Rhodesia.** N: Ganderowe Falls, 43·5 km. N.W. of Copper Queen, Gokwe, c. 700 m., fr. 3.vii.1961, *Leach* 1163 (SRGH). S: Sabi-Lundi confluence, Chitsa's Kraal, 250 m., fr. 5.vi.1950, *Chase* 2311 (BM; K; LISC; SRGH). **Nyasaland.** N: Karonga, c. 500 m., fr. 27.ii.1955, *Williamson* 184 (BM). **Mozambique.** N: Quissanga, between Maate and Metuge, fl. & fr. 3.x.1948, *Barbosa* 2331 (LISC; LMJ). Z: Mopeia and Régulo Changalaze, fl. & fr. 2.viii.1947, *Barbosa & Carvalho* 3810 (LMJ). T: R. Mazoe, near Dique, 330 m., fl. & fr. 21.ix.1948, *Wild* 2577 (SRGH). MS: Maríngué, c. 200 m., fr. 20.vi.1950, *Chase* 2553 (BM; SRGH). SS: Massangena, fr. vii.1932, *Smuts* P. 375 (K; PRE). LM: between Manhiça and Xinavane, fr. 9.vii.1947, *Barbosa* 249 (LISC).

Tropics and subtropics of the world. Mainly in riverine forest from 0–1300 m.

Var. **microcarpum** (Kunth) Bl. in Rumphia, **3**: 185 (1847).—Klotzsch in Peters, loc. cit.—Radlk., tom. cit.: 391 (1932). Type from S. America.

Cardiospermum microcarpum Kunth, Nov. Gen. & Sp. Pl. **5**: 104 (1821).—E. Mey. in Drège, Zwei Pflanz.-Docum.: 170 (1843) (" microspermum "). Type as above.

Cardiospermum truncatum A. Rich., Tent. Fl. Abyss. **1**: 101 (1847). Type from Ethiopia.

Bechuanaland Prot. N: Chobe-Zambesi confluence, fr. 11.iv.1955, *E.M. & W.* 1466 (BM; LISC; SRGH). **N. Rhodesia.** B: Kabompo mouth, 1050 m., fr. 27.v.1954, *Gilges* 386 (PRE). W: 1·6 km. E. of Ndola, Itawa Dambo, 1330 m., fl. & fr., 18.ii.1953, *Draper* 41 (K; LISC; SRGH). S: Livingstone, fl. i.1929, *Young* 1641 (BM). **S. Rhodesia.** N: Urungwe, S. edge of Nyasau Peak, Kariba, fl. & fr. 14.iii.1959, *Chase* 7059 (K; SRGH). W: Wankie, Lobangwe, c. 800 m., fl. & fr. iv.1918, *Eyles* 1300 (BM; K; SRGH). S: Sabi Valley Experimental Station, fl. & fr. ii.1960., *Soane* 270 (SRGH). **Mozambique.** T: near R. Chueza, fl. 21.vi.1949, *Barbosa & Carvalho* 3209 (LISC; LMJ). LM: Namaacha, fr. 9.vii.1947, *Barbosa* 700 (LISC; SRGH).

Tropics and subtropics of the world. Mainly in riverine forest from 0–1350 m.

2. **Cardiospermum corindum** L., Sp. Pl. ed. 2: 526 (1762).—Gibbs in Journ. Linn. Soc., Bot. **37**: 436 (1906).—Eyles in Trans. Roy. Soc. S. Afr. **5**: 405 (1916).—Radlk. in Engl., Pflanzenr. IV, 165, **1**, 2: 397 (1932).—Steedman, Trees etc. S. Rhod.: 43 (1933).—Exell & Mendonça, C.F.A. **2**, 1: 75 (1954). Type from Brazil.

Cardiospermum pechuelii Kuntze in Jahrb. Königl. Bot. Gart. Berl. **4**: 262 (1886). Type from SW. Africa.

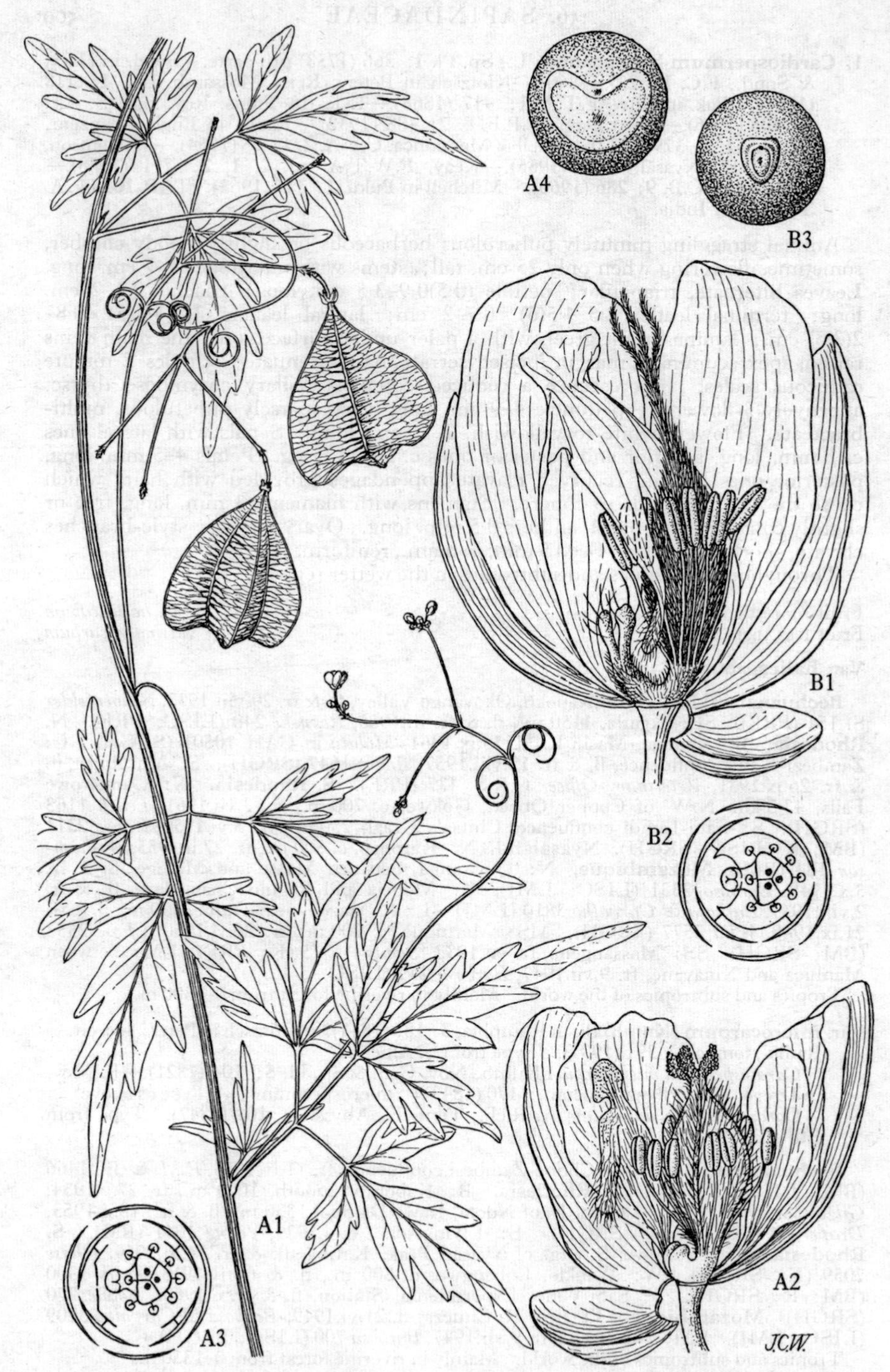

Tab. 101. A.—CARDIOSPERMUM HALICACABUM. A1, branch with flowers and fruit (×1); A2, flower with two petals removed and sepals displayed (×15); A3, diagram of flower; A4, seed (×6); *Chase* 2311. B.—CARDIOSPERMUM GRANDIFLORUM. B1, flower with two petals removed and sepals displayed (×9); B2, diagram of flower; B3, seed (×4·5); *Banda* 177.

Cardiospermum corindum forma *pechuelii* (Kuntze) Radlk. in Mart., Fl. Bras. **13,** 3: 448 (1897). Type as above.

Cardiospermum alatum Bremek. & Oberm. in Ann. Transv. Mus. **16:** 422 (1935). Type: Bechuanaland Prot., Ngami, *van Son* in Herb. Transv. Mus. 29000 (PRE, holotype).

Annual herbaceous or slightly woody climber, densely puberulous; internodes 4–8(11) cm. long. Leaves pinnate to biternate; petiole 0·5–3(4·5) cm. long; rhachis 1–3 cm. long. Leaflets with lowest pair ternate, terminal one 2–4 × 1–2 (2·3) cm., lateral ones 2–3·5 × 1–2 cm., lamina with upper surface bright green and under surface grey-green with raised venation, puberulous, apex obtuse and apiculate, margin crenate to lobed, base attenuate. Inflorescence an axillary many-branched numerous-flowered corymbose thyrse; peduncle c. 6 cm. long, pubescent, multibracteate. Flowers white to creamy yellow, 4–6 mm. long. Sepals with lateral ones c. 1 mm. long, anterior and posterior ones c. 3 mm. long, ovate. Petals 4–5 mm. long, 2 anterior ones ovate with small petaloid appendages, 2 posterior ones ovate bearing recurved appendages with clavate apices covered with patent hairs which cause them to cohere. Stamens with the filaments of the inner ones slightly shorter; filaments free. Ovary hirsute. Fruit 2–3·6 cm. long, 3-gonous or 3-quetrous. Seed 5 mm. in diam., globose; hilum reniform.

Caprivi Strip. Singalamwe, fl. 1.i.1959, *Killick & Leistner* 3248 (SRGH). **Bechuanaland Prot.** N: Francistown, fl. i.1926, *Rand* 45 (BM). **N. Rhodesia.** B: Sesheke, fl. 25.i.1952, *White* 1958 (BM; FHO; K). C: Lusaka, fl. 26.i.1957, *Noak* 74 (SRGH). **S. Rhodesia.** W: Matopos, fl. & fr. iii.1918, *Eyles* 999 (BM; SRGH). E: Birchenough Bridge, fl. & fr. i.1938, *Obermeyer* 2432 (SRGH). S: 3·2 km. from Beitbridge on West Nicholson Road, fl. 15.ii.1955, *E.M. & W.* 411 (BM; LISC; SRGH).

Widespread in tropical and subtropical regions; in our area confined mainly to the Limpopo, Lundi-Sabi and Upper Zambezi river valleys. Margins of rivers and lower altitude savannas.

Radlkofer (loc. cit.) divides this species into 14 forms which are difficult to separate satisfactorily. He places all the specimens cited from our area in forma *pechuelii* (loc. cit.) which, if distinct, seems to be confined to SW. Africa.

3. **Cardiospermum grandiflorum** Sw., Prodr.: 64 (1788).—Radlk. in Engl., Pflanzenr. IV, 165, **1,** 2: 372 (1932).—Exell & Mendonça, C.F.A. **2,** 1: 74 (1954).—Keay, F.W.T.A. ed. 2, **1,** 2: 711 (1958).—Hauman, F.C.B. **9:** 285 (1960). TAB. **101** fig. B. Type from Jamaica.

Cardiospermum hirsutum Willd. in L., Sp. Pl. ed. 4, **2,** 1: 467 (1799). Type from Guinea.

Cardiospermum elegans Kunth, Nov. Gen. Sp. Pl. **5:** 99, t. 439 (1821). Type from S. America.

Cardiospermum grandiflorum forma *hirsutum* (Willd.) Radlk. in Sitz.-Ber. Bayer. Akad. **8:** 260 (1878). Type as for *C. hirsutum*.

Cardiospermum grandiflorum forma *elegans* (Kunth) Radlk., loc. cit.—White, F.F.N.R.: 224 (1962). Type as for *C. elegans*.

Annual slightly woody climber; stems densely to sparsely hirsute with long brown crisped hairs (c. 2 mm. long) or glabrous; internodes 4–7·6 cm. long. Leaves biternate; petiole 2·1–7·3 cm. long, pubescent; leaflets with the terminal one 4·5–8·4 × 1·8–3·7 cm. and lateral ones (4)4·5–8·7(11·7) × 1·6–3·3 cm.; lamina ovate, upper surface bright green, sparsely pubescent, lower surface tomentose with occasional long brown hairs on the veins, or both surfaces almost glabrous, apex acute, margin serrate, base attenuate; stipules minute caducous scales. Inflorescence an axillary many-flowered corymbose thyrse; peduncle 8–15(20) cm. long. Flowers white or yellow, 7–11 mm. long; pedicels bearing numerous brown scale-like bracts. Sepals with 2 lateral ones c. 2 mm. long and anterior and posterior ones c. 7 mm. long, ovate, petaloid. Petals with 2 anterior and 2 posterior ones ovate bearing petaloid appendages c. 7 mm. long, and joined laterally, those of the posterior petals being apically recurved, passing across the disk, clavate and bearing patent hairs at the tips causing the latter to cohere. Disk produced posteriorly into 2 corniform glands and anteriorly into a short androgynophore (c. 1 mm. long). Stamens with free filaments, Ovary hirsute; style-branches filiform, hirsute. Fruit 4·4–6·6 cm. long, 3-gonous. Seeds 5 mm. in diam., globose; hilum oblong.

N. Rhodesia. W: Kitwe, fl. & fr. 16.ii.1961, *Matushi* 106 (FHO; K; SRGH). **S. Rhodesia.** E: Umtali; fl. & fr. 4.iv.1947, *Chase* 367 (BM; K; SRGH). **Nyasaland.** S: Zomba Mt., fl. & fr. 22.ii.1956, *Banda* 177 (BM; LISC; SRGH).

Tropical America and tropical Africa. Riverine forests.

Specimens from our area belong mainly to forma *hirsutum*. *Fanshawe* 3318 cited by White (loc. cit.) as forma *elegans* shows, on investigation of the specimen at Kew, some hairs characteristic of forma *hirsutum*. Some specimens (e.g. *Banda* 177) show great variation even in a single plant, the apex of the stem showing the characters of forma *hirsutum* and the lower portions those of forma *elegans*. The typical form (forma *grandiflorum*) is said to be confined to tropical America.

3. PAULLINIA L.

Paullinia L., Sp. Pl. **1**: 365 (1753); Gen. Pl. ed. 5: 170 (1754).

Woody climbers usually with paired spirally coiled tendrils arising at the apex of the peduncle; stems cable-like appearing to be composed of several coalescent axes (see t. 102 fig. 3). Leaves stipulate, imparipinnate; leaflets 2-jugate. Inflorescence a paniculate or racemoid thyrse. Flowers spuriously polygamous, zygomorphic. Sepals 5, unequal, the 2 inner ones connate at the base. Petals 4, unequal, each with a large cucullate scale with an apical appendage. Disk unilateral, with 4 glands. Stamens 8, excentric, reduced or absent in the ♀ flowers; filaments shortly connate at the base. Ovary 3-locular, loculi 1-ovulate; style 3-fid; pistillode very reduced in the ♂ flowers. Fruit capsular, 3-gonous or 3-winged; loculi 1-seeded; dehiscence septicidal. Seed with a fleshy arillode.

Genus of c. 150 species mainly in tropical America.

Paullinia pinnata L., Sp. Pl. **1**: 366 (1753) emend.—Bak. in Oliv., F.T.A. **1**: 419 (1868).—Bak. f. in Trans. Linn. Soc., Ser. 2, **4**: 8 (1894); in Journ. Linn. Soc., Bot. **40**: 47 (1911).—Gibbs in Journ. Linn. Soc., Bot. **37**: 436 (1906).—R.E.Fr. in Wiss. Ergebn. Schwed. Rhod.-Kongo-Exped. **1,** 1: 130 (1914).—Brenan, T.T.C.L.: 560 (1949).—Wild, Guide Fl. Vict. Falls: 151 (1953).—Exell & Mendonça, C.F.A. **2,** 1: 73 (1954).—Keay, F.W.T.A., ed. 2, **1,** 2: 710, fig. 196 (1958).—Hauman, F.C.B. **9**: 283 (1960).—White, F.F.N.R.: 225 (1962). TAB. **102.** Syntypes from the West Indies and Brazil.

Liane or scandent shrub; branchlets pubescent. Leaves up to 30 cm. long (usually c. 12 cm.); stipules 4–5 × 1–1·5 mm., narrowly triangular, pubescent; petiole up to 10 cm. long, winged, pubescent; rhachis with wing up to 8 mm. broad; petiolules 1–3 mm. long, pubescent; leaflet-lamina up to 16 × 9 cm., ovate or elliptic or obovate, pubescent or pilose on the nerves but otherwise nearly glabrous, apex usually acuminate, margin dentate to subentire, base cuneate; lateral nerves 7–9 pairs. Inflorescence an axillary paniculate thyrse, floriferous portion usually c. 5 cm. long, tendrils c. 2 cm. long when coiled (sometimes absent); peduncle up to 11 cm. long (occasionally very short), minutely pubescent. Flowers white; pedicels up to 4 mm. long. Sepals 1·5–3 mm. long, the larger ones elliptic, the smaller ones subcircular. Petals 4 × 2 mm., narrowly obovate, unguiculate, with a large ciliate scale attached to the claw. Stamens unequal; filaments 2–4 mm. long, pilose. Ovary 3-gonous, shortly stipitate, appressed-pubescent; style 3 mm. long; pistillode 1 mm. long in ♂ flowers. Fruit up to 28 × 12 mm., narrowly obconic, 3-gonous, apiculate, shortly stipitate, glabrous. Seed blackish, 12–15 × 6–8 mm.

N. Rhodesia. B: Kabompo, fl. 10.xi.1952, *Gilges* 269 (K; SRGH). N: Kawambwa, fl. 26.viii.1957, *Fanshawe* 3649 (K). W: Mwinilunga Distr., Tshikundulu stream, fl. & fr. 6.x.1952, *Angus* 611A (FHO; K). S: Katombora, fl. & fr. 18.x.1947, *Morze* 42 (FHO). **S. Rhodesia.** E: Melsetter, 320 m., fl. & fr. 25.xl.1955, *Drummond* 5026 (K; SRGH). **Nyasaland.** N: Kaziwiziwe R., 1500 m., fl. 8.i.1959, *Richards* 10550 (K; SRGH). S: Zomba Distr., fl. 1937, *Clements* 647 (FHO; K). **Mozambique.** N: Amaramba, between Cuamba and Macía, fl. & fr. 14.x.1942, *Mendonça* 830 (LISC). Z: Milange, fl. 10.x.1942, *Torre* 4575 (LISC). MS: between Moribane and Quedas do Revuè, fl. x.1953, *Pedro* 4185 (K; LMJ).

Widespread in tropical America, tropical Africa and Madagascar. Evergreen and mixed forests.

Said to be a fish poison.

This species is almost certain to occur in the western division of S. Rhodesia at or near

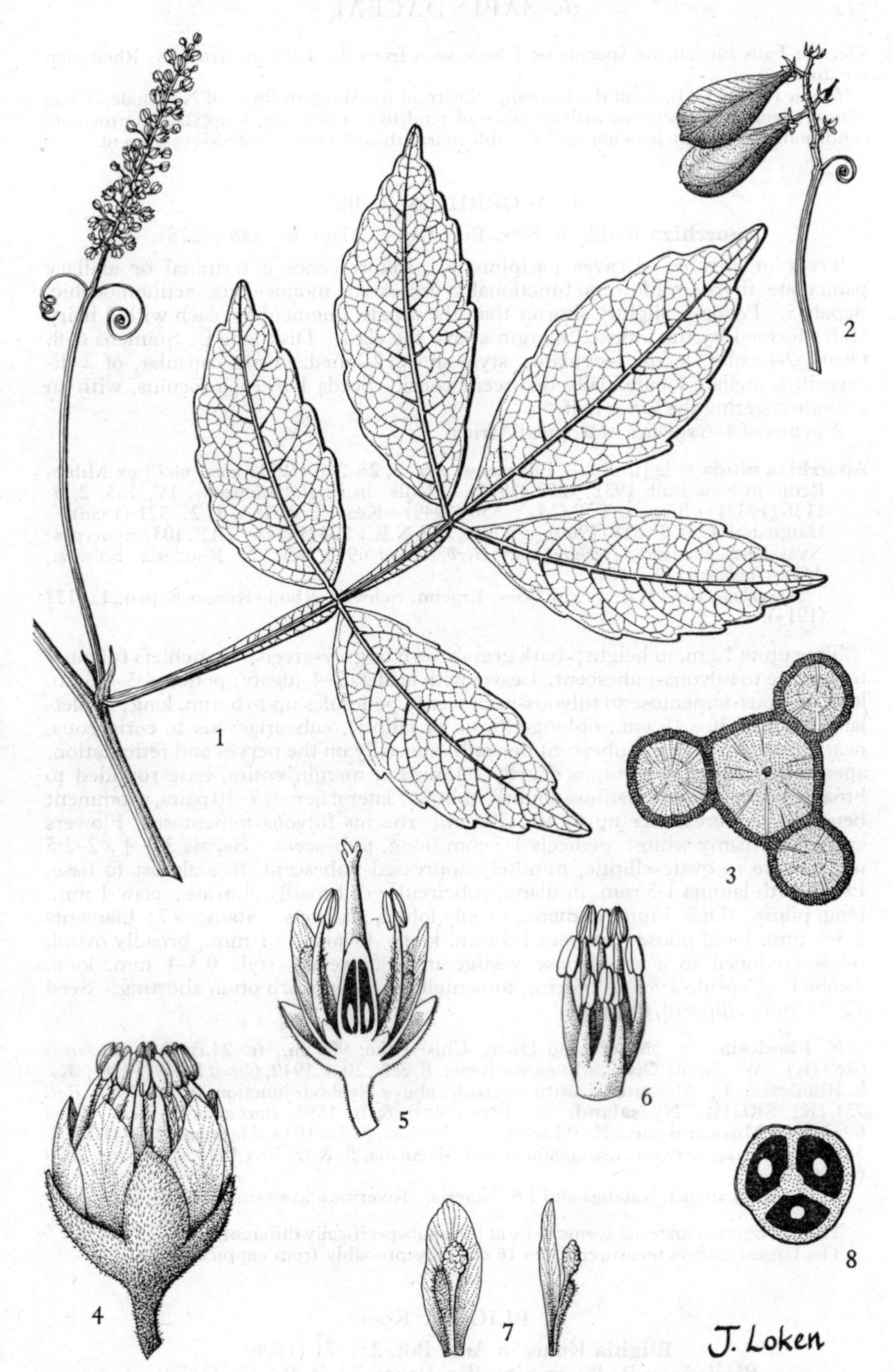

Tab. 102. PAULLINIA PINNATA. 1, leaf and inflorescence ($\times \frac{2}{3}$) *Clements* 647; 2, infructescence ($\times \frac{2}{3}$) *Chase* 979; 3, transverse section of old stem ($\times \frac{2}{3}$) *Chase* 979; 4, flower ($\times 7$); 5, vertical section of flower ($\times 7$); 6, androecium ($\times 7$); 7, front and side view of petal ($\times 7$); 8, transverse section of ovary ($\times 7$). 4–8 from *Clements* 647.

Victoria Falls but all the specimens I have seen from the Falls are from N. Rhodesian territory.

Holmes 1003 (FHO), from the Kabompo Distr. of the Western Prov. of N. Rhodesia, has almost sessile inflorescences with no trace of tendrils. There are, however, intermediate conditions, the peduncle being very variable in length and tendrils not always present.

4. **APORRHIZA** Radlk.

Aporrhiza Radlk. in Sitz.-Ber. Bayer. Akad. **8**: 338 (1878).

Trees or shrubs. Leaves paripinnate. Inflorescence a terminal or axillary paniculate thyrse. Flowers functionally unisexual, monoecious, actinomorphic. Sepals 5. Petals 5, slightly shorter than the sepals, unguiculate, each with 2 hairy scales formed by the inflexed margin above the claw. Disk lobed. Stamens 6–8. Ovary 2-locular, loculi 1-ovulate; style short, 2-lobed. Fruit capsular, of 2 divergent scutellate loculicidally dehiscent cocci. Seeds 1 in each loculus, with an arillode covering the lower half.

A genus of 4–6 species in tropical Africa.

Aporrhiza nitida Gilg [in Engl., Pflanzenw. Afr. **3,** 2: 280 (1921) *nom. nud.*] ex Milne-Redh. in Kew Bull. **1931**: 272 (1931).—Radlk. in Engl., Pflanzenr. IV, 165, **2,** 6: 1136 (1933).—Brenan, T.T.C.L.: 556 (1949)—Keay, F.W.T.A. **1,** 2: 721 (1958).—Hauman, F.C.B. **9**: 318 (1960).—White, F.F.N.R.: 224 (1962). TAB. **103**. Syntypes: Nyasaland, Blantyre, *Buchanan* in *Herb. Wood* 6992 (K); N. Rhodesia, Solwezi, *Milne-Redhead* 1119 (K).

Aporrhiza sp.—R.E. Fr. in Wiss. Ergebn. Schwed. Rhod.-Kongo-Exped. **1**: 131 (1914).

Tree up to 12 m. in height; bark grey-brown or grey-green; branchlets fulvous-tomentose to fulvous-pubescent. Leaves with leaflets 2–4-jugate; petiole 4·5–6·5 cm. long, fulvous-tomentose to fulvous-pubescent; petiolules up to 6 mm. long; leaflet-lamina up to 36 × 16 cm., oblong-elliptic or elliptic, subcoriaceous to coriaceous, nearly glabrous above, pubescent beneath especially on the nerves and reticulation, apex rather blunt, sometimes shortly acuminate, margin entire, base rounded to broadly cuneate and sometimes slightly oblique; lateral nerves 7–10 pairs, prominent beneath. Inflorescence up to 40 × 20 cm.; rhachis fulvous-tomentose. Flowers cream or creamy white; pedicels 1·5 mm. long, pubescent. Sepals 3·5–4 × 2–2·5 mm., ovate to ovate-elliptic, minutely appressed-pubescent, free almost to base. Petals with lamina 1·5 mm. in diam., subcircular or broadly obovate; claw 1 mm. long, pilose. Disk 3 mm. in diam., deeply lobed, glabrous. Stamens 7; filaments 3·5–4 mm. long, pilose; anthers 1·3 mm. long. Ovary 1 × 1 mm., broadly ovoid, pilose (reduced to a subglobose vestige in ♂ flowers); style 0·5–1 mm. long, 2-lobed. Capsule 1·8 × 2–2·3 cm., tomentellous, 1 mericarp often aborting. Seed 12 × 6 mm., ellipsoid, glabrous.

N. Rhodesia. N: Mporokoso Distr., Chisi Lake, 900 m., fr. 24.ix.1956, *Richards* 6269 (K). W: Ndola Distr., Mpongwe Boma, fl. & fr. 20.ix.1949, *Grout* 10/49 (FHO; K). **S. Rhodesia.** E: Melsetter, Lusitu riverside, above Nyahodi junction, fr. 7.ix.1958, *Ball* 751 (K; SRGH). **Nyasaland.** S: Blantyre, fl. & fr. 1895, *Buchanan* in *Herb. Wood* 6992 (K). **Mozambique.** Z: Massingire, Metalola, fl. 2.x.1944, *Mendonça* 2320 (LISC). MS: Mossurize, between Espungabera and Macuiana, fl. & fr. 30.x.1944, *Mendonça* 2679 (LISC).

Also in Tanganyika, Katanga and ? S. Nigeria. Riverine and swamp forests.

The S. Nigerian material seems to be at least subspecifically different.
The largest leaflets measured (36 × 16 cm.) are probably from coppice.

5. **BLIGHIA** Konig

Blighia Konig in Ann. Bot. **2**: 571 (1806).
Phialodiscus Radlk. in Sitz.-Ber. Bayer. Akad. **9**: 479 (1879).

Large trees or shrubs. Leaves paripinnate; leaflets 1–5-jugate. Inflorescence an axillary usually racemoid thyrse (often partially racemose). Flowers spuriously polygamous, usually dioecious, actinomorphic. Sepals 5, connate. Petals 5, pouchlike owing to the fusion of the margins of the scales. Disk lobed. Stamens 8–10. Ovary 3-gonous, 3-locular; loculi 1-ovulate. Fruit a 3-lobed, narrowly to

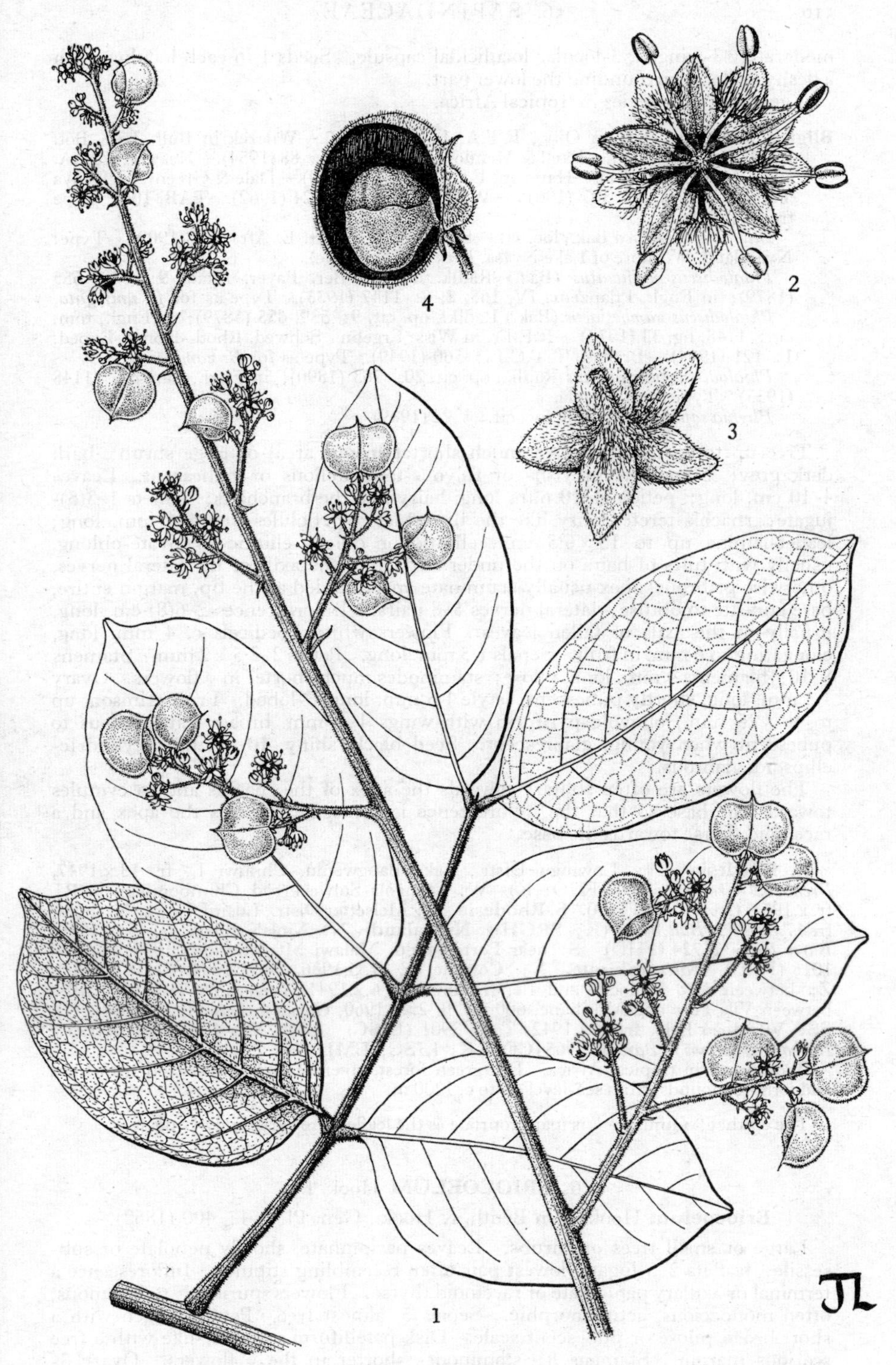

Tab. 103. APORRHIZA NITIDA. 1, leaf and inflorescence (× ⅔) *Grout* 10/49; 2, male flower (× 6) *Grout* 10/49; 3, calyx (× 6) *Grout* 10/49; 4, one coccus of fruit opened to show seed (× 2) *Duff* 242/37.

moderately 3-winged, 3-locular loculicidal capsule. Seeds 1 in each loculus, with a fleshy arillode surrounding the lower part.

A genus of 6–7 species in tropical Africa.

Blighia unijugata Bak. in Oliv., F.T.A. **1**: 427 (1868).—Wilczek in Bull. Jard. Bot. Brux. **21**: 159 (1951).—Exell & Mendonça, C.F.A. **2**, 1: 88 (1954).—Keay, F.W.T.A. ed. 2, **1**, 2: 723 (1958).—Hauman, F.C.B. **9**: 316 (1960).—Dale & Greenway, Kenya Trees and Shrubs: 507 (1961).—White, F.F.N.R.: 224 (1962). TAB. **104**. Type from Nigeria.

Blighia zambesiaca Bak., loc. cit.—Sim, For. Fl. Port. E. Afr.: 32 (1909). Type: Nyasaland, W. shore of Lake Nyasa, *Kirk* (K, holotype).

Phialodiscus unijugatus (Bak.) Radlk. in Sitz.-Ber. Bayer. Akad. **9**: 539, 655 (1879); in Engl., Pflanzenr. IV, 165, **2**, 6: 1147 (1933). Type as for *B. unijugata*.

Phialodiscus zambesiacus (Bak.) Radlk., op. cit. **9**: 539, 655 (1879); in Engl., tom. cit.: 1148, fig. 33 (1933).—R.E.Fr. in Wiss. Ergebn. Schwed. Rhod.-Kongo-Exped. **1**: 121 (1914).—Brenan, T.T.C.L.: 560 (1949). Type as for *B. zambesiaca*.

Phialodiscus plurijugatus Radlk., op. cit. **20**: 263 (1890); in Engl., tom. cit.: 1148 (1933). Type from Angola.

Blighia sapida sensu Sim, loc. cit.: t. 22 (1909).

Tree up to 25 m. tall (often much shorter in our area) or large shrub; bark dark grey; branchlets greyish- or fulvous-tomentellous or pubescent. Leaves 4–10 cm. long; petiole 4–10 mm. long, hairy like the branchlets; leaflets 1–3(5)-jugate; rhachis terete, hairy like the branchlets; petiolules up to 4 mm. long; leaflet-lamina up to 15 × 6·5 cm., elliptic or oblong-elliptic or ovate-oblong, usually with tufts of hairs on the under surface in the axils of the lateral nerves, otherwise glabrous, apex usually acuminate and rounded at the tip, margin entire, base cuneate to obtuse; lateral nerves 5–8 pairs. Inflorescences 2–6(8) cm. long, usually in the axils of fallen leaves. Flowers white; pedicels c. 4 mm. long, elongating to 6 mm. in fruit. Sepals 1·5 mm. long. Petals 2–2·5 × 2 mm. Stamens 8–10; filaments 4 mm. long, pilose; staminodes much shorter in ♀ flowers. Ovary 1·5 mm. long, shortly pubescent; style 1·5 mm. long, 3-lobed. Fruit crimson, up to 40 × 25 mm., obovoid-pyriform with wings 1–3 mm. broad, tomentellous to pubescent when young, glabrescent. Seed black, shiny, 10 × 6 mm., cylindric-ellipsoid, glabrous.

The flowers are often solitary towards the apex of the rhachis and in cymules towards the base so that the inflorescence is a raceme towards the apex and a racemoid thyrse towards the base.

N. Rhodesia. N: Luwingu Distr., Lake Bangweulu, Chiluwi I., fr. 13.x.1947, *Brenan & Greenway* 8098 (FHO; K). W: Chingola-Solwezi road, Chimbombomene R., fr. x.1951, *Grout* 3/51 (FHO). **S. Rhodesia.** E: Melsetter Distr., Lusitu, Lower Nyahodi, fr. 17.ix.1959, *Ball* 811A (K; SRGH). **Nyasaland.** N: Njola's village, fr. 15.ix.1929, *Burtt Davy* 21724 (FHO). S: near Port Herald, Malawi Mt., st. viii-ix.1937, *Topham* 1016 (FHO). **Mozambique.** N: Corrane, fr. 17.xi.1936, *Torre* 1004 (COI; LISC). Z: between Alto Molocúè and Ile, fr. (immat.) 6.x.1941, *Torre* 3596 (LISC). MS: between Vila Pery and Moribane, 600 m., fl. 2.xii.1960, *Gomes e Sousa* 4628 (COI; K). SS: Vila João Belo, fr. 21.i.1942, *Torre* 3901 (LISC). LM: Vila Luiza, fl. 3.x.1957, *Barbosa & Lemos* in *Barbosa* 7965 (COI; K; LISC; LMJ).

Widespread in tropical Africa. Evergreen forest, riverine forest, grassland with trees and termite mounds, from sea-level up to c. 1200 m.

The epithet " unijugata " is inappropriate as the leaflets are very rarely 1-jugate.

6. **ERIOCOELUM** Hook. f.

Eriocoelum Hook. f. in Benth. & Hook., Gen. Pl. **1**, 1: 400 (1862).

Large or small trees or shrubs. Leaves paripinnate, shortly petiolate or subsessile; leaflets 2–5-jugate, lowest pair often resembling stipules. Inflorescence a terminal or axillary paniculate or racemoid thyrse. Flowers spuriously polygamous, often monoecious, actinomorphic. Sepals 5, almost free. Petals 5, each with a short broad pilose or pubescent scale. Disk patelliform or collar-like with a free scarious margin. Stamens 8; staminodes shorter in the ♀ flowers. Ovary 3-locular, densely hairy, loculi 1-ovulate; pistillode subglobose in the ♂ flowers. Fruit capsular, 3-gonous, 3-locular, loculicidally dehiscent, valves lanuginous at the base within. Seeds 1 per loculus, ellipsoid, with an arillode at the base.

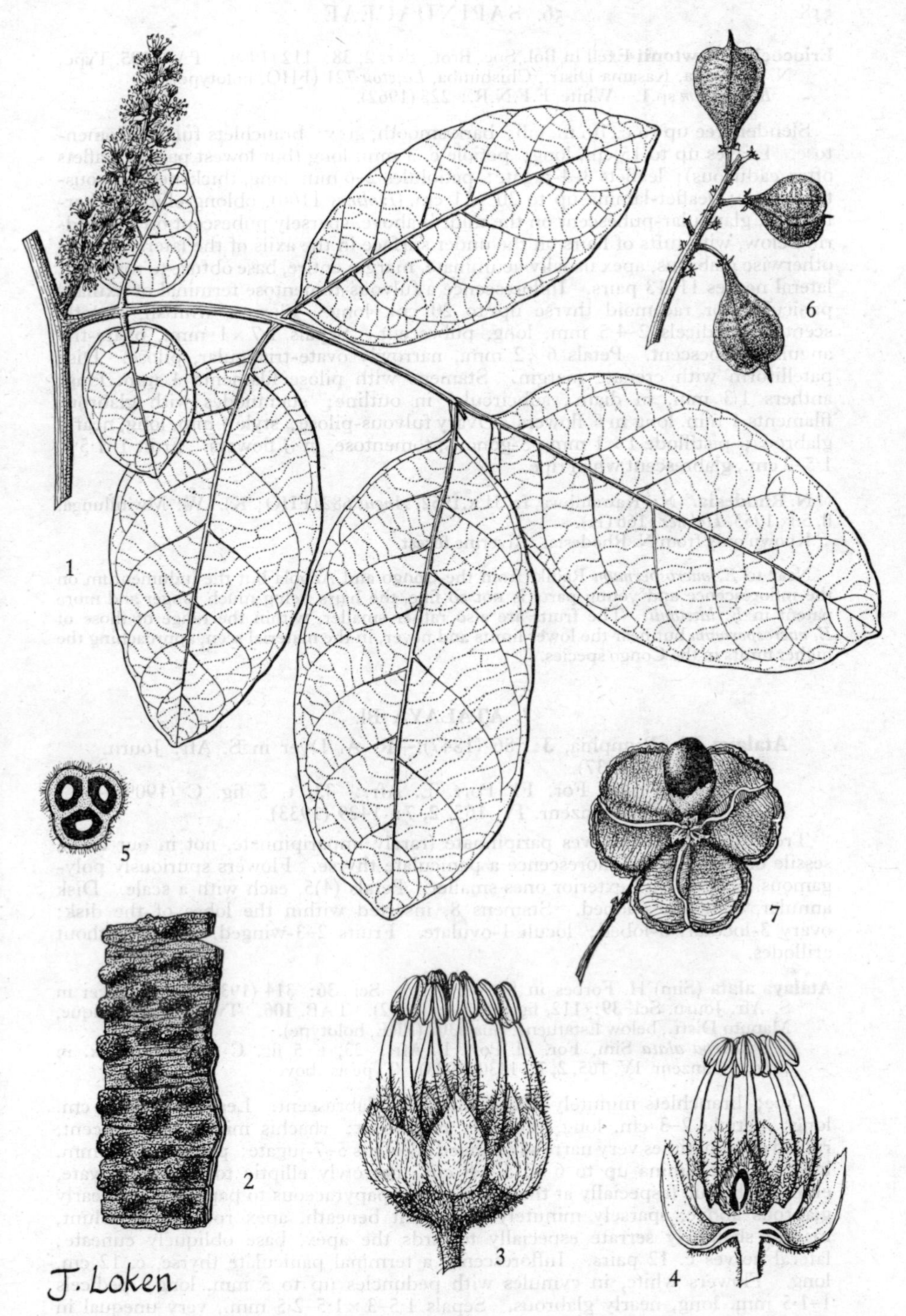

Tab. 104. BLIGHIA UNIJUGATA. 1, leaf and inflorescence (× ⅔) *Barbosa* s.n.; 2, bark showing warts (× ⅔) *Chase* 7400; 3, flower (× 7) *Barbosa* s.n.; 4, vertical section of flower (× 7) *Barbosa* s.n.; 5, transverse section of ovary (× 7) *Barbosa* s.n.; 6, infructescence (× ⅔) *Torre* 3485; 7, ripe fruit showing seed (× ⅔) *Torre* 3901.

Eriocoelum lawtonii Exell in Bol. Soc. Brot., Sér. 2, **38**: 112 (1964). TAB. **105**. Type: N. Rhodesia, Kasama Distr., Chishimba, *Lawton* 721 (FHO, holotype).
Eriocoelum sp.1.—White, F.F.N.R.: 225 (1962).

Slender tree up to c. 10. m. tall; bark smooth, grey; branchlets fulvous-tomentose. Leaves up to 15 cm. long; petiole c. 3 mm. long (but lowest pair of leaflets often caducous); leaflets 3–4-jugate; petiolules 2–6 mm. long, thickened, fulvous-tomentose; leaflet-lamina up to 30 × 11 cm. (*Holmes* 1166), oblong-elliptic, chartaceous, glandular-pubescent on the midrib above, sparsely pubescent on the midrib below, with tufts of hairs on the under surface in the axils of the lateral nerves, otherwise glabrous, apex usually acuminate, margin entire, base obtuse to rounded; lateral nerves 11–13 pairs. Inflorescence a fulvous-tomentose terminal or axillary paniculate or racemoid thyrse up to 20 cm. long. Flowers whitish, sweetly scented; pedicels 2–4·5 mm. long, pubescent. Sepals 1·7 × 1 mm., ovate-triangular, pubescent. Petals 6 × 2 mm., narrowly ovate-triangular, pilose. Disk patelliform with crenate margin. Stamens with pilose filaments 4 mm. long; anthers 1·3 mm. in diam., subcircular in outline; staminodes with glabrous filaments 1 mm. long in ♀ flowers. Ovary fulvous-pilose; style 4 mm. long, nearly glabrous; pistillode 1 × 1 mm., 3-gonous, tomentose, in ♂ flowers. Fruit 1–1·5 × 1·5–2 cm., glabrescent when ripe.

N. Rhodesia. N: Kawambwa, fr. 31.x.1952, *Angus* 682 (FHO; K). W: Mwinilunga, fl. 5.ix.1955, *Holmes* 1166 (K).
Known only from N. Rhodesia. Riverine forest.

Near to *E. microspermum* Radlk. from the Congo and Angola but the indumentum on the inflorescence and young parts is not so fine, the hairs being much longer and more patent in *E. lawtonii*. The fruits are also rather smaller, within the range of those of *E. microspermum* but near the lower limits and never, in the material seen, approaching the higher limits in the Congo species.

7. **ATALAYA** Bl.

Atalaya Bl., Rumphia, **3**: 86 (1847).—R. A. Dyer in S. Afr. Journ. Sci. **34**: 214 (1937).
Diacarpa Sim, For. Fl. Port. E. Afr.: 33, t. 5 fig. C (1909).—Radlk. in Engl., Pflanzenr. IV, 165, **2**, 7: 1429 (1933).

Trees or shrubs. Leaves paripinnate (rarely imparipinnate, not in our area), sessile or petiolate. Inflorescence a paniculate thyrse. Flowers spuriously polygamous. Sepals 5, 2 exterior ones smaller. Petals (4)5, each with a scale. Disk annular, sometimes lobed. Stamens 8, inserted within the lobes of the disk; ovary 3-locular, 3-lobed; loculi 1-ovulate. Fruits 2–3-winged. Seeds without arillodes.

Atalaya alata (Sim) H. Forbes in S. Afr. Journ. Sci. **36**: 314 (1939)—R. A. Dyer in S. Afr. Journ. Sci. **39**: 112, fig. 04, 2–2a (1942). TAB. **106**. Type: Mozambique, Maputo Distr., below Estatuene, *Sim* 6307 (PRE, holotype).
Diacarpa alata Sim, For. Fl. Port. E. Afr.: 33, t. 5 fig. C (1909).—Radlk. in Engl., Pflanzenr. IV, 165, **2**, 7: 1430 (1933). Type as above.

Tree; branchlets minutely pubescent, soon glabrescent. Leaves up to 15 cm. long; petiole 2–3 cm. long, minutely pubescent; rhachis minutely pubescent, ridged or sometimes very narrowly winged; leaflets 5–7-jugate; petiolules 1–2 mm. long; leaflet-lamina up to 6 × 1·5–1·8 cm., narrowly elliptic to narrowly ovate, elliptic, oblique (especially at the base), thinly papyraceous to papyraceous, nearly glabrous above, sparsely minutely pubescent beneath, apex rounded or blunt, margin shallowly serrate especially towards the apex, base obliquely cuneate; lateral nerves c. 12 pairs. Inflorescence a terminal paniculate thyrse, c. 12 cm. long. Flowers white, in cymules with peduncles up to 5 mm. long; pedicels 1–1·5 mm. long, nearly glabrous. Sepals 1·5–3 × 1·5–2·5 mm., very unequal in size, elliptic to subcircular, ciliate. Petals 2·5 × 2 mm., broadly triangular, pilose, margins incurved at the base to form a pouch-like scale. Disk 2·5 mm. in diam., glabrous. Stamens with filaments 3·5 mm. long, pilose towards the base, and apiculate anthers 1·2 mm. long. Fruit a 2(3)-winged samara; wing laterally elongated, 1·2 × 3 cm., glabrous.

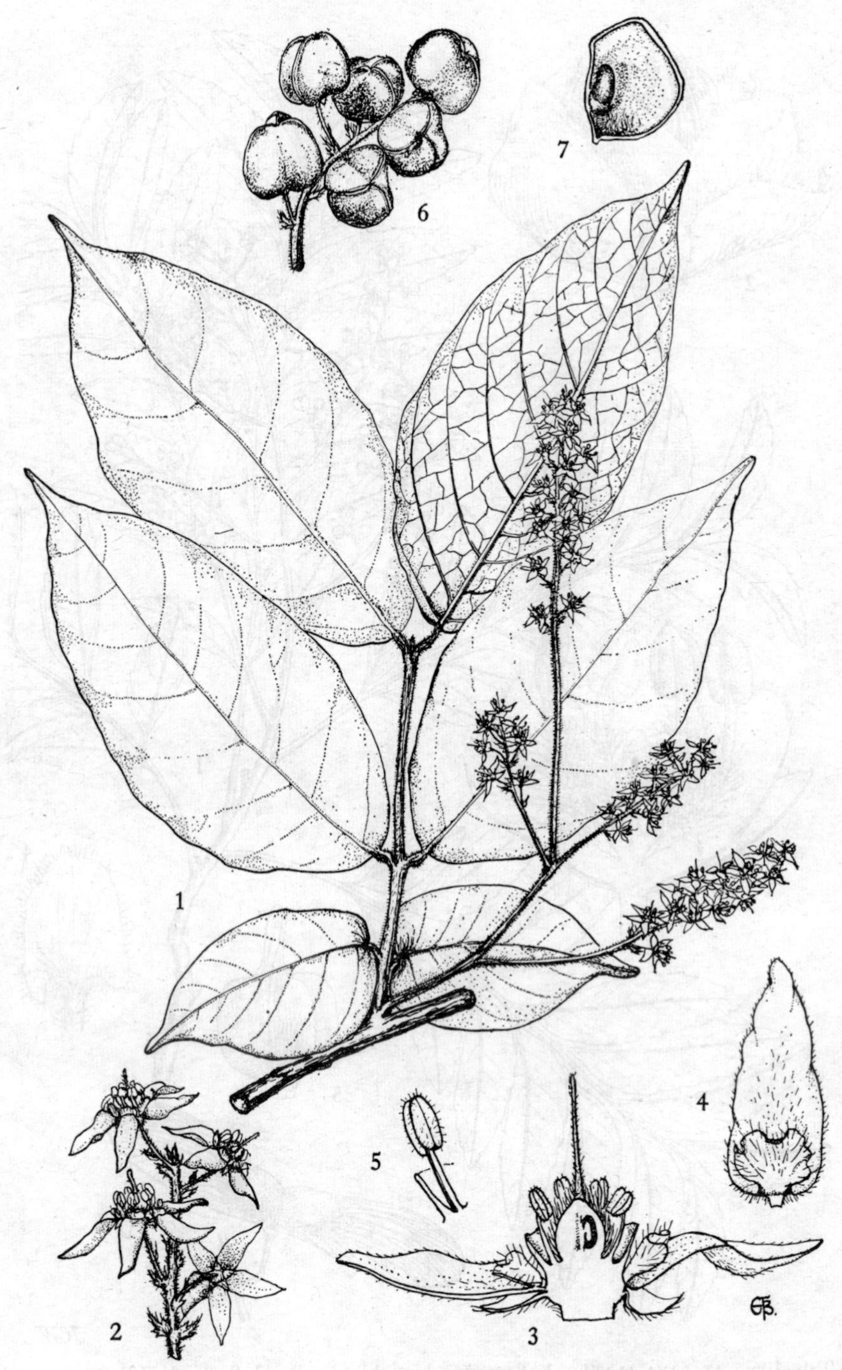

Tab. 105. ERIOCOELUM LAWTONII. 1, leaf and inflorescence (× ½); 2, part of inflorescence (× 1½); 3, vertical section of flower (× 5); 4, petal (× 6); 5, stamen (× 6); 6, infructescence (× ⅔); 7, one loculus of fruit opened to show seed (× 1) all from *Lawton* 721.

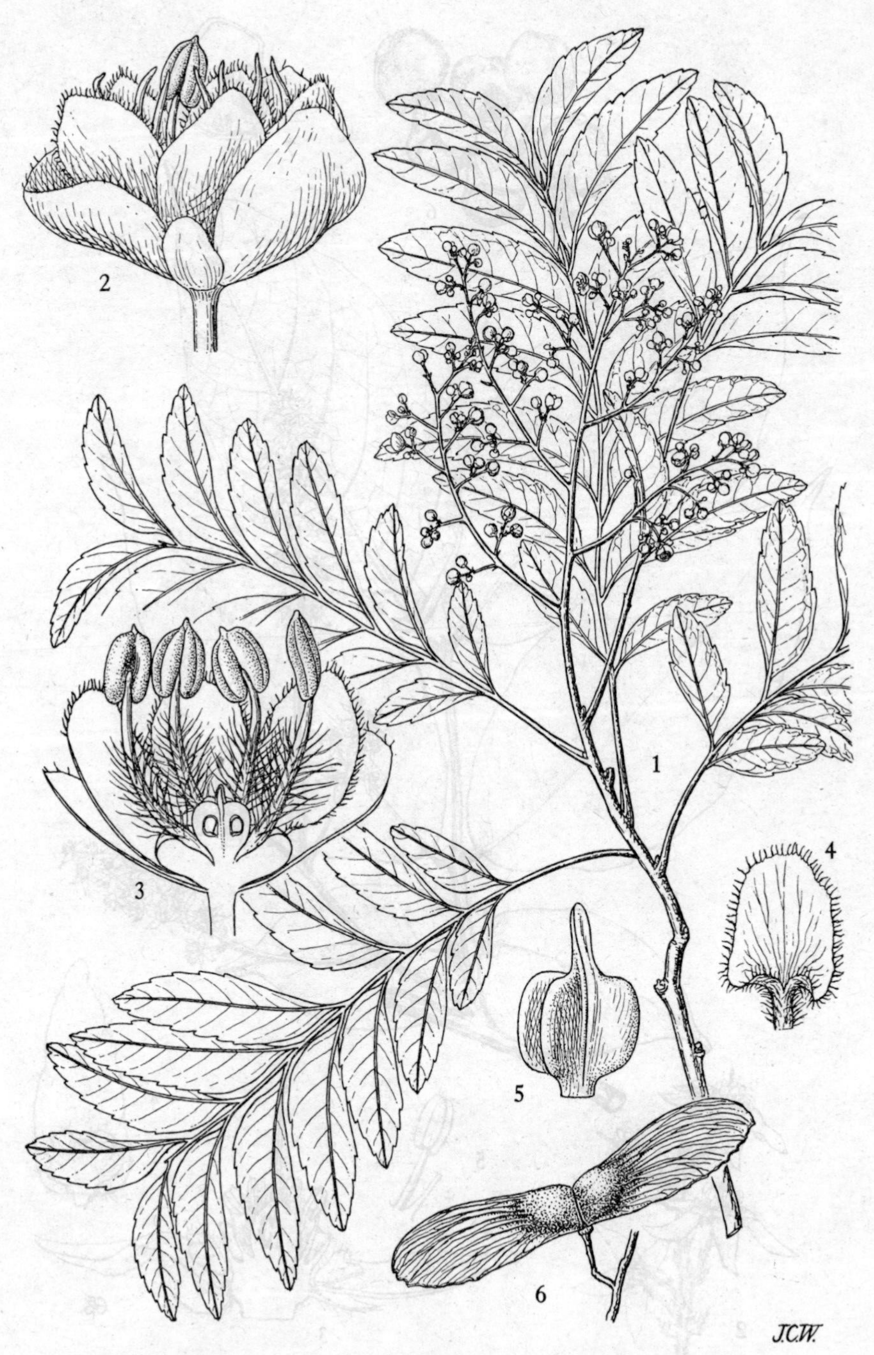

Tab. 106. ATALAYA ALATA. 1, flowering branchlet ($\times \frac{2}{3}$); 2, flower ($\times 12$); 3, vertical section of flower ($\times 12$); 4, petal ($\times 9$); 5, ovary ($\times 24$); 6, fruit ($\times 1$) all from *Balsinhas* 175.

Mozambique. LM: Libombos, Mazimiane Farm, fr. 19.iii.1948, *Gomes e Sousa* 3700 (K; LISC); Maputo, near Goba, fl. 15.xi.1940, *Torre* 2042 (LISC).
Also in Zululand and Natal. In *Androstachys* thickets and along streams.

8. DEINBOLLIA Schumach.

Deinbollia Schumach. in Kongel. Dansk. Vid. Selskr. Naturv. Math. Afh. **4**: 16 (1829).

Trees or shrubs or rarely shrublets. Leaves normally paripinnate, petiolate to subsessile; leaflets 2–11-jugate, alternate or subopposite. Inflorescence an axillary or terminal paniculate thyrse. Flowers usually dioecious (in our species), usually spuriously polygamous, actinomorphic. Sepals 5, the 2 external ones smaller. Petals 5, ciliate, each with a hairy deeply 2-lobed scale at the base. Disk ± annular, surrounding the stamens. Stamens (8)12–30; staminodes shorter in ♀ flowers. Ovary (2)3(5)-lobed, (2)3(5)-locular; loculi 1-ovulate; style gynobasic; pistillode vestigial in ♂ flowers. Fruit of (2)3(5) cocci (of which 1 to 4 usually abort), subglobose, indehiscent, somewhat fleshy or leathery. Seeds 1 in each coccus, without arillodes.
A genus of 30–40 species in tropical and S. Africa, Madagascar and Mascarenes.

Plant a shrub or small tree; pedicels tomentose to pilose; fruit tomentose to pilose:
  Petiole 3–9 cm. long:
    Sepals 4·5–5·5 mm. long - - - - - - - - 1. *oblongifolia*
    Sepals 2·5–4 mm. long:
      Rhachis of inflorescence fuscous-tomentellous (in dried specimens) to sparsely pubescent with paler usually somewhat silvery flower-buds; leaflets 3–8-jugate, distinctly petiolulate - - - - - 2. *borbonica*
      Rhachis of inflorescence fulvous-tomentose or densely fulvous-pubescent (in dried specimens); flower-buds scarcely silvery; leaflets 3–5-jugate:
        Upper surface of leaflet with rather inconspicuous reticulation; leaflets subcoriaceous, distinctly petiolulate, cuneate at the base; sepals 2·5–3 mm. long 3. *nyasica*
        Upper surface of leaflet with conspicuous reticulation; leaflets chartaceous, subsessile or very shortly petiolulate, rounded to subcordate at the base; sepals 3·5–4 mm. long - - - - - - - 4. *nyikensis*
  Petiole not more than 1 cm. long or leaf subsessile (with a small pair of leaflets at the base of the rhachis); leaf-rhachis usually narrowly winged - 5. *xanthocarpa*
Plant a shrublet up to 30 cm. tall; pedicels glabrous; fruit glabrous - 6. *fanshawei*

1. **Deinbollia oblongifolia** (E. Mey.) Radlk. in Sitz.-Ber. Bayer. Akad. **8**: 299 (1878); in Engl., Pflanzenr. IV, 165, **1**, 3: 673 (1932). Type from Natal.
Rhus oblongifolia E. Mey. in Drège, Zwei Pflanz.-Docum.: 156, 159 (1843). Type as above.
Hippobromus oblongifolius (E. Mey.) Drège in Linnaea, **19**: 614 (1847). Type as above.
Sapindus oblongifolius (E. Mey.) Sond. in Harv. & Sond., F.C. **1**: 240 (1860).—Sim. For. Fl. Port. E. Afr.: 32 (1909). Type as above.

Shrub up to c. 1·5 m. tall; branchlets glabrous or soon glabrescent. Leaves up to 30 cm. long, petiolate; petiole up to 9 cm. long, somewhat sparsely pubescent (more densely so in the furrows) or glabrous; rhachis terete, often ribbed but scarcely winged, glabrous or sparsely pubescent; leaflets 5–7-jugate; petiolules up to 3 mm. long; leaflet-lamina up to 15 × 5 cm., elliptic to oblong-elliptic, glabrous or nearly so, apex acute (rarely long-acuminate) to blunt, margin entire, base cuneate to rounded; lateral nerves 12–15 pairs. Inflorescence terminal, up to 35 cm. long, fulvous-tomentose or fulvous-pubescent. Flowers cream-coloured, in short-stalked or subsessile cymules; pedicels 1·5–3·5 mm. long, tomentose. Sepals 4·5–5·5 × 3–3·5 mm., elliptic to broadly elliptic, very densely fuscous- or silvery-pubescent. Petals 4·5–5·5 × 2·5 mm., elliptic, ciliate. Disk glabrous. ♂ flowers not seen; staminodes 14–16 in ♀ flowers with filaments 1 mm. long. Ovary 3-lobed, tomentose; style 3·5 mm. long. Fruit whitish-yellow; cocci 10–15 × 8–10 mm., subglobose or ellipsoid, sparsely pubescent at first, glabrescent. Seeds up to 10 × 8 mm., subglobose to obovoid or ellipsoid, glabrous.

Mozambique. Z: Maganja da Costa, fl. vi.1946, *Pedro* 1519 (LMJ). SS: Bilene, Macia, fl. 20.vi.1950, *Gomes e Sousa* 3998 (K; PRE). LM: Vila Luísa, fr. 1.x.1957,

Barbosa & Lemos 7888 (K; LMJ); Inhaca I., fr. 16.ix.1954, *Barbosa & Balsinhas* 5569 (BM; LMJ).
Also in Natal. Coastal thickets.

2. **Deinbollia borbonica** Scheff. in Flora, **52**: 306 (1869).—Radlk. in Engl., Pflanzenr IV, 165, **1**, 3: 674 (1932) pro parte excl. specim. *Buchanan* 1181. Type from Réunion.
Deinbollia borbonica forma *glabrata* Radlk. in Sitz.-Ber. Bayer. Akad. **8**: 369, 370 (1878); in Engl., tom. cit.: 675 (1932) pro parte excl. specim. *Buchanan* 1181.—Brenan T.T.C.L: 556 (1949). Type not specified.

Small tree or shrub; branchlets at first tomentellous, glabrescent. Leaves up to 30 cm. long; petiole 4–6 cm. long; leaflets 3–8-jugate; petiolules up to 4 mm. long; rhachis terete, sometimes ridged, glabrous or tomentellous; leaflet-lamina 5–14 × 2–5 cm., oblong to elliptic, chartaceous, sparsely pubescent or nearly glabrous, apex acute to rounded, margin entire, base cuneate; lateral nerves up to 14 pairs. Inflorescence up to 45 cm. long; rhachis fuscous-tomentellous to sparsely pubescent. Flowers in shortly stalked cymules; pedicels 1 mm. long, pubescent. Sepals 3–4 × 2·5–3 mm., subcircular to ovate-triangular, fuscous-pubescent. Petals 3·5–4·5 × 2·5 mm., elliptic, pilose-ciliate. Disk glabrous (in our area). Stamens 12–15; filaments 2 mm. long, pilose; anthers 1 mm. long. Ovary 3-lobed, 3-locular, tomentose. Fruit 1(2)-coccous by abortion; coccus 12 × 10 mm., obovoid, tomentose at first, glabrescent.

Mozambique. N: Metangula, fl. 25.v.1948, *Pedro & Pedrógão* 3899 (LMJ).
From Somaliland to northern Mozambique, Réunion and Zanzibar. In open woodlands on sandy soils.
Radlkofer divides the species into 4 forms and places the only specimen cited from our area, *Buchanan* 1181 from Nyasaland, in his forma *glabrata* (loc. cit.). This specimen belongs, in my opinion, to *D. nyikensis* Bak.

3. **Deinbollia nyasica** Exell in Bol. Soc. Brot., Sér. 2,**38**: 111 (1964). Type: Nyasaland, Cholo, *Hornby* 2915 (K, holotype; PRE).

Large tree; branchlets fulvous-tomentose. Leaves c. 20 cm. long; petiole 3–3·5 cm. long, stout; leaflets 3–5-jugate; petiolules up to 5 mm. long; rhachis densely pubescent to tomentose; leaflet-lamina up to 16 × 8·5 cm., elliptic, subcoriaceous, glabrous above, very sparsely pubescent beneath, apex shortly and bluntly acuminate, margin entire, base cuneate; midrib impressed above; lateral nerves 8–10 pairs; reticulation rather inconspicuous on upper surface. Inflorescence up to 25 cm. long, fulvous-tomentose, terminal or axillary. Flowers immature. Sepals 2·5–3 mm. long, tomentellous.

Nyasaland. S: Cholo, near Mboma stream, fl. (immat.) viii.1943, *Hornby* 2915 (K; PRE).
Known only from the type locality. Ecology unknown.
Nearest to *D. nyikensis* from which it differs by the characters given in the key to the species.

4. **Deinbollia nyikensis** Bak. in Kew Bull. **1897**: 249 (1897).—Radlk. in Engl. Pflanzenr. IV, 165, **1**, 3: 677 (1932).—Burtt Davy & Hoyle, N.C.L.: 69 (1936). Type: Nyasaland, Nyika Plateau, *Whyte* (K, holotype).
Deinbollia borbonica forma *glabrata* sensu Radlk., tom. cit.: 675 (1932) pro parte quoad specim. *Buchanan* 1181.

Tree; branchlets fulvous-pubescent when young. Leaves c. 20 cm. long; petiole up to 7 cm. long; leaflets 3–5-jugate, subsessile or very shortly petiolulate; leaflet-lamina up to 14 × 5 cm., oblong-elliptic, chartaceous, nearly glabrous above, sparsely pubescent beneath, apex usually rounded to obtuse, margin entire, base usually rounded to subcordate; lateral nerves 7–9 pairs. Inflorescence up to 30 cm. long, terminal or axillary, fulvous-tomentose. Sepals 3·6–3·8 mm. long, fuscous- or fulvous-tomentose. Petals 3·5–3·8 mm. long. Stamens c. 17 (8 *fide* Baker). Fruit with 1–2 cocci developing and with a very short (up to 0·7 mm. long) glabrous torus; cocci up to 18 × 13 mm., ellipsoid, fulvous-tomentose.

Nyasaland. N: unlocalized, fl. 1891, *Buchanan* 1181 (K). **Mozambique.** N: banks of Lake Nyasa, 600 m., fr., *Gomes e Sousa* 1509 (K).
Also in Tanganyika. Ecology unknown.

5. **Deinbollia xanthocarpa** (Klotzsch) Radlk. in Sitz.-Ber. Bayer. Akad. **8**: 308 (1878); in Engl., Pflanzenr. IV, 165, **1,** 3: 676 (1932).—Burtt Davy & Hoyle, N.C.L.: 69 (1936).—Brenan in Mem. N.Y. Bot. Gard. **8,** 3: 240 (1953).—*White*, F.F.N.R.: 224 (1962). TAB. **107**. Type: Mozambique, Rios de Sena, *Peters* (B, holotype †).

Sapindus xanthocarpa Klotzsch in Peters, Reise Mossamb. Bot. **1**: 119 (1862).—Bak. in Oliv., F.T.A. **1**: 431 (1868).—Sim, For. Fl. Port. E. Afr.: 32 (1909). Type as above.

Deinbollia marginata Radlk. in Engl., tom. cit.: 673 (1932).—Burtt Davy & Hoyle, N.C.L.: 69 (1936). Type: Nyasaland, Zomba, *Sharpe* 10 (K, holotype).

Small tree up to 6–7 m. tall or bush; branchlets tomentose or densely pubescent, eventually glabrescent. Leaves usually paripinnate (rarely imparipinnate) shortly petiolate to subsessile (lowest pair of leaflets, at or near the base of the rhachis, resembling stipules); rhachis usually narrowly winged, sometimes marginate or ribbed, pubescent; leaflets 3–9-jugate, subsessile or very shortly petiolulate; leaflet-lamina up to 8(10) × 3(5·5) cm., usually oblong to narrowly oblong, sometimes oblong-elliptic or elliptic, chartaceous, pubescent to shortly pilose mainly on the nerves and reticulation of the under surface, apex often obtuse or retuse but sometimes acute and apiculate, margin entire and often undulate, base cuneate; lateral nerves 6–12 pairs. Inflorescence c. 20 cm. long, terminal, once–twice-branched, often fuscous-tomentose especially when young. Flowers white, in shortly stalked cymules; pedicels c. 1·5 mm. long, fulvous- to fuscous-tomentose. Sepals 4·5–6·5 × 2·2–2·5 mm., elliptic, silvery-pubescent outside except where covered by imbrication. Petals 4·5–6 × 4–5·5 mm., elliptic, pilose-ciliate. Disk glabrous. Stamens 16; anthers 1·8 mm. long (1·5 mm. long in staminodes of ♀ flowers); filaments 3 mm. long (2 mm. long in staminodes), pilose. Ovary 3 (5)-lobed, tomentose; style 4 mm. long. Fruit 1–2-coccous; cocci yellow, 12–14 × 6–9 mm., obovoid to subglobose, tomentose, glabrescent, edible.

N. Rhodesia. C: Chingombe, fl. & fr. 27.ix.1957, *Fanshawe* 3745 (K). **S. Rhodesia.** N: Gokwe Distr., Gasani R, fl. 11.ix.1949, *West* 2985 (SRGH). E: Melsetter Distr., Sabi Valley, fl. 17.ix.1953, *Chase* 5071 (LISC; SRGH). S: Ndanga Distr., fl. 29.viii.1958, *Phelps* 243 (PRE; SRGH). **Nyasaland.** C: Salima Bay, fl. 22.ix.1935, *Galpin* 15015 (K; PRE). S: Mlanje, Njobru Valley, fr. 16.xi.1955, *Jackson* 1763 (COI; FHO; PRE). **Mozambique.** N: Ilha Mareli, fl. 25.viii.1946, *Gouveia & Pedro* in *Pedro* 1810 (LMJ). Z: between Régulo Simogo and Campo, fl. 30.viii.1949, *Barbosa & Carvalho* 3908 (LISC; LMJ). T: Tete, Tomo-Mazoe road, 365 m., fl. 21.ix.1948, *Wild* 2563 (K; SRGH). MS: Gorongosa, R. Pungué, fl. 25.viii.1958, *Chase* 6977 (PRE; SRGH). SS: between Jangamo and Inharrime, fl. 20.ix.1948, *Myre & Carvalho* 256 (LISC).

Also in the Transvaal. Thickets, kopjes and riverine formations.

Gouveia & Pedro in *Pedro* 1810, cited above, has some imparipinnate leaves.

Much the most common species in our area and usually distinguishable by its subsessile leaves and narrowly winged leaf-rhachis (the latter character also occurring in *D. fanshawei*).

6. **Deinbollia fanshawei** Exell in Bol. Soc. Brot., Sér. 2, **38**: 112 (1964). Type: N. Rhodesia, Mongu, fl. & fr. 21.ix.1962, *Fanshawe* 7048 (FHO, holotype; K).

Shrublet up to 30 cm. in height; branchlets densely appressed-pubescent to tomentose, glabrescent. Leaves up to 15 cm. long; petiole up to 5 cm. long, nearly glabrous, narrowly winged; leaflets 2–4-jugate, sessile or subsessile; rhachis with wings up to 2 mm. broad; leaflet-lamina up to 9 × 2·5 cm., oblong-elliptic to narrowly oblong-elliptic, chartaceous to subcoriaceous, nearly glabrous, apex blunt to acute and often mucronate, margin entire, base cuneate; lateral nerves 9–11 pairs, rather prominent below. Inflorescence 2–4 cm. long, terminal or leaf-opposed; rhachis tomentose. Flowers white; pedicels 1·5 mm. long, glabrous; flower-buds nearly glabrous. Sepals 5, 4·5 × 4 mm., subcircular, minutely ciliate, otherwise glabrous. Petals 5, 4·5 × 3 mm., rhombic, pilose on the margin. Stamens with filaments c. 2·5 mm. long (perhaps immature), pilose. Ovary (in an apparently ⚥ flower) 3-lobed, one lobe containing an apparently well-developed ovule, glabrous; pistillode 3-lobed in ♂ flowers. Fruit with 1–2 cocci developing; cocci c. 10 mm. in diam., subglobose, glabrous. Ripe seed not seen.

N. Rhodesia. B: Senanga, fl. 27.vii.1962, *Fanshawe* 6979 (BM; FHO). Known only from Barotseland. Kalahari Sand woodland.

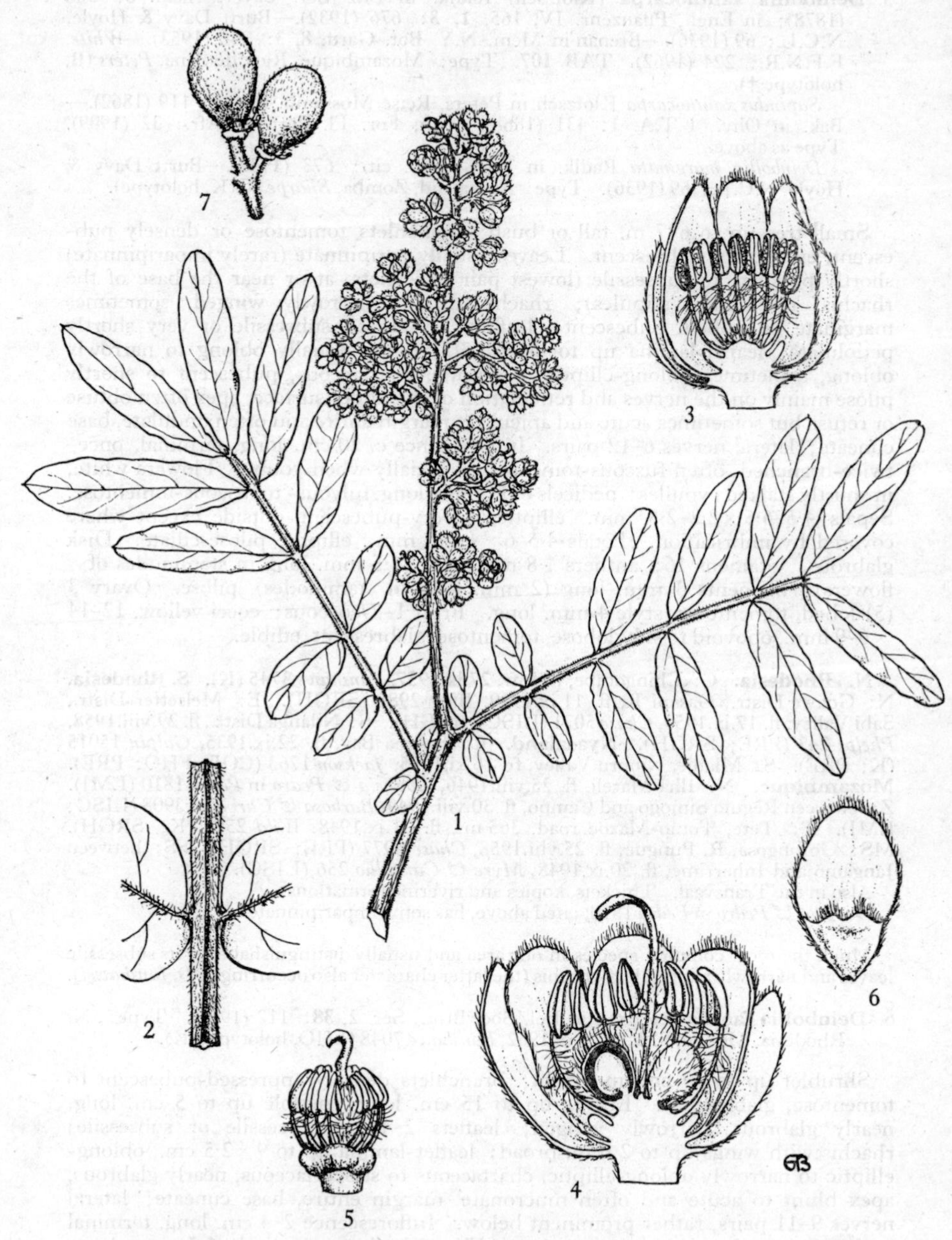

Tab. 107. DEINBOLLIA XANTHOCARPA. 1, flowering branchlet ($\times \frac{2}{3}$) *Chase* 6977; 2, winged leaf-rhachis ($\times 7$) *Brenan* 7814; 3, vertical section of male flower ($\times 5$) *Brenan* 7814; 4, vertical section of female flower ($\times 5$) *Chase* 6977; 5, female flower with sepals and petals removed ($\times 3$) *Chase* 6977; 6, petal ($\times 5$) *Chase* 6977; 7, fruits ($\times 1$) *McGregor* 71/51.

Apparently smaller than any other known species of *Deinbollia* and differing from the other species in our area by the glabrous ovary, fruit and pedicel. The glabrous pedicel is articulated to a tomentose branch of the rhachis and the fruit later breaks off at this point with the pedicel attached to it as a short stipe. Very few mature flowers were available for examination: one of these seemed to have a well-developed ovule in one of the ovary-lobes although the stamens were also well developed.

9. APHANIA Bl.

Aphania Bl., Bijdr. Fl. Ned. Ind. **5**: 236 (1825).

Medium-sized trees or shrubs, sometimes subscandent. Leaves usually paripinnate, rarely simple (not in our area) or tending to be imparipinnate, very shortly petiolate; leaflets 1–5-jugate. Inflorescence a paniculate or racemoid thyrse, axillary or terminal on short shoots. Flowers spuriously polygamous, usually monoecious. Sepals 5, 2 exterior ones smaller. Petals (2–4)5, each with a small hairy scale at the base. Disk patelliform, 5-crenate or 5-lobed, glabrous. Stamens 5–8, inserted inside the disk; filaments pilose. Ovary 2(3)-locular; loculi 1-ovulate. Fruit drupaceous, 2(3)-coccous (often 1-coccous by abortion). Seed without arillode.

A genus of c. 20 species in tropical Africa and tropical Asia.

Aphania senegalensis (Juss. ex Poir.) Radlk., Sapind. Holl.-Ind.: 21, 69 (1877–78); in Sitz.-Ber. Bayer. Akad. **8**: 238 (1878); in Engl., Pflanzenr. IV, 165, **1**, 3: 703 (1932).—Brenan, T.T.C.L.: 555 (1949).—Exell & Mendonça, C.F.A. **2**, 1: 81 (1954).—Keay, F.W.T.A. ed. 2, **1**, 2: 716 (1958).—Hauman, F.C.B. **9**: 343 t. 36 (1960).—Dale & Greenway, Kenya Trees and Shrubs: 507 (1961). TAB. **108.** Type from Senegal.

Sapindus senegalensis Juss. ex Poir. in Lam., Encycl. Méth. Bot. **6**: 666 (1804). —Bak. in Oliv., F.T.A. 1: 430 (1868). Type as above.

Medium-sized tree or shrub; bark grey; branchlets fulvous-setulose, soon glabrescent. Leaves paripinnate; petiole 1–5(7) cm. long; leaflets 1–2-jugate; petiolules 3–5 mm. long; leaflet-lamina 10–20 × 4–7 cm., narrowly elliptic, subcoriaceous, glabrous, apex usually blunt and sometimes shortly acuminate, margin entire, base cuneate. Inflorescence up to 15 cm. long, paniculate, terminal, fulvous-setulose. Flowers with slender pedicels 1–2 mm. long in ♂ flowers and more robust and 5 mm. long in ♀ flowers. Sepals c. 2·5 × 2 mm., ovate, ciliate. Petals 5, ovate, ciliolate but otherwise glabrous, each with an emarginate pilose scale at the base. Stamens (6)8; filaments 3–4 mm. long; staminodes 2 mm. long in ♀ flowers. Style 2 mm. long, subterminal. Fruit reddish-coloured, of 2 cocci (often 1 by abortion); cocci up to 18 × 12 mm., ellipsoid, glabrous. Seed 12 mm. long, ovoid, glabrous.

Mozambique. N: Nangororo, fr. 28.x.1959, *Gomes e Sousa* 4489 (K; SRGH). MS: Vila Machado, st. 20.iv.1948, *Garcia* 938 (LISC).

From Senegambia to Angola, Sudan, Abyssinia, Uganda, Kenya, Tanganyika and Mozambique. Usually in evergreen forest but ecology unknown in our area.

Fruit edible but the seed is said to be poisonous.

Owing to insufficiency of material from our area (where the species appears to be rare) much of the description is from Hauman (loc. cit.).

10. PANCOVIA Willd.

Pancovia Willd. in L., Sp. Pl. ed. 4, **2**, 1: 280, 285 (1799–1800).

Trees or shrubs. Leaves paripinnate, petiolate; leaflets 2–12-jugate. Inflorescence a racemoid or paniculate thyrse, axillary or caulinary or flowers fasciculate. Flowers spuriously polygamous, monoecious or dioecious, zygomorphic. Sepals 4–5, unequal, ± connate. Petals 3–4, unguiculate, each with a 2-lobed basal scale. Disk unilateral. Stamens 6–9 in an excentric bundle, reduced in ♀ flowers. Ovary 3-locular; loculi 1-ovulate; pistillode much reduced or absent in ♂ flowers. Fruit fleshy, indehiscent; carpels almost completely connate or joined only at the base (1 or 2 often aborting). Seed elliptic, laterally compressed, without arillode.

A genus of 10–12 species in tropical and subtropical Africa.

Pancovia golungensis (Hiern) Exell & Mendonça, C.F.A., **2**, 1: 85 (1954). TAB. **109.** Type from Angola.

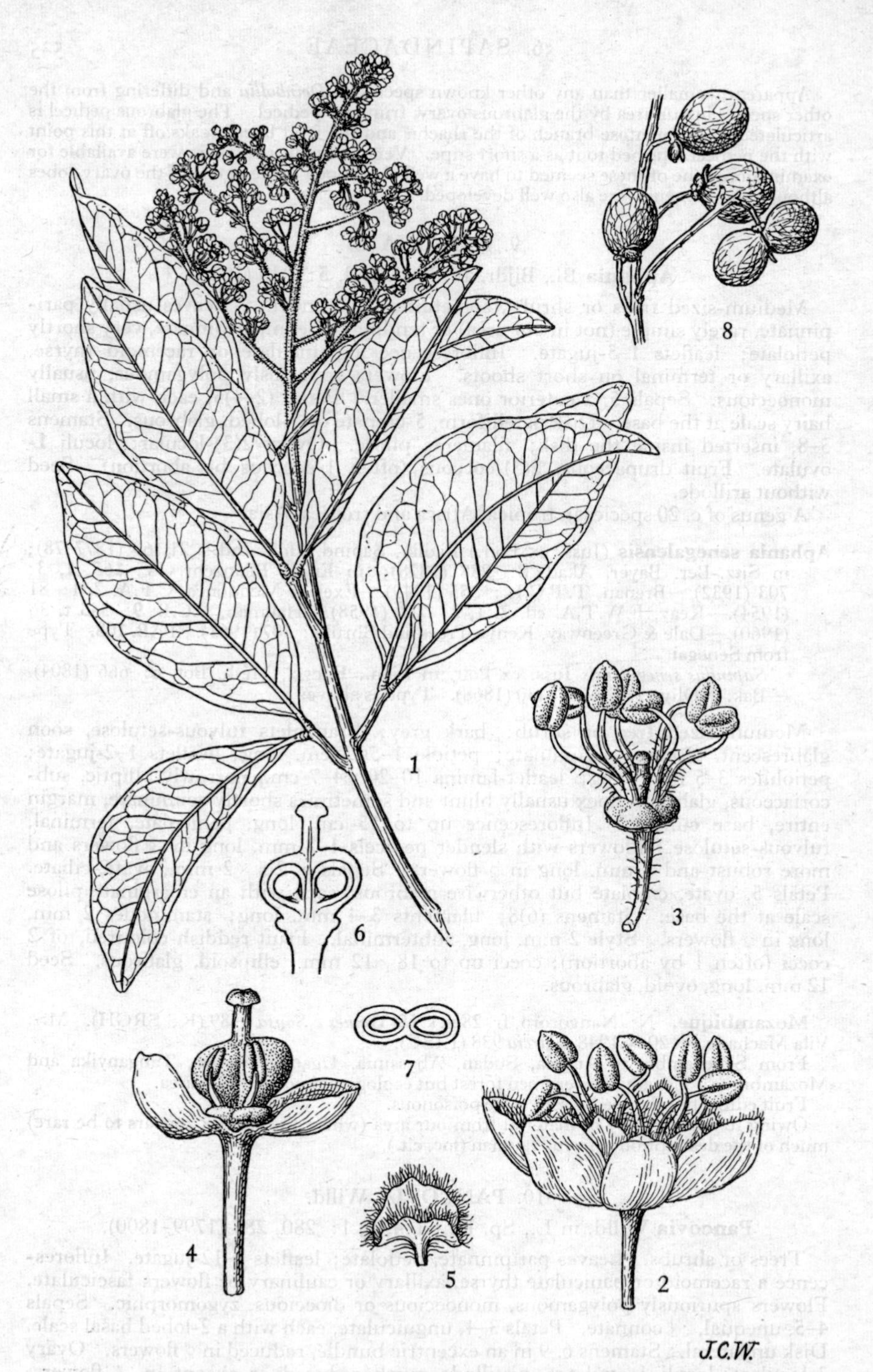

Tab. 108. APHANIA SENEGALENSIS. 1, flowering branchlet (× ⅔); 2, male flower (× 6); 3, male flower with sepals and petals removed (× 6); 4, female flower with 3 sepals and petals removed (× 6); 5, petal (× 6); 6, vertical section of ovary (× 6); 7, transverse section of ovary (× 6); 8, fruits (× ⅔). No. 1 from *Warnecke* 43, 2–7 from *Warnecke* 383, 8 from *Hoyle* 539.

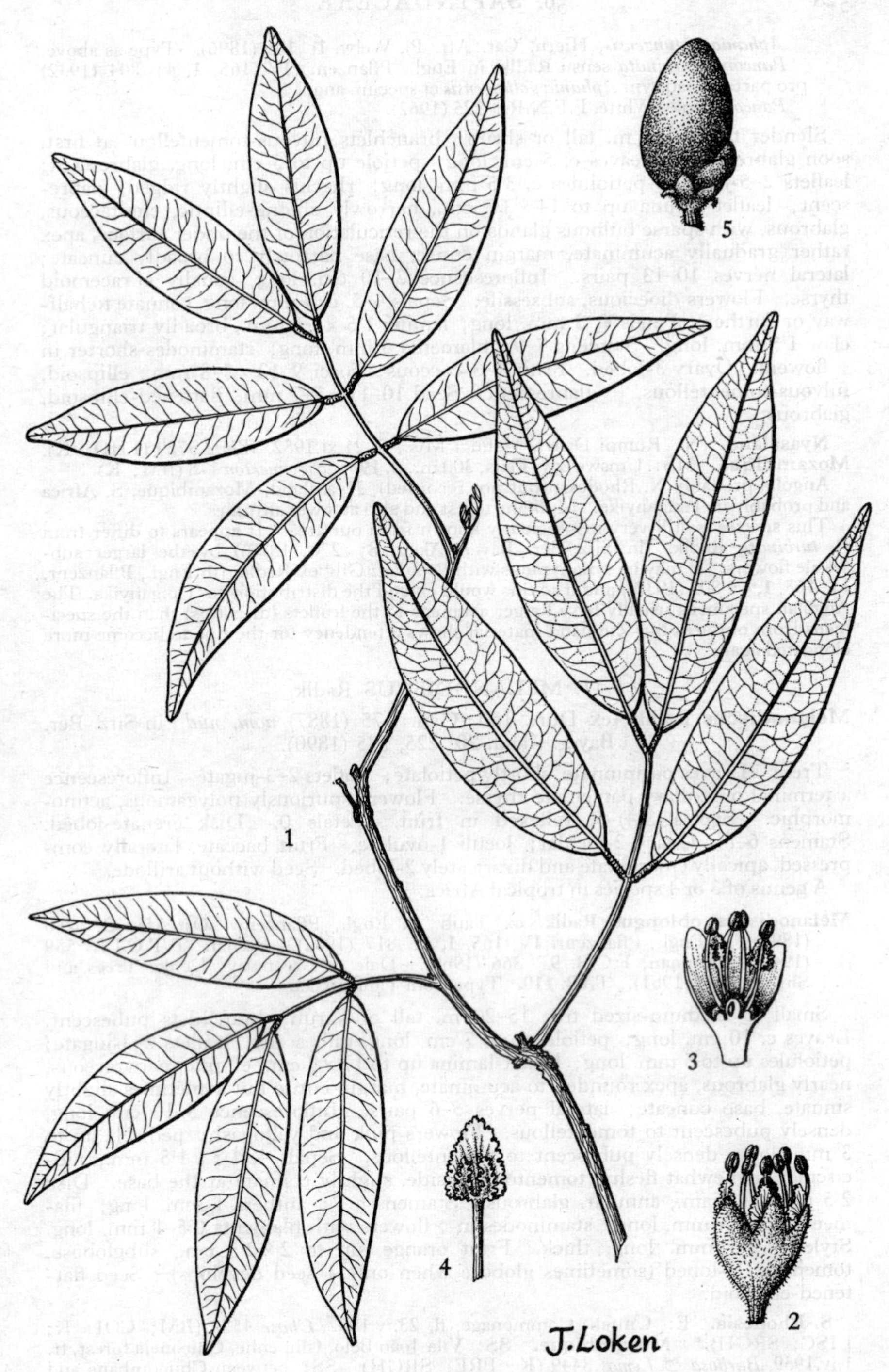

Tab. 109. PANCOVIA GOLUNGENSIS. 1, branchlet (× ⅔) *White* 3729; 2, flower (×3); 3, vertical section of flower (×3); 4, petal (×6); 5, fruit (×2) *Welwitsch* 4516. Nos. 2–5 from sketches by M. E. Church.

Aphania golungensis, Hiern, Cat. Afr. Pl. Welw. **1**: 169 (1896). Type as above.
Pancovia turbinata sensu Radlk. in Engl., Pflanzenr. IV, 165, **1**, 4: 804 (1932) pro parte quoad syn. *Aphania golungensis* et specim. angol.
Pancovia sp.—White, F.F.N.R.: 225 (1962).

Slender tree c. 15 m. tall or shrub; branchlets fulvous-tomentellous at first, soon glabrescent. Leaves c. 5 cm. long; petiole up to 3 cm. long, glabrescent; leaflets 2–5-jugate; petiolules c. 3·5 mm. long; rhachis slightly ridged, glabrescent; leaflet-lamina up to 14 × 3·5 cm., narrowly oblong-elliptic, chartaceous, glabrous, with sparse bulbous glands on the reticulation of the under surface, apex rather gradually acuminate, margin entire, base narrowly to broadly cuneate; lateral nerves 10–13 pairs. Inflorescence 3–10 cm. long, usually a racemoid thyrse. Flowers dioecious, subsessile. Sepals 4–5, c. 4 mm. long, connate to halfway or further. Petals 4, 3 mm. long; lamina 1·5 × 1·5 mm., broadly triangular; claw 1·5 mm. long. Stamens 7–9; filaments 5 mm. long; staminodes shorter in ♀ flowers. Ovary 3-lobed. Fruit 1–3-coccous; cocci 9–12 × 5–8 mm., ellipsoid, fulvous-tomentellous, ± glabrescent. Seed 10–11 × 5–7 mm., flattened-ellipsoid, glabrous.

Nyasaland. N: Rumpi Distr., Mafingi Mts., fl. 21.xi.1952, *White* 3729 (FHO; K). **Mozambique.** MS: Umswirizwi Flats, 300 m., fl. 1906, *Swynnerton* 178 (BM; K).

Angola, probably N. Rhodesia (but not recorded), Nyasaland, Mozambique, S. Africa and probably in Tanganyika. Montane forest and also at lower altitudes.

This species is still very insufficiently known from our area. It appears to differ from *P. turbinata* Radlk. (in Sitz.-Ber. Bayer. Akad. **8**: 270 (1878)) by the larger subsessile flowers. It may be synonymous with *P. holtzii* Gilg ex Radlk. (in Engl., Pflanzenr. IV, 165, **1**, 4: 802 (1932)) and if so this would extend the distribution to Tanganyika. The Angolan specimens usually have longer acumens to the leaflets (drip-tips) than the specimens from our area but Zululand material shows a tendency for the tips to become more elongated again.

11. MELANODISCUS Radlk.

Melanodiscus Radlk. [ex Dur., Ind. Gen.: 75 (1887) *nom. nud.*] in Sitz.-Ber. Bayer. Akad. **20**: 225, 285 (1890).

Trees. Leaves paripinnate, shortly petiolate; leaflets 2–3-jugate. Inflorescence a terminal or axillary paniculate thyrse. Flowers spuriously polygamous, actinomorphic. Sepals 3(4)–5, reflexed in fruit. Petals 0. Disk crenate-lobed. Stamens 6–8. Ovary 2-locular; loculi 1-ovulate. Fruit baccate, laterally compressed, apically emarginate and divaricately 2-lobed. Seed without arillode.

A genus of 3 or 4 species in tropical Africa.

Melanodiscus oblongus Radlk. ex Taub. in Engl., Pflanzenw. Ost-Afr. **C**: 250 (1895); in Engl., Pflanzenr. IV, 165, **1**, 4: 817 (1932).—Brenan, T.T.C.L.: 559 (1949).—Hauman, F.C.B. **9**: 366 (1960).—Dale & Greenway, Kenya Trees and Shrubs: 515 (1961). TAB. **110**. Type from Tanganyika.

Small to medium-sized tree 15–20 m. tall or shrub; branchlets pubescent. Leaves c. 10 cm. long; petiole up to 7 cm. long, pubescent; leaflets 2–3-jugate; petiolules up to 5 mm. long; leaflet-lamina up to 17 × 7 cm., elliptic, papyraceous, nearly glabrous, apex rounded to acuminate, margin entire but sometimes slightly sinuate, base cuneate; lateral nerves 5–6 pairs. Inflorescence 5–10 cm. long, densely pubescent to tomentellous. Flowers pink and yellowish; pedicels up to 3 mm. long, densely pubescent to tomentellous. Sepals 4, 1·5 × 1·5 mm., subcircular, somewhat fleshy, tomentose outside, slightly connate at the base. Disk 2·5 mm. in diam., annular, glabrous. Stamens 6–7; anthers 1 mm. long; filaments up to 4 mm. long; staminodes in ♀ flowers with filaments 0·5–1 mm. long. Style c. 2·5 mm. long, thick. Fruit orange, up to 2 × 2·5 cm., subglobose, tomentose, 2-lobed (sometimes globose when only 1 seed develops). Seed flattened-ellipsoid.

S. Rhodesia. E: Umtali, Commonage, fl. 23.iv.1952, *Chase* 4523 (BM; COI; K; LISC; SRGH).* **Mozambique.** SS: Vila João Belo, Chipenhe, Chiconela forest, fr. 1.iv.1959, *Barbosa & Lemos* 8449 (K; PRE; SRGH). SS: between Chucumbane and Zumane, fr. 7.ii.1948, *Torre* 7292 (LISC).

Also in the Congo, Kenya and Tanganyika. Evergreen woodlands up to c. 1300 m.

* Specimen just received: **Nyasaland.** S: Malawi Hills, st. 25 v. 1963, *Chapman* 2105 (FHO; SRGH).

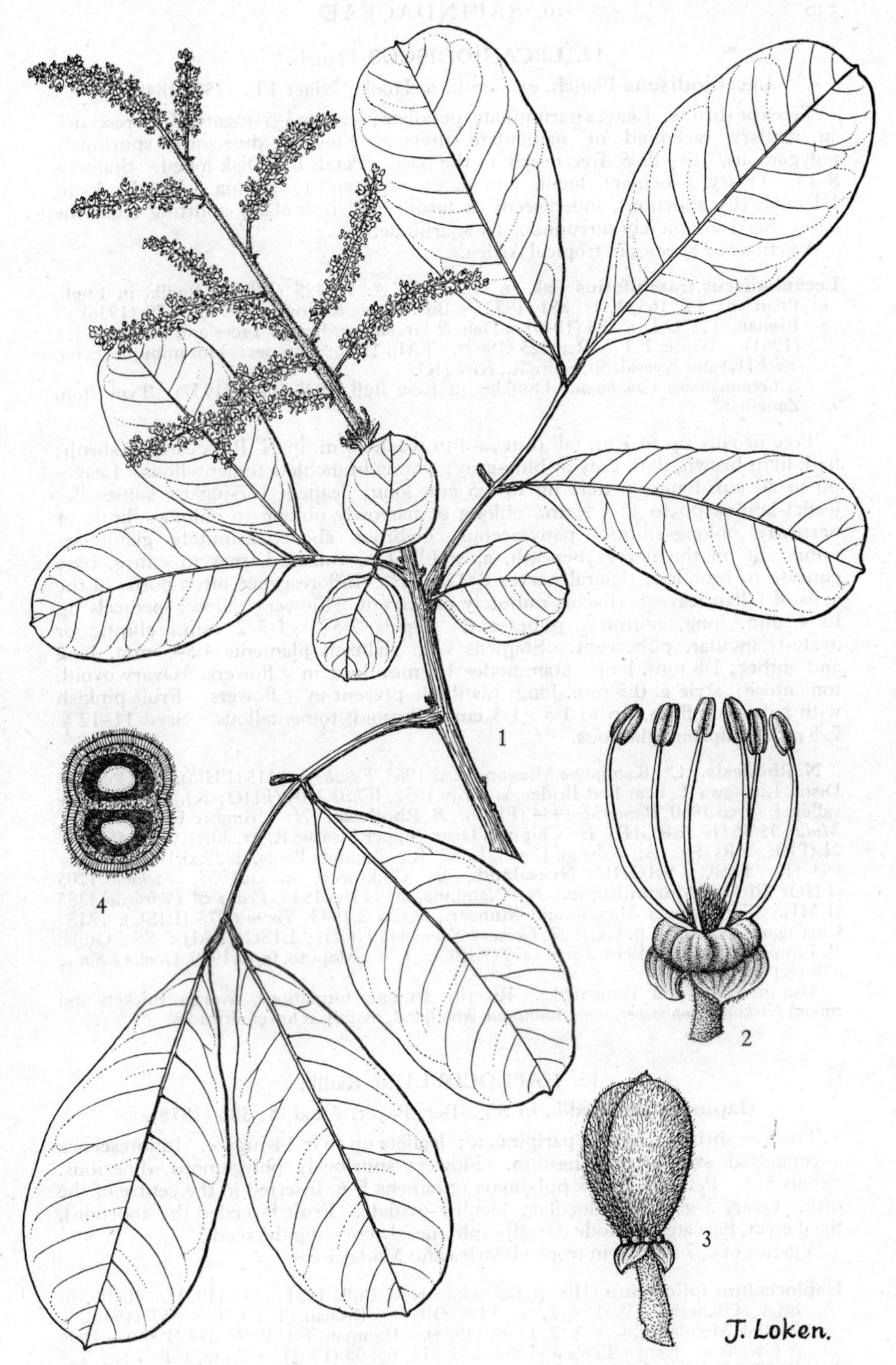

Tab. 110. MELANODISCUS OBLONGUS. 1, flowering branchlet (× ⅔) *Chase* 4523; 2, male flower (× 5) *Chase* 4419; 3, young fruit (× 3) *Chase* 4697; 4, transverse section of fruit (× 3) *Chase* 4697.

12. **LECANIODISCUS** Planch.

Lecaniodiscus Planch. ex Benth. in Hook., Niger Fl.: 250 (1849).

Trees or shrubs. Leaves paripinnate, petiolate; leaflets 3–7-jugate. Inflorescence an axillary racemoid or paniculate thyrse. Flowers dioecious, spuriously polygamous. Sepals 5, free nearly to the base. Petals 0. Disk lobed. Stamens 8–13. Ovary 3-locular; loculi 1-ovulate; style short; stigma 3-lobed. Fruit 1-locular (by abortion), indehiscent or tardily and irregularly splitting from the base. Seed completely surrounded by anarillode.

A genus of 2 species in tropical Africa.

Lecaniodiscus fraxinifolius Bak. in Oliv., F.T.A. **1**: 429 (1868).—Radlk. in Engl., Pflanzenr. IV, 165, **1,** 4: 881 (1932).—Burtt Davy & Hoyle, N.C.L.,: 70 (1936).—Brenan, T.T.C.L.: 558 (1949).—Dale & Greenway, Kenya Trees and Shrubs: 514 (1961).—White, F.F.N.R.: 225 (1962). TAB. **111**. Syntypes: Mozambique, Sena, *Kirk* (K) and Nyasaland, Shire R., *Kirk* (K).

Lecaniodiscus vaughaniae Dunkley in Kew Bull. **1937**: 469 (1937). Type from Zanzibar.

Tree usually up to 7 m. tall (but said to reach 20 m. in N. Rhodesia) or shrub; bark light brown, dark grey or blue-grey; young branchlets tomentellous. Leaves up to 30 cm. long; petiole up to 5·5 cm. long; leaflets 3–7-jugate, subsessile; leaflet-lamina up to 11 × 4 cm., oblong or narrowly oblong to oblong-elliptic or narrowly oblong-elliptic, papyraceous, glabrous above, minutely glandular-pubescent on the nerves beneath, apex blunt to rounded, margin entire, base cuneate to rounded; lateral nerves 8–12 pairs. Inflorescence often borne in the axils of fallen leaves; rhachis minutely pubescent. Flowers yellow; pedicels up to 4 mm. long, minutely pubescent. Sepals 2·5–3 × 1·5–2 mm., elliptic or ovate-triangular, pubescent. Stamens with glabrous filaments 4·5–5 mm. long and anthers 1·8 mm. long; staminodes 1·5 mm. long in ♀ flowers. Ovary ovoid, tomentose; style c. 0·5 mm. long; pistillode present in ♂ flowers. Fruit pinkish with pale blue flesh, up to 1·5 × 1·3 cm., ellipsoid, tomentellous. Seed 11–12 × 7–8 mm., ellipsoid, glabrous.

N. Rhodesia. C: Katondwe Mission, 12.xi.1963, *Fanshawe* 8115 (FHO). E: Petauke Distr., Luangwa R., near Beit Bridge, st. 17.iv.1952, *White* 2404 (FHO; K). S: Gwembe valley, fr. 1.xii.1960, *Bainbridge* 434 (FHO). **S. Rhodesia.** N: Urungwe Distr., Kariba, *Mullin* 95/56 (K; SRGH). E: Chipinga Distr., upper Msaswi R., fr. 30.v.1955, *Mowbray* 21 (PRE; SRGH). S: Ndanga Distr., Lundi R., Chipinda Pools, fl. 29.xi.1959, *Goodier* 684 (K; LISC; SRGH). **Nyasaland.** S: Chikwawa, st. vii.1955, *Jackson* 1705 (FHO; PRE). **Mozambique.** N: Nampula, st. 11.vi.1948, *Pedro & Pedrógão* 4385 (LMJ). T: between Mandiè and Mungari, fr. 26.xi.1943, *Torre* 6078 (LISC). MS: Cheringoma, Conduè, fl. 6.x.1957, *Gomes e Sousa* 4439 (COI; LISC; LMJ). SS: Guijá, R. Limpopo, fr. 12.xii.1940, *Torre* 2379 (LISC). LM: Maputo, fr. ix.1930, *Gomes e Sousa* 346 (K; LISC).

Also in Kenya and Tanganyika. Riverine fringing formations, riverine thickets and mixed *Kirkia-Colophospermum-Adansonia* woodland, mainly at lower altitudes.

13. **HAPLOCOELUM** Radlk.

Haplocoelum Radlk. in Sitz.-Ber. Bayer. Akad. **8**: 336 (1878).

Trees or shrubs. Leaves paripinnate; leaflets up to c. 14-jugate. Inflorescence a congested axillary polychasium. Flowers spuriously polygamous, dioecious. Sepals 5–6. Petals 0. Disk pulvinate. Stamens 5–6, inserted in the centre of the disk. Ovary 3-gonous, 3-locular; loculi 1-ovulate. Fruit 1-seeded (by abortion). Seed erect, baccate; arillode dorsally split, nearly covering the seed.

A genus of c. 7 species in tropical Africa and Madagascar.

Haplocoelum foliolosum (Hiern) Bullock in Kew Bull. **1931**: 353 (1931).—Radlk. in Engl., Pflanzenr. IV, 165, **2,** 8: 1498 (1934).—Brenan, T.T.C.L.: 557 (1949).—Exell & Mendonça, C.F.A. **2,** 1: 88 (1954).—Hauman, F.C.B. **9**: 374 (1960).—Dale & Greenway, Kenya Trees and Shrubs: 512, fig. 93 (1961).—White, F.F.N.R.: 225 (1962). TAB. **112** fig. A. Type from Angola (Huila).

Balsamea ? foliolosa Hiern, Cat. Afr. Pl. Welw. **1**: 126 (1896). Type as above.

Pistaciopsis dekindtiana Engl., Bot. Jahrb. **32**: 126 (1902). Type from Angola (Huila).

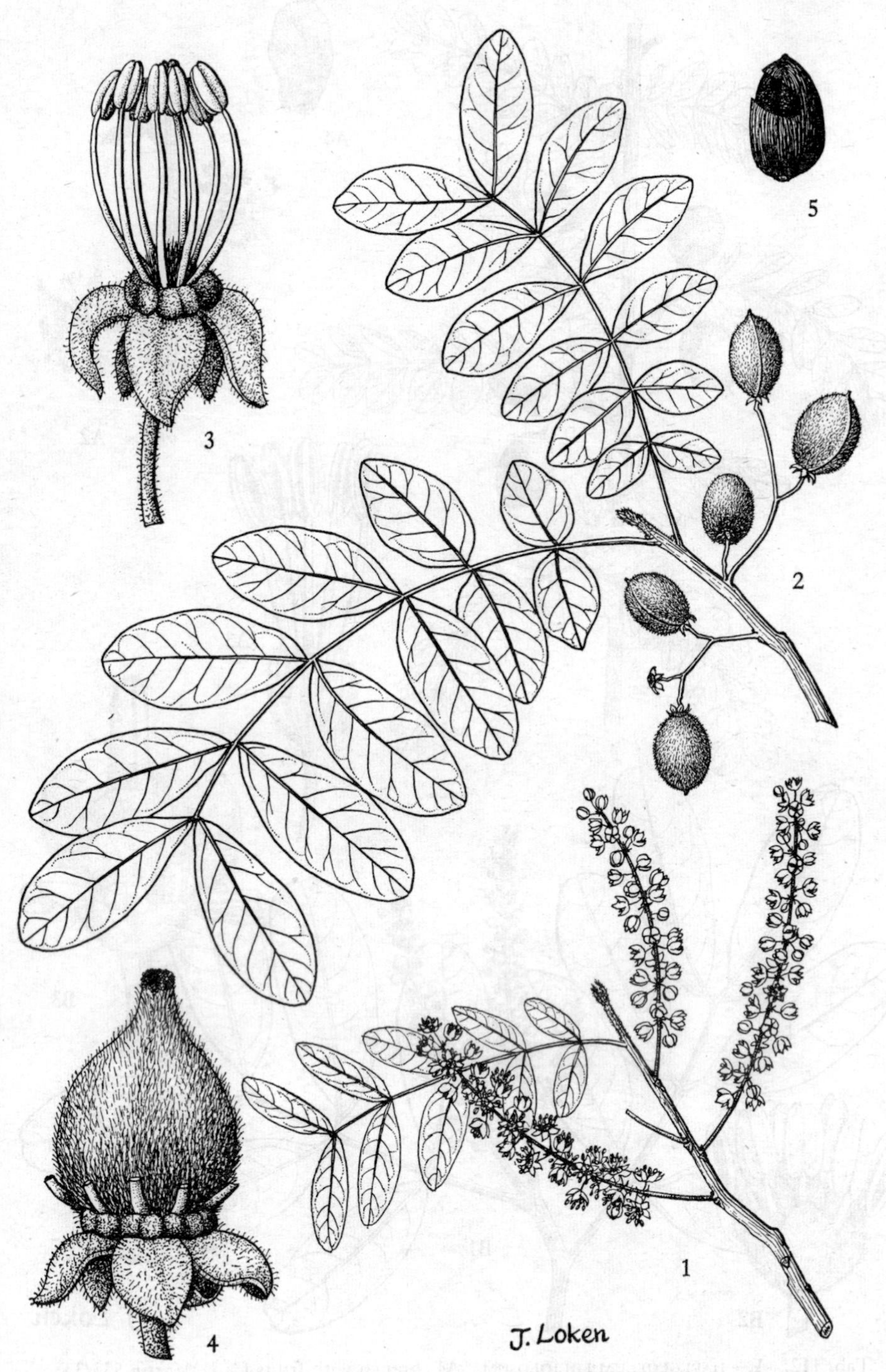

Tab. 111. LECANIODISCUS FRAXINIFOLIUS. 1, flowering branchlet (× ⅔) *Gomes e Sousa* 4440; 2, fruiting branchlet (× ⅔) *Phelps* 111; 3, male flower (× 5) *Mullin* 95/56; 4, young fruit developing (× 5) *Gomes e Sousa* 4440; 5, seed (× 1½) *Phelps* 111.

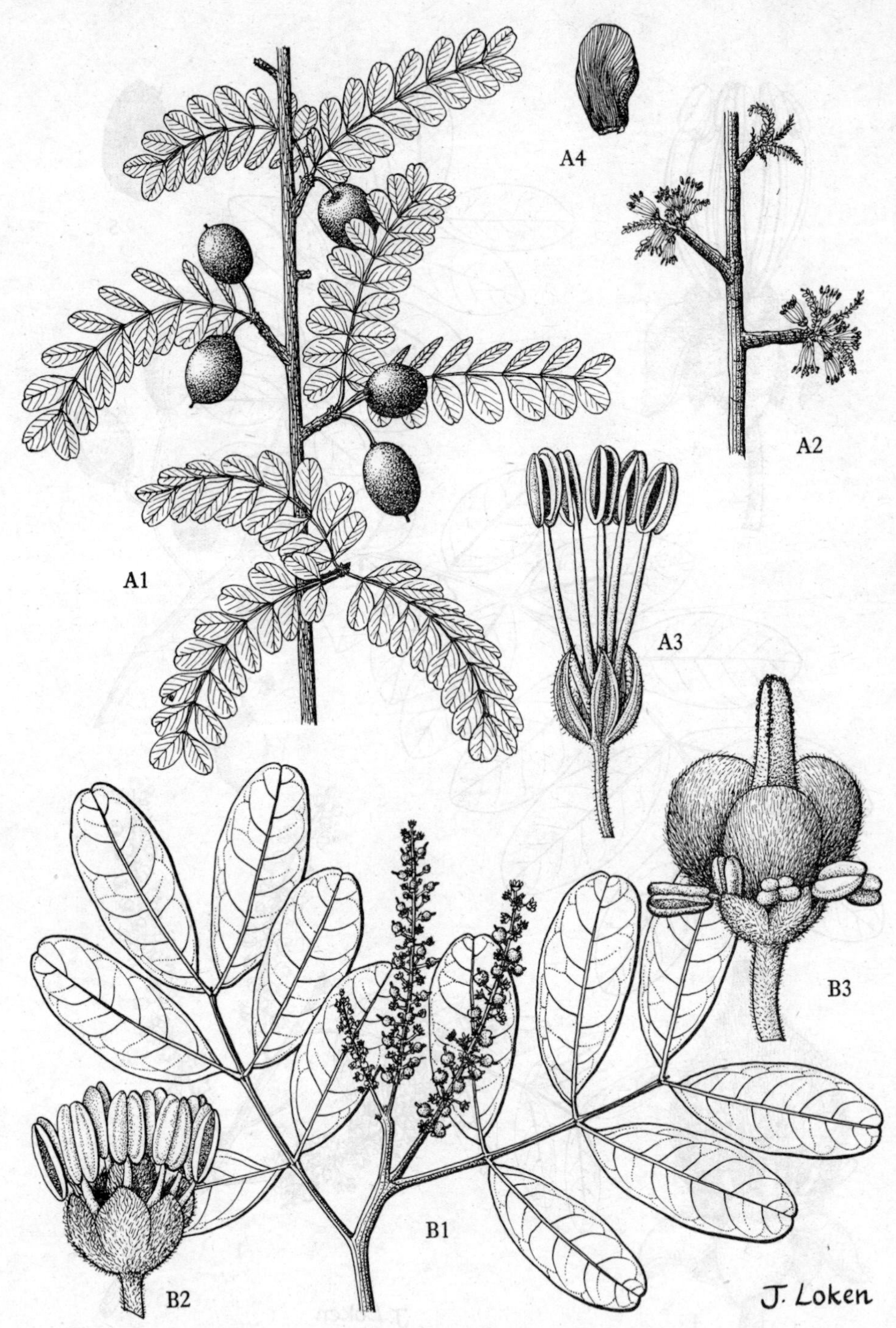

Tab. 112. A.—HAPLOCOELUM FOLIOLOSUM. A1, branch with fruits ($\times \frac{2}{3}$) *Martin* 533/33; A2, flowers and young leaves ($\times \frac{2}{3}$) *Martin* 413/32; A3, male flower ($\times 7$) *Martin* 413/32; A4, seed ($\times 1\frac{1}{2}$) *Martin* 533/33. B.—STADMANIA OPPOSITIFOLIA SUBSP. RHODESICA. B1, branchlet with inflorescences ($\times \frac{2}{3}$) *Chase* 4733; B2, male flower ($\times 7$) *Chase* 4733; B3, young fruit developing ($\times 7$) *Chase* 4733.

Haplocoelum dekindtianum (Engl.) Radlk. in Engl. & Prantl, Nat. Pflanzenfam. Nachtr. III: 204 (1907); in Engl., Pflanzenr. IV,165, **1**, 4: 885 (1932).—R.E.Fr. in Wiss. Ergebn. Schwed. Rhod.-Kongo-Exped. **1**: 131 (1914). Type as above.

Small tree or shrub; bark smooth, grey; wood reddish; branchlets dark greyish-brown, pubescent. Leaves 2–10 cm. long, often borne on short shoots; rhachis pubescent, often narrowly winged; petiole 3–8 mm. long, pubescent; leaflets 3–14-jugate; petiolules up to 0·5 mm. long, or leaflets subsessile; leaflet-lamina up to 15 × 9 mm., obliquely oblong, chartaceous, sparsely pubescent mainly on the midrib, apex usually emarginate, margin entire, base ± truncate and attached to the petiolule almost at the corner; lateral nerves numerous for the size of the leaflet and rather closely spaced. Flowers often appearing before the leaves, cream-coloured; pedicels 1–3 mm. long in ♂ flowers and 5–6 mm. long in ♀ flowers (*fide* Hauman), articulated at the base, pubescent. Sepals 5, 2 × 1 mm., oblong, pubescent. Disk 0·8 mm. in diam. Stamens 5; filaments 4 mm. long, glabrous; anthers 1·8–2 mm. long, apiculate; staminodes in ♀ flowers very short. Ovary 3-gonous; style 1 mm. long. Fruit orange to red, c. 15 × 15 mm., subglobose, glabrescent when mature, apiculate at the apex; stipe up to 7 mm. long. Seed 10 × 6 mm., ellipsoid.

N. Rhodesia. N: Abercorn Distr., Kalambo, 1220 m., st. 15.v.1936, *Burtt* 6003 (BM; K). W: Balovale Distr., near Chavuma, fr. 12.x.1952, *White* 3471 (BM; FHO; K). S: Kalomo Distr., Siburu Forest, 940 m., fr. 30.i.1963, *Bainbridge* 701 (FHO). **S. Rhodesia.** N: Urungwe Distr., Msukwe R., fl. 18.xi.1953, *Wild* 4189 (K; SRGH). W: Sebungwe Distr., st. 28.v.1947, *West* 2323 (SRGH). C: Sebakwe R., fr. xii.1948, *Hodgson* 38/48 (SRGH). **Mozambique.** N: Mogincual, st. 26.vi.1948, *Pedro & Pedrógão* 4684 (LMJ). T: Boroma, R. Zambeze, st. 16.vii.1950, *Chase* 2645 (BM; LISC; SRGH).

Also in Angola, Katanga and Tanganyika. In *Isoberlinia* woodlands, in *Combretum-Pterocarpus* woodlands, in *Baikiaea* forest, in thickets, and frequent on termite mounds.

14. STADMANIA Lam.

Stadmania Lam., Tabl. Encycl. **2**, 2: 441, t. 312 (1793).

Trees. Leaves paripinnate. Inflorescence a terminal or axillary racemoid thyrse. Flowers functionally ♂ or ♀, monoecious. Sepals 5, ± connate. Petals 0. Disk 5-crenate. Stamens (6)8, inserted within the disk. Ovary 3-sulcate, 3-locular; loculi 1-ovulate; ovules basal. Fruit (1)2–3-coccous. Seed covered by an arillode.

A genus of 1 species from eastern tropical Africa, S. Africa, Madagascar and Réunion.

Stadmania oppositifolia Poir. in Lam., Encycl. Méth. **7**: 376 (1806). Type from Réunion.

Stadmania sideroxylon DC., Prodr. **1**: 615 (1824).—Radlk. in Engl., Pflanzenr. IV, 165, **1**, 5: 1009 (1933) (" Stadmannia ").—Dale & Greenway, Kenya Trees and Shrubs: 516 (1961). *Nom. illegit.* Type as above.

Melanodiscus venulosus Bullock ex Dale & Greenway, op. cit.: 518 (1961) *nom. nud.*

Subsp. **rhodesica** Exell in Bol. Soc. Brot., Sér. 2, **38**: 114 (1964). TAB. **112** fig. B. Type: S. Rhodesia, Umtali Distr., near Bazeley Bridge, *Chase* 4733 (K; LISC; SRGH, holotype).

Tree up to 20 m. tall; branchlets whitish-grey, tomentellous, glabrescent. Leaves up to 15 cm. long; petiole up to 5·5 cm. long, tomentellous; leaflets (1) 2–3-jugate; leaflet-lamina up to 8·5 × 3·5 cm., elliptic to narrowly elliptic, chartaceous to subcoriaceous, somewhat shiny and nearly glabrous above except towards the base and on the midrib, tomentellous to minutely pubescent beneath, apex rounded and usually emarginate, margin entire, base cuneate to rounded; lateral nerves 6–10 pairs, rather inconspicuous. Inflorescence up to 10 cm. long; rhachis tomentellous. Flowers cream-coloured, 3–6-fasciculate on the rhachis; pedicels up to 3 mm. long, tomentellous. Sepals 1 × 1 mm., connate for ⅔ of their length, tomentellous. Disk very small. Stamens with very short filaments and anthers 1 mm. long. Ovary tomentellous; style 2 mm. long, fleshy, scarcely lobed at the apex; pistillode in centre of ♂ flower. Fruit 1–3-coccous; coccus c. 12 mm. in diam., subglobose, sericeous-tomentellous. Seed c. 10 mm. in diam., subglobose.

S. Rhodesia. E: Odzi R., 56 km. S. of Umtali, fr. 5.ii.1962, *Pole Evans* 6297 (PRE). S: Nuanetsi Distr., Shirugwe Hill, 4·8 km. S. of Mateke Hills, st. 5.v.1958, *Drummond* 5584 (SRGH).

Also in the Transvaal.

Differs from subsp. *oppositifolia* (from Kenya, Tanganyika, Madagascar and Réunion) in having shorter inflorescences, smaller leaves, leaves tomentellous to minutely pubescent on the leaflets beneath (especially on the midrib towards the base) and in the indumentum of ± patent hairs (appressed in subsp. *oppositifolia*). The species seems to divide into a northern and southern subspecies, very similar but slightly different and with a considerable gap in the distribution.

15. MACPHERSONIA Bl.

Macphersonia Bl., Rumphia, **3**: 156 (1847).

Trees or shrubs. Leaves abruptly bipinnate, Inflorescence a paniculate or racemoid thyrse. Flowers dioecious, actinomorphic. Sepals 5, petaloid, free almost to the base. Petals 5, small, unguiculate, each with 2 linear scales at the base formed by infolding of the margin. Disk annular, glabrous. Stamens 7–8. Ovary (2)3-locular; loculi 1-ovulate; pistillode very small in ♂ flowers. Fruit baccate, usually 1–2-locular and 1–2 seeded (by abortion). Seed with a thin arillode.

About 8 species in Madagascar and the Comoro Is. of which 1 extends to eastern tropical Africa.

Macphersonia hildebrandtii O. Hoffm., Sert. Pl. Madagasc.: 14 (1881).—Radlk. in Sitz.-Ber. Bayer. Akad. **20**: 247 (1890); in Engl., Pflanzenr. IV, 165, **1,** 4: 891 (1932).—Brenan, T.T.C.L.: 559 (1949). TAB. **113.** Type from Madagascar.

Small tree or shrub, 2–4 m. tall; branchlets pubescent and pilose. Leaves up to 20–25 cm. long, with 3–6 pinnae; rhachis pilose; petiole usually less than 5 mm. long; leaflets 5–10(15)-jugate, subsessile; leaflet-lamina up to 13 × 6 mm., obliquely oblong, nearly glabrous, apex usually emarginate, margin entire, base obliquely truncate; lateral nerves numerous for the size of the leaflet, closely spaced, rather inconspicuous. Inflorescence an axillary racemoid thyrse up to 15 cm. long. Flowers with glabrous pedicels up to 2–3 mm. long. Sepals 2 × 2 mm., subcircular, ciliate, slightly connate at the base. Petals 1 × 1 mm., obovate-triangular, unguiculate, pilose. Disk 1·2 mm. in diam., annular, glabrous, with crenulate or lobed margin. Stamens 8; filaments 4 mm. long, glabrous; anthers 0·9 mm. long. No ♀ flowers seen; pistillode very small in ♂ flowers. Fruit 13 mm. in diam. (*ex descr.*), subglobose, glabrous.

Mozambique. Z: Mocuba, Namagoa, fl. ix.1944, *Faulkner* 30 (BM; K; PRE). MS: Mossurize, R. Lucite, fl. 19.x.1953, *Pedro* 4335 (LMJ). SS: Inhambane, near Morrumbene, fl. 22.x.1937, *Gomes e Sousa* 2056 (BM; K).

Also in Madagascar, Comoro Is., Zanzibar and Tanganyika. Along streams.

16. FILICIUM Thw. ex. Hook. f.

Filicium Thw. ex. Hook f. in Benth. & Hook., Gen. Pl. **1**: 325 (1862); Enum. Pl. Zeyl.: 408 (1864).

Trees or shrubs. Leaves paripinnate with winged rhachis. Inflorescence a terminal or axillary paniculate thyrse. Flowers functionally ♂ or ♀. Sepals 5, persistent. Petals 5. Disk lanate. Stamens 5, inserted between the lobes of the disk, sterile in ♀ flowers. Ovary 2-locular; loculi 1-ovulate; ovule pendent from the apex of the loculus; pistillode in ♂ flowers. Fruit drupaceous, 1–2-locular. Seeds 1–2, without arillode.

A genus of 3–4 species in tropical Asia, possibly in Fiji, and in tropical Africa.

Filicium decipiens (Wight & Arn.) Thw., Enum. Pl. Zeyl.: 408 (1864).—Radlk. in Engl., Pflanzenr. IV, 165, **2,** 7: 1427 (1933).—Brenan, T.T.C.L.: 557 (1949).—Dale & Greenway, Kenya Trees and Shrubs: 511 (1961). TAB. **114.** Type from southern India.

Rhus decipiens Wight & Arn., Prodr. Fl. Penins. Ind. Or. **1**: 172 (1834). Type as above.

Usually a small tree up to 6–7 m. tall or shrub; bark variously described as smooth and greyish-brown or rough and dull chocolate-brown; branchlets

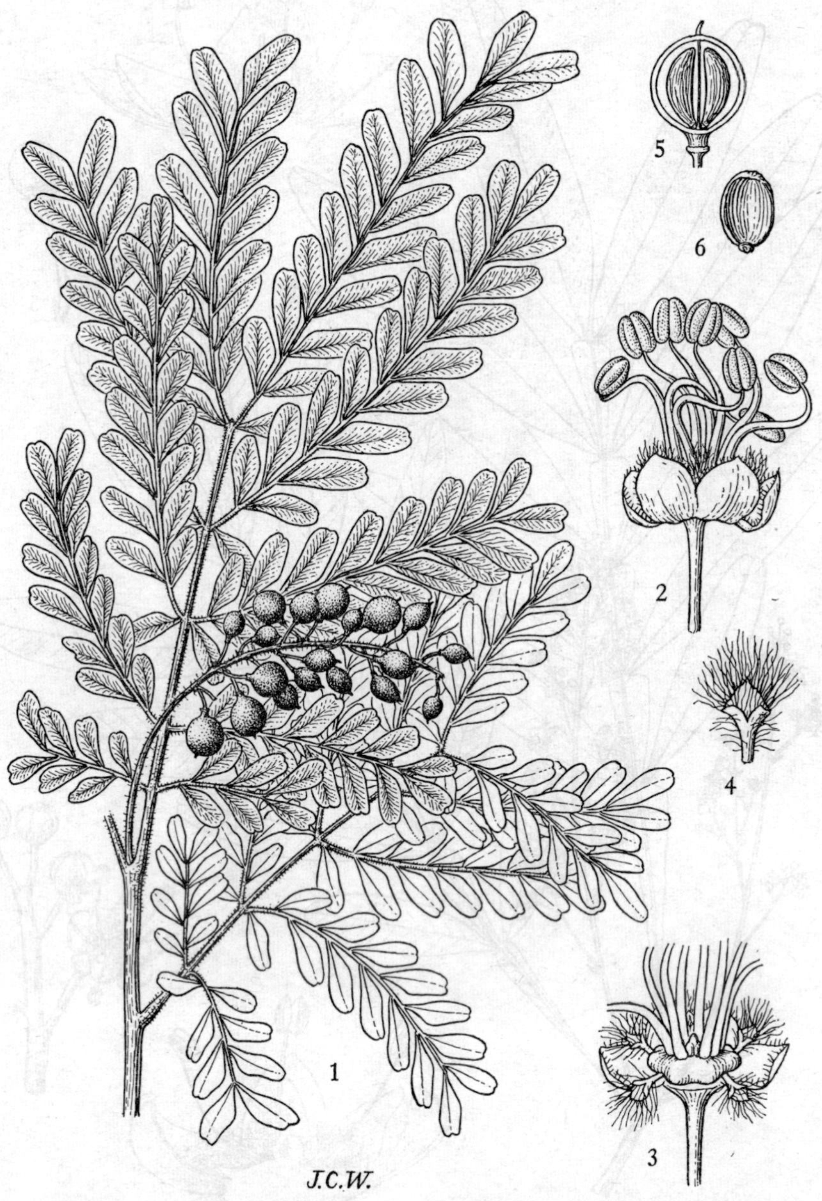

Tab. 113. MACPHERSONIA HILDEBRANDTII. 1, fruiting branchlet ($\times \frac{2}{3}$) *Gomes e Sousa* 2056 and *Hildebrandt* 1240; 2, male flower (×6) *Faulkner* 30; 3, male flower with one sepal removed to show petals and disk (×6) *Faulkner* 30; 4, petal (×10) *Faulkner* 30; 5, vertical section of fruit (×2) *Faulkner* 30; 6, seed (×2) *Faulkner* 30.

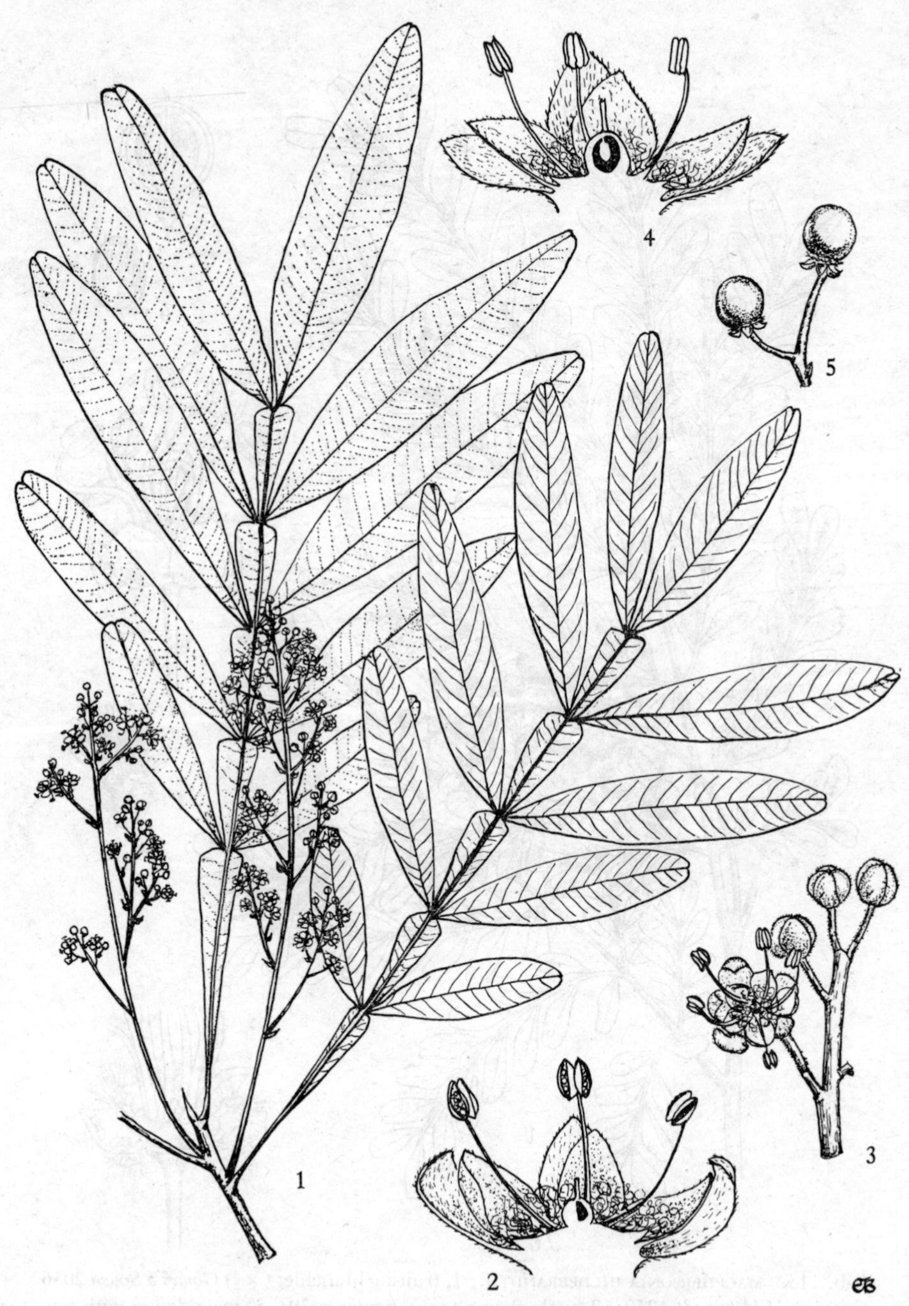

Tab. 114. FILICIUM DECIPIENS. 1, flowering branchlet ($\times \frac{2}{3}$) *Farquhar & Finn* in GHS 26611; 2, vertical section of male flower ($\times 16$) *Wild* 5275; 3, male flower and flower-buds ($\times 5$) *Wild* 5275; 4, vertical section of female flower ($\times 5$) *Farquhar* 10/49; 5, fruits ($\times 1$) *Goldsmith* in GHS 68162.

glabrous. Leaves up to 35 cm. long; rhachis glabrous, rather broadly winged with wing up to c. 15 mm. broad; petiole up to 10 cm. long, glabrous, winged towards the apex; leaflets up to 11-jugate, sessile; leaflet-lamina up to 15 × 3 cm., very narrowly oblong to oblong-elliptic, chartaceous, glandular-punctate and shiny above, otherwise glabrous, apex acute to rounded, margin entire, base cuneate to narrowly cuneate; lateral nerves numerous. Inflorescence up to 40 cm. long. Sepals 1·5 × 1·1 mm., ovate, ciliolate, free almost to the base. Petals 1·5 × 1·2 mm., ovate, ciliolate. Disk 2 mm. in diam. Stamens with filaments 3 mm. long and anthers 0·8 mm. long. Fruit green and dark red, 6(9) × 5(7) mm., conical to subglobose, glabrous.

S. Rhodesia. E: Melsetter Distr., Msapa Gap, fl. xii.1949, *Farquhar & Finn* in GHS 26611 (SRGH). **Nyasaland.** S: Namasi, fl. xi–xii.1899 *Cameron* 76 (K). **Mozambique.** MS: Gorongosa, R. Morombodzi, st. 8.x.1944, *Mendonça* 2390 (LISC).

Also in southern India, Ceylon, Fiji (*fide* Radlkofer), Ethiopia, Kenya and Tanganyika. Riverine forest.

J. D. Chapman informs me that this species also occurs in the Northern and Central Provinces of Nyasaland.

In spite of the presence of petals and the 1-ovulate loculi of the ovary the affinity, according to Radlkofer (loc. cit.), is rather with *Zanha* than with the genera adjacent to it in the somewhat artificial key to the genera (p. 496) from which it differs in having the ovule pendent from the apex of the loculus, a character which would be of practical inconvenience if used as a major separation in the key.

17. **ZANHA** Hiern

Zanha Hiern, Cat. Afr. Pl. Welw. **1**: 128 (1896).

Dialiopsis Radlk. [in Busse, Ber. Forschungsr. Deutsch-Ostafr.: 21 (1902) *nom. nud.*] in Engl. & Prantl, Nat. Pflanzenfam. Nachtr. III: 207 (1907).

Medium-sized trees. Leaves paripinnate; leaflets 3–7-jugate. Inflorescence a short to very short paniculate thyrse usually borne on a short shoot. Flowers dioecious, actinomorphic. Sepals 4–5(6). Petals 0. Disk small. Stamens (3) 4–7, coiled in bud; staminodes of ♀ flowers anantherous. Ovary 2-locular, loculi 2-ovulate; style short; stigma 2-lobed; pistillode absent in ♂ flowers. Fruit 1-locular (by abortion), ± drupaceous. Seed 1, pendulous from the apex of the loculus; testa provided with stomata; arillode absent.

A genus of two species in tropical Africa.

Fruit tomentellous; mature leaves pubescent at least on the nerves beneath or if lamina (rarely) almost glabrous then rhachis always densely pubescent towards the base
1. *africana*

Fruit glabrous; mature leaves glabrous; leaf-rhachis glabrous or very sparsely pubescent when young - - - - - - - - - - - - 2. *golungensis*

1. **Zanha africana** (Radlk.) Exell, comb. nov. TAB. **115** fig. A. Syntypes from Tanganyika (*Busse* 785 (B†) and 785a (B†)).

Dialiopsis africana Radlk. [in Busse, Ber. Forschungsr. Deutsch-Ostafr.: 21 (1902) *nom. nud.*] in Engl. & Prantl, Nat. Pflanzenfam. Nachtr. III: 208 (1907); in Engl., Pflanzenr. IV, 165, **2**, 7: 1419 (1933).—Burtt Davy & Hoyle, N.C.L.: 70 (1936).—Brenan, T.T.C.L.: 557 (1949).—Palgrave, Trees of Central Afr.: 403 cum phot. et tab. (1957).—Dale & Greenway, Kenya Trees and Shrubs: 509 (1961).—White, F.F.N.R.: 224 (1962). Syntypes as above.

Tree up to 10 m. tall or shrub; bark grey, scaling in large flakes; branchlets tomentellous when young. Leaves up to 15 cm. long; rhachis fulvous-tomentose; petiole up to 6 cm. long, fulvous-tomentose; leaflets 3–5-jugate, subsessile; leaflet-lamina up to 8 × 4 cm., elliptic to oblong-elliptic, chartaceous to subcoriaceous, tomentose when young, later usually pubescent on both surfaces but rarely glabrescent except on some of the nerves below, apex blunt to rounded, margin crenate to crenate-serrate or sometimes almost entire, base rounded to subcordate; lateral nerves 5–8 pairs. Inflorescence a congested fulvous-tomentose subglobose paniculate thyrse c. 2 cm. in diam.; peduncle up to 12-20 mm. long in ♀ plants, fulvous-tomentose. Flowers greenish, sweet-scented; pedicels 2–2·5(4) mm. long,

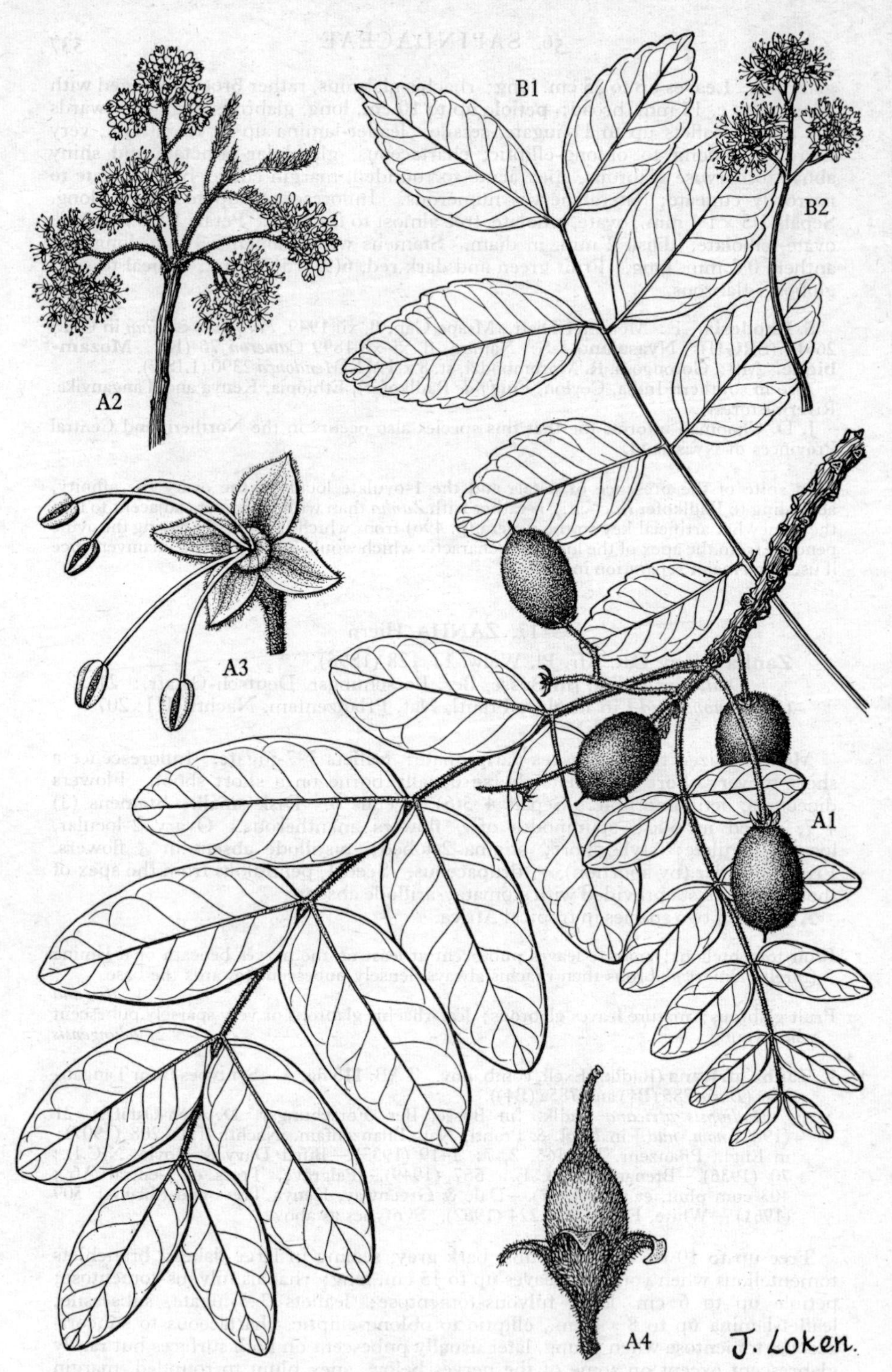

Tab. 115. A.—ZANHA AFRICANA. A1, branchlet with fruits (× ⅔) *Chase* 950 and *Angus* 1786; A2, inflorescences and young leaves (× ⅔) *Chase* 951; A3, male flower (×4) *Chase* 951; A4, young fruit (×4) *Chase* 947; B.—ZANHA GOLUNGENSIS. B1, leaf (× ⅔) *Chase* 1540; B2, inflorescence (× ⅔) *Chase* 536.

tomentose. Sepals 4–6, 2–2·5 × 1·5 mm., ovate-oblong, tomentose outside, glabrous within, connate at the base. Disk 1·3 mm. in diam., annular. Stamens 4–7; filaments 4 mm. long, glabrous; anthers 0·5–0·8 mm. long. Ovary pubescent to tomentellous; style 1·5–2 mm. long. Fruit orange or yellow, up to 3 × 2 cm., obovoid-ellipsoid to subglobose, tomentose, often beaked by the persistent style. Seed 1·5 × 1 cm., ellipsoid.

N. Rhodesia. B: Ndayo Forest, fr. 2.i.1939, *Martin* 927/39 (FHO). W: Ndola West Forest Reserve, fr. 29.xii.1951, *White* 1829 (FHO; K). C: Mt. Makulu, fr. 16.iv.1957, *Angus* 1786 (FHO; PRE; SRGH). E: Lundazi, st. 6.ix.1929, *Stevenson* 74 (FHO). S: Mazabuka Distr., Siamambo Forest Reserve, st. vii.1952, *White* 3876 (FHO). **S. Rhodesia.** N: Mtoko Distr., 16 km. E. of Mtoko, fr. 28.xii.1950, *Whellan* 495 (K; SRGH). W: Matabeleland, st. iii.1929, *Pardy* 5012 (FHO; SRGH). C: S. of Salisbury, fr. 18.xii.1945, *McGregor* 24/45 (FHO). E: Umtali Commonage, 1100 m., fl. 30.x.1951, *Chase* 4152 (SRGH). S: Buhera Distr., 1065 m., fr. xi.1953, *Davies* 608 (SRGH). **Nyasaland.** N: Mzimba Distr., fr. 21.i.1959, *Adlard* 292 (FHO; SRGH). **Mozambique.** MS: Gorongosa, near R. Mucosa, 200 m., st. 1936, *Gomes e Sousa* 4324 (FHO; K; PRE).

Also in S. Angola and Tanganyika. Woodland, often on granite ridges or kopjes, occasionally in riverine forest.

See note under *Z. golungensis*.

2. **Zanha golungensis** Hiern, Cat. Afr. Pl. Welw. **1**: 128 (1896).—Radlk. in Engl., Pflanzenr. IV, 165, **2**, 7: 1421 (1933).—Brenan, T.T.C.L.: 561 (1949).—Exell & Mendonça, C.F.A. **2**, 1: 92 (1954).—Keay, F.W.T.A. ed. 2, **1**, 2: 725 (1958).—Hauman, F.C.B. **9**: 361 (1960).—White, F.F.N.R.: 225 (1962). TAB. **115** fig. B. Type from Angola (Cuanza Norte).
Talisiopsis oliviformis Radlk. in Engl. & Prantl, Nat. Pflanzenfam. Nachtr. III: 208 (1907). Type from Togo Republic.

Tree up to 18 m. tall or shrub; bark reddish; branchlets glabrous. Leaves up to 20 cm. long; rhachis glabrous or very sparsely pubescent when young; petiole up to 7 cm. long, glabrous; leaflets 3–7-jugate, subsessile or with petiolules up to 2 mm. long; leaflet-lamina up to 10(16) × 4(6) cm., elliptic to oblong-elliptic, papyraceous to chartaceous, glabrous, apex blunt or rounded and often acuminate, margin crenate or remotely and shallowly dentate or entire, base cuneate; lateral nerves 5–10 pairs. Inflorescence a congested glandular-pubescent subglobose paniculate thyrse 1·5–2 cm. in diam. (♀ inflorescence fewer-flowered than the ♂); peduncle up to 6 cm. long, sparsely pubescent. Flowers (♂) with pedicels 2 mm. long. Sepals 4–5, 2–2·5 × 1·5 mm., ovate-triangular, pubescent outside, slightly connate at the base. Disk 1 mm. in diam., annular. Stamens 4–5; filaments 4 mm. long; anthers 0·8 mm. long; staminodes in ♀ flowers anantherous. Fruit yellow or pink, up to 20 × 15 mm., ellipsoid or globose-ellipsoid, apiculate, glabrous.

N. Rhodesia. N: Luwingu, fl. 13.x.1947, *Brenan & Greenway* 8092 (FHO; K). **S. Rhodesia.** W: Mafungabusi, st. 25.vi.1947, *Goldsmith* 19/47 (FHO; SRGH). E: near Umtali, fr. 29.xii.1945, *McGregor* 14/45 (FHO). **Nyasaland.** S: Zomba Distr., fl. v.1937, *Townsend* 670 (FHO). **Mozambique.** MS: Chimoio, R. Vanduzi, fr. 19.ix.1944, *Mendonça* 2509 (LISC).

Widespread in tropical Africa. Evergreen and mixed forests and woodland with nearly closed canopy; up to c. 1400 m.

In my opinion there is no doubt that this and the preceding species both belong to the same genus and that *Dialiopsis* is a synonym of *Zanha*. Not only do they agree almost completely in all the usual characters but they also agree in 4 characters unusual in the family one of which is in fact unique among Angiosperms. These are: (1) stamens coiled in the bud; (2) staminodes of ♀ flowers anantherous; (3) pistillode absent in ♂ flowers; (4) testa of seed provided with stomata. J.R. Metcalfe, of the Jodrell Laboratory, Kew, has kindly confirmed the last character in both species and thus verified Radlkofer's original observations. As far as can be ascertained there is no other record of stomata on a seed-coat. It is of course possible that they may be found in other genera if they are carefully looked for but it is unlikely that Radlkofer missed them if they are present in other Sapindaceae.

Radlkofer was able to separate the two genera only on the size and shape of the cells of the pericarp. The two species are in fact sometimes difficult to separate in the absence of flowers or fruit and where they overlap (at Umtali and in the Manica e Sofala Prov. of Mozambique) it is possible that they may hybridize.

18. PAPPEA Eckl. & Zeyh.

Pappea Eckl. & Zeyh., Enum. Pl. Afr. Austr. Extratrop. **1**: 53 (1834–35?).

Trees or shrubs. Leaves simple. Inflorescence an axillary racemoid thyrse. Flowers dioecious, actinomorphic. Sepals 5, connate to form a cupuliform calyx. Petals (4)5(6), each with 2 hairy scales at the base. Disk annular. Stamens 8–10; anthers with an apical gland. Ovary 3-locular; loculi 1-ovulate. Fruit capsular, 3-coccous (often reduced to 1 by abortion). Seed with a lobed arillode.

A genus of probably only 1 variable species in tropical and S. Africa.

Pappea capensis Eckl. & Zeyh., Enum. Pl. Afr. Austr. Extratrop. **1**: 53 (1834–35?).—Sim, For. Fl. Port. E. Afr.: 33 (1909).—Radlk. in Engl., Pflanzenr. IV, 165, **1,** 5: 1013 (1933).—Steedman, Trees etc. S. Rhod.: 43 (1933).—Burtt Davy & Hoyle, N.C.L.: 70 (1936).—Dale & Greenway, Kenya Trees and Shrubs: 516, fig. 94 (1961).—White, F.F.N.R.: 225 (1962). TAB. **116**. Syntypes from S. Africa.

Baccaurea capensis Spreng. in Flora, **12,** 1, Beibl.: 3 (1829) *nom. nud.*

Sapindus pappea Sond. in Harv. & Sond., F.C. **1**: 241 (1860) *nom. illegit.* Type as above.

Pappea radlkoferi Schweinf. ex Radlk. in Engl. & Prantl, Nat. Pflanzenfam. **3,** 5: 334 (1896). Type from Ethiopia.

Pappea ugandensis Bak. f. in Journ. Linn. Soc., Bot. **37**: 138 (1905).—Radlk. in Engl, Pflanzenr. IV, 165, **1,** 5: 1014 (1933).—Burtt Davy & Hoyle, N.C.L.: 70 (1936).—Brenan, T.T.C.L.: 560 (1949).—Hauman, F.C.B. **9**: 381 (1960). Type from Uganda.

Pappea fulva Conrath in Kew Bull. **1908**: 221 (1908).—Radlk., tom. cit.: 1016 (1933).—Steedman, Trees etc. S. Rhod.: 44 (1933). Type from the Transvaal.

Pappea capensis var. *radlkoferi* (Schweinf. ex Radlk.) Schinz in Viertel. Naturf. Gesell. Zür. **4**: 400 (1909).—Burtt Davy & Hoyle, loc. cit.—O.B. Mill., B.P.C.L.: 36 (1948). Type as for *P. radlkoferi*.

Pappea radlkoferi var. *angolensis* Schlecht. in Engl., Bot. Jahrb. **50**, Suppl.: 422 (1914).—Exell & Mendonça, C.F.A. **2,** 1: 91 (1954). Type from Angola (Huila).

Tree up to 10 m. tall with spreading crown; branchlets tomentellous, glabrescent. Leaves petiolate; petiole up to 15 mm. long, tomentellous; lamina up to 10 × 6 cm., oblong to narrowly oblong, chartaceous to subcoriaceous, shortly pubescent above especially on the nerves, tomentellous beneath, apex rounded and usually shortly mucronate, margin entire to sharply serrate, base rounded to subcordate; lateral nerves 10–20 pairs, branching near the margin. Inflorescence up to 16 cm. long; rhachis tomentellous. Flowers yellowish or greenish, in fascicles or subsessile cymules on the rhachis; pedicels up to 4 mm. long. Sepals 1·3 mm. long, connate to above the middle. Petals 0·7–1 mm. in diam., subcircular, unguiculate. Disk 2 mm. in diam. Stamens 8–10 (reduced in size in ♀ flowers); filaments up to 3 mm. long; anthers 1 mm. long. Ovary 3-gonous, tomentellous; style up to 5 mm., appressed-pubescent. Fruit (usually reduced to 1 coccus) c. 10 mm. in diam., subglobose, tomentellous. Seed black, shiny, c. 7 mm. in diam., subglobose, nearly surrounded by the arillode.

Bechuanaland Prot. N: Tsessebe, fl. 13.i.1960, *Leach & Noel* 8 (K; SRGH). SE: Gaberones road, c. 32 km. N. of Lobatsi, fl. 20.ii.1955, *McConnell* in GHS 68921 (SRGH). **N. Rhodesia.** N: Mpika, fl. 1.ii.1955 *Fanshawe* 1919 (FHO; SRGH). C: Lusaka, st. 19.v.1934, *Miller* 23 (FHO). E: Fort Jameson, fr. 1.vi.1958, *Fanshawe* 4492 (FHO). S: Livingstone, fr. 4.vi.1958, *Fanshawe* 5724 (FHO; K). **S. Rhodesia.** W: Bulawayo, fl. 1908, *Chubb* 44 (BM). C: Salisbury, 1465 m., st. 22.v.1927, *Eyles* 4924 (FHO; K). E: Umtali Distr., Maranka Reserve, fl. 10.ii.1953, *Chase* 4760 (BM; K; LISC; SRGH). S: Victoria Distr., fl. & fr. 1909, *Monro* 988 (BM). **Nyasaland.** C: Dedza Distr., Bunda Hill, fl. 26.iii.1961, *Chapman* 1179 (FHO; SRGH). S: Zomba Distr., fl. iii.1932, *Clements* 243 (FHO). **Mozambique.** T: Vila Mousinho, fr. 16.vii.1949, *Andrada* 1774 (COL; LISC). MS: between Mambone and Quicuaxa, fl. 2.ix.1942, *Mendonça* 113 (LISC). SS: Inhambane, near Vilanculos, fl. 1.x.1942, *Mendonça* 68 (LISC). LM: Sábiè, near Moamba, fl. 1.xii.1942, *Mendonça* 1541 (LISC).

Eastern and southern tropical Africa and S. Africa. Savanna and riverine woodland, termite mounds and often on sandy soil.

Apparently one very variable species with all transitions occurring.

This is the " Indaba Tree " of Government House, Bulawayo.

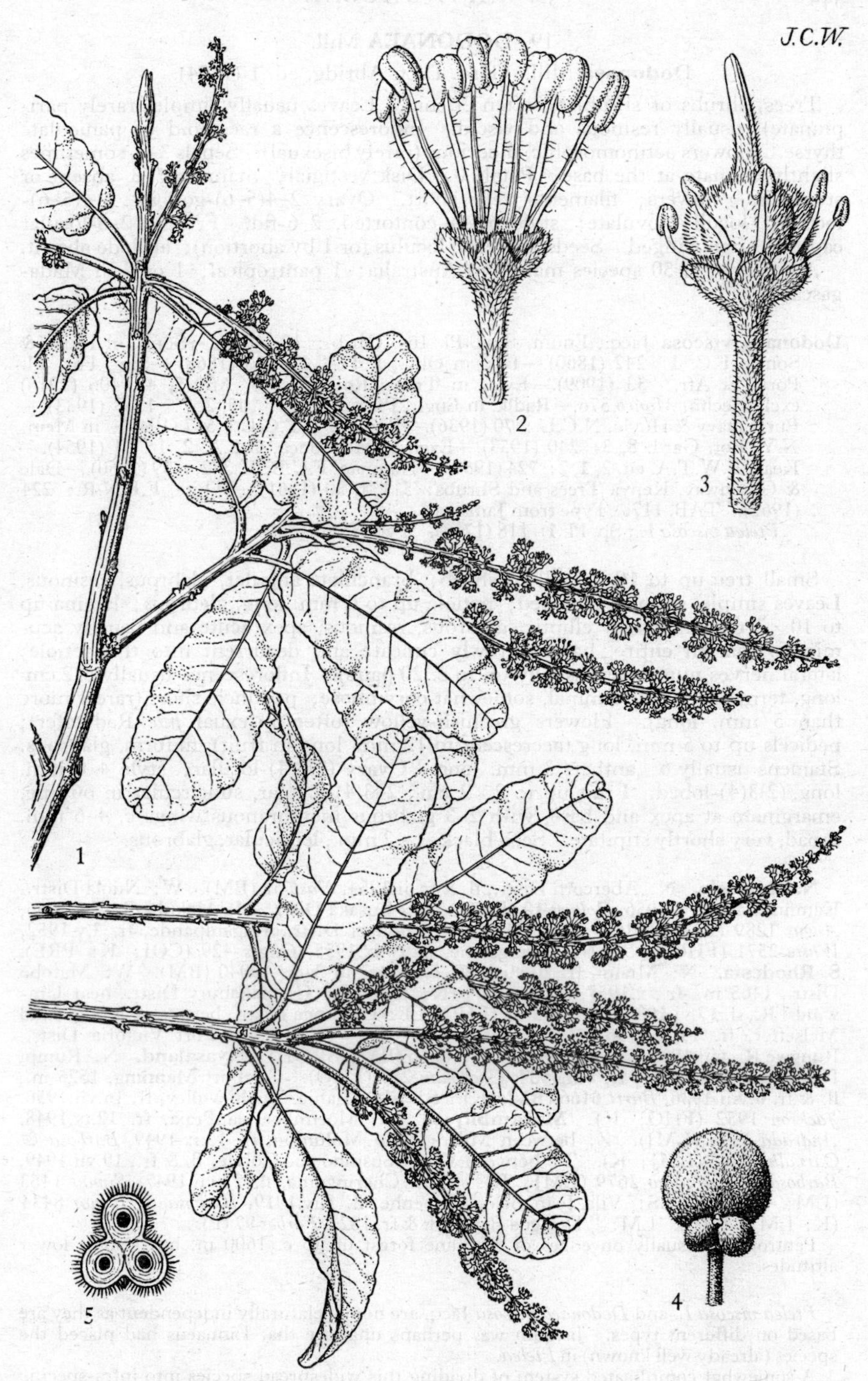

Tab. 116. PAPPEA CAPENSIS. 1, flowering branch (× ⅔) *Chase* 633; 2, male flower (× 8) *Chase* 633; 3, female flower (× 8) *Fanshawe* 5724; 4, fruit (× 2) *Brenan* 7764; 5, transverse section of fruit (× 2) *Brenan* 7764.

19. **DODONAEA** Mill.

Dodonaea Mill., Gard. Dict. Abridg. ed. **1** (1754).

Trees, shrubs or shrublets, often ericoid. Leaves usually simple (rarely paripinnate), usually resinous and viscid. Inflorescence a racemoid or paniculate thyrse. Flowers actinomorphic, dioecious (rarely bisexual). Sepals 3–7, sometimes slightly connate at the base. Petals 0. Disk vestigial. Stamens 5–8, absent or sterile in ♀ flowers; filaments very short. Ovary 2–4(5–6)-gonous, 2–4(5–6)-locular; loculi 2-ovulate; style often contorted, 2–6-fid. Fruit a 2–6-locular capsule, often winged. Seeds 2 in each loculus (or 1 by abortion); arillode absent.

A genus of c. 50 species mainly in Australia; 1 pantropical; 1 or 2 in Madagascar.

Dodonaea viscosa Jacq., Enum. Syst. Pl. Ins. Carib.: 19 (1760).—Sond. in Harv. & Sond., F.C. **1**: 242 (1860).—Bak. in Oliv., F.T.A. **1**: 433 (1868).—Sim, For. Fl. Port. E. Afr.: 33 (1909).—Eyles in Trans. Roy. Soc. S. Afr. **5**, 4: 406 (1916) excl. specim. *Monro* 576.—Radlk. in Engl., Pflanzenr. IV, 165, **2**, 7: 1363 (1933).—Burtt Davy & Hoyle, N.C.L.: 70 (1936).—Brenan, T.T.C.L.: 557 (1949); in Mem. N.Y. Bot. Gard. **8**, 3: 240 (1953).—Exell & Mendonça, C.F.A. **2**, 1: 91 (1954).—Keay, F.W.T.A. ed. 2, **1**, 2: 724 (1958).—Hauman, F.C.B. **9**: 382, t. 39 (1960).—Dale & Greenway, Kenya Trees and Shrubs: 511, t. 30 (1961).—White, F.F.N.R.: 224 (1962). TAB. **117**. Type from Jamaica.

Ptelea viscosa L., Sp. Pl. **1**: 118 (1753). Type from India.

Small tree up to 10 m. tall or shrub; branchlets angular, glabrous, resinous. Leaves simple, spirally arranged; petiole up to 6 mm. long, glabrous; lamina up to 10 × 3 cm., narrowly elliptic, glabrous, resinous, apex acute and usually acuminate, margin entire, base narrowly cuneate and decurrent into the petiole; lateral nerves numerous (usually up to c. 20 pairs). Inflorescence usually c. 2 cm. long, terminal or subterminal, somewhat corymbose; peduncle short (rarely more than 5 mm. long). Flowers greenish-yellow (often bisexual *fide* Radlkofer); pedicels up to 5 mm. long (accrescent to 12 mm. long in fruit), filiform, glabrous. Stamens usually 6; anthers 3 mm. long. Ovary (2)3(4)-locular; style 4–6 mm. long, (2)3(4)-lobed. Fruit up to 2 × 2 cm., 2–3(4)-locular, subcircular in outline, emarginate at apex and base, with 2–3 glabrous membranous wings c. 4–6 mm. broad, very shortly stipitate. Seed black, 3 × 2 mm., lenticular, glabrous.

N. Rhodesia. N: Abercorn Distr., fl. & fr. ii.1954, *Nash* 52 (BM). W: Ndola Distr., Kamfinsa, fl. 11.vi.1956, *Holmes* 19 (FHO). C: Lusaka Distr., Mt. Makulu, fl. 15.v.1956, *Angus* 1289 (BM; FHO; K; PRE). E: Lundazi Distr., Kagampande, fr. 3.v.1952, *White* 2571 (FHO; K). S: Livingstone, fl. 10.ix.1955, *Gilges* 429 (COI; K; PRE). **S. Rhodesia.** N: Miami, fr. 20.vii.1930, *Goldberg* in *Moss* 19040 (BM). W: Matobo Distr., 1465 m., fr., x.1957, *Miller* 4569 (K; SRGH). C: Salisbury Distr., near Umwindsi R., fl. 17.vi.1933, *Pardy* 95/33 (FHO). E: Chipinga Distr., between Chipinga and Melsetter, fr. 19.viii.1961, *Lord Methuen* 107 (K; PRE). S: Fort Victoria Distr., Rungwe R., viii.1932, *Cuthbertson* in GHS 6066 (BM; SRGH). **Nyasaland.** N: Rumpi Distr., Misuku Distr., fr. viii.1954, *Chapman* 219 (FHO). C: Fort Manning, 1525 m., fl. & fr. 6.viii.1936, *Burtt* 6166 (BM; K). S: Mt. Mlanje, Tuchi Valley, fr. 16.vii.1956, *Jackson* 1952 (FHO; K). **Mozambique.** N: Mocímboa da Praia, fr. 12.ix.1948, *Andrada* 1346 (LMJ). Z: between Milange and Molumbo, fr. 13.ix.1949, *Barbosa & Carvalho* 4063 (LMJ; K). T: between Vila Mousinho and Zóbuè, fl. & fr., 19.vii.1949, *Barbosa & Carvalho* 3679 (LMJ; K). MS: Cheringoma, fl. 22.vii.1947, *Simão* 1463 (LM; SRGH). SS: Vila João Belo, Chipenhe, fl. 1.iv.1919, *Barbosa & Lemos* 8434 (K; LMJ; PRE). LM: " Delagoa Bay ", fl. & fr. 1822, *Forbes* 97 (P).

Pantropical, usually on edges of montane forest up to c. 1600 m. but also at lower altitudes.

Ptelea viscosa L. and *Dodonaea viscosa* Jacq. are nomenclaturally independent as they are based on different types. Jacquin was perhaps unaware that Linnaeus had placed the species (already well known) in *Ptelea*.

A somewhat complicated system of dividing this widespread species into infra-specific taxa (varieties, forms and subforms) was elaborated by Radlkofer (op. cit.: 1368 et seq.) The only specimen in that work cited from our area appears under "specimina quoad formam indeterminata ". Brenan (loc. cit. 1953) says that the *Brass* specimens from Nyasaland are referable to var. *vulgaris* forma *burmanniana* (DC.) Radlk. as defined by Radlkofer.

Tab. 117. DODONAEA VISCOSA. 1, flowering branch ($\times \frac{2}{3}$) *Chase* 2182; 2, male flower ($\times 6$) *Chase* 2182; 3, female flower ($\times 6$) *Chase* 1671; 4, vertical section of female flower ($\times 6$) *Chase* 1671; 5, infructescence ($\times \frac{2}{3}$) *Swynnerton* s.n.; 6, seed ($\times 4$) *Swynnerton* s.n.

57. MELIANTHACEAE

By F. White

Trees or shrubs. Indumentum of simple hairs. Leaves alternate, imparipinnate or 3-foliolate. Stipules present, usually large, either in pairs at the base of the petiole or fused and intrapetiolar (*Bersama*). Inflorescence of conspicuous terminal or axillary racemes. Flowers bisexual in appearance but often unisexual and then apparently dioecious or polygamous, slightly to markedly zygomorphic, usually 4–5-merous; sepals and petals dissimilar. Sepals 4–5, shorter or longer than the petals, united at the base, with lobes imbricate. Petals 4–5, free, imbricate, unequal, unguiculate. Disk extrastaminal, annular-pentagonal or variously unilateral. Stamens 4–5(8), free or connate at the base; anthers 2-thecous, dehiscing longitudinally. Ovary superior, 4–5-locular, with basal or axile placentation; ovules 1–4 per loculus; style 1, stigma capitate. Fruit a papery or woody loculicidal capsule. Seeds large, with copious endosperm; aril present or absent.

A family of 2 genera confined to Africa. A third genus, *Greyia* Hook. & Harv., was formely included in the Melianthaceae but is now generally given family rank. *G. sutherlandii* Hook. & Harv. is grown for ornament in S. Rhodesia.

BERSAMA Fresen.

Bersama Fresen., Mus. Senckenb. **2**: 280, t. 17 (1837).

Trees or shrubs. Leaves imparipinnate or 3-foliolate; rhachis terete or narrowly to broadly winged; leaflets entire to coarsely dentate; stipules intrapetiolar, papery, persistent or not. Inflorescence a dense raceme. Flowers unisexual (or possibly polygamous) but with well-developed vestiges of the opposite sex, slightly zygomorphic. Sepals 4–5, much shorter than the petals, persistent, the 2 anterior ones ± connate or completely fused. Petals 5, spathulate, densely hairy. Disk annular-pentagonal or unilateral forming 2* or 4 sides of a pentagon. Stamens 4–5(8); filaments densely hairy, flattened and connate at the base; anthers dorsifixed; antherodes smaller than the anthers and without pollen. Ovary 4–5-locular, narrowly ovoid-conic, densely hairy, gradually narrowed into the style; ovules 1 per loculus, basal; stigma capitate; pistillode similar to the fertile gynoecium but much smaller and shorter, with very small vestigial ovules and lacking a stigma. Fruit a woody capsule opening by 4–5 thick valves, smooth to densely echinate. Seeds with a conspicuous waxy aril.

About 10 species, confined to Africa. The S. African species (as well as *B. lucens* and *B. swynnertonii* from the SE. of our area) are distinct and easy to distinguish: the tropical species are excessively variable and difficult to classify. 54 species have been described from tropical Africa. Most of these have been united with the type-species, *B. abyssinica*, by Verdcourt (in Kew Bull. **5**, 2: 233–244 (1950)). When this is done, *B. abyssinica* becomes a remarkably variable species, but one which cannot conveniently be split into segregate species. Some of its variation is sufficiently correlated with geographical distribution to permit the recognition of subspecies, 3 of which occur in our area. Much of the remaining variation is too sporadic or too localized, in my opinion, to justify the recognition of named varieties as is done by Verdcourt (in Kew Bull. **10**, 4: 600 (1956); and in F.T.E.A. Melianth. (1958)).

All the flowers examined from our area belong to one of two types: (*a*) with stamens much longer than the gynoecium, the anthers dehiscing and producing abundant pollen, the ovary at first sight appearing to be functional and with what appear to be ovules, but scarcely half the size of that of " b " and lacking a capitate stigma; (*b*) with stamens much shorter than the gynoecium, the " anthers " not dehiscing and producing no pollen, the gynoecium much longer than that of " a " and the stigma capitate. The flowers of " b " are functionally female. Whether

* Expressed by the sign ⌃ in future references.

those of " a " are functionally male or bisexual can only be decided by observation in the field.

The plant formerly called *Bersama mossambicensis* Sim has been removed by Verdcourt to a new genus, *Pseudobersama*, in the Meliaceae (see Fl. Zamb. **2,** 1: 304 (1963))

Capsule smooth; aril covering about ½ of seed; flowers large, petals (10)11–22 mm. long; leaf-rhachis often winged; leaflets often hairy beneath - - - 1. *abyssinica*
Capsule rugose; aril covering ⅓ of seed or less; flowers small, petals up to 9 mm. long; leaf-rhachis not winged; leaflets glabrous beneath:
Leaflets acuminate, (5)7–9, margin not thickened and not undulate; aril covering basal 1/5 of seed - - - - - - - - - - - 2. *swynnertonii*
Leaflets rounded or emarginate, 3–5(9), margin thickened and undulate; aril covering about ⅓ of seed - - - - - - - - - - 3. *lucens*

1. **Bersama abyssinica** Fresen. in Mus. Senckenb. **2**: 281, t. 17 (1837).—Bak. in Oliv., F.T.A. **1**: 434 (1868).—Bak. f. in Journ. of Bot. **45**: 20 (1907).—Verdcourt in Kew Bull. **5**, 2: 236 (1950); F.T.E.A. Melianth.: 2, t. 1 (1958).—Exell & Mendonça, C.F.A., **2,** 1: 93 (1954).—Keay, F.W.T.A. ed. 2, **1,** 2: 726 (1958).—Toussaint, F.C.B. **9**: 388, t. 9 (1960).—White, F.F.N.R.: 226 (1962). Type from Ethiopia.

Usually a small or medium-sized tree 6–12 m. tall, but sometimes a shrub c. 3 m. tall or a larger tree up to 25 m. tall. Leaves usually less than 50 cm. long, imparipinnate; rhachis not winged to broadly winged; leaflets 5–10-jugate, opposite, up to 14 × 5 cm., sessile or shortly petiolulate, very variable in shape, ovate or ovate-oblong to narrowly lanceolate or lanceolate-elliptic, apex acute to acuminate, margin entire to deeply serrate, base cuneate or rounded or rarely cordate, glabrous to densely hairy beneath. Racemes up to 35 cm. long. Flowers creamy-white or yellowish or tinged with pink, subsessile or with pedicels up to 1 cm. long. Calyx c. 6 mm. long, densely puberulous to tomentellous. Petals 10–22 mm. long, tomentellous. Disk ⌒-shaped to annular-pentagonal. Capsule yellowish or reddish, c. 2·5 × 2·5 cm., not rugose, tomentose at first, sometimes glabrescent. Seeds bright red with a yellow cup-shaped aril, c. 11 × 8 mm.

Subsp. **abyssinica.**—Verdcourt in Kew Bull. **5**, 2: 237 (1950); F.T.E.A. Melianth.: 4, t. 1 fig. 10–12 (1958).—Brenan in Mem. N.Y. Bot. Gard. **8,** 3: 241 (1953).

Disk annular-pentagonal or forming 4 sides of a pentagon; leaf-rhachis not winged; leaflets entire, glabrous; calyx (in our area) nearly always tinged with pink; petals (10)11–12 mm. long.

Nyasaland. C: Dedza, fl. xi.1958, *Jackson* 2255 (FHO; SRGH). S: Cholo Mt., fl. ix.1946, *Brass* 17797 (K; SRGH).
Also in Ethiopia, the Congo, Uganda, Kenya and Tanganyika. In evergreen forest, especially at the edges, and in secondary forest, 1200–1600 m.

Subsp. **englerana** (Gürke) F. White, comb. et stat. nov. Type from Tanganyika.
Bersama englerana Gürke in Engl., Bot. Jahrb. **14**: 307, t. 5 (1891). Type as above.
Bersama abyssinica subsp. *paullinioides* var. *englerana* (Gürke) Verdcourt in Kew Bull. **10**, 4: 600 (1956); F.T.E.A. Melianth.: 6, t. 1 fig. 1–9 (1958).—Keay, F.W.T.A., ed. 2, **1,** 2: 726 (1958). Type as above.

Disk ⌒-shaped; leaf-rhachis narrowly to broadly winged; leaflets entire to deeply serrate, glabrous beneath or with a few white hairs on the midrib and lateral nerves; calyx densely velutinous with golden hairs or (exceptionally) tinged with pink; petals (10)11–18 mm. long.

N. Rhodesia. B: Mongu, fl. xi.1959, *Gilges* 782 (SRGH). N: Abercorn, Lake Chila, fl. xi.1952, *Angus* 753 (FHO; K). W: Mwinilunga, 16 km. W. of Kakoma, fl. ix.1952, *Angus* 560 (FHO; K). C: Broken Hill, fl. xi.1947, *Brenan & Greenway* in *Brenan* 7851 (BM; FHO; K). **S. Rhodesia.** E: Umtali, fl. xi.1933, *McGregor* 2/33 (FHO). **Nyasaland.** C: Dedza, fl. xii.1957, *Adlard* 259 (K; SRGH). S: Ncheu Escarpment, fl. xi.1960, *Chapman* 1030 (COI; FHO; K; SRGH). **Mozambique.** T: between Furancungo and Angónia, fl. x.1943, *Torre* 6031 (LISC). MS: Gorongosa, near Vila Paiva, fl. x.1961, *Gomes e Sousa* 4732 (COI; K; LMJ).

Also in Nigeria, Ethiopia, Sudan Republic, Uganda, Kenya, Tanganyika and Angola. In riverine forest and (in N. Rhodesia) in evergreen thicket and forest patches on well-drained soils (" mateshi "); sometimes in fire-protected *Brachystegia* woodland; in our area usually at lower altitudes than the other two subspecies.

Intermediates between subsp. *abyssinica* and subsp. *englerana* occur scattered throughout the northern half of N. Rhodesia (e.g. *Denning* 23 from Mwinilunga, *White* 3195 from Ndola, *Angus* 875 from Mpika and *Richards* 11495 from Abercorn). They all have the typical disk of subsp. *englerana* but a wingless leaf-rhachis. Within our area, the calyx of subsp. *abyssinica* is almost without exception suffused with pink, whereas that of subsp. *englerana* is golden-velutinous. As the calyx of these intermediates is of the latter type, they appear to be more closely related to subsp. *englerana*. Sterile and fruiting specimens cannot be fully identified. Thus, *Simão* 1545 from Manica e Sofala, the only specimen seen from Mozambique with a wingless leaf-rhachis, could equally well belong to subsp. *abyssinica* or subsp. *englerana*.

The roots are said to be poisonous and are often taken by people wishing to commit suicide. They were found to be toxic by the Govt. Analyst, N. Rhodesia (SRGH 31259).

Subsp. **nyassae** (Bak. f.) F. White, comb. et stat. nov. Type: without precise locality, *Buchanan* 280 (BM, holotype).

Bersama nyassae Bak. f. in Journ. of Bot. **45**: 19 (1907); in Journ. Linn. Soc., Bot. **40**: 48 (1911).—Steedman, Trees etc. S. Rhod.: 44 (1933). Type as above.

Bersama abyssinica subsp. *paullinioides* var. *nyassae* (Bak. f.) Verdcourt in Kew Bull. **10**, 4: 600 (1956); F.T.E.A. Melianth.: 6 (1958). Type as above.

Bersama maschonensis Gürke in Engl., Bot. Jahrb. **40**: 88 (1907).—Eyles in Trans. Roy. Soc. S. Afr. **5**: 406 (1916). Type: S. Rhodesia, Umtali, *Engler* 3141 (B, holotype †).

Bersama schreberifolia v. Brehm. in Engl., Bot. Jahrb. **54**: 411 (1917). Type: Mozambique, Mossurize Distr., Jihu, R. Zona, *Swynnerton* 40 (B. syntype †; BM; K; SRGH); Mossurize Distr., R. Xinica, *Swynnerton* 1369 (B, syntype †; BM; K).

Bersama myriantha Gilg & v. Brehm., tom. cit.: 412 (1917). Type from Tanganyika.

Bersama zombensis Dunkley in Kew Bull. **1937**: 470 (1937). Syntypes: Nyasaland, Zomba, 1934, *Clements* 503 (FHO; K, holotype); 1933, *Clements* 334 (FHO, paratype); 1934, *Clements* 406 (K, paratype); 1935, *Clements* 577 (FHO; K, paratype).

Bersama abyssinica subsp. *paullinioides* sensu Brenan in Mem. N.Y. Bot. Gard. **8**, 3: 241 (1953).—Pardy in Rhod. Agr. Journ. **51**: 490 (1954).

Bersama abyssinica subsp. *paullinioides* var. *ugandensis* sensu Verdcourt, F.T.E.A. Melianth.: 6 (1958) quoad syn. *B. myriantha* tantum.

Disk ⌃-shaped; leaf-rhachis broadly winged; leaflets usually drying blackish on the upper surface, entire to deeply serrate, characteristically densely pubescent to tomentose beneath with long spreading golden hairs; nerves and veins often impressed on the upper surface; calyx densely velutinous with golden hairs; petals 12–21 mm. long.

N. Rhodesia. N: Lundazi, Nyika Plateau, headwaters of Shire R., fr. v.1952, *White* 2787 (FHO; K). **S. Rhodesia.** E: Umtali, fl. xi.1952, *Chase* 4683 (BM; COI; K; LISC; SRGH). **Nyasaland.** N: Vipya Plateau, Mzumara Forest Reserve, fr. iii.1953, *Chapman* 76 (FHO). C: Dedza, Lengwe, fr. vii.1961, *Chapman* 1414 (SRGH). S: Kirk Range, Chisongole, fl. xii.1960, *Chapman* 1064 (SRGH). **Mozambique.** N: Macondes, between Mueda and Chomba, fr. iv.1960, *Gomes e Sousa* 4565 (COI; FHO; K; LMJ; SRGH). Z: Gúruè, R. Malema, fr. viii.1949, *Andrada* 1841 (COI; LISC). T: Macanga, Vila Luso, fr. xii.1942, *Andrada* 67 (LISC); MS: Mossurize, fl. xi.1943, *Torre* 6146 (LISC).

Also in Tanganyika and the Congo. In evergreen (including riverine) forest, especially at the edges, in secondary forest and sometimes on termite mounds; (650)1000–2250 m.

Verdcourt (in F.T.E.A. Melianth.: 6 (1958)), who treats it as a variety of subsp. *paullinioides*, says that this variant is not well-defined. In fact it is the most distinct of all the many forms of *B. abyssinica*. In the eastern part of our area it is the most widespread and prevalent of the three subspecies recognised here. Its most distinctive feature is the golden tomentum on the lower surface of the leaf. In a few specimens (especially old fruiting ones) the hairs are sparse but in these cases the other differential characters permit easy identification.

2. **Bersama swynnertonii** Bak. f. in Journ. of Bot. **45**: 14 (1907); in Journ. Linn. Soc., Bot. **40**: 48 (1911).—Eyles in Trans. Roy. Soc. S. Afr. **5**: 406 (1916).—Verdcourt in Kew Bull. **5**, 2: 244 (1950). FRONTISP. Type: S. Rhodesia, Chipete forest patch, *Swynnerton* 9 (BM, holotype; K; SRGH).

Evergreen tree up to 22 m. tall; bark on bole dark brown, rough, on branches smooth. Leaves up to 20 cm. long, imparipinnate; rhachis not winged; leaflets

(2)3–4-jugate, opposite, distinctly petiolulate, up to 8 × 3·4 cm., ovate or lanceolate to obovate or oblanceolate, apex acuminate, margin entire, base cuneate, glabrous beneath, with venation prominent on both surfaces. Racemes up to 18 cm. long, slender. Flowers purplish-pink, with pedicels c. 6 mm. long. Calyx c. 5 mm. long, tomentellous. Petals c. 9 mm. long, tomentellous. Disk ⌒-shaped. Capsule c. 1·5 × 1·75 cm., tomentose at first, glabrescent in patches, rugose. Seeds bright red with a yellow cushion-shaped aril, c. 8 × 5 mm.

S. Rhodesia. E: Melsetter, Chisengu Forest Reserve, fl. xii.1955, *Pardy* 4/55 (FHO; K; SRGH).

Not known elsewhere. In patches of evergreen forest, especially at edges and in kloof forest and riverine forest; 1200–1800 m.

3. **Bersama lucens** (Hochst.) Szyszyl., Polypet. Disc. Rehm. **2**: 50 (1888).—Wood & Evans, Natal Pl. **1**: 71, t. 88 (1899).—Sim, For. & For. Fl. Col. Cap. Good Hope: 155 (1907).—Bak. f. in Journ. of Bot. **45**: 13 (1907).—Phillips in Bothalia, **1**: 37 (1921).—Verdcourt in Kew Bull. **5**, 2: 244 (1950). Type from Natal.
Natalia lucens Hochst. in Flora, **24**: 663 (1841).—Sond. in Harv. & Sond., F.C. **1**: 369 (1860). Type as above.

Evergreen shrub up to 3 m. tall. Leaves c. 9 cm. long, imparipinnate or 3-foliolate; rhachis not winged; leaflets 1–2(4)-jugate, opposite, subsessile, obovate or obovate-elliptic or rarely elliptic, apex rounded or subacute or very rarely acute, margin thickened and undulate, base cuneate, glabrous beneath, with venation prominent on both surfaces. Racemes up to 15 cm. long. Flowers greenish-cream, with pedicels c. 6 mm. long. Calyx c. 4 mm. long, tomentellous. Disk ⌒-shaped. Capsule c. 2 × 2 cm., tomentellous, rugose. Seeds bright red with a yellow cup-shaped aril, c. 9 × 6 mm.

Mozambique. LM: Namaacha, fr. v.1957, *Carvalho* 232 (LM).

Also in the Transvaal, Natal, Swaziland and Cape Prov. Rock-crevices.

Imperfectly Known Species

Bersama sp.

Tree up to 25 m. tall. Leaves (from a sapling) up to 30 cm. long, imparipinnate; rhachis with broad serrate wings; leaflets c. 4-jugate, margin very deeply serrate, glabrous beneath except for a few hyaline hairs on the nerves. Capsule (picked from the ground) similar to that of *B. abyssinica*.

S. Rhodesia. E: Chirinda Forest, fr. x.1947, *Wild* 2220 (COI; K; SRGH).

This plant, of which there are about 6 gatherings, all from the Eastern Division of S. Rhodesia, has been identified as *B. acutidens* Welw. ex. Hiern, an Angolan species, by Verdcourt (in Kew Bull. **5**, 2: 243 (1950)), but there is no convincing evidence that it belongs there. Similar juvenile foliage with deeply serrate leaflets occurs sporadically throughout the range of *B. abyssinica* subsp. *englerana*, which is otherwise known from S. Rhodesia only by a single gathering. Leaves from mature trees and flowers are needed to establish the identity more closely.

58. PTAEROXYLACEAE

By F. White and B. T. Styles

Aromatic trees or shrubs with indistinct oil-cavities in the younger parts. Leaves opposite or alternate, pari- or imparipinnate, without stipules. Flowers dioecious or polygamous, actinomorphic, 4–5-merous; sepals and petals dissimilar. Sepals small, 4–5, variously connate or almost free, the lobes imbricate or with open aestivation, not concealing the corolla in the bud. Petals 4–5, free, valvate or imbricate. Stamens 4–5, free, alternating with the petals; filaments glabrous. Disk well-developed, broadly annular, intrastaminal. Ovary superior, 2–5-locular, with axile placentation; ovules 1–2 per loculus, descending, with adaxial rhaphe. Fruit a capsule, the carpels separating from a persistent central column and dehiscing along the adaxial suture and also near the apex of the abaxial suture. Seeds with a long terminal wing.

A small family with two genera. *Ptaeroxylon* is confined to Africa and *Cedrelopsis* Baill. to Madagascar. Sonder (in Harv. & Sond., F.C. **1**: 242 (1860)) gave family rank to *Ptaeroxylon*, placing it immediately after the Sapindaceae but nevertheless expressing doubt as to its true affinity. J. D. Hooker (in Benth. & Hook., Gen. Pl. **1**: 411 (1862)) and many subsequent authors placed it in Sapindaceae. Radlkofer (in Sitz.-Ber. Bayer. Akad. **20**: 160 (1890)), the monographer of the latter family, transferred *Ptaeroxylon* from Sapindaceae to Meliaceae, where Harms (in Engl. & Prantl, Nat. Pflanzenfam., ed. 1, **3** (4): 270 (1896); ed. 2, **19b1**: 48 (1940)), with some hesitation, allowed it to remain. Mauritzon (in Bot. Notis. **1936**: 198 (1936)) has shown that the ovule-structure of *Ptaeroxylon* is very similar to that of *Sapindus*, and offers no grounds for the removal of *Ptaeroxylon* from the Sapindaceae. More recently, Leroy (in Compt. Rend. Acad. Sci. Fr. **248**: 106 (1959) and in Journ. Agric. Trop. Bot. Appl. **7**: 456 (1960)) has discussed its relationships and concluded that, together with *Cedrelopsis*, it should constitute the family Ptaeroxylaceae, related to but distinct from the Sapindaceae. As *Ptaeroxylon* shares some important characters with Rutaceae, Meliaceae and Sapindaceae, but differs in equally important characters from each, Leroy's treatment seems to us to be fully justified, though the exact position of *Ptaeroxylon* must remain, to a certain extent, a matter of opinion. Jenkin (unpublished thesis, Oxford University, 1961) found that *Ptaeroxylon* has smaller and more numerous vessels in the wood than all known Meliaceae and differs in other important respects, but agrees closely with certain Rutaceae and Sapindaceae. The pollen grains of *Ptaeroxylon* and *Cedrelopsis* (Pennington, unpublished) differ from those of all known Meliaceae in having a conspicuous prominent reticulum. Among related families this occurs in *Pometia* (Sapindaceae) and *Ailanthus* (Simaroubaceae), but in other features their grains are different.

PTAEROXYLON Eckl. & Zeyh.

Ptaeroxylon Eckl. & Zeyh., Enum. Pl. Afr. Austr. Extratrop. **1**: 54 (1834–35?).

Leaves opposite, paripinnate. Flowers dioecious, in contracted thyrses. Sepals 4, free almost to the base, with open aestivation. Petals 4, imbricate. Stamens 4. Ovary laterally compressed, 2-locular; loculi 1-ovulate; style slightly more than half as long as the ovary; stigmas 2, large, capitate, spreading. Capsule splitting into 2 persistent bilobed valves; central column breaking up into a number of fibrous strands.

1 species confined to the southern half of Africa.

Ptaeroxylon obliquum (Thunb.) Radlk. in Sitz.-Ber. Bayer. Akad. **20**: 165 (1890).—Gürke in Engl., Pflanzenw. Ost-Afr. **C**: 232 (1895).—Siebenlist, Forstwirtsch. in Deutsch-Ostafr: 98 (1914).—Engl., Pflanzenw. Afr. **3,** 1: 800, t. 376 (1915).—Chalk & al. in Chalk & Burtt Davy, For. Trees Brit. Emp. **3**: 56, t. 8, fig. 11 (1935).—Brenan, T.T.C.L.: 318 (1949).—Gomes e Sousa, Dendrol. Moçamb. **2**: 114 (1949); Dendrol. Mozamb. **1**: 197 cum photogr. et tab. (1951).—Exell & Mendonça, C.F.A. **1,** 2: 306 (1951).—O.B. Mill. in Journ. S. Afr. Bot. **18**: 39 (1952). TAB. **118**. Type from S. Africa.

Rhus obliqua Thunb., Fl. Cap. **2**: 224 (1818).—DC., Prod. **2**: 68 (1825). Type as above.

Ptaeroxylon utile Eckl. & Zeyh., Enum. Pl. Afr. Austr. Extratrop. **1**: 54 (1834–35?).—Harv., Thes. Cap. **1**: 11, t. 17 (?1859).—Sond. in Harv. & Sond., F.C. **1**: 243 (1860).—Sim, For. & For. Fl. Col. Cape Good Hope: 166, t. 31 (1907). Type from S. Africa (Cape Prov.).

Ptaeroxylon utile forma *robustum* Szyszyl., Polypet. Disc. Rehm.: 48 (1888) ("Pteroxylon"). Type from the Transvaal.

Shrub or small to medium-sized tree up to 15 m. tall, usually deciduous; bole up to 0·3 m. in diam. at breast height, rarely more; bark whitish-grey and smooth at first, later darker with longitudinal fissures. Leaves densely puberulous when young, the hairs sometimes persisting on the petiole and rhachis and less densely so on the lamina; rhachis (+petiole) up to 12 cm. long, flattened and slightly winged, usually ending in a short appendage; leaflets 3–7-jugate, subsessile, opposite, rarely subopposite, exceptionally alternate, leaflet-lamina up to 5 × 2·4 cm., usually

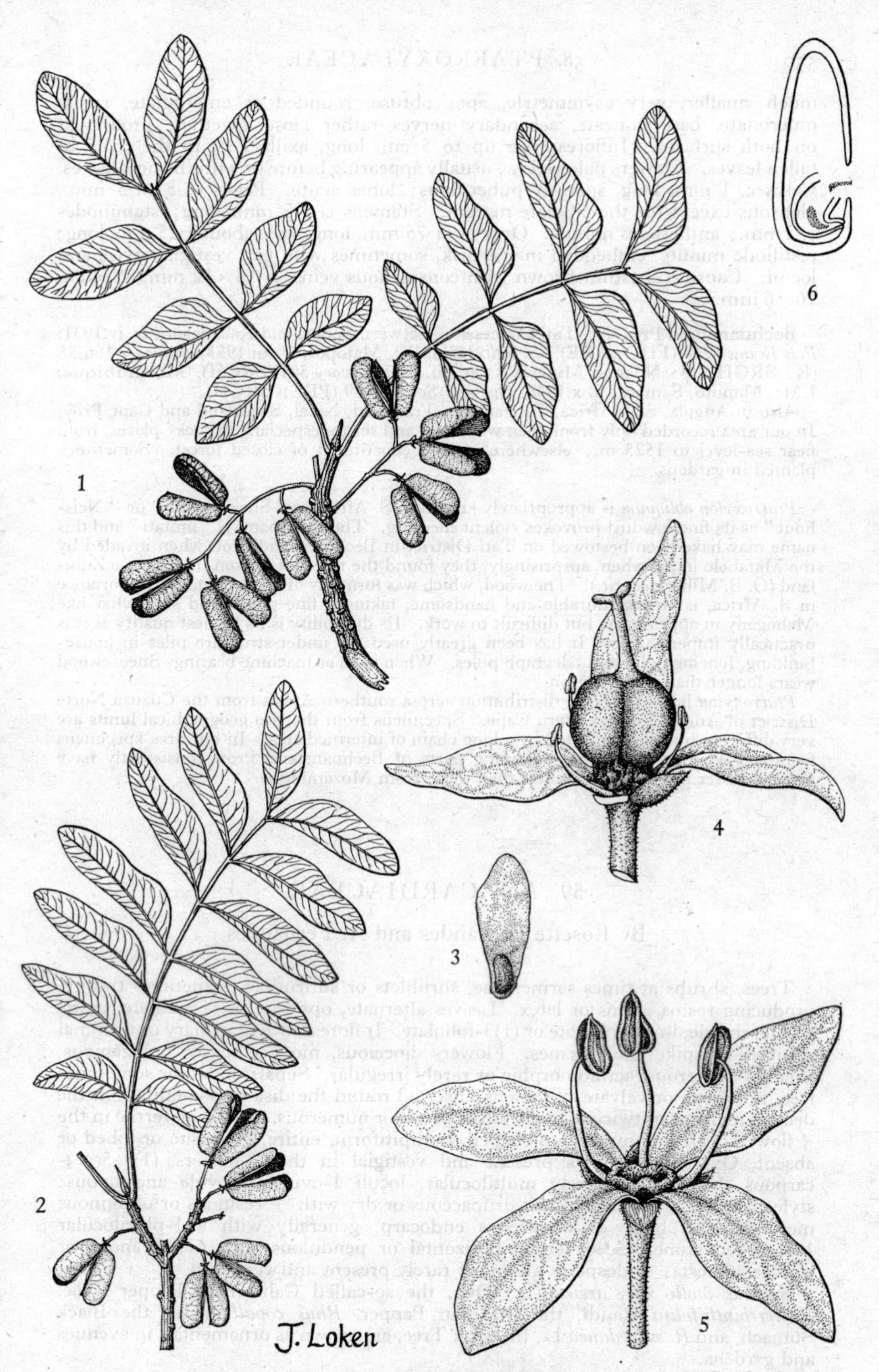

Tab. 118. PTAEROXYLON OBLIQUUM. 1, branchlet with leaves and fruit (× ⅔), *Armitage* 186/55; 2, branchlet with leaf and fruits (× ⅔), *Barbosa* 721 (from Maputo, Mozambique); 3, seed (× 1⅓) *S. Afr. For. Dept.* 5533; 4, female flower with one sepal and one petal removed (× 7), *Galpin* 8097; 5, male flower with one stamen removed (× 7), *Gomes e Sousa* 3869; 6, ovule in vertical section (× 10) after Mauritzon.

much smaller, very asymmetric, apex obtuse, rounded or emarginate, rarely mucronate, base cuneate, secondary nerves rather close together, prominent on both surfaces. Inflorescence up to 5 cm. long, axillary or in the axils of fallen leaves. Flowers pale yellow, usually appearing before or with the new leaves. Calyx c. 1 mm. long, sparsely puberulous; lobes acute. Petals c. 5 × 1·5 mm., glabrous except for the ciliolate margin. Stamens c. 3·5 mm. long; staminodes c. 2 mm.; antherodes minute. Ovary c. 1·75 mm. long, style about 1·25 mm. long; pistillode minute, embedded in the disk, sometimes with two vestigial styles and loculi. Capsule chestnut-brown with conspicuous veins, c. 18 × 12 mm. Seed c. 16 × 6 mm.

Bechuanaland Prot. N: Tati Concession, between Sebina and Kalamakati, fr. iv.1931, *Pole Evans* 3264 (FHO; PRE). **S. Rhodesia.** W: Matopos, fr. xii.1955, *Armitage* 186/55 (K; SRGH). S: Ndanga, Mutewa R., st. vii.1959, *Savory* 501 (SRGH). **Mozambique.** LM: Maputo, Santaca, fl. x.1948, *Gomes e Sousa* 3869 (PRE; SRGH).

Also in Angola, SW. Africa, Tanganyika, Transvaal, Natal, Swaziland and Cape Prov. In our area recorded only from open woodland and scrub, especially in rocky places, from near sea-level to 1525 m.; elsewhere often a constituent of closed forest. Sometimes planted in gardens.

Ptaeroxylon obliquum is appropriately known in S. Africa as " Sneezewood " or " Neishout " as its fine sawdust provokes violent sneezing. The Zulu name is " umtati " and this name may have been bestowed on Tati District in Bechuanaland Prot. when invaded by the Matabele impis when, surprisingly, they found the tree so far from their native Zululand (O. B. Miller, *loc. cit.*). The wood, which was formerly of great economic importance in S. Africa, is strong, durable and handsome, taking a fine polish and somewhat like Mahogany in appearance, but difficult to work. Its durability is its highest quality as it is practically imperishable. It has been greatly used for under-structure piles in house-building, fencing poles and telegraph poles. When used as machine bearings Sneezewood wears longer than brass or iron.

Ptaeroxylon has a scattered distribution across southern Africa from the Cuanza Norte District of Angola to the eastern Cape. Specimens from the two geographical limits are very different but are connected by a long chain of intermediates. In our area specimens from the Matopos Hills and adjoining parts of Bechuanaland Prot. consistently have fewer broader more obtuse leaflets than those from Mozambique.

59. ANACARDIACEAE

By Rosette Fernandes and A. Fernandes

Trees, shrubs at times sarmentose, shrublets or suffrutices, sometimes thorny, producing resins, gums or latex. Leaves alternate, opposite or verticillate, exstipulate, simple, imparipinnate or (1)3-foliolate. Inflorescence of axillary or terminal panicles or spike-like racemes. Flowers dioecious, monoecious or polygamous, small, 3–5-merous, actinomorphic or rarely irregular. Sepals united or sometimes free, imbricate or valvate. Stamens, inserted round the disk or sometimes on the disk, as many as or twice as many as the petals or numerous, usually all fertile in the ♂ flowers. Disk annular, cup-shaped or stipitiform, entire, crenulate or lobed or absent. Ovary, sometimes present and vestigial in the ♂ flowers, (1)3–5(∞)-carpous, 1–6-locular, rarely multilocular, loculi 1-ovulate; ovule anatropous; styles free or ± connate. Fruit drupaceous or dry with ± resinous or oleaginous mesocarp and bony or coriaceous endocarp, generally with a 1-plurilocular 1–3-seeded stone. Seeds erect, horizontal or pendulous with membranous or coriaceous testa; endosperm absent or rarely present and very thin.

Schinus molle var. *areira* (L.) DC., the so-called Californian Pepper Tree, *S. terebinthifolius* Raddi, the Brazilian Pepper, *Rhus copallina* L., the Black Sumach, and *R. succedanea* L., the Wax Tree, are grown as ornamentals in avenues and gardens.

Leaves simple:
Style and stigma 1; ♂ flowers with 1(2) fertile stamens and staminodes ± developed; fruits large; leaf-lamina glabrous; flowers polygamous:

Pedicel and receptacle swollen and fleshy in fruit, finally becoming larger than the fruit; fruit not fleshy, reniform, compressed; stamens 7–10 (usually only 1 fertile) with the filaments connate in a ring at the base - - **1. Anacardium**

Pedicel and receptacle not fleshy; drupe ovoid or subglobose, pulpy; stamens 5(10–12), 1(2–5) fertile, the others ± reduced, with the filaments not connate in a ring - - - - - - - - - - - **2. Mangifera**

Styles 3, sometimes slightly connate at the base; ♂ flowers with 5 stamens all fertile; drupes black and shining, generally somewhat compressed, up to 13 mm. broad; leaf-lamina ± hairy; flowers dioecious - - - - - - **9. Ozoroa**

Leaves compound (rarely 1-foliolate):

Hairs on the inflorescences (and usually on the young leaves) stellate; flowers 4-merous; stamens 8; styles 3–4, persistent in fruit - - - - **6. Lannea**

Hairs (if present) simple:

Leaves 3-foliolate (very rarely 5-foliolate); drupes up to 8 mm. in diam. (in African species), globose or compressed; flowers (4)5(6)-merous with imbricate petals; ovary usually 1-locular and 1-seeded; styles 3; flowers dioecious **10. Rhus**

Leaves pinnate or, if 1–3-foliolate, drupes ellipsoid or ovoid:

Style 1, with a lobed or capitate stigma; flowers 5-merous with valvate petals (in our area); ovary 1-locular and 1-seeded - - - - **7. Sorindeia**

Styles 2–4 or stigmas sessile:

Fruit 1-locular and 1-seeded; petals 4(5), valvate; stamens 4(5); ovary hirsute or pubescent; styles 3–4 or stigmas sessile; drupes globose, ovoid or turbinate, generally 1 cm. or more long - - - - - **8. Trichoscypha**

Fruit 2–4-locular; petals imbricate:

Flowers 4–5-merous with the sepals connate below; stamens 7–10; fruits 4–5-locular with 2 loculi fertile - - - - - - **5. Harpephyllum**

Flowers 3–4(5)-merous; sepals free:

Fruit 1(2)-seeded; stamens 6–8(10) with rounded anthers; panicles ample, lax - - - - - - - - - - - **4. Pseudospondias**

Fruit 2–3-seeded; stamens (8–10)15–25(30) with oblong anthers; inflorescences spike-like - - - - - - - **3. Sclerocarya**

1. ANACARDIUM L.

Anacardium L., Sp. Pl. **1**: 383 (1753); Gen. Pl. ed. 5: 180 (1754).

Shrubs or trees. Leaves alternate, sessile or petiolate, simple, entire, coriaceous, generally oblong-obovate. Flowers polygamo-dioecious, in corymbose terminal or subterminal panicles. Sepals 5, imbricate, slightly connate at the base. Petals 5, imbricate, reflexed, caducous. Stamens 7–10, unequal, 1(2) fertile, the others sterile; filaments connate at the base; anthers dorsifixed. Disk absent. Ovary free, sessile, obovoid or obcordate, 1-locular, with the ovule ascendent; style 1, with a capitate stigma. Fruit ± obliquely reniform, compressed, on the swollen pedicel; mesocarp oleaginous. Seeds reniform with thick cotyledons.

Anacardium occidentale L., Sp. Pl. **1**: 383 (1753).—Engl. in A. & C. DC., Mon. Phan. **4**: 219 (1883); Pflanzenw. Afr. **3**, 2: 175, fig. 86 (1921).—Oliv., F.T.A. **1**: 443 (1868).—Schinz, Pl. Menyharth.: 60 (1905).—Brenan, T.T.C.L.: 32 (1949).—Exell & Mendonça, C.F.A. **2**, 1: 96 (1954).—Williamson, Useful Pl. Nyasal.: 17 (1955).—Mogg in Macnae & Kalk, Nat. Hist. Inhaca I. Moçamb.: 148 (1958).—Van der Veken, F.C.B. **9**: 7 (1960).—White, F.F.N.R.: 208 (1962). Type a plant cultivated in Europe.

A shrub or tree up to 10 m. tall, with rather thick glabrous striate branchlets. Leaves glabrous; petiole 1–3 cm. long and c. 2 mm. broad at the base, hemicylindric; lamina slightly discolorous, shining, 6·5–18 × 3·8–10 cm., broadly obovate to oblong-obovate, obtuse, rounded or emarginate at the apex, cuneate or rounded at the base, coriaceous to subcoriaceous; midrib not raised above, very prominent below, lateral nerves 9–17 pairs, slightly raised, reticulation slightly visible. Inflorescence 10–25 cm. long, with reddish to blackish ascendent puberulous branches; pedicels c. 0·5 mm. long, puberulous; bracts 5–10 × 2–5 mm., ovate-lanceolate, densely greyish-puberulous outside, glabrous inside. Calyx segments 4–5 × 1–2·2 mm., lanceolate to oblong-ovate, greyish-puberulous outside. Petals yellowish-white to pale reddish, 7–13 × 1–1·7 mm., linear-lanceolate, puberulous on both faces. Stamens unequal, 1–2 with filaments 6–9 mm. long, the others with filaments 2–3 mm. long and sterile anthers. Ovary of the ⚥ flowers obliquely obovoid, puberulous; style c. 4 mm. long. Fruit greenish-yellow,

2–3 × 1·5–2·5 × 1·5 cm., laterally umbilicate, with a very thickened and fleshy pedicel, 6–7·5 × 4·5 cm.

N. Rhodesia. E: Chadiza, fl. 1.ii.1958, *Robson* 799 (K). **S. Rhodesia.** E: Umtali, fr. 4.iv.1957, *Chase* in GHS 75835 (SRGH). **Mozambique.** N: Monapo, fl. 22.x.1952, *Barbosa & Veloso* 5174A (LISC). Z: Pebane, fl. 9.ix.1950, *Munch* 252 (SRGH). T: Massanga, fl. & fr. 25.ix.1948, *Wild* 2628 (K; SRGH). MS: Mossurize, Maringa, fl. 30.vi.1950, *Chase* 2486 (BM; LISC; SRGH). SS: Bilene-Macia, fl. 8.vii.1948, *Torre* 8047 (K; LISC). LM: Marracuene, fr. 29. i. 1952, *Barbosa & Veloso* 4617 (LISC); Inhaca I., fl. & fr. 16.xii.1956, *Mogg* 26803 (J).

Native of tropical America. Cultivated and sometimes naturalized in warm countries. Fruit-pedicels and seeds (Cashew Nut; caju, castanha de caju in Portuguese) edible.

2. MANGIFERA L.

Mangifera L., Sp. Pl. **1**: 200 (1753); Gen. Pl. ed. 5: 93 (1754).

Trees. Leaves alternate, petiolate, entire, with the lateral nerves arched and anastomosing near the margin; petiole hemicylindric, thickened at the base. Flowers polygamous, in terminal or subterminal panicles. Sepals 5 (rarely 4), free or nearly so, imbricate. Petals 4–5, imbricate, inserted between the disk-lobes, with 1–5 somewhat prominent nerves, the median one prolonged like a crest at the base of the internal face. Stamens 5–10, 1(2–5) fertile, the others sterile; filaments filiform; anthers ovoid. Disk 4–5-lobulate or vestigial. Ovary subglobose, 1-locular, 1-ovulate; style subterminal or lateral, incurved, with a single stigma. Drupe subreniform or ovoid with a fleshy mesocarp and a fibrous-woody endocarp. Seed oblong-ovoid, compressed, with a chartaceous testa; embryo with cotyledons plano-convex, sometimes lobed and the radicle ascendent.

Mangifera indica L., Sp. Pl. **1**: 200 (1753).—Oliv., F.T.A. **1**: 442 (1868).—Engl. in A. & C. DC., Mon. Phan. **4**: 198 (1883); Pflanzenw. Afr. **3,** 2: 174, fig. 85 (1921).—Schinz, Pl. Menyharth.: 60 (1905).—Brenan, T.T.C.L.: 35 (1949).—Exell & Mendonça, C.F.A. **2,** 1: 96 (1954).—Williamson, Useful Pl. Nyasal.: 78 (1955).—Mogg in Macnae & Kalk, Nat. Hist. Inhaca I. Moçamb.: 148 (1958).—Van der Veken, F.C.B. **9**: 8 (1960).—White, F.F.N.R.: 211 (1962). Type from India.

A tree 10–30 m. high; branchlets fulvous-puberulous, glabrescent when older. Leaves glabrous; petiole 2–4·5 cm. long, striate; lamina concolorous, green, (9)10–33 × (2)2·5–8 cm., oblong or oblong-lanceolate, obtuse to acute or acuminate at the apex, cuneate at the base, coriaceous; midrib prominent on both sides, mainly below, lateral nerves and reticulation raised. Panicle up to 30 cm. long, pyramidal, with the axis tomentose or puberulous. ♂ and ⚥ flowers in the same panicle; pedicels 2–4 mm. long, puberulous. Sepals 5, green with whitish margin, 2–2·5 × 1–1·5 mm., sericeous-pilose outside. Petals 5, whitish with the 3–7 nerves reddish, 3–5 × 1–1·5 cm. Stamens 1(2) fertile with filament 4–5 mm. long; staminodes 3(4), very short. Disk c. 2 mm. high, thick. Ovary 1·5 mm. long, depressed-globose; style c. 2 mm. long, lateral, opposite to the fertile stamen. Drupe very variable in shape and size, 8–25 × 7–10 cm.

N. Rhodesia. N: 27 km. NW. of Abercorn, fl. 19.vii.1930, *Hutchinson & Gillett* 3945 (BM; K). C: Lusaka Forest Nursery, fl. 8.viii.1952, *White* 3038 (K). **S. Rhodesia.** E: Chipinga, Rupisi Hot Springs, fl. immat. 26.vii.1960, *Weir* in CAH 7308 (K). **Mozambique.** N: mouth of R. Lúrio, fl. 21.viii.1948, *Barbosa* 1843 (LISC; LMJ). SS: Inhambane, fl. ix.1937, *Gomes e Sousa* 2022 (COI; K). LM: Manhiça, fl. 4.ix.1945, *Pedro* 30 (SRGH); Inhaca I., fl. 11.vii.1957, *Mogg* 27249 (J).

Native of E. tropical Asia. Extensively cultivated in all warm regions. The fruits (Mango), very much appreciated, are eaten fresh or as a conserve. There are many cultivated varieties.

3. SCLEROCARYA Hochst.

Sclerocarya Hochst. in Flora, **27,** Bes. Beil.: 1 (1844).

Small or medium-sized trees. Leaves alternate, imparipinnate, crowded at the apex of the branchlets, petiolate; leaflets petiolulate. Flowers unisexual (sometimes bisexual?), shortly pedicellate, in axillary or terminal spike-like inflorescences (the ♀ ones sometimes reduced), simple or with a few basal branches. Sepals and petals 4–5, imbricate. ♂ flowers: stamens (8–10)15–25(30) with subulate filaments; anthers oblong, dorsifixed; disk patelliform; pistillode absent. ♀ flowers slightly larger than the ♂ ones, with sterile stamens ± developed; ovary subglobose, immersed in the disk, 2–3(4)-locular; styles 2–3, short, lateral; stigmas capitate. Drupe obovoid-subglobose with fleshy mesocarp; endocarp thick, bony, 3-operculate at the apex, opercules ovate. Seeds 2–3, obclavate, compressed, with papyraceous brownish testa; cotyledons plano-convex.

African and Madagascan genus with 2 species.

Sclerocarya caffra Sond. in Linnaea, **23**: 26 (1850); in Harv. & Sond., F.C. **1**: 525 (1860).—Oliv., F.T.A. **1**: 449 (1868).—Engl. in A. & C. DC., Mon. Phan. **4**: 257 (1883); Pflanzenw. Afr. **3,** 2: 180, fig. 88 (1921).—Burkill in Johnston, Brit. Centr. Afr.: 241 (1897).—Schinz, Pl. Menyharth.: 60 (1905).—Gibbs in Journ. Linn. Soc., Bot. **37.** 436 (1906).—Monro in Proc. & Trans. Rhod. Sci. Ass. **8**: 72 (1908).—Eyles in Trans. Roy. Soc. S. Afr. **5**: 401 (1916).—Burtt Davy, F.P.F.T. **2**: 491, fig. 78 (1932).—Steedman, Trees etc. S. Rhod.: 40, cum phot. et t. 38 (1933).—Burtt Davy & Hoyle, N.C.L.: 29 (1936).—Pole Evans in Mem. Bot. Surv. S. Afr. **21**: 25 (1948).—O. B. Mill., B.C.L.: 35 (1948); in Journ. S. Afr. Bot. **18**: 48 (1952).—Gomes e Sousa, Dendrol. Moçamb.: 217, cum tab. (1948).—Brenan, T.T.C.L.: 38 (1949).—Pardy in Rhod. Agric. Journ. **50,** 6: 463 (1953).—Williamson, Useful Pl. Nyasal.: 106 (1955).—Wild, Guide Fl. Vict. Falls: 152 (1953).—Palgrave, Trees of Central Afr.: 9, fig. p. 10, 11 (1956).—Mogg in Macnae & Kalk, Nat. Hist. Inhaca I. Moçamb.: 148 (1958).—White, F.F.N.R.: 214 (1962). TAB. **119.** Syntypes from Natal and the Transvaal.

Sclerocarya birrea sensu Exell & Mendonça, C.F.A. **2,** 1: 130 (1954).—sensu Van der Veken, F.C.B. **9**: 67 (1960).—sensu Palmer & Pitman, Trees S. Afr.: 292 cum fig. (1961).

Sclerocarya caffra var. *dentata* Engl. in A. & C. DC., Mon. Phan. **4**: 258 (1883). Type from the Transvaal.

Sclerocarya caffra var. *oblongifoliolata* Engl., Pflanzenw. Ost-Afr. **C**: 243 (1895); Pflanzenw. Afr. **3,** 2: 180 (1921).—Brenan, T.T.C.L.: 38 (1949). Type from Zanzibar.

A thick-boled laxly branched tree, up to 18 m. tall; bark greyish, widely reticulate; branches spreading. Leaves 7–13(17)-foliolate; petiole and rhachis 15–30 cm. long, hemicylindric, canaliculate above, glabrous; leaflets discolorous, 3–9·5(11) × 1·5–3·5(6) cm., ovate, oblong-elliptic or elliptic, acuminate with the acumen up to 10 × 4 mm. or sometimes acuminate-caudate, margin entire or sometimes dentate-serrate on suckers, asymmetric and slightly cuneate or rounded at the base, petiolulate, the lateral petiolules 0·5–3 cm. long, the terminal one up to 5 cm. long, membranous to subcoriaceous, glabrous; midrib prominent on both sides, lateral nerves distinct above, impressed or slightly raised below, reticulation very close and nearly inconspicuous; ♂ inflorescences axillary and terminal, 7–17 cm. long, with puberulous axis; bracts 2 × 1·5 mm., ovate, obtuse, puberulous or glabrous. ♂ flowers: sepals c. 2 mm. long and broad; petals yellow to purplish-pink (red in bud), 4–6 × 3–4 mm., oblong-ovate, obtuse; filaments c. 3 mm. long; anthers 1–1·5 mm. ♀ inflorescences subterminal, shorter than the ♂ ones and with fewer flowers; axis and pedicels thickened in the fruiting state; sepals and petals ± as in the ♂ flowers; staminodes 15–25; ovary subglobose. Drupe yellow, 3–3·5 cm. in diam., with very juicy mesocarp; stone 2–3 × 2·5 cm., thick, obovoid. Seed c. 1·5–2 cm. × 0·4–0·8 cm.

Caprivi Strip: E. of Cuando R., st. x.1945, *Curson* 1179 (PRE). **Bechuanaland Prot.** N: Nata R., near Madsiara Drift, st. 1.v.1957, *Drummond* 5290 (K; SRGH). SE: 0.8 km. E. of Pharing, near Kanye, ♂ fl. x.1947, *Miller* B/516 (K). **N. Rhodesia.** B: near Senanga, st. 2.viii.1952, *Codd* 7357 (K; PRE). N: Chiengi, ♂ fl. 12.x.1949, *Bullock* 1240 (K; SRGH). S: c. 110 km. up stream from Kariba Gorge, ♂ fl. ii.1957, *Scudder* 46 (SRGH). **S. Rhodesia.** N: Mtoko, Suskwe, ♂ fl. 28.x.1953, *Phelps* 70 (K; LISC; LMJ; SRGH). W: Matopo Hills, ♂ fl. & fr. x.1905, *Gibbs* 261 (BM). C: between Gatooma and Mondoro Reserve, st. 28.ii.1956, *Cleghorn* 163 (SRGH). E: Sabi R., fr. 9.xi.1906, *Swynnerton* 1210 (BM). S: Bikita, 24 km. E. of Moodie's Pass, ♂ fl. 12.x.1959, *Leach* 9472 (M; SRGH). **Nyasaland.** N: Loanga R., fr., *Nicholson* (K). C: Dowa Distr., Chitala, ♂ fl. 28.x.1941, *Greenway* 6375 (EA; K). S: Zomba, fr. xii.1915, *Purves* 230 (K). **Mozambique.** N: between Imala and Muite, ♂ fl. 24.x.1948, *Barbosa* 2562 (LISC; LM; LMJ). Z: Mocuba, Namagoa, ♂ fl. & fr. x.1944, xii.1945, 3.vi.1946, *Faulkner* 2 (COI; EA; K; LISJC; PRE; SRGH). T: Moatize, st. 8.v.1948, *Mendonça* 4141 (LISC). MS: Báruè, between Mandiè and Mungari, ♂ fl. 30.x.1941, *Torre* 3713 (COI; K; LISC; LMJ; SRGH); Gorongosa, fr. 12.xi.1963,

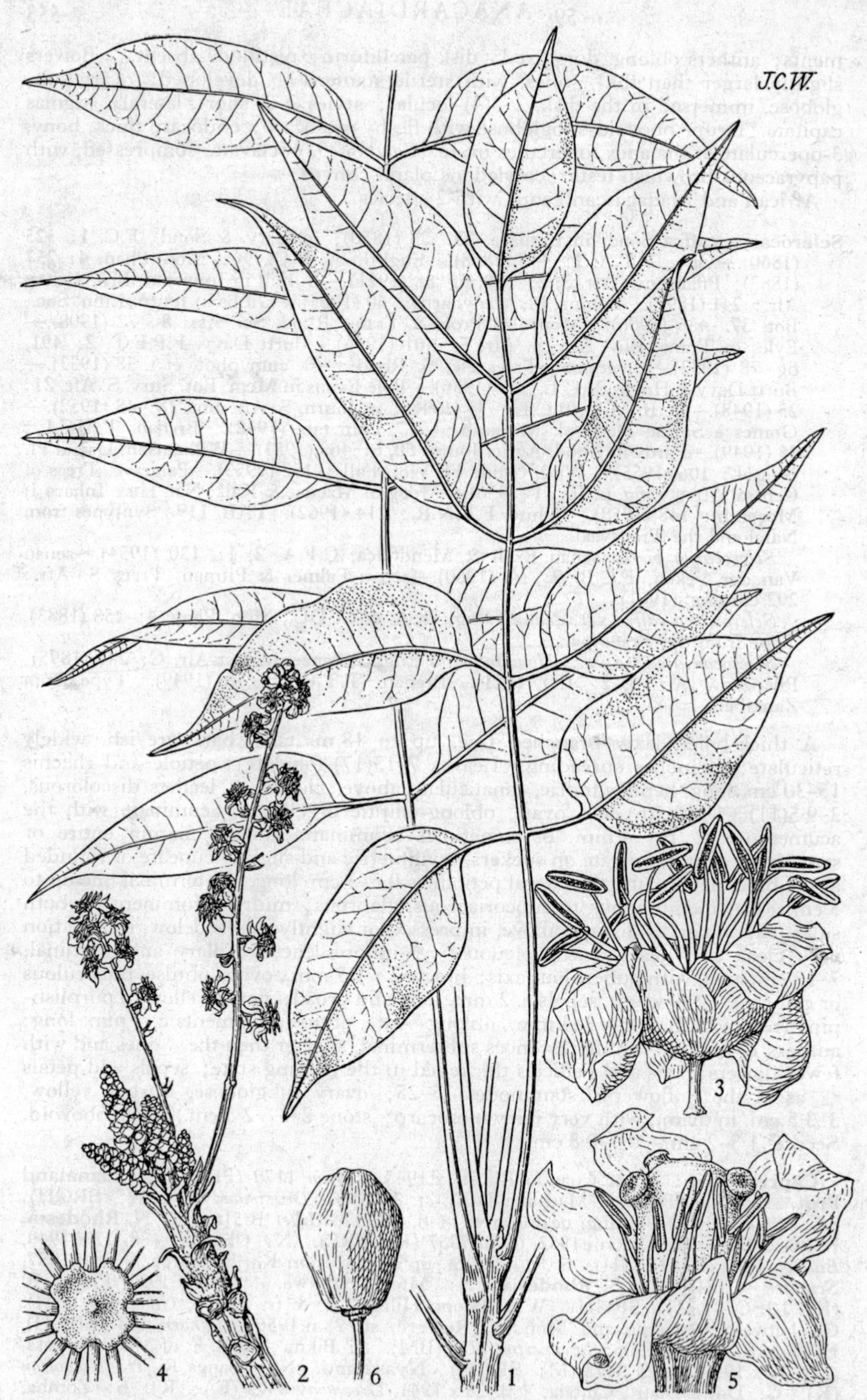

Tab. 119. SCLEROCARYA CAFFRA. 1, upper part of a leafy branch ($\times\frac{2}{3}$); 2, upper part of flowering male branch ($\times\frac{2}{3}$); 3, male flower ($\times 6$); 4, disk of male flower from above ($\times 8$); 5, female flower with a very young fruit ($\times 6$); 6, fruit ($\times\frac{2}{3}$).

Torre & Paiva 9211 (COI; K; LISC; LMJ; SRGH). SS: between Saúte and Funhalouro, st. 19.v.1941, *Torre* 2695 (COI; FHO; LISC; LMJ). LM: Marracuene, Bobole, fr. 9.iii.1946, *Gomes e Sousa* 3369 (EA; K; LISC; P); Inhaca I., ♂ fl. 29.ix.1958, *Mogg* 28383 (J).

Also in Angola, Katanga, Kenya, Tanganyika, Zanzibar, Madagascar, SW. Africa, Transvaal and Natal. Widespread in woodlands and savannas of several types on various types of soils.

4. PSEUDOSPONDIAS Engl.

Pseudospondias Engl. in A. & C. DC., Mon. Phan. **4**: 258 (1883).

Shrubs or trees. Leaves alternate, imparipinnate, the lateral leaflets asymmetric at the base, petiolulate, with the secondary nerves curved. Flowers dioecious in axillary, much branched and lax many-flowered panicles. ♂ flowers: sepals 3–4(5), imbricate; petals 3–4(5), imbricate; stamens 6–8(10), the episepalous slightly longer, inserted below the annular crenulate disk; vestigial ovary 3–4-lobulate, at the centre of the disk. ♀ flowers: ± similar to the ♂ ones, with small staminodes; ovary globose, 3–4(5)-locular; styles 3–4(5), subterminal, very short. Drupe oblong-obovoid with resinous mesocarp and woody endocarp, 3–4(5)-operculate at the apex, usually with only 1 loculus fertile. Seeds 1(2) oblong; testa thin; cotyledons plano-convex.

An African genus with 2 species. Fruit edible, very sweet.

Pseudospondias microcarpa (A. Rich.) Engl. in A. & C. DC., Mon. Phan. **4**: 259 (1883); Pflanzenw. Afr. **3,** 2: 181 (1921).—Brenan, T.T.C.L.: 36 (1949).—Exell & Mendonça, C.F.A. **2,** 1: 128 (1954).—Van der Veken, F.C.B. **9**: 44, t. 6 (1960).—White, F.F.N.R.: 211 (1962). TAB. **120.** Type from Senegal.

Spondias microcarpa A. Rich. in Guill., Perr. & Rich., Fl. Senegamb. Tent. **1**: 151 (1831), t. 40 (1832).—Oliv., F.T.A. **1**: 448 (1868). Type as above.

A tall shrub with the main stems stout and spreading, or more often a tree up to 35 m. high; bole usually short, 3–18 m. tall and up to 2 m. in diam., twisted, strongly buttressed; bark greyish-yellow, falling off in large flakes. Leaves 5–17-foliolate; petiole and rhachis 10–50 cm. long, striate, glabrous, rarely pubescent; petiole dorsally convex, flat above; rhachis subcylindric; leaflets opposite or alternate, discolorous, sometimes shining, 5–20 × 3–8 cm., oblong-ovate to elliptic, the basal ones smaller, acuminate, the acumen obtuse, the terminal one symmetrically cuneate at the base, the others very asymmetric, entire, papyraceous to coriaceous, glabrous or rarely with the midrib pubescent below; petiolules 0·3–1 cm. long, grooved, glabrous or pubescent; midrib prominent mainly beneath, lateral nerves and venation scarcely raised on the upper surface, slightly prominent below. Panicles up to as long as the leaves or longer, with the axis and branches dull-brownish-red, puberulous. ♂ flowers: sepals 4(5), 0·8–1 mm. long, ovate, obtuse, glabrous; petals whitish, 1·5–2·2 × 1–1·25 mm., elliptic; stamens 8(10), with very short filaments and 0·3–0·5 mm. long anthers. ♀ flowers: ovary globose, 4(5)-locular; styles 4. Drupe blue-black when ripe, 1·5–2·5 × 1–1·8 cm., broadly ellipsoid when fresh; stone oblong-ovoid to obovoid, 4(5)-gonal, 4(5)-locular (usually only 1 loculus fertile), with 4(5) opercules near the apex. Seed 10–15 × 2–4 mm., oblong.

N. Rhodesia. N: Puta, ♂ fl. 17.viii.1958, *Fanshawe* 4708 (FHO; K); Lunzua R. Escarpment, SW. of Lake Tanganyika, ♂ fl. 21.vii.1930, *Hutchinson & Gillett* 3981 (BM; K).

From Senegal to Angola, Sudan, Congo, Uganda and Tanganyika. In riverine forest.

5. HARPEPHYLLUM Bernh. ex Krauss

Harpephyllum Bernh. ex Krauss in Flora, **27**: 349 (1844).

Trees. Leaves alternate, imparipinnate, aggregated at the ends of the branches. Flowers dioecious in axillary panicles. ♂ flowers: calyx 4–5-partite, the segments imbricate; petals 4(5), longer than the calyx, imbricate; stamens 7–10; filaments inserted below the disk, subulate at the apex, flattened at the base; anthers oblong-ovate; vestigial ovary at the centre of the cup-shaped crenulate disk. ♀ flowers with perianth similar to that of the ♂; staminodes conspicuous; ovary rhomboid

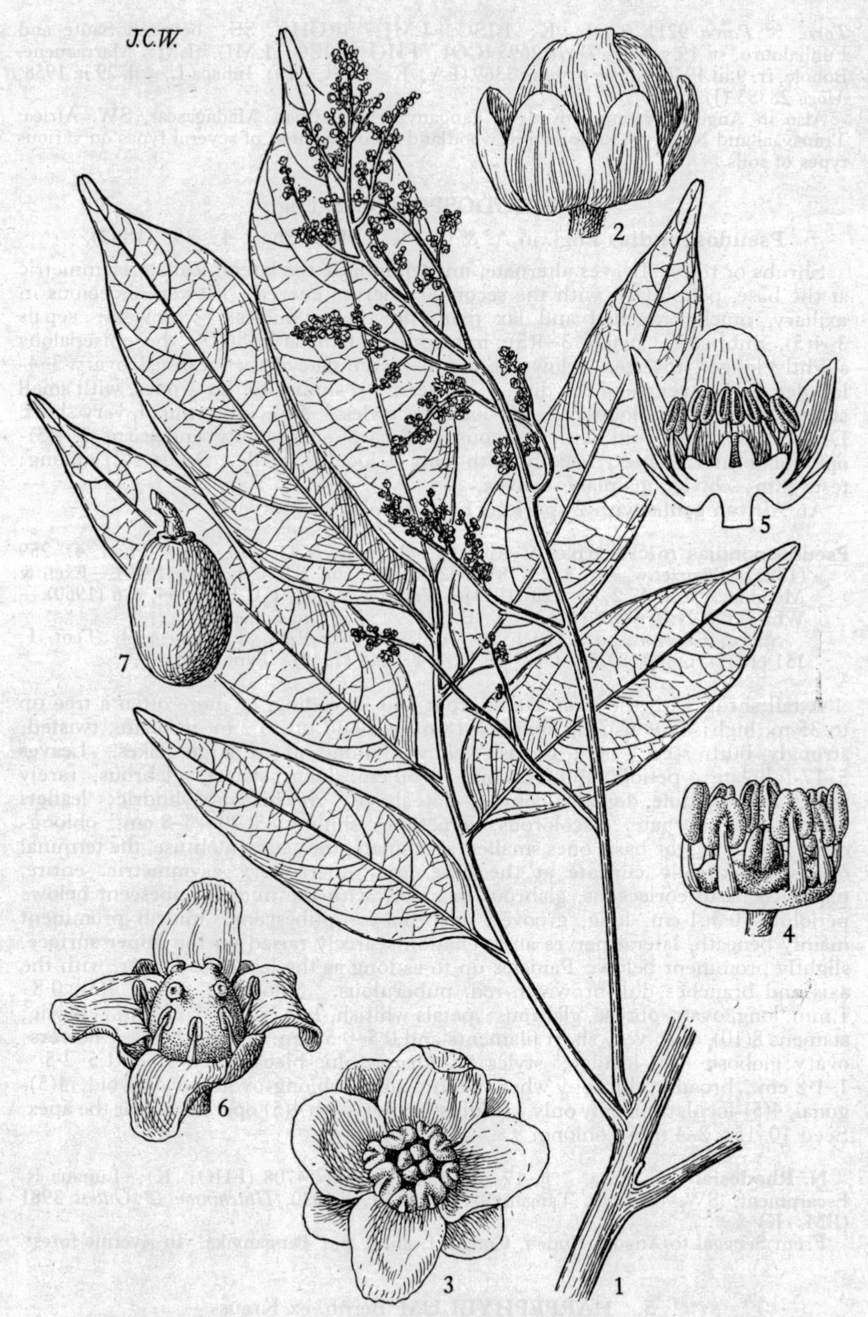

Tab. 120. PSEUDOSPONDIAS MICROCARPA. 1, male flowering branch ($\times \frac{2}{3}$); 2, male flower-bud ($\times 16$); 3, male flower from above ($\times 10$); 4, male flower, sepals and petals removed ($\times 16$); 5, vertical section of male flower ($\times 10$); 1–5 from *Hutchinson & Gillett* 3981; 6, female flower with young fruit ($\times 10$) *Harris* H190; 7, fruit ($\times 2$) *Scott-Elliot* 7504.

with 4–5 sessile stigmas. Drupe oblong-obovoid with subfleshy mesocarp and bony 4-locular endocarp; stone 4-locular with only 2 loculi fertile, each 1-seeded.
Monospecific genus from southern Africa.

Harpephyllum caffrum Bernh. ex Krauss in Flora, **27**: 349 (1844).—Sond. in Harv. & Sond., F.C. **1**: 525 (1860).—Engl. in A. & C. DC., Mon. Phan. **4**: 283 (1883); Pflanzenw. Afr. **3**, 2: 181 (1921).—Burtt Davy, F.P.F.T. **2**: 493 (1932).—Palmer & Pitman, Trees S. Afr.: 288, fig. 289 (1961). TAB. **121**. Type from S. Africa.

An evergreen tree up to 15 m. tall; stem 45–75 cm. in diam., with dark-brown thick rough bark; branchlets nodose due to the leaf-scars, glabrous. Leaves 9–17-foliolate; petiole and rhachis 10·5–30 cm. long, glabrous; petiole 4–10 cm. long, furrowed above, convex below; rhachis narrowly winged; leaflets pale green, 4–10 × 1·3–2·5 cm., opposite, the terminal lanceolate to ovate, acuminate at the apex and symmetrically cuneate at the base, the lateral ones lanceolate-falcate, with a very acute apex, cuneate and abruptly contracted at the petiolule-like base, coriaceous, glabrous; midrib slightly raised above, somewhat prominent beneath, lateral nerves 6–8, arcuate, conspicuous above, nearly invisible beneath. Panicles shorter than the leaves, with axis and branches glabrous. ♂ flowers: calyx-lobes 1 mm. long; petals whitish or yellowish, 3 × 1 mm., oblong-ovate, concave, margin revolute; stamens with filaments 2·5 mm. long and anthers c. 1 mm. long; disk glabrous. Drupe red, 2·5–3 × 1·2–1·7 cm.; stone 2·5 × 1·2 cm., oblong-ovate. Seed compressed.

S. Rhodesia. S. Fort Victoria, cultivated in the Glenlivet Garden, st. 29.vi.1952, *Munch* 379 (SRGH). **Mozambique.** LM: Namaacha Falls, ♂ fl. 22.ii.1955, *E.M. & W.* 547 (BM; LISC; SRGH); Namaacha Falls, fr. 22.viii.1944, *Torre* 6934 (COI; FHO; K; LISC; LMJ; SRGH).
Also in St. Helena (FHO), probably introduced, Cape Prov., Transvaal and Natal. In riverine forest.

6. LANNEA A. Rich.

Lannea A. Rich. in Guill., Perr. & Rich., Fl. Senegamb. Tent. **1**: 153 (1831) *nom. conserv.*

Shrublets, suffrutices, shrubs or trees, with the young parts and inflorescences stellately tomentose. Leaves alternate, imparipinnate or 3-foliolate or rarely 1-foliolate, usually clustered at the end of the branches; leaflets entire, sessile or petiolulate. Panicles terminal or axillary, ± branched and pyramidal or spike-like (the lateral branches very short), often arising before the leaves. ♂ flowers: calyx 4-partite, the segments imbricate; petals 4, imbricate; stamens 8 with the filaments subulate inserted below the disk; anthers ovate or sagittate, dorsifixed; disk ± cup-shaped, 8-crenulate with a 4-cleft vestigial ovary at the centre. ♀ flowers: perianth similar to that of the ♂ flowers; staminodes short; ovary ovoid or subglobose, 4-locular, with 2–3 abortive loculi; ovule pendent from a long funicle attached at the apex of the loculus; styles 3–4, subterminal; stigmas ± capitate. Drupe obovoid or ovoid, ± compressed or subglobose; mesocarp thin; endocarp woody, the surface unequally ridged and alveolate, with 1–2 ovate opercules at the apex. Seeds ± reniform, compressed; cotyledons plano-convex.
Genus with about 40 species in tropical Africa and tropical Asia.

Trees or shrubs:
Drupes tomentose; leaflets (3)5–21, up to 3 cm. long, elliptic to ovate, obtuse, at first with ± sparse stellate hairs mixed with short simple stiff ones, later glabrescent above and densely white-tomentose beneath - - - - - 1. *humilis*
Drupes glabrous; adult leaflets usually larger; simple hairs, if present, longer:
Leaflets with or without stellate hairs but these not in a dense tomentum; simple hairs usually present in tufts in the axils of the lateral nerves and sometimes scattered on the nerves and lamina:
Inflorescences arising at the same time as the leaves, solitary in the leaf-axils, unbranched or with some basal branches; simple hairs of the lateral nerve-axils long and usually abundant - - - - - - 4. *stuhlmannii*
Inflorescences arising before the leaves, spike-like, crowded at the top of ± short branches:
Leaflets without stellate hairs, very asymmetric at the base, decurrent into the petiolule and narrowly acuminate at the apex; inflorescence-axis and

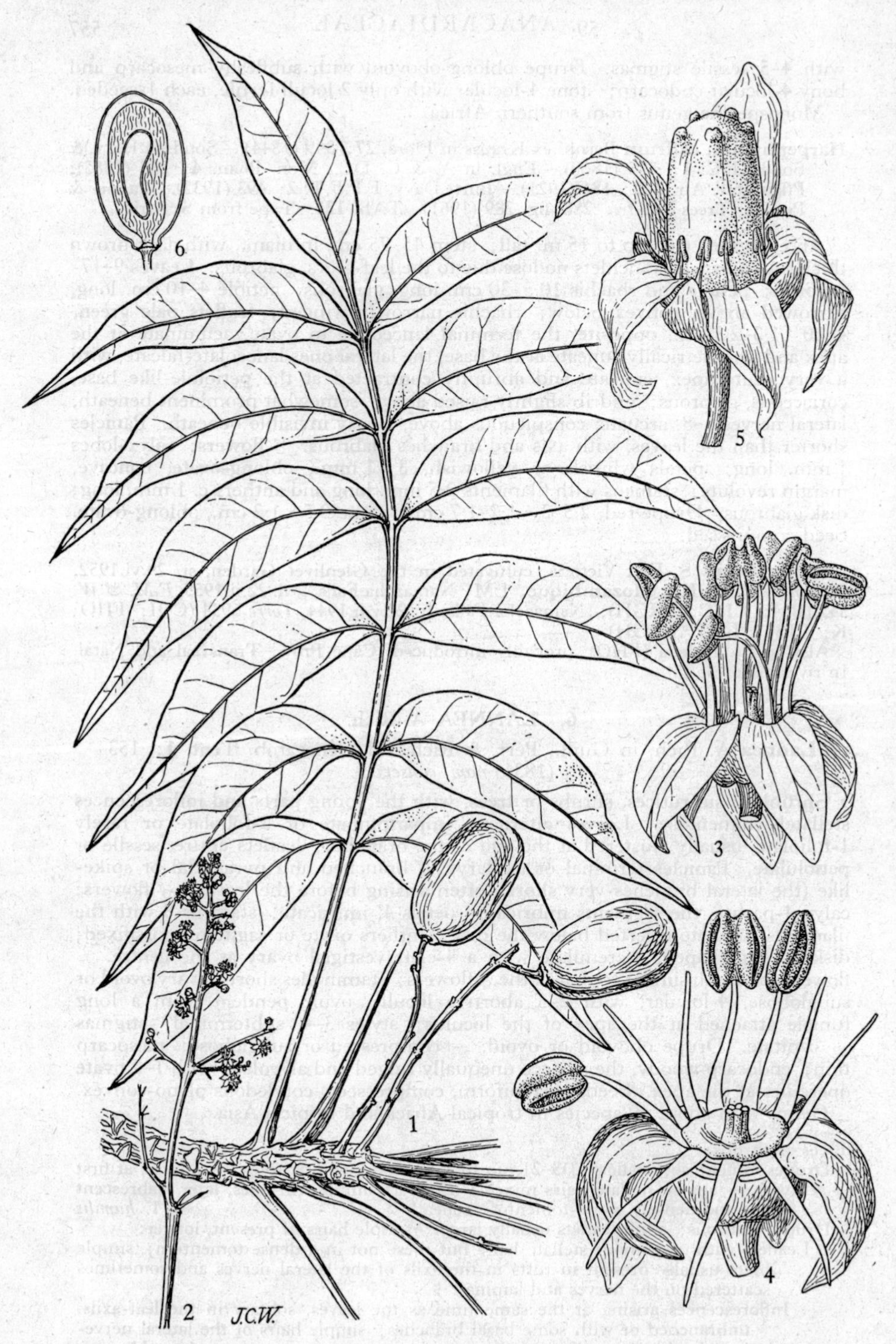

Tab. 121. HARPEPHYLLUM CAFFRUM. 1, upper part of a fruiting branch (× 2/3) *Torre* 6934; 2, male inflorescence (× 2/3) *Pedro* 5017; 3, male flower (× 10) *Pedro* 5017; 4, same with one petal and 7 anthers removed (× 12) *Pedro* 5017; 5, female flower with young fruit (× 8) *Pole Evans* H18263; 6, vertical section of drupe (× 2/3) *Torre* 6934.

fructiferous pedicels thick, with sparse stellate hairs; drupe ellipsoid, 10–11 × 6–7 mm. - - - - - - - - 6. *asymmetrica*

Leaflets not so asymmetric at the base, with a broader and more obtuse acumen; inflorescence-axis and fructiferous pedicels slender, stellate-tomentose; drupe ovoid to subglobose, 7–9 × 6–7 mm. - - - - - - 5. *antiscorbutica*

Leaflets with a dense tomentum of stellate hairs mainly below, at least when young; simple hairs, if present, not in tufts at the lateral-nerve axils:

Tomentum of the under surface of the adult leaflets felted, whitish; lateral nerves and reticulation visible beneath, darker than the tomentum, glabrous or with sparse stellate hairs; leaflets up to 10·5 × 5·5 cm., usually smaller; inflorescence-axis covered by a greyish indumentum - - - - - - 2. *discolor*

Tomentum of the under surface of the adult leaflets not felted, rusty-ochraceous; lateral nerves and reticulation ± raised below, densely covered by the indumentum; leaflets generally larger than above, up to 15·5 × 7·5 cm., more attenuate at the apex; inflorescence-axis covered by a fulvous indumentum 3. *schimperi*

Shrublets up to 1·2 m. high, with simple or few-branched stems:

Leaves 1-foliolate; lamina broadly ovate or elliptic, up to 19·5 × 15 cm., densely fulvous- to whitish-tomentose below; inflorescences axillary, solitary, arising at the same time as the leaves - - - - - 7. *katangensis*

Leaves 3-∞-foliolate or, if 1-foliolate, then plants not as above:

Stems virgate, 0·6–0·9 m. long; inflorescences very dense, in fascicles along the stem, arising just after leaf-fall; leaflets membranous to papyraceous, contracted into an acumen up to 1 cm. long - - - - 8. *virgata*

Stems up to 0·3 m. long (usually shorter), tortuous; inflorescences less dense, borne at the stem-base just before the leaves; rootstock and the long trailing roots woody; leaflets obtuse or shortly acuminate:

Leaflets (3)5–11, narrowly to broadly elliptic, 2·5–7 × 1·5–4 cm.; lateral nerves and reticulation not or scarcely raised beneath; drupe 9–12 × 6–9 mm., ovoid-oblong - - - - - - - - - - 9. *gossweileri*

Leaflets (1)3–7, elliptic to subcircular, 9–20 × 4–12 cm.; lateral nerves and reticulation raised beneath; drupe 9–11 × 8–9 mm., ovoid - - 10. *edulis*

1. **Lannea humilis** (Oliv.) Engl. in Engl. & Prantl, Nat. Pflanzenfam. Nachtr. I zu II–IV: 213 (1897); Pflanzenw. Afr. **3,** 2: 184 (1921).—Brenan, T.T.C.L.: 34 (1949).—Van der Veken, F.C.B. **9**: 56 (1960).—White, F.F.N.R.: 210 (1962). Syntypes from Ethiopia and N. Central Africa.

Odina humilis Oliv., F.T.A. **1**: 447 (1868).—Engl. in A. & C. DC., Mon. Phan. **4**: 271 (1883). Syntypes as above.

Odina tomentosa Engl., Bot. Jahrb. **15**: 103 (1893); Pflanzenw. Ost-Afr. **C**: 244 (1895); Pflanzenw. Afr., loc. cit.—Brenan, T.T.C.L.: 34 (1949). Type from E. Africa.

Lannea tomentosa (Engl.) Engl. in Engl. & Prantl, Nat. Pflanzenfam. Nachtr. I zu II–IV: 213 (1897). Type as above.

A shrub or small tree up to 6 m. tall; bole thick; crown flat or spreading; branches brownish-grey, cylindric, striate, lenticellate, glabrous; branchlets densely whitish-tomentose with small rigid-armed stellate hairs. Leaves (3)5–21-foliolate, scattered along the terminal twigs or ± crowded on the short floriferous lateral branches; petiole and rhachis 3–15 cm. long, subcylindric, canaliculate on the upper side, ± densely stellate-tomentose; leaflets discolorous, dark green and with ± sparse stellate hairs mixed with minute simple ones on the upper surface, densely white-stellate-tomentose to sparsely so (in aged leaves) below, coriaceous; terminal leaflet 2–3 × 1–1·8 cm., oblong to obovate, obtuse at the apex, ± cuneate at the base with a petiolule 0·2–1 cm. long; lateral leaflets 1–3 × 0·8–1·3 cm., ovate, oblong or elliptic, obtuse at the apex, unequally rounded or subcordate at the base, subsessile to very shortly petiolulate; midrib sunk above, prominent below, lateral nerves nearly invisible above, raised below. Inflorescences spike-like, unbranched or with a few branches, 2–5 cm. long, clustered on short lateral branchlets, with the flowers glomerate; axis and branches densely whitish-tomentose; pedicels 1–4 mm. long. Calyx-segments c. 1·5 × 1·5 mm., ovate, tomentose; petals yellowish, 2·5–4·5 × 1·5–2 mm., unguiculate. Drupe greyish-tomentose, 9–13 × 5–8 mm., oblong, somewhat compressed and oblique.

N. Rhodesia. N: Bulaya, ♂ fl. 9.x.1958, *Fanshawe* 4891 (BR; FHO; K). E: Petauke, Luembe, fr. 13.xii.1958, *Robson* 944 (K; LISC). **S. Rhodesia.** N: Mtoko, Chazarini, ♂ fl. ii.1958, *Brayne* 13 (SRGH).

Also in Senegal, Nigeria, Cameroon, Chari [*Chevalier* 9117 (COI)], Sudan, Rwanda, Uganda, Ethiopia, Kenya and Tanganyika. Woodlands of several types and in swamps.

2. **Lannea discolor** (Sond.) Engl. in Engl. & Prantl, Nat. Pflanzenfam. Nachtr. I zu II–IV: 213 (1897); in Sitz. Königl. Preuss. Akad. Wiss. Berl. **1906,** 52: 894 (1906); Pflanzenw. Afr. **3,** 2: 185 (1921).—Monro in Proc. Trans. Rhod. Sci. Ass. **8**: 71 (1908).—Eyles in Trans. Roy. S. Afr. **5**: 401 (1916).—Burtt Davy, F.P.F.T. **2**: 493 (1932).—Steedman, Trees etc. S. Rhod.: 39, t. 36 (1933).—Burtt Davy & Hoyle, N.C.L.: 29 (1936).—Hoyle & Jones in Kew Bull. **2**: 86 (1947).—O. B. Mill., B.C.L.: 34 (1948); in Journ. S. Afr. Bot. **18**: 46 (1952).—Codd. Trees and Shrubs Kruger Nat. Park: 104, fig. 100a, b (1951).—Pardy in Rhod. Agric. Journ. **48**, 6: 506 (1951).—Williamson, Useful Pl. Nyasal.: 74 (1955).—Van der Veken, F.C.B. **9**: 56 (1960).—Palmer & Pitman, Trees S. Afr.: 287, fig. p. 288, phot. 116 (1961).—White, F.F.N.R.: 209 (1962). Type from the Transvaal.
Odina discolor Sond. in Linnaea, **23**: 25 (1850); in Harv. & Sond., F.C. **1**: 504 (1860).—Engl. in A. & C. DC., Mon. Phan. **4**: 272 (1883). Type as above.
Odina sp.—Steedman in Proc. & Trans. Rhod. Sci. Ass. **24**: 15, t. 14 (1925).
Odina schimperi sensu Gibbs in Journ. Linn. Soc., Bot. **37**: 436 (1906).—Monro, tom. cit.: 72 (1908).

Tree up to 15 m. tall with rounded crown and upright bole 10–30 cm. in diam.; bark on bole pale grey, shallowly and irregularly fissured, exfoliating at the base, smooth and grey-purple on the upper branches; floriferous branches ± short, very rugose, ± stellate-tomentose; leafy branchlets, young petioles, rhachis and juvenile leaflets densely grey- or pinkish-to rusty-reddish-tomentose. Leaves appearing after the flowers, 5–11-foliolate; petiole and rhachis 13–35 cm. long, semicylindric, slightly canaliculate on the upper side, glabrescent; leaflets very discolorous (dark green or ± deep reddish-brown to blackish and glabrescent above, pale fulvous-greyish-tomentose beneath), 2·5–10·5 × 1·5–5·5 cm., opposite or sub-opposite, oblong-ovate, elliptic or oblong-elliptic, rarely broadly ovate to sub-circular, acute to obtuse or nearly rounded at the apex, the terminal one symmetric, cuneate at the base and with petiolule 1–2·5 cm. long, the lateral ones ± asymmetric at the base and subsessile or with compressed petïolules up to 3 mm. long; midrib, lateral nerves and reticulation somewhat sunk above, slightly prominent below, not so tomentose as the lamina or glabrous and darker than the latter. Inflorescences generally unbranched, spike-like, 2·5–23 cm. long, rarely with a few short basal branchlets, crowded at the apices of short densely stellate-tomentose branches; flowers with pedicels 1–3·5 mm. long, in dense bundles. Calyx-segments ovate, obtuse, glabrous or with a few stellate hairs at the base. Petals creamy to bright tawny-yellow, 3–5 × 1·3–1·6 mm., oblong. Drupe reddish to purple, 9–15 × 7–10 × 4–5 mm., ovoid to subglobose, compressed.

Bechuanaland Prot. SE: Kanye, Pharing, fr. 14.xi.1948, *Hillary & Robertson* 515 (PRE). **N. Rhodesia.** B: Balovale, fr. 14.x.1952, *Angus* 633 (BM; BR; FHO). N: half way to Kalambo Falls, st. 13.v.1936, *Burtt* 6066 (BM; K). C: Lusaka Arboretum, st. 2.iii.1952, *White* 2151 (FHO; K). E: Chikowa Mission to Jumbe, ♂ fl. 13.x.1958, *Robson* 83 (K; LISC). S: Mazabuka, Mapanza, Choma, ♂ fl. 28.ix.1958, *Robinson* 2893 (K; M; PRE; SRGH). **S. Rhodesia.** N: Sebungwe, Kariangwe, st. 9.xi.1951, *Lovemore* 112 (LISC; SRGH), fr. 21.ix.1951, *Lovemore* 112A (K; SRGH). W: Matobo, st. x.1953, *Miller* 1930 (K; SRGH); Matobo, Farm Besna Kobila, ♂ fl. ix.1957, *Miller* 4539 (K; SRGH). C: Salisbury, st. 1931, *Trapnell* 688 (K). E: Umtali, Commonage, ♀ fl. & fr. 2.x.1948, *Chase* 1650 (BM; K; SRGH), ♂ fl. 29.ix.1954, *Chase* 5299 (BM; K; SRGH). S: Belingwe, ♀ fl. & fr. 23.ix.1922, *Eyles* 3671 (BOL; SRGH). **Mozambique.** T: Mt. Zóbuè, ♂ fl. & fr. 3.x.1942, *Mendonça* 588 (COI; K; LISC; LMJ; PRE; SRGH). MS: Bárue, Vila Gouveia, Mungári road, fr. 18.ix.1942, *Mendonça* 333 (BR; COI; LISC). LM: between Moamba and Boane, st. 2-xii.1940, *Torre* 2186 (K; LISC; SRGH).

Also in the Congo, Swaziland and the Transvaal. In open woodlands of several types, especially on rocky slopes, or on sandy soil (White, loc. cit.). Fruit edible with a pleasant grape-like flavour.

3. **Lannea schimperi** (Hochst. ex A. Rich.) Engl. in Engl. & Prantl, Nat. Pflanzenfam. Nachtr. I zu II–IV: 213 (1897); Pflanzenw. Afr. **3,** 2: 186 (1921). —Burtt Davy & Hoyle, N.C.L.: 29 (1936).—Hoyle & Jones in Kew Bull. **2**: 80 (1947).—Brenan, T.T.C.L.: 34 (1949).—Williamson, Useful Pl. Nyasal.: 75 (1955).—Van der Veken, F.C.B. **9**: 57, t. 7 (1960).—White, F. F.N.R.: 210 (1962). Syntypes from Ethiopia.
Odina schimperi Hochst. ex A. Rich., Tent. Fl. Abyss. **1**: 140 (1847).—Oliv., F.T.A. **1**: 445 (1868).—Engl. in A. & C. DC., Mon. Phan. **4**: 269 (1883). Syntypes as above.

Tree usually 5–10(15) m. tall, with spreading crown and short bole; bark grey to nearly black, rather rough; floriferous branches very rugose. Leaves appearing after the flowers and the fruit, 5–11(13)-foliolate; petiole and rhachis 8–33 cm. long, dorsally convex, canaliculate above, at first densely pinkish-rusty-tomentose, glabrescent with age; leaflets 6·3–15·5 × 4·2–7·5 cm., elliptic, oblong-ovate to ovate, the basal ones somewhat shorter and proportionately broader, acute or obtuse at the apex, the terminal one symmetric, acute and up to 3·5 cm. long petiolulate, the lateral ones asymmetric, rounded, truncate or subcordate at the base and sessile or 1·5–3 mm. long petiolulate, all at first densely pink-rusty-tomentose on both surfaces, then discolorous on drying (almost black and with ± sparse stellate hairs above, persistently rusty-tomentose below), finally reddish-brown or dull brown and glabrous above and laxly tomentose and rusty-ochraceous beneath; midrib a little prominent above, rather so beneath, lateral nerves and reticulation sunk in the upper face, raised on the under surface, concealed by a tomentum as dense as that of the lamina. Panicles spike-like, crowded at the top of short branches, the ♂ ones up to 22 cm. long, the ♀ ones up to 8 cm. long, with tomentose axis. Flowers in dense bundles; pedicels 1–3 mm. long, tomentose; calyx-segments c. 1·5 mm. long, ovate to subcircular, ciliolate, covered by stellate hairs or almost glabrous; petals greenish to bright yellow, 3·5–5 × 1·5–2·5 mm., oblong-ovate, fragrant. Drupe red, 7–10 × 4–6 mm., obliquely ovoid.

Mozambique. N: Malema, Mutuáli, road to Lioma, ♂ fl. 30.ix.1953, *Gomes e Sousa* 4133 (COI; LMJ). Z: Alto Molócuè, road to Mugena, ♂ fl. 24.viii.1949, *Andrada* 1896 (LISC).

Without the leaves we cannot refer the specimens cited above (and others) to the varieties but, on account of the geographical distribution, it seems probable that they belong to var. *stolzii*.

Leaflets with only stellate caducous hairs - - - - - - var. *schimperi*
Leaflets with simple persistent hairs mixed with stellate-ones on the under surface var. *stolzii*

Var. **schimperi**

N. Rhodesia. B: Mankoya, st. 22.ii.1952, *White* 2114A (FHO). N: 1·6 km. E. of Mpulungu, on shore of Lake Tanganyika, ♂ fl. 17.xi.1952, *Angus* 780 (FHO). **Mozambique.** Z: Mocuba, st. 20.x.1949, *Faulkner* 481 (K). MS: between Matarara do Lucite and Goonda, near Chiboma, st. 21.x.1953, *Pedro* 4348 (LMJ).

Widespread in tropical Africa. Savannas and woodlands, sometimes on termite mounds.

Var. **stolzii** (Engl. & v. Brehm.) R. & A. Fernandes in Bol. Soc. Brot., Sér. 2, **38**: 145 (1965). Type: Tanganyika, Kyimbila, Mbaka, *Stolz* 1603 (BR; K, lectotype; P).
Lannea stolzii Engl. & v. Brehm. in Engl., Bot. Jahrb. **54**: 325 (1917).—Engl., Pflanzenw. Afr. **3**, 2: 185 (1921). Syntypes from Tanganyika (Kyimbila Distr.).

N. Rhodesia. N: Kasama, Mbesuma Ranch, fol. & fr. 20.x.1961, *Astle* 980 (SRGH). **Nyasaland.** S: Mlanje, st. viii–ix.1937, *Topham* 1008 (FHO). **Mozambique.** N: Malema, Mutuáli, near road to Lúrio, ♂ fl. 21.ix.1953, fol. 21.ii.1954, *Gomes e Sousa* 4123 (COI; K; LMJ; P). Z: Mocuba, Namagoa, ♂ fl. 1.ix.1944, st. xi.1944, *Faulkner* 64 (BM; EA; K; PRE). MS: Chimoio, Gondola, Mupindanganga, st. 5.ii.1948, *Garcia* 75 (K; LISC; SRGH).

Also in Tanganyika. Woodlands of several types.

Lannea discolor vel **L. schimperi**

N. Rhodesia. B: Balovale, Sansongu, ♂ fl. 1.viii.1952, *Gilges* 165 (K; PRE; SRGH). C: between Luangwa Bridge and Rufunsa R., ♂ fl. 6.ix.1947, *Brenan* 7823 (K). **Nyasaland.** C: Lilongwe Research Sta., ♂ fl. 26.xi.1951, *Jackson* 578 (K). S: Mlanje, between Likabula and Tuchila R., ♂ fl. 9.viii.1957, *Chapman* 403 (BM; K). **Mozambique.** T: between Tete and Zóbuè, ♂ fl. 8.ix.1941, *Torre* 3376 (COI; LISC). MS: between Vila Machado and Vila Pery, ♂ fl. 22.ix.1943, *Torre* 5924 (COI; LISC).

L. schimperi is very similar to *L. discolor*, but distinguishable by its leaves, generally larger, with a less pale under surface and lateral nerves and reticulation concealed below by a tomentum as dense as that of the lamina. When the leaves are lacking, it is almost impossible to distinguish the one species from the other from herbarium material.

Lannea discolor × schimperi ?

From studying *L. discolor* and *L. schimperi* in the field, F. White, A. Angus and A. C. Hoyle have arrived at the conclusion that the areas of these species overlap in certain regions and that hybrids occur. According to their opinion, the following specimens from N. Rhodesia are such hybrids. The leaves, however, are like those of *L. discolor*.

N. Rhodesia. W: c. 19 km. NNE. of Solwezi, Chafugoma Hill, ♀ fl. & fr. 6.ix.1949, *Hoyle* 1193, 1193A, ♂ fl. 1204, fol. 1205, fr. 1206, fr. 1207 (all FHO).

4. **Lannea stuhlmannii** (Engl.) Engl. in Engl. & Prantl, Nat. Pflanzenfam. Nachtr. I zu II–IV: 214 (1897); Pflanzenw. Afr. **3,** 2: 184 (1921).—Eyles in Trans. Roy. Soc. S. Afr. **5**: 273 (1916).—Burtt Davy & Hoyle, N.C.L.: 29 (1936).—Brenan, T.T.C.L.: 35 (1949).—Williamson, Useful Pl. Nyasal.: 75 (1955).—Van der Veken, F.C.B. **9**: 52 (1960).—White, F. F.N.R.: 210 (1962). No specimen cited under the specific epithet.

Odina stuhlmannii Engl., Pflanzenw. Ost-Afr. **C**: 244 (1895). See above.

Odina stuhlmannii var. *oblongifoliolata* Engl., loc. cit. Syntypes from Nubia, Zanzibar, Mozambique, Kilimanjaro and Massaihochland.

Odina stuhlmannii var. *brevifoliolata* Engl., loc. cit. Syntypes from Zanzibar and Tete.

Lannea kirkii Burtt Davy in Kew Bull. **1921**: 51 (1921); F.P.F.T. 2: 493 (1932).—Codd, Trees and Shrubs Kruger Nat. Park: 105, fig. 100c (1951).—Palmer & Pitman, Trees S. Afr.: 288 cum fig. (1961). Type from the Transvaal.

A shrub or more frequently an irregularly branched tree 3–15(18) m. high; branches wide, spreading, lightly foliaged; bark light- to dark-brown or grey, reticulate, exfoliating in ± rectangular scales up to 10 cm. long; branchlets sparsely stellate-tomentose, soon glabrous. Leaves 3–9(13)-foliolate, rarely 1-foliolate, crowded at the apices of the branchlets; petiole and rhachis 5–22 cm. long, subcylindric, canaliculate on the upper side, glabrous or with a few stellate hairs; leaflets somewhat discolorous, subcoriaceous, with sparse minute glandular hairs and ± dense stellate ones when young and tufts of white simple hairs in the nerve-axils below, sometimes also with simple hairs on the nerves and lamina, the terminal one 2–9 × 1·7–6 cm., ovate to broadly ovate or subcircular, symmetric at the base with a petiolule 0·7–2·5 cm. long, the lateral ones 3·2–10 × 2·2–5·3 cm., ovate or elliptic to broadly elliptic, asymmetric at the base, subsessile or with petiolules up to 0·5 cm. long, all obtuse to ± acuminate, with blunt acumen at the apex, margin entire; midrib sunk above, prominent beneath, lateral nerves 4–8 pairs, slightly raised on both surfaces, reticulation usually impressed beneath. Inflorescences arising with the leaves, 2–20(40) cm. long, axillary, simple and spike-like or with a few short (exceptionally up to 13 cm. long) branches at the base; pedicels 0·5–4 mm. long, in scattered fascicles, with stellate caducous hairs as also the inflorescence-axis. Calyx-segments c. 1·5 mm. long, semicircular, ciliate-denticulate, glabrous or stellate-hairy. Petals greenish-yellow to yellow, 3–4·5 × 1·25–2·5 mm., oblong-ovate, concave. Drupe red or brown, 8–12 × 6–8 × 3–4 mm., oblong-ellipsoid, compressed.

Mature leaflets without simple hairs on the nerves and lamina - - var. *stuhlmannii*
Mature leaflets with ± dense simple hairs on the nerves and lamina as well as on the petiole, rhachis and inflorescences - - - - - - var. *tomentosa*

Var. **stuhlmannii**

Odina wodier var. *brevifolia* Engl. in A. & C. DC., Mon. Phan. **4**: 268 (1883). Type: Mozambique, Shupanga, *Kirk* (K, holotype).

Lannea wodier var. *brevifolia* (Engl.) Eyles in Trans. Roy. Soc. S. Afr. **5**: 401 (1916). Type as above.

N. Rhodesia. B: Senanga, Lilengo Forest Reserve, fr. 6.ii.1952, *White* 2021 (BR; FHO; K). **S. Rhodesia.** W: Mashaba ♀ fl. & fr., *Fynn* 1/37 (FHO). E: Umtali, Darlington, ♂ fl. 4.i.1949, *Chase* 1426 (BM; K; SRGH). S: Ndanga, north bank of Lundi R., ♂ fl. 8.xi.1959, *Goodier* 621 (COI; K; SRGH). **Nyasaland.** S: Cheruza, Port Herald, fr. 13.vii.1916, *Purves* 267 (K). **Mozambique.** N: between Nacala and Fernão Veloso, ♀ fl. & fr. 14.x.1948, *Barbosa* 2408 (LISC; LM; LMJ). Z: Mocuba, Namagoa, ♂ fl., ♀ & fr. xi.1944, *Faulkner* 336 (BM; K; PRE; SRGH; UPS). MS: Mossurize, between Espungabera and Chibababa, ♂ fl. 10.xi.1943, *Torre* 6142 (COI; K; LISC; LMJ; SRGH). SS: Guijá, near Caniçado, ♂ fl. 12.xii.1940, *Torre* 2358 (K; LISC). LM: Sábiè, around Moamba, fr. 16.ii.1948, *Torre* 7334 (BR; LISC; LM); Inhaca I., fl. immat. 3.xi.1962, *Mogg* 30084 (J).

Also in Kenya, Uganda, Tanganyika, Zanzibar and the Transvaal. In river valleys forests, woodlands of several types (including coastal), savannas, on termite mounds, etc.

Var. **tomentosa** Dunkley in Kew Bull. **1937**: 471 (1937).—Wild, Guide Fl. Vict. Falls: 152 (1953).—White, F.F.N.R.: 210 (1962). Type: N. Rhodesia, Bombwe, *Martin* 357/32 (K, holotype).

Bechuanaland Prot. N: Sigara Pan, c. 48 km. W. of mouth of Nata R., st. 26.iv.1957, *Drummond & Seagrief* 5242 (SRGH). SE: Sofala, ♀ fl. & fr. xii.1940, *Miller* B/257 (PRE). **N. Rhodesia.** B: Nangweshi, st. 28.vii.1952, *Codd* 7211 (BM; K; PRE). E: Petauke Distr., Luembe, ♀ fl. & fr. 13.xii.1958, *Robson* 938 (K; LISC). S: Gwembe Valley, 14·5 km. S. of Sinazezi, st. 3.iv.1952, *White* 2604 (K). **S. Rhodesia.** N: Urungwe, Kariba, near Sanyati R., fr. 20.i.1956, *Phelps* 108 (BR; K; SRGH). W: Matopos, fl. immat. x.1930, *Eyles* 6623 (K; SRGH). C: Hartley, Poole Farm, ♂ fl. 4.xii.1947, *Hornby* 2926 (K; SRGH). E: Melsetter, Hot Springs Hotel, fr. 30.xii.1948, *Chase* 1427 (BM; K; SRGH). S: Beitbridge, ♂ fl. x.1959, *Davies* 2617 (SRGH). **Nyasaland.** C: Chitala-Salima road, at bridge over Chitala R., fr. 13.ii.1959, *Robson* 1581 (K; LISC). S: Zomba, ♀ fl. & fr. 1936, *Clements* 754 (FHO; K). **Mozambique.** T: Boroma, ♂ fl. i.1891, *Menyhart* 749 (K).

Known until now only from our area. Ecology as for the type.

5. **Lannea antiscorbutica** (Hiern) Engl., Bot. Jahrb. **24**: 499 (1898).—Exell & Mendonça, C.F.A. **2**, 1: 132 (1954).—Van der Veken, F.C.B. **9**: 53, phot. 1 (1960). Syntypes from Angola.

Calesiam antiscorbutica Hiern, Cat. Afr. Pl. Welw. **1**: 178 (1896). Syntypes as above.

Lannea stuhlmannii sensu White pro parte quoad specim. *White* 2061, 3477, 3589 et *Holmes* 437.

A shrub or tree up to 15 m. tall; trunk erect, covered by greyish bark; old branches rugose and glabrous; branchlets brownish-grey to almost black, smooth or striate, glabrous or with a few stellate hairs. Leaves (3)5–9(11)-foliolate; petiole and rhachis 4–25(43) cm. long, subterete, ± canaliculate above, glabrous or very sparsely stellate-hairy; leaflets subconcolorous or, when dried, almost black above and brown beneath, 4–14(18) × 1·8–6·7(8·2) cm., lanceolate-elliptic or ovate to elliptic or oblong, membranous to papyraceous, the young ones with a narrow acumen, resinous and covered by minute pinkish or whitish glandular hairs mixed or not with stellate ones, when adult glabrous on the lamina except for tufts of simple hairs in the nerve-axils, attenuate to the apex or ± abruptly acuminate, the acumen 0·5–2 cm. long and somewhat broad and blunt; terminal leaflet symmetric, with petiolule 2–4 cm. long, the lateral ones unequally rounded or subcuneate at the base and with petiolule 0·1–0·6 cm. long; midrib and lateral nerves slender, not or slightly raised on both surfaces, reticulation visible below in the oldest leaves. Inflorescences spike-like, arising before the leaves, crowded at the apices of the branchlets; axis 2·5–10 cm. long, pinkish-salmon-stellate-tomentose; pedicels 0·2–2·5 mm. long. Calyx-segments c. 1 mm. long, ovate, entire, obtuse, with few stellate hairs or glabrous. Petals c. 3 × 1·5 mm., oblong-ovate, unguiculate. Drupe 7–9(12) × 6–7 mm., irregularly ovoid.

N. Rhodesia. B: Kalabo, fr. 16.xi.1959, *Drummond & Cookson* 6514 (COI; K; SRGH). N: Samfya, ♂ fl. 3.x.1953, *Fanshawe* 352 (BR; K; SRGH). W: Ndola, fr. 19.x.1954, *Fanshawe* 1624 (BR; EA; K; LISC; SRGH). **Mozambique.** LM: Maputo, Salamanga, fr. 28.xi.1947, *Mendonça* 3563 (COI; K; LISC; LMJ; SRGH).

Also in Angola and the Congo. Woodlands on alluvial soil, Kalahari Sand and termite mounds.

6. **Lannea asymmetrica** R. E. Fr. in Wiss. Ergebn. Schwed. Rhod.-Kongo-Exped. **1**: 126 (1914).—Engl., Pflanzenw. Afr. **3**, 2: 184 (1921).—Van der Veken, F.C.B. **9**: 52 (1960).—White, F.F.N.R.: 209 (1962). Type: N. Rhodesia, Kalambo, *Fries* 1374 (UPS).

A small tree 5 m. tall or shrub, with greyish glabrous grooved branches. Leaves 5–7-foliolate; petiole and rhachis up to at least 22 cm. long, subterete, canaliculate above, glabrous; young leaflets dark green to reddish-brown, up to 12 × 4 cm., lanceolate, attenuate or contracted at the apex into a very narrow acumen up to 1 cm. long, translucent when dried, membranous, glabrous on both surfaces (except for dense tufts of white short somewhat crisped simple hairs in the axils of the lateral nerves beneath), the terminal one symmetric, cuneate and with petiolule up to 1·5–2·5 cm. long, the lateral ones very unequal, with the proximal part decurrent into a petiolule up to 1 cm. long. Mature leaves not seen. Inflorescences spike-like, arising before the leaves, crowded at the apices of the branchlets;

flowers not seen. Fructiferous-axis up to 6 cm. long with sparse stellate hairs, thick; fructiferous pedicels c. 2 mm. long, thick. Drupe blackish, 10–11 × 6–7 mm. oblong-ellipsoid, compressed.

N. Rhodesia. N: Lunzua R., ♂ fl. 20.vii.1930, *Hutchinson & Gillett* 3967 (K); Abercorn, near Lake Tanganyika 1·6 km. E. of Mpulungu, st. 16.xi.1952, *White* 3670 (FHO; K).
Also in Katanga. Open woodlands on rocky places or escarpments.

7. **Lannea katangensis** Van der Veken in Bull. Jard. Bot. Brux. **29**: 239 (1959); F.C.B. **9**: 58 (1960). TAB. **122** fig. A. Type from Katanga.
Lannea sp. 2.—White, F.F.N.R.: 211 (1962).

Suffrutex or shrublet up to 1·20 m. tall; stems arising from a woody rootstock, cylindric, at first densely fulvous- to greyish-stellate-tomentose, finally glabrescent. Leaves 1-foliolate; petioles (0·3)1–5 cm. long, subterete, narrowly canaliculate above; lamina (3·5)6–19·5 × (2·5)4·5–15 cm., ovate to broadly elliptic, rounded at the apex, margin entire or undulate, truncate to cordate at the base, subcoriaceous, at first subconcolorous (softly and densely fulvous-stellate-tomentose on both surfaces) then discolorous (dark green with ± dense dull-yellowish stellate hairs above and fulvous- to whitish-tomentose below), finally glabrous or almost so above, whitish-yellow- to greyish-tomentose below; midrib, lateral and tertiary nerves impressed above, raised but covered by the tomentum beneath as is also the reticulation. Inflorescences arising with the leaves, 1·5–7·5 cm. long, axillary, solitary, spike-like, with tomentose axis. Flowers fasciculate, nearly sessile. Calyx-segments c. 1 × 0·75 mm., ovate, with stellate hairs on the outside. Petals greenish-cream, c. 2·5 mm. long, oblong-ovate, contracted at the base. Drupe green tinged with red (unripe), 9–11 × 7–8 mm., oblong, compressed.

N. Rhodesia. N: Mpika, fr. 10.ii.1955, *Fanshawe* 2044 (FHO; K). E: Lundazi, Tigone Dam, ♂ fl. 19.xi.1958, *Robson* 658 (K; LISC), fr. 19.xi.1958, 658A (K; LISC). **Nyasaland.** N: Mzimba Distr., st. vi–vii.1942, *Barker* 479 (EA).
Also in Katanga. Woodlands, sometimes in rocky places.

8. **Lannea virgata** R. & A. Fernandes in Bol. Soc. Brot., Sér. 2, **38**: 146, t. 2–4 (1965). Type: N. Rhodesia, Kasempa, *Fanshawe* 6685 (FHO, holotype; K).

A many-stemmed suffrutex. Stems unbranched or somewhat branched, upright, salmon-pink, 60–90 cm. tall, cylindric, striate, stellate-tomentose to glabrescent. Leaves arising before the flowers, 3–7-foliolate; petiole and rhachis 13–18 cm. long, subterete, slender, tomentose to glabrescent; lateral and terminal leaflets similar, (4·5)6–10·5 × (2·5)3·2–4·5 cm., roundish, ovate to oblong, contracted at the apex into an acute acumen up to 1 cm. long, rounded or cordate, symmetric or a little asymmetric and petiolulate at the base (lateral petiolules up to 9 mm., the terminal one up to 3 cm. long), membranous to subcoriaceous, at first ± densely rufous-tomentose with slender, long-armed stellate hairs on both surfaces, then glabrescent above and ± arachnoid-tomentose beneath, the margin somewhat incurved in the old leaves; midrib and the arched lateral nerves visible but not prominent above, slightly raised below as also the venation, covered with a tomentum somewhat denser than that of the lamina. Inflorescences 1·5–6 cm. long, spike-like, very compact, densely clustered above the leaf-scars along the leafless stems, with salmon-pink-stellate-tomentose axis. ♂ flowers in bundles; pedicels 0–1·5 mm. long, tomentose. Calyx-segments flushed with red, ovate, obtuse, tomentose. Petals yellow flushed with red, c. 3·5 mm. long, oblong, obtuse, constricted at the base. Drupe reddish, 9–10 × 6–7 × 3 mm., oblong, asymmetric, compressed.

N. Rhodesia. B: Balovale, ♂ fl. vi.1952, *Gilges* 119 (SRGH). W: Katuba, st. 12.xii.1960, *Fanshawe* 5840A (FHO; K), 6521 (FHO; K); Kasempa, ♂ fl. 20.viii.1961, *Fanshawe* 6689 (FHO; K); Kasempa, fr. 15–30.vi.1953, *Holmes* 1098 (FHO).
Known up to the present only from N. Rhodesia. Woodlands, sometimes by dambos (seasonal swamps) and on termite mounds.

9. **Lannea gossweileri** Exell & Mendonça in Bol. Soc. Brot., Sér. 2, **26**: 280, t. **5** (1952); C.F.A. **2**, 1: 134 (1954).—Van der Veken, F.C.B. **9**: 62 (1960). Type from Angola.

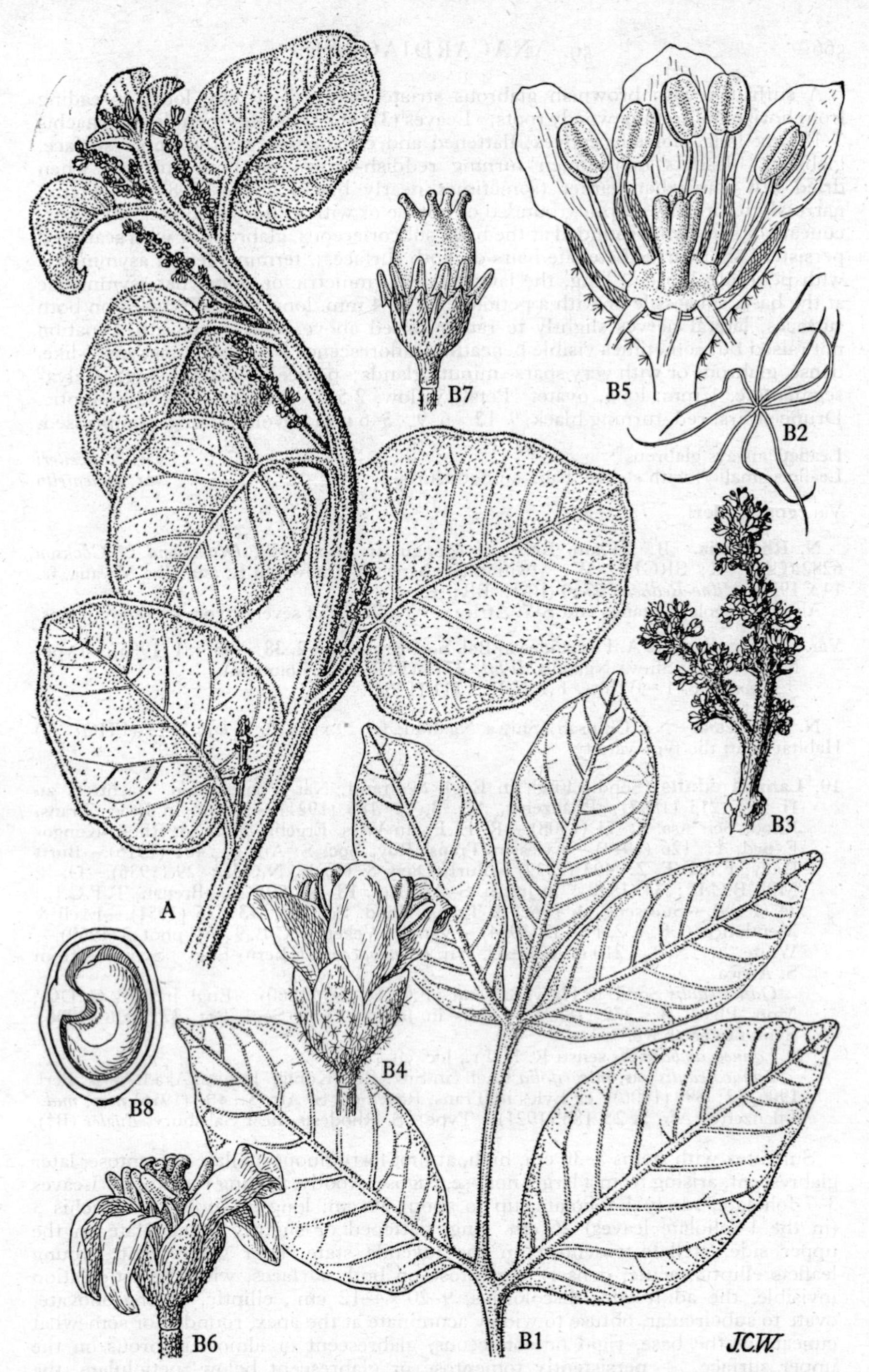

Tab. 122. A.—LANNEA KATANGENSIS. Terminal part of flowering branch (× ⅔) *Robson* 658. B.—LANNEA EDULIS var. EDULIS. B1, upper part of leaf (× ⅔) *Norman* R7; B2, stellate hair (× 300) *Richards* 11472; B3, male inflorescence (× ⅔) *Richards* 11472; B4, male flower (× 8) *Richards* 11472; B5, vertical tangential section of the male flower (× 12) *Richards* 11472; B6, female flower (× 8) *Norman* R7; B7, female flower with very young fruit after petal-fall (× 6) *Norman* R7; B8, vertical section of the drupe (× 2) *Norman* R7.

A suffrutex with brownish glabrous striate stems, 5–30 cm. long, spreading from horizontal trailing woody roots. Leaves (3)5–11-foliolate; petiole and rhachis 3–18 cm. long, dorsally convex, flattened and canaliculate on the upper surface, glabrous; leaflets light green turning reddish-brown on both surfaces when dried but somewhat darker (sometimes nearly black) above, 3–8 × 1·5–4 cm., narrowly to broadly elliptic, rounded or obtuse or with a short acumen at the apex, cuneate or somewhat rounded at the base, subcoriaceous, glabrous or with scattered persistent short-armed stellate hairs on both surfaces; terminal leaflet asymmetric with petiolule 1–5 cm. long, the lateral ones symmetric or somewhat asymmetric at the base, subsessile or with a petiolule up to 4 mm. long; midrib raised on both surfaces, lateral nerves slightly to rather raised above, less so beneath, venation not raised but sometimes visible beneath. Inflorescence 1·5–3 cm. long, spike-like, dense, glabrous or with very sparse minute glands; pedicels 1–2 mm. long. Calyx-segments c. 1 mm. long, ovate. Petals yellow, 2·5–3 × 1·5 mm., obovate-elliptic. Drupe at first red, turning black, 9–12 × 6–9 × 5–6 mm., ovoid-oblong, compressed.

Leaflets always glabrous - - - - - - - - var. *gossweileri*
Leaflets smaller, with stellate hairs on both surfaces - - - var. *tomentella*

Var. **gossweileri**

N. Rhodesia. B: 24 km. ENE. of Mongu, fr. 10.xi.1959, *Drummond & Cookson* 6282 (COI; K; SRGH). W: Mwinilunga, between Kamwezhi R. and Cha Mwana, fr. 14.x.1937, *Milne-Redhead* 2760 (BM; BR; K).
Also in Angola, Katanga and SW. Africa. Woodlands of several types on sandy plains.

Var. **tomentella** R. & A. Fernandes in Bol. Soc. Brot., Sér. 2, **38**: 145, t. 1 (1965). Type: N. Rhodesia, Shiwa Ngandu, *White* 3775 (FHO; K, holotype).
Lannea sp. 1.—White, F.F.N.R.: 211 (1962).

N. Rhodesia. N: Chinsali, Shiwa Ngandu, fr. 29.xi.1952, *White* 3775 (FHO; K) Habitat as in the type variety.

10. **Lannea edulis** (Sond.) Engl. in Engl. & Prantl, Nat. Pflanzenfam. Nachtr. I zu II–IV: 213 (1897); Pflanzenw. Afr. **3,** 2: 185 (1921).—Monro in Proc. Trans. Rhod. Sci. Ass. **8**: 71 (1908).—R. E. Fr. in Wiss. Ergebn. Schwed. Rhod.-Kongo-Exped. **1**: 126 (1914).—Eyles in Trans. Roy. Soc. S. Afr. **5**: 401 (1916).—Burtt Davy, F.P.F.T. **2**: 493 (1932).—Burtt Davy & Hoyle, N.C.L.: 29 (1936).—O. B. Mill, B.C.L.: 34 (1948); in Journ. S. Afr. Bot. **18**: 46 (1952).—Brenan, T.T.C.L.: 34 (1949).—Suesseng. in Proc. & Trans. Rhod. Sci. Ass. **43**: 35 (1951).—Exell & Mendonça, C.F.A. **2,** 1: 133 (1954).—Van der Veken, F.C.B. **9**: 61, phot. 3 (1960).—White, F.F.N.R.: 210 (1962) excl. syn. *L. ambacensis* (Hiern) Engl. Syntypes from S. Africa.
Odina edulis Sond. in Harv. & Sond., F.C. **1**: 503 (1860).—Engl. in A. & C. DC., Mon. Phan. **4**: 272 (1883).—Gibbs in Journ. Linn. Soc., Bot. **37**: 436 (1906). Syntypes as above.
Lannea ambacensis sensu R. E. Fr., loc. cit.
Lannea edulis var. *integrifolia* Engl. [in Sitz.-Ber. Königl. Preuss. Akad. Wiss. Berl. **1906,** 52: 890 (1906); ex Eyles in Trans. Roy. Soc. S. Afr. **5**: 401 (1916) *nom. nud.*] Pflanzenw. Afr. **3,** 2: 186 (1921). Type: S. Rhodesia, near Salisbury, *Engler* (B†).

Suffrutex with stems 3–30 cm. high, at first ferruginous-stellate-tomentose, later glabrescent, arising from a large, nodose, rugose, woody, trailing rootstock. Leaves 3–7-foliolate, rarely 1-foliolate, up to about 30 cm. long; petiole and rhachis 3 (in the 1-foliolate leaves)–26 cm. long, flattened or slightly canaliculate on the upper side, stellate-tomentose in the juvenile state, later glabrescent; young leaflets elliptic, acute, densely tomentose on both surfaces, with the reticulation invisible, the adult ones discolorous, 9–20 × 4–12 cm., elliptic, oblong-obovate, ovate to subcircular, obtuse to widely acuminate at the apex, rounded or somewhat cuneate at the base, rigid or coriaceous, glabrescent or almost glabrous on the upper surface, ± persistently tomentose or glabrescent below, petiolulate, the petiolules compressed, canaliculate, the terminal one 0·5–5 cm. long, the lateral ones up to 0·4 cm. long; midrib sunk or slightly raised above, strongly prominent beneath, lateral nerves (7)8–10 on each side, visible above, raised below, reticulation ± visible on the upper face, very prominent beneath, mainly in the oldest leaves. Inflorescences almost at ground level, coming up some weeks after burnings, before the leaves or with the young leaves, composed of panicles, with the 3–10 cm. long axis, the very short lateral branches and the pedicels covered by minute red glands and stellate hairs; pedicels 1–3 mm. long. Calyx segments red in dry state, ovate, obtuse, glabrescent, with or without some glands. Petals yellowish to pinkish, 2–3 × 1·3–1·5 mm., elliptic, somewhat unguiculate. Drupe bright to deep red, 9–11 × 8–9 × 6–7 mm., ovoid, compressed.

Leaflets up to 20 × 12 cm.; plant covered by dense tomentum - - var. *edulis*
Leaflets up to 10 × 5 cm.; plant glabrescent or glabrous - - - var. *glabrescens*

Var. **edulis.** TAB. **122** fig. B.

N. Rhodesia. B: Sesheke, ♀ fl. & fr., *Gairdner* 140 (K). N: Abercorn–Mpulungu road, c. 16 km. from Mpulungu, ♂ fl. 26.ix. 1959, *Richards* 11472 (K). W: Matonchi Farm, ♀ fl. & fr. 1.ix.1930, *Milne-Redhead* 1019 (BR; K). C: c. 96 km. from Mumbwa on the road to Broken Hill, fr. 19.ix.1947, *Brenan* 7890 (EA; FHO; K). E: Chadiza, fr. 28.xi.1958, *Robson* 759 (K; LISC). S: Choma, Siamambo Forest Reserve, st. vii. 1952, *White* 3885 (FHO), 3886 (FHO). **S. Rhodesia.** N: Lomagundi, near Banket, ♀ fl. 20.ix.1959, *Leach* 9385 (SRGH). W: Matopos, Mtsheleli Valley, fr. 22.xi. 1951, *Plowes* 1326 (K; SRGH). C: Hartley, Poole Farm, st. 7.iv.1954, *Wild* 4559 (K; LISC; SRGH). E: Umtali Commonage, ♀ fl. 20.i.1949, *Chase* 1662 (BM; SRGH). S: Buhera, fr. xi.1953, *Davies* 613 (SRGH). **Nyasaland.** C: Dedza, Chongoni, fr. 22.xi.1960, *Chapman* 1056 (FHO; LISC; SRGH). S: Mlanje, near Likabula, fr. x.1929, *Clements* 142 (FHO). **Mozambique.** N: Ribáuè, Posto Agrícola, ♀ fl. 5.ix.1942, *Mendonça*.1271 (LISC). Z: Milange, fr. 13.xi.1942, *Mendonça* 1431c (K; LISC). T: Macanga, Furancungo, at 20 km. on Angónia road, ♀ fl. & fr. 29.ix.1942, *Mendonça* 543 (LISC). MS: Manica, Posto de Mavita, st. 30.viii.1947, *Pimenta* 7 (LISC).

Also in Ubangi, Angola, Congo, Uganda, Tanganyika and the Transvaal. In open woodlands of several types, open grassy plains, dambos (swamps), termite mounds, burnt ground, etc.

Var. **glabrescens** (Engl.) Burtt Davy, F.P.F.T. **2**: 494 (1932). Type from the Transvaal.
Odina edulis var. *glabrescens* Engl. in A. & C. DC., Mon. Phan. **4**: 272 (1883). Type as above.
Lannea edulis var. A.—White, F.F.N.R.: 211 (1962).

N. Rhodesia. B: 36·8 km. W. of Mankoya, st. 19.vii.1961, *Angus* 2994 (FHO; K; SRGH). W: Mwinilunga aerodrome, st. 6.vi.1934, *Duff* 186/34 (FHO). **S. Rhodesia.** C: Umvuma, Mtao Forest Reserve, fr. 26.x.1952, *Mullin* 29/52 (SRGH).

Also in the Transvaal. Habitat as for the type variety.

MATERIAL INSUFFICIENT

Lannea sp. A

N. Rhodesia. W: Chingola, ♂ fl. 24.ix.1955, *Fanshawe* 2461 (K).
The young leaves of this specimen recall those of *L. asymmetrica* but the material is insufficient to decide.

Lannea sp. B

Nyasaland. C: near Lake Nyasa, fr. ix.1859, *Kirk* (K).
This poor specimen may belong to *L. schimperi* but the material (inflorescence-axis and fruit) is insufficient.

Lannea sp. C

Mozambique. N: Cabo Delgado, Montepuez, fr. 17.x.1942, *Mendonça* 909 (LISC).
The young leaves and inflorescence-axis recall those of *L. antiscorbutica*, but the material is insufficient to decide.

Lannea sp. D

Mozambique. N: Palma, 2 km. S. of R. Rovuma and 16 km. from Nangade, alt. 220 m., fr. 18.iv.1964, *Torre & Paiva* 12146 (LISC).
Probably a new species near *L. stuhlmannii*.

7. SORINDEIA Thou.

Sorindeia Thou., Gen. Nov. Madag.: 23 (1806).

Shrubs, sometimes sarmentose, or trees. Leaves alternate, imparipinnate or rarely 1-foliolate; leaflets petiolulate; tertiary nerves generally collected into an oblique nerve, directed towards the angle between the midrib and the lateral nerves. Flowers dioecious, 5-merous, in axillary or terminal branched panicles. ♂ flowers: calyx ± cupuliform, shallowly 5-lobulate or -dentate; petals valvate, or sometimes imbricate, longer than the calyx; stamens 10–20, filaments subulate, inserted below and on the disk, anthers dorsifixed, introrse; disk crenulate, glabrous; pistillode absent. ♀ flowers: perianth similar to the ♂; staminodes usually 5, small; ovary free, ovoid, 1-locular, the ovule pendent, apical; style 1 with a 3-lobed stigma, persisting on the fruit. Drupe ellipsoid or asymmetrically ovoid; mesocarp thin, fleshy; endocarp chartaceous or woody. Seed ellipsoid.

Genus with about 40 species in tropical and subtropical Africa and Madagascar.

Tertiary nerves not collected into an oblique nerve directed towards the angle between the midrib and the lateral nerves; adult leaflets membranous to papyraceous; panicles arising on the old branches, just above the leaf scars - 1. *madagascariensis*

Tertiary nerves collected into an oblique nerve directed towards the angle between the midrib and the lateral nerves; panicles axillary and terminal:

Petals with an inner apical cuneiform thickening; leaflets coriaceous with a very undulate thickened margin:

Calyx-lobes roundish, ciliolate; petals pale-yellowish; leaves (1)3–7(9)-foliolate with the petiole and rhachis up to 20 cm. long - - - - 2. *katangensis*

Calyx-lobes subacute, not ciliolate; petals dark-reddish-purple; leaves 9–11-foliolate with the petiole and rhachis up to 30 cm. long - - 3. *undulata*

Petals without an inner apical cuneiform thickening; leaflets not so coriaceous and not so undulate at the margin as above:

Petals of the ♂ flower subacute, c. 3·5 × 1·5 mm.; anthers not apiculate, c. 1·5 mm. long; drupe c. 1 cm. long; panicles ample, up to 70 cm. long; leaflets up to 22 × 11·5 cm. - - - - - - - 4. *juglandifolia*

Petals of the ♂ flower acute, c. 4 × 2 mm.; anthers apiculate, c. 2 mm. long; drupe c. 2 × 1·2 cm.; panicles smaller; leaflets up to 14·3 × 7–8 cm. - 5. *rhodesica*

1. **Sorindeia madagascariensis** Thou., Gen. Nov. Madag.: 24 (1806).—Engl. in A. & C. DC., Mon. Phan. **4**: 300 (1883); Pflanzenw. Afr. **3,** 2: 190 (1921).—Engl. & Krause in Engl., Bot. Jahrb. **46**: 337 (1912).—Perrier, Fl. Madagasc., Anacardiaceae: 26 (1946).—Meikle in Mem. N.Y. Bot. Gard. **8,** 3: 244 (1953). Type from Madagascar.

Sorindeia obtusifoliolata Engl., Pflanzenw. Ost-Afr. **C**: 244 (1895); Pflanzenw. Afr. **3,** 2: 190 (1921).—Engl. & Krause in Engl., Bot. Jahrb. **46**: 337 (1912).—Burtt Davy & Hoyle, N.C.L.: 29 (1936).—Brenan, T.T.C.L.: 28 (1949). Syntypes from Tanganyika.

An evergreen tree up to 10 m. tall, much branched, glabrous everywhere except for the sometimes minutely puberulous pedicels. Leaves imparipinnate; petiole and rhachis 10–32 cm. long, terete, striate, whitish to greyish or brownish, the petiole rather swollen at the base; leaflets 7–13, light green, concolorous or somewhat discolorous, up to 34 × 13 cm., membranous to chartaceous (rarely subcoriaceous), subopposite or usually alternate, the terminal one symmetric, oblong or obovate, ± cuneate at the base and with petiolule 1–2·3 cm. long, the median ones up to 24 × 9 cm., narrowly oblong, oblong or elliptic, asymmetric at the base as also the rather smaller basal ones, petiolulate (with petiolules 0·5–12 mm. long, somewhat swollen, wrinkled, canaliculate), all rounded or obtuse to ± acuminate at the apex, entire and not thickened at the undulate margin; lateral nerves arcuate near the margin, visible but not prominent above, prominulous below, tertiary nerves prominulous beneath (as also the reticulation), not collected into a nerve directed towards the axils of the lateral nerves. Panicles 20–95 cm. long, generally arising from the older branches below the leafy region, inserted just above the leaf-scars, pendent, ample, lax, with divaricate branches; pedicels of the ♂ flowers c. 6 mm. long, slender, articulated near the apex, with minute bracts at the base, those of the ♀ flowers a little thicker and shorter. Flower-bud globose, not apiculate. Calyx brownish-red, c. 1·5 × 2·5 mm., shallowly lobed, the lobules apiculate or subobtuse. Petals pale reddish-orange to dull red outside, dull yellow inside, 4·5 × 1·5–2 mm., fleshy, oblong, acute. Drupe bright yellow, up to 2·5 × 1·3 cm., with stipe up to 1 cm. long, ellipsoid, acute, apiculate.

Nyasaland. N: Deep Bay, fr., *Lewis* 94 (FHO). C: Dedza, Mua-Livulezi Forest Reserve, ♂ fl. 30.iv.1953, *Adlard* 40 (K; SRGH). S: Mlanje Mt., from Tuchila to Likabula, fr. 23.ix.1957, *Chapman* 437 (FHO; K). **Mozambique.** N: Porto Amélia, R. Nangororo, Ridi, ♂ fl. 25.x.1959, *Gomes e Sousa* 4487 (COI; K; PRE). Z: Lugela-

Mocuba, Namagoa, ♀ fl. & fr., *Faulkner* 213 (BR; COI; K; LISC; LISJC; PRE; SRGH).

Also in Tanganyika, Zanzibar and Mascarene Is. In riverine or other forest types and in woodland.

Fruit edible with a pleasant flavour.

2. **Sorindeia katangensis** Van der Veken in Bull. Jard. Bot. Brux. **29**: 245 (1959) pro parte excl. specim. *Fanshawe* 3503, 3647 et 4060; F.C.B. **9**: 98 (1960). TAB. **123.** Type from Katanga.
Sorindeia lemairei sensu White, F.F.N.R.: 214 (1962).

Weak-stemmed shrub or tree up to 10 m. tall; branches dark grey to brownish, striate; bark of the bole flaking into thin pieces of 5 × 2·5 cm. Leaves imparipinnate, (1)3–7(9)-foliolate; petiole in the 1-foliolate leaves 1·5–2 cm. long, petiole and rhachis in the multifoliolate ones 2·5–20 cm. long, greyish-brown to yellowish, terete, striate, glabrous or ± pilose; leaflets light green, turning olive-green when dried, concolorous or nearly so, glabrous, except sometimes for the base of the midrib, ± rigidly coriaceous, the terminal 6·5–20 × 3–10 cm., obovate-elliptic to broadly elliptic, acute or ± cuneate at the base and petiolulate, the lateral ones 5·5–15 × 2·7–7·7 cm., oblong-elliptic or ovate, rounded or subacute at the base and petiolulate (petiolules 0·4–1 cm. long, canaliculate, somewhat swollen and wrinkled), ± rounded or usually shortly acuminate at the apex (acumen obtuse), and with very undulate pale-cartilaginous thickened margin; midrib prominent above and below; lateral nerves and reticulation whitish or lighter than the lamina, slightly prominent above, more so beneath, the tertiary nerves somewhat raised beneath as also the venation, collected into an oblique nerve directed towards the lateral-nerve-axils. Panicles up to 40 × 27 cm., axillary and terminal, pyramidal, lax, with ± pilose axis and branches, finally glabrescent; pedicels 1·3 mm. long, articulated near the middle. Flower-bud ovoid, not apiculate. Calyx 1·5–2 × c. 3 mm., cupuliform, lobulate, the lobules c. 0·5 mm. long, rounded, ciliolate. Petals cream-yellow, oblong-elliptic, acute, fleshy, with an inner apical cuneiform thickening and longitudinally cristate along the median line on the inner side, glabrous. Anthers c. 1 mm. long. Drupe blackish-violet at maturity, 15–20 × 12–15 × 8–10 mm., globose-ellipsoid, asymmetric.

N. Rhodesia. W: Chingola, fr. 14.x.1955, *Fanshawe* 2507 (K; SRGH); Mwinilunga, 6·4 km. N. of Kalene Hill Mission, fr. 22.ix.1952, *White* 3323 (FHO; K).

Also in Katanga. In riverine and other evergreen forests.

3. **Sorindeia undulata** R. & A. Fernandes in Bol. Soc. Brot., Sér. 2, **38**: 149, t. 8 (1965). Type: N. Rhodesia, Chienge, *Fanshawe* 4739 (FHO; K, holotype).

Tree up to 9 m. high; branchlets greyish, striate. Leaves 9–11-foliolate, up to 41 cm. long; petiole and rhachis up to 30 cm. long, puberulous; leaflets alternate, discolorous (greyish-green above, greenish beneath), coriaceous, glabrous, contracted at the apex into an acumen 5–11 mm. long, with the margin undulate and thickened, petiolulate (petiolule c. 7 mm. long, canaliculate above), the terminal one 11 × 5 cm., elliptic, symmetric at the base, the intermediate ones 9–10 × 3·8–4·2 cm., ovate to oblong, very asymmetric at the base, the basal ones 5–6 × 3–4 cm., broadly ovate; midrib and lateral nerves (9–11 on each side) straw-coloured, prominulous above, prominent below, reticulation visible on both surfaces but more so beneath. Panicles reddish, 20 cm. long (or more?), axillary, lax. Flower-bud purplish, subglobose. Calyx-lobes 0·5 × c. 1·5 mm., subacute. Petals dark reddish-purple, c. 3 × 1·75 mm., with an apical cuneiform thickening on the inner face. Stamens 13–16; anthers c. 1·25 mm. long, oblong. ♀ flowers and fruit unknown.

N. Rhodesia. N: Chienge, ♂ fl. 18.viii.1958, *Fanshawe* 4739 (FHO; K).

Known only from N. Rhodesia. In riverine forest.

4. **Sorindeia juglandifolia** (A. Rich.) Planch. ex Oliv., F.T.A. **1**: 440 (1868).—Keay in Bull. Jard. Bot. Brux. **26,** 2: 209 (1956); F.W.T.A. ed. 2, **1,** 2: 737 (1958). Type from Senegal.
Dupuisia juglandifolia A. Rich. in Guill., Perr. & Rich., Fl. Senegamb. Tent. **1**: 148, t. 38 (1831). Type as above.

A sarmentose shrub or small tree up to 9 m. tall, glabrous except sometimes for the inflorescence; branchlets striate, sometimes densely lenticellate. Leaves

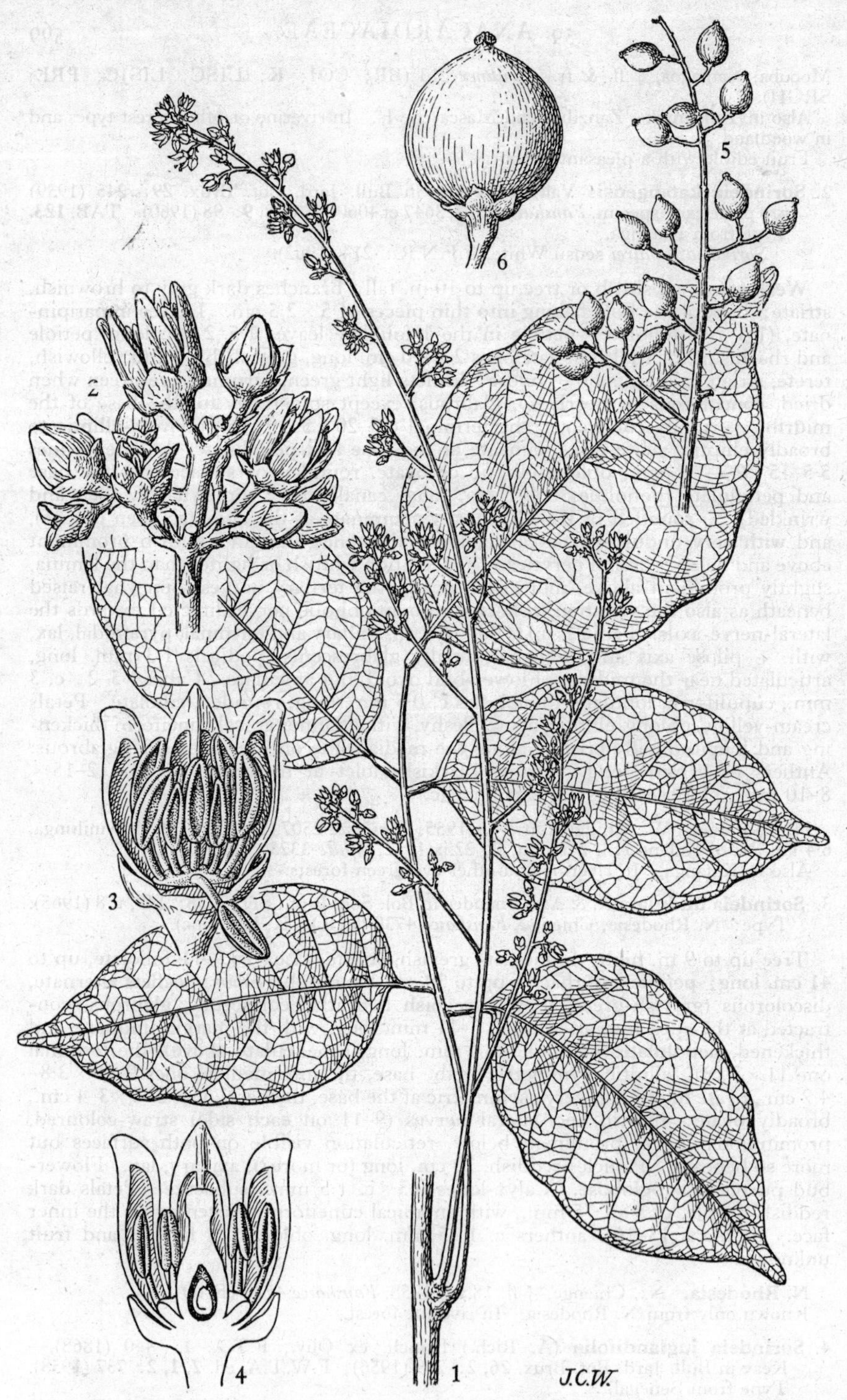

Tab. 123. SORINDEIA KATANGENSIS. 1, leaf and male panicle (× ⅔) *Angus* 426; 2, terminal part of male panicle (×4) *Angus* 426; 3, male flower, front sepals and petals removed (×8) *Angus* 426; 4, vertical section of female flower (×8) *Angus* 426; 5, terminal part of young infructescence (× ⅔) *Holmes* 976; 6, young fruit (×3) *Holmes* 976.

(1)3–9-foliolate; petiole and rhachis 10–30 cm. long, subcylindric, slightly compressed above, striate; leaflets concolorous, subopposite or alternate, obovate, ovate, elliptic or oblong-elliptic, subcoriaceous to coriaceous, shortly broadly and obtusely acuminate at the apex, entire, slightly thickened and undulate at the margin, rounded or cuneate at the base, petiolulate (with petiolule dull purple, 0·3–1 cm. long, thick, wrinkled, furrowed), usually large, the terminal one up to 22 × 11·5 cm., the lateral ones somewhat smaller, decreasing in size to the leaf base; midrib prominulous above, very prominent beneath, lateral nerves 6–10 on each side, slender, straw-coloured in a dry state, ascendent-arcuate, raised above and more so beneath; tertiary nerves collected into an oblique nerve directed towards the lateral-nerve-axils, prominulous on both surfaces as also the lax venation. Panicles up to 70 cm. long and with basal branches up to 45 cm. long, pyramidal, lax, sparsely pilose or glabrous, the ♂ ones usually longer than the ♀ ones; pedicels slender, up to 3·5 mm. Flowers whitish or yellowish, tinged with red; bud globose, obtuse. Calyx 2 × 2·5 mm., very shallowly 5-lobulate, glabrous. Petals 3–5 × 1·5 mm., oblong, subacute, glabrous, not thickened internally at the apex. Filaments c. 1 mm. long; anthers oblong, 1·5 mm. long. Drupe yellow, c. 1 cm. long, broadly ellipsoid.

Mozambique. LM: Maputo, between Polana and Costa do Sol, fr. 29.xi.1942, *Mendonça* 1511 (COI; K; LISC; LMJ).
Also in West tropical Africa and Ubangi-Chari. In the dense coastal bush of the dunes.

5. **Sorindeia rhodesica** R. & A. Fernandes in Bol. Soc. Brot., Sér. 2, **38**: 147, t. 5–7 (1965). Type: N. Rhodesia, Fort Rosebery, *Angus* 260 (BM; BR; FHO, holotype; K).
Sorindeia juglandifolia sensu White, F.F.N.R.: 214 (1962).
Sorindeia katangensis Van der Veken in Bull. Jard. Bot. Brux. **29**: 245 (1959) pro parte quoad specim. *Fanshawe* 3503, 3647 et 4060.

Scrambling shrub 4 m. high or small tree up to 9 m. tall. Leaves usually 7–9-foliolate; petiole and rhachis 16–35 cm. long, terete, not much thickened; leaflets light green, subconcolorous, papyraceous to subcoriaceous, glabrous, contracted at the apex into a rather narrow acumen up to 1·8 cm. long, a little thickened and undulate at the margin, broadly elliptic to ovate, the terminal one up to 14·3 × 7·8 cm., symmetric at the base, the lateral ones up to 14 × 7 cm., symmetric or asymmetric, all rounded or somewhat cuneate at the base and with petiolule 0·5–1 cm. long, not much thickened, canaliculate, not wrinkled; lateral nerves and the narrow reticulation slightly raised above, more so below; tertiary nerves collected in an oblique one directed towards the lateral nerve-axil. Panicles 7–34 cm. long, terminal and axillary, much branched; pedicels up to 2 mm. long; flower-bud purple-red, globose, obtuse. ♂ flower: calyx c. 2·5 × 3·5 mm., very shallowly lobulate; petals pink edged with cream outside, cream inside, c. 4 × 2 mm., oblong, acute, very fleshy, not thickened at the apex; stamens 12–16; filaments 0·75 mm. long; anthers 2 mm. long, oblong, apiculate. ♀ flowers: calyx 2·5 × 3 mm., glabrous, cupuliform; petals as in the ♂ flower but narrower; staminodes 5, c. 2 mm. long; ovary 1·75 mm. in diam., glabrous, ovoid, attenuate into a style 1–1·5 mm. long. Drupe orange to red, c. 2 × 1·2 cm., ellipsoid.

N. Rhodesia. N: Kawambwa, ♂ fl. 22.viii.1957, *Fanshawe* 3503 BR; EA; K); Abercorn, Chinakila, near Senga Hill, fr. 25.x.1959, *Lawton* 650 (FHO).
Known only from the Northern Prov. of N. Rhodesia. Evergreen forests.

8. TRICHOSCYPHA Hook. f.

Trichoscypha Hook. f. in Benth. & Hook., Gen. Pl. **1**: 423 (1862).

Lianes, shrubs or trees. Leaves alternate, imparipinnate; leaflets petiolulate. Flowers dioecious, in many-flowered panicles, the ♀ ones usually shorter and with thicker axis than in the ♂. Calyx shortly cupuliform, with 4(5) triangular-ovate lobes, valvate in bud. Petals 4(5), patent, later reflexed, valvate, those of the ♀ flowers slightly larger. Stamens 4(5), smaller and sterile in ♀ flowers; filaments filiform; anthers dorsifixed. Disk cupuliform, ferruginous-hirsute or glabrous. Ovary ovoid, 1-locular, 1-ovulate; styles 3–4, reflexed or erect, compressed, with

2-lobed stigmas (rarely sessile). Drupe ovoid, turbinate or subglobose, appressed-setulose or glabrous; exocarp and mesocarp ± fleshy; endocarp thin, coriaceous or crustaceous. Seed ovoid with very thick cotyledons.

Leaflets with a very undulate, not thickened margin; anthers obtuse at the apex, c. 1 mm. long; disk 1·5 mm. in diam., densely rubro-pilose - - 1. *ulugurensis*
Leaflets with a thickened usually not undulate margin; anthers narrowing gradually to an acute apex; disk 1 mm. in diam., with the hairs shorter and not so dense as above 2. *silveirana*

1. **Trichoscypha ulugurensis** Mildbr. in Notizbl. Bot. Gart. Berl. **11**: 1071 (1934).— Brenan, T.T.C.L.: 39 (1949). TAB. **124**. Type from Tanganyika.

A tree up to 10·5 m. tall; branchlets brownish, striate, lenticellate. Leaves up to 50 cm. long, (5)7–11-foliolate; petiole and rhachis brownish, up to 38 cm. long, glabrous or shortly appressed-pilose, striate; leaflets opposite or subopposite or alternate, 5·5–13 × 3–5·5 cm., subcoriaceous, elliptic or oblong-lanceolate, subequal, acuminate at the apex, with an often falcate acumen c. 1 cm. long, conspicuously undulate at the margin, obtuse or sometimes acute and asymmetric at the base, petiolulate, petiolule 3–7 mm. long, canaliculate above, convex and rugose beneath; midrib impressed and sometimes appressed-pilose above, raised beneath; lateral nerves 10–12 pairs, arcuate and anastomosed near the margin, slightly raised above, more so beneath as also the reticulation. Panicles up to 30 cm. long, terminal, pyramidal, many-flowered, with the axis and branches densely rufous-pilose; pedicels short. ♂ flowers: calyx c. 1 mm. long, 4-dentate; petals 3 × 2 mm.; stamens with filaments up to 3 mm. long and broad ellipsoid anthers c. 1 mm. long; disk 1·5 mm. in diam., densely reddish-pilose. ♀ flowers like the ♂ but a little larger; staminodes with filaments c. 1 mm. long and anthers c. 0·5 mm. long; ovary 3–4 mm. long, ovoid, densely appressed-reddish-pilose; styles 3, c. 1 mm. long, reflexed on the ovary. Drupe unknown.

S. Rhodesia. E: Chimanimani Mts., Martin Forest Reserve, ♀ fl. 15.viii.1948, *McGregor* 37/48 (FHO; K; SRGH). **Nyasaland.** S: Mlanje Mt., Malosa stream, ♀ fl. 14.vii.1958, *Chapman* 610 (FHO; K; SRGH).
Also in Tanganyika. In evergreen forests.

2. **Trichoscypha silveirana** Exell & Mendonça in Bol. Soc. Brot., Sér. 2, **26**: 278, t. 3 (1952); C.F.A. **2,** 1: 124, t. 30 (1954).—Van der Veken, F.C.B. **9**: 80 (1960).— White, F.F.N.R.: 214 (1962). Type from Angola.

Shrub or tree up to 15 m. tall, with smooth bole; bark of branchlets dull brown, fissured, lenticellate, glabrous. Leaves (1)2–6-foliolate; petiole and rhachis reddish-brown, 15–35 cm. long, sparsely appressed-puberulous (the hairs whitish, very short); petiole 2·5–10 cm. long, plano-convex; rhachis subcylindric, striate; leaflets concolorous, 6–17 × 3–8 cm., oblong, oblong-elliptic or elliptic, unequal (the basal ones shorter and sometimes ovate), abruptly or gradually acuminate, the acumen up to 1·5 × 0·5 cm., obtuse and sometimes slightly notched at the apex, with the margin somewhat thickened, revolute and sometimes undulate, subacute or rounded and symmetric at the base, subcoriaceous to coriaceous, glabrous except for the midrib minutely whitish-hirsute above and appressed-brownish-puberulous below; midrib impressed above, very prominent beneath; lateral nerves and the very lax reticulation nearly invisible or prominulous above, prominent beneath; petiolules c. 0·5 cm. long, canaliculate. Panicles up to 30 × 15 cm., terminal, with the cylindric axis much thickened in fruiting state, as also the densely lenticellate and sparsely appressed-brown-reddish-puberulous branches; pedicels 1–3 mm. long, very thick, densely puberulous. ♂ flowers ± similar to the ♀ flowers: calyx-lobes c. 2 mm. × 2·5 mm., shortly ovate; petals c. 2·5–3 × 1·5–2 mm., oblong-ovate, obtuse, dark-veined; ovary densely dull-reddish-hispid; styles 3, compressed, reflexed; staminodes c. 2 mm. long. Drupe dull purple, 1–1·8 × 0·8–1·4 cm., ovoid-globose, ± puberulous.

N. Rhodesia. N: Kawambwa, fr. 23.viii.1957, *Fanshawe* 3555 (EA; K; PRE; SRGH). W: Mwinilunga, fr. 25.i.1955, *Holmes* 1206 (K).
Also in Angola and Katanga. In riverine forests.

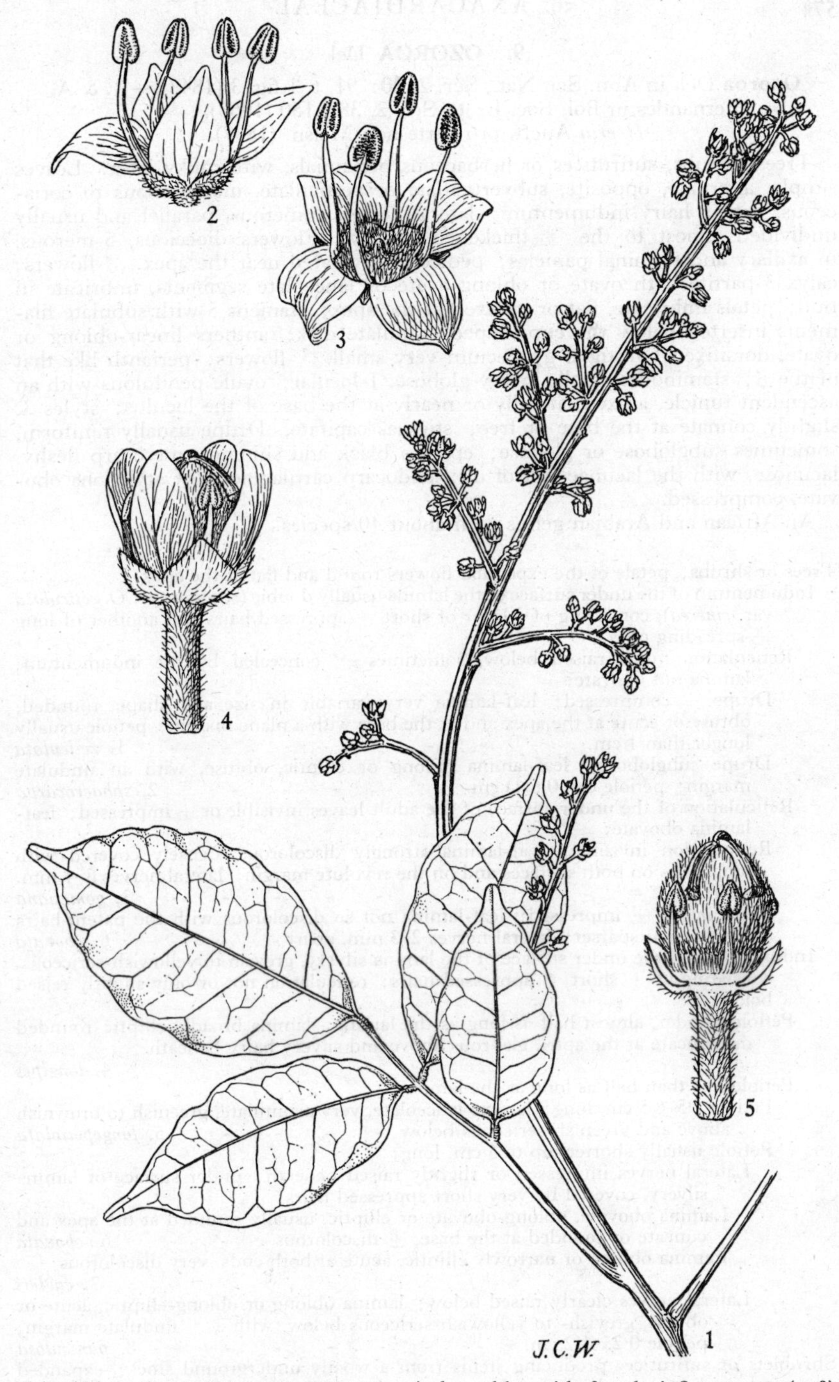

Tab. 124. TRICHOSCYPHA ULUGURENSIS. 1, branchlet with female inflorescence (×⅔) *Chapman* 610; 2, male flower (×6) *Drummond & Hemsley* 1750; 3, same showing disk (×6) *Drummond & Hemsley* 1750; 4, female flower with very young fruit (×6) *Chapman* 610; 5, female flower after petal-fall and with the front sepal removed (×6) *Chapman* 610.

9. OZOROA Del.

Ozoroa Del. in Ann. Sci. Nat., Sér. 2, **20**: 91, t. 1 fig. 3 (1843).—R. & A. Fernandes in Bol. Soc. Brot., Sér. 2, **38**: 150 (1965).
Heeria Auctt. pro parte non Meisn. (1837).

Trees, shrubs, suffrutices or herbaceous perennials, with milky latex. Leaves simple, alternate, opposite, subverticillate or verticillate, membranous to coriaceous, with a hairy indumentum; lateral nerves numerous, parallel and usually undivided almost to the ± thickened margin. Flowers dioecious, 5-merous, in axillary and terminal panicles; pedicels articulated near the apex. ♂ flowers: calyx 5-partite with ovate or oblong-ovate or lanceolate segments, imbricate in bud; petals imbricate, flat or inflexed at the apex; stamens 5 with subulate filaments inserted below the cup-shaped crenulate disk; anthers linear-oblong or ovate, dorsifixed; vestigial gynoecium very small. ♀ flowers: perianth like that of the ♂; staminodes small; ovary globose, 1-locular; ovule pendulous with an ascendent funicle, affixed laterally or nearly at the base of the loculus; styles 3, slightly connate at the base or free; stigmas capitate. Drupe usually reniform, sometimes subglobose or globose; epicarp black and shining; mesocarp fleshy, lacunose, with the lacunae full of oil; endocarp cartilaginous. Cotyledons obovate, compressed.

An African and Arabian genus with about 40 species.

Trees or shrubs; petals of the expanded flowers round and flat at the apex:
Indumentum of the under surface of the lamina usually double (see however *O. reticulata* var. *cinerea*), consisting of a layer of short ± appressed hairs and another of long ± spreading ones:
Reticulation much raised below, sometimes ± concealed by the indumentum; lamina not obovate:
Drupe ± compressed; leaf-lamina very variable in size and shape, rounded, obtuse or acute at the apex and at the base with a plane margin; petiole usually longer than 1 cm. - - - - - - - - - 1. *reticulata*
Drupe subglobose; leaf-lamina oblong or elliptic, obtuse, with an undulate margin; petiole 0·3–0·6(1) cm. - - - - - - 2. *sphaerocarpa*
Reticulation of the under surface of the adult leaves invisible or ± impressed; leaf-lamina obovate:
Reticulation invisible; leaf-lamina strongly discolorous, densely covered with long hairs on both surfaces and on the revolute margin; lateral nerves 3–4 mm. apart - - - - - - - - - - - - - 3. *gomesiana*
Reticulation ± impressed; leaf-lamina not so discolorous with the patent hairs shorter and sparser; lateral nerves 2–3 mm. apart - - - - 6. *obovata*
Indumentum of the under surface of the lamina silvery, greyish to yellowish-sericeous, consisting of ± short ± appressed hairs; reticulation not or only slightly raised below:
Petiole slender, almost half as long as the lamina; lamina broadly elliptic rounded or truncate at the apex, glabrous above and silvery hairy beneath 4. *longipes*
Petiole less than half as long as the lamina:
Petiole 2·5–6·5 cm. long; lamina lanceolate, very acuminate, greenish to brownish above and greenish-sericeous below - - - - 5. *longepetiolata*
Petiole usually shorter, up to 3 cm. long:
Lateral nerves impressed or slightly raised beneath; under surface of lamina silvery, covered by very short appressed hairs:
Lamina obovate, oblong-obovate or elliptic, usually rounded at the apex and cuneate or rounded at the base, ± discolorous - - 6. *obovata*
Lamina oblong or narrowly elliptic, acute at both ends, very discolorous 7. *engleri*
Lateral nerves clearly raised below; lamina oblong or oblong-elliptic, acute or obtuse, greyish- to yellowish-sericeous below, with a ± undulate margin; petiole 0·25–1·2 cm. - - - - - - - - - 8. *paniculosa*
Shrublets or suffrutices producing stems from a woody underground stock; expanded petals with a triangular inflexed apex; petiole usually shorter than 1 cm.:
Leaves not acicular:
Indumentum of the under surface of the leaf-lamina double, consisting of a layer of short ± appressed hairs and another of long ones:
Reticulation neither raised nor visible beneath; lamina oblong-elliptic or broadly elliptic to ovate, membranous, very discolorous (dark brown and glabrous above and densely ferruginous- or olivaceous-hairy beneath); lateral nerves

(2)3–5(7) mm. apart, slightly raised on the under surface
9. *nigricans* var. *nigricans*

Reticulation raised below, at times concealed by the indumentum:

Leaf-lamina linear, 6–11 × 1–1·6 cm.; petiole 1–3 mm. long; lateral nerves 2·5–3·5 mm. apart - - - - - - - - - 10. *bredoi*

Leaf-lamina not linear; petiole longer:

Leaf-lamina subcoriaceous, not shining above, with a flat not very thickened margin - - - - - - - - - - - 11. *kassneri*

Leaf-lamina coriaceous, shining above, with a much thickened cartilaginous undulate margin - - - - - - - - - 12. *nitida*

Indumentum of the under surface of the leaf-lamina usually simple, consisting of dense appressed hairs; reticulation not or only slightly raised below:

Leaf-lamina neither linear nor lanceolate, somewhat discolorous:

Leaf-lamina oblong-elliptic or oblong-obovate, (4·2)7–14 × 2–6 cm., with the margin much thickened; lateral nerves 3–5(7) mm. apart; petiole 1–8 mm. long - - - - - - - - - - - - 13. *marginata*

Leaf-lamina elliptic or obovate, 4·5–8 × 2·5–4·5 cm. with the margin less thickened; lateral nerves 2–3·5 mm. apart; petiole 1–1·2 cm. long
14. *viridis*

Leaf-lamina linear or lanceolate, conspicuously discolorous:

Lateral nerves 2–3(3·5) mm. apart, forked near the margin; lamina up to 5·5 cm. broad, slightly or not shining above - - - - - - 15. *pwetoensis*

Lateral nerves 1–1·5 mm. apart, not branched until the margin; lamina up to 2 cm. broad, shining above - - - - - - - 16. *homblei*

Leaves acicular, with revolute margins, the uppermost 5–10-verticillate
17. *stenophylla*

1. **Ozoroa reticulata** (Bak. f.) R. & A. Fernandes in Bol. Soc. Brot., Sér. 2, **38**: 195 (1965). Type: S. Rhodesia, Bulawayo, *Rand* 64 (BM, lectotype).

Heeria insignis var. *reticulata* Bak. f. in Journ. of Bot. **37**: 428 (1899).—Eyles in Trans. Roy. Soc. S. Afr. **5**: 401 (1916).—Van der Veken, F.C.B. **9**: 11 (1960). Syntypes: S. Rhodesia, Bulawayo, *Rand* 64 (BM), 307 (BM).

Heeria reticulata (Bak. f.) Engl. in Sitz.-Ber. Preuss. Akad. Wiss. Berl. **1906,** 52: 878, 891 (1906); Pflanzenw. Afr. **3,** 2: 197 (1921).—Monro in Proc. & Trans. Rhod. Sci. Ass. **8**: 71 (1908).—Eyles in Trans. Roy. Soc. S. Afr. **5**: 402 (1916).—Steedman, Trees etc. S. Rhod.: 39 (1933).—Brenan, T.T.C.L.: 33 (1949).—Pardy in Rhod. Agr. Journ. **48,** 6: 504, cum photo (1951).—Codd, Trees and Shrubs Krueger Nat. Park: 105, fig. 101 c. (1951).—Palgrave, Trees of Cent. Afr.: 1, fig. p. 2, 3 (1956).—White, F.F.N.R.: 208 (1962). Syntypes as above.

A much-branched tree up to 15 m. tall or sometimes a shrub; branches brownish, ochraceous or greyish, cylindric or grooved, rarely angled, puberulous to ± densely yellowish-villous, the oldest glabrescent and lenticellate. Leaves alternate or in whorls of 3; petiole subterete, flattened at the base, grooved to the apex on the upper side, densely villous to puberulous; lamina nearly concolorous to very discolorous, very variable in size and shape (see subspecies and varieties), usually 2·5–4 times as long as broad, obtuse or acute and mucronate at the apex, rounded or acute at the base, subcoriaceous to coriaceous; upper surface glabrous or puberulous to villous only on the nerves or ± velvety; under surface ± densely villous, usually with double indumentum; midrib impressed above, much raised below; lateral nerves ± raised below; reticulation ± prominent beneath, concealed or not by the indumentum. Panicles terminal and axillary, up to 17 cm. long, much branched, many-flowered, with the axis and branches puberulous to ± densely villous; bracts 3–4(10) mm. long, subulate; pedicels 1–1·5 mm. long, villous. Sepals 1·5–3·7 × 0·75–1·5 mm., ovate or ovate-triangular, somewhat acute, externally ± villous. Petals whitish or yellowish, 2·2–4 × 1–2 mm., oblong, obtuse and flat at the apex, dorsally appressed-pilose. Ovary compressed. Drupes black, shining, 7–8 × 9–11 mm., transversely ellipsoid, compressed, wrinkled.

A very polymorphic species, the polymorphism probably due mainly to mutation and hybridization. Subspp. *foveolata* and *grandifolia* may prove to be distinct species.

Reticulation of the leaf-lamina lax and not much raised below; small trees or shrubs:

Lateral nerves and reticulation of the old leaves completely covered by the very dense indumentum; under layer of short hairs usually concealed by a layer of long soft ones; lateral nerves ± arcuate; trees, rarely shrubs - - subsp. *reticulata*

Lateral nerves and reticulation of the old leaves generally not completely covered by the indumentum; under layer of the short hairs usually not concealed by the layer of the long stiff ones; leaf-lamina slightly to very discolorous when dried (brownish to dark brownish above, ochraceous, brownish to whitish beneath), 7·5–21·5 × 3·9–9·8 cm., broadly ovate or elliptic, rounded at the apex and base; lateral nerves straight, almost at right angles to the midrib; shrubs, rarely small trees subsp. *grandifolia*

Reticulation of the leaf-lamina close and much raised below; small trees, rarely shrubs (or sometimes suffrutices?) - - - - - - - subsp. *foveolata*

Subsp. **reticulata**

Heeria pulcherrima sensu Eyles in Trans. Roy. Soc. S. Afr. **5**: 402 (1916).

Heeria insignis sensu Steedman in Proc. & Trans. Rhod. Sci. Ass. **26**: 7, t. 8 (1927); Trees, etc. S. Rhod.: 38 (1933).

Leaf-lamina flat or nearly so at the margin; lateral nerves and veins not or slightly raised above:

Leaf-lamina dark-green, reddish-brown to dark brown (when dried), usually not shining above, 4–17(21) × 1·7–4·8(5·4) cm., lanceolate, narrowly elliptic, elliptic-oblong to broadly elliptic; lateral nerves 2–4(6) mm. apart; petiole 0·5–2(2·5) cm. long - - - - - - - - - - - - - var. *reticulata*

Leaf-lamina olive-green and shining above, 5–19 × 2·5–6·5(8), ovate-oblong, acute or subobtuse at the apex; lateral nerves (3·5)4–6(7) mm. apart; petiole 2–3 cm. long var. *nyasica*

Leaf-lamina very conspicuously bullate towards the margin, with the latter crisped, thickened and recurved, 6–17·7 × 2–5·5 cm., usually elliptic or oblong-elliptic, subchartaceous; lateral nerves (3·5)5–8 mm. apart, ± impressed above as also the reticulation; petiole 0·6–1·5 cm. - - - - - - - - var. *crispa*

Var. **reticulata.** TAB. **125.**

Heeria paniculosa sensu Eyles in Trans. Roy. Soc. S. Afr. **5**: 401 (1916).

Heeria sp.—Eyles, tom. cit.: 402 (1916).

Heeria insignis var. *lanceolata* sensu Suesseng. in Proc. & Trans. Rhod. Sci. Ass. **43**: 35 (1951).

Bechuanaland Prot. N: Tsessebe, ♀ fl. & fr. 13.i.1960, *Leach & Noel* 2 (COI; SRGH). SE: Hillside, ♂ fl. xii.1944, *Martineau* 715 (SRGH). **N. Rhodesia.** S: Mazabuka, fr. 20.viii.1929, *Burtt Davy* 20749 (FHO; K). **S. Rhodesia.** N: Urungwe, ♂ fl. 17.xii.1957, *Goodier* 469 (LISC; SRGH). W: Nyamandhlovu, near Chesa Valley Pasture Station, fr. 9.i.1954, *Plowes* 1652 (K; LISC; SRGH). C: Umvuma, fr. 10.iii. 1947, *Acheson* 1/47 (FHO; SRGH); Que Que Reserve, ♂ fl. 28.xi.1952, *McLean* 13/52 (SRGH). E: Umtali, Zimunya's Reserve, fr. 8.iv.1956, *Chase* 6066 (BM; K; LMJ; SRGH). S: Belingwe, Lundi Reserve, fr. 24.iv.1954, *Plowes* 1723 (SRGH).

Woodlands of several types, savannas, rocky hillsides and outcrops.

The specimens " **N. Rhodesia.** C: Makulu Research Station, c. 16 km. S. of Lusaka, ♂ fl. 5.xi.1956, *Simwanda* 80 (COI; FHO; SRGH); S: Choma, fr. 24.viii.1961, *Bainbridge* 576 (FHO); Mazabuka, Bweli Estate, ♀ fl. i.1930, *Browne* 37/30 (FHO); near Choma, st. 16.viii.1929, *Burtt Davy* 20639 (FHO); Livingstone to Choma, st. 15.viii.1929, *Burtt Davy* 20681 (FHO); Livingstone, Machili Survey, fr. 25.iii.1933, *Martin* 658 (FHO); Mazabuka, Yates Jones Farm, near Choma, fr. 27.i.1960, *White* 6493 (FHO) " are somewhat intermediate, particularly in the close reticulation, between subspp. *reticulata* and *foveolata*. We therefore suppose that they may be hybrids. If so, there is in these plants dominance of *reticulata* characters.

Var. **nyasica** R. & A. Fernandes in Bol. Soc. Brot., Sér. 2, **38**: 169, t. 26–28 (1965). Type: Nyasaland, Mlanje Mt., between Likabula and Palombe, *Chapman* 500 (BM; BR; FHO; K, holotype).

Rhus insignis sensu Burkill in Johnston, Brit. Centr. Afr., App. II: 241 (1897).

Heeria mucronata sensu Burtt Davy & Hoyle, N.C.L.: 29 (1936).—Williamson, Useful Pl. Nyasal.: 64 (1955) pro parte (" muconata ").

Heeria mucronifolia sensu Burtt Davy & Hoyle, N.C.L.: 29 (1936).

Heeria pulcherrima sensu Burtt Davy & Hoyle, loc. cit., pro parte.

Nyasaland. S: Mlanje Mt., between Likabula and Palombe, fr., *Chapman* 500 bis (FHO; K); Blantyre Distr., Matope, ♂ fl. 9.i.1956, *Jackson* 1785 (BM; BR; FHO; K); Zomba, ♀ fl. & fr. xi.1905; *Purves* 224 (K).

In montane woodlands of several types.

Barbosa & Carvalho in *Barbosa* 3390 (K; LISC; LMJ) from Mozambique, Tete, near Fingoè is somewhat intermediate between vars. *nyasica* and *crispa* and may be a hybrid.

Var. **crispa** R. & A. Fernandes in Bol. Soc. Brot., Sér. 2, **38**: 171, t. 29–30 (1965). Type: S. Rhodesia, Salisbury, *Eyles* 873 (BM, holotype; SRGH).

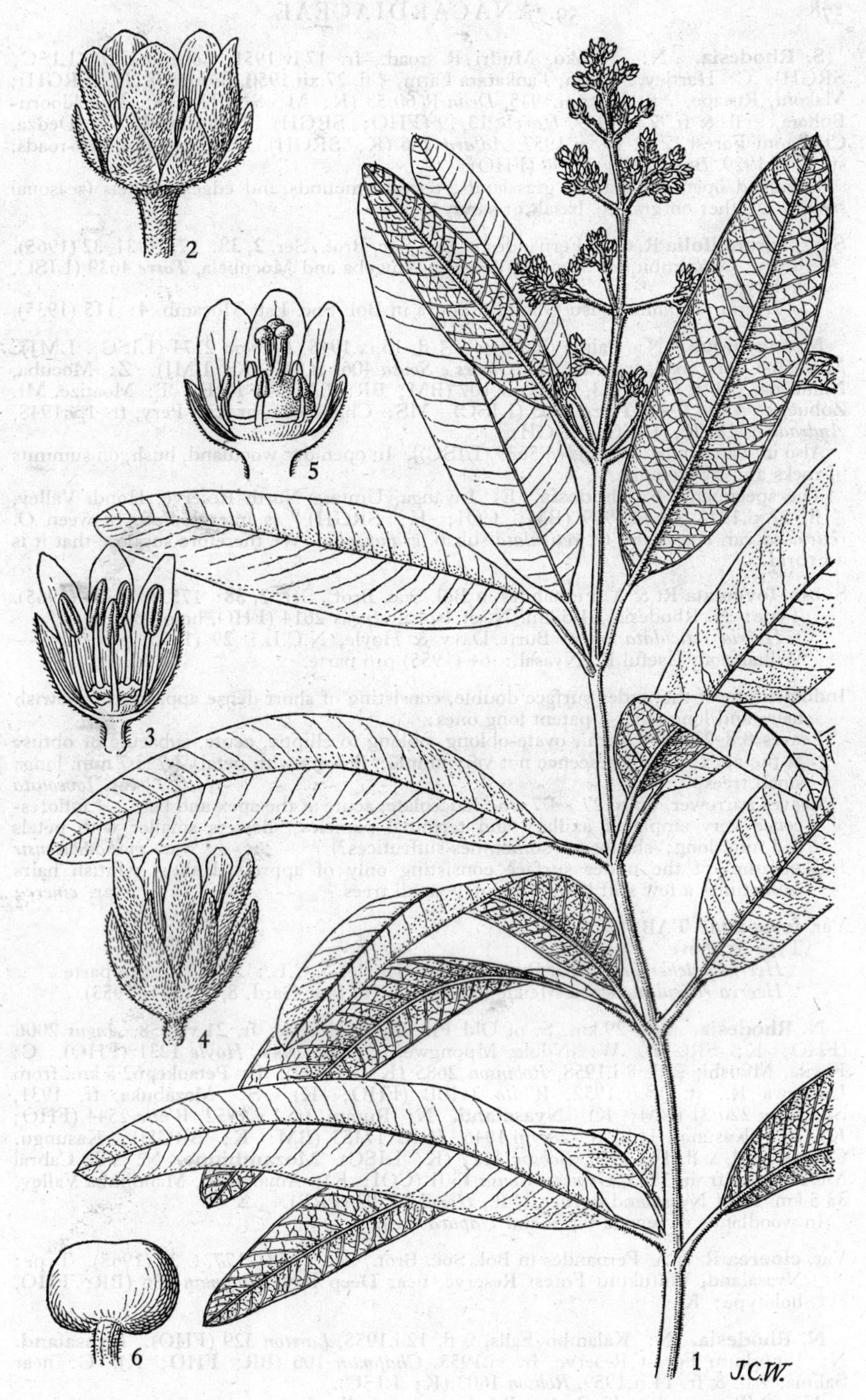

Tab. 125. OZOROA RETICULATA subsp. RETICULATA var. RETICULATA. 1, female flowering branch (× ⅔) *Leach & Noel* 2; 2, male flower (× 6) *Howden* 3/58; 3, same, vertical tangential section (× 6) *Howden* 3/58; 4, female flower (× 6) *Leach & Noel* 2; 5, vertical tangential section of female flower (× 6) *Leach & Noel* 2; 6, fruit (× 2) *Plowes* 1723.

S. Rhodesia. N: Mtoko, Mudri R. road., fr. 17.iv.1951, *Lovemore* 16 (LISC; SRGH). C: Hartley, Norton, Tankatara Farm, ♂ fl. 27.xii.1950, *Birkett* 38 (K; SRGH); Makoni, Rusape, ♀ fl. & fr. iii.1935, *Dehn* R 60/53 (K; M; SRGH). S: Enkeldoorn-Buhera, ♀ fl. & fr. 7.xi.1949, *Harvie* 13/49 (FHO; SRGH). **Nyasaland.** C: Dedza, Chongoni Forest, ♀ fl. 19.xii.1957, *Adlard* 266 (K; SRGH). S: Nfunzi, Cross-roads, st. 11.ix.1929, *Burtt Davy* 21590 (FHO).

In mixed open woodlands, grasslands, termite mounds and edges of vleis (seasonal swamps), either on granite, basalt or sandy soil.

Subsp. **grandifolia** R. & A. Fernandes in Bol. Soc. Brot., Sér. 2, **38**: 173, t. 31–32 (1965). Type: Mozambique, Mocuba, between Mugeba and Mocubela, *Torre* 4639 (LISC, holotype).

Heeria reticulata sensu Gomes e Sousa in Bol. Soc. Est. Moçamb. **4**: 115 (1935).

Mozambique. N: Palma (Tungue), ♂ fl. 16.ix.1948, *Barbosa* 2174 (LISC; LMJ); Mutuáli-Lioma road, fr. 9.iii.1953, *Gomes e Sousa* 4061 (COI; K; LMJ). Z: Mocuba, Namagoa, ♀ fl. & fr. x.1943, *Faulkner* 107 (BM; BR; EA; K; PRE). T: Moatize, Mt. Zóbuè, st. 18.vi.1941, *Torre* 2905 (LISC). MS: Chimoio, near Vila Pery, fr. 1.v.1948, *Andrada* 1218 (K; LISC; SRGH).

Also in Tanganyika [*Schlieben* 5639 (LISC)]. In open dry woodland, bush, on summits of rocks and by streams.

The specimen "**S. Rhodesia.** E: Inyanga, Umtasa North Reserve, Hondi Valley, ♂ fl. 17.xi.1948, *Chase* 989 (BM; COI; K; SRGH)" is intermediate between *O. reticulata* var. *crispa* and *O. reticulata* subsp. *grandifolia.* We therefore suppose that it is a hybrid.

Subsp. **foveolata** R. & A. Fernandes in Bol. Soc. Brot., Sér. 2, **38**: 175, t. 33–34 (1965). Type: N. Rhodesia, 51·5 km. N. of Isoka, *Angus* 2614 (FHO, holotype; K).

Heeria reticulata sensu Burtt Davy & Hoyle, N.C.L.: 29 (1936) pro parte.—Williamson, Useful Pl. Nyasal.: 64 (1955) pro parte.

Indumentum of the under surface double, consisting of short dense appressed yellowish hairs and long stiff ± patent long ones:
- Leaves 8·8–23 × 3–8·6 cm., ovate-oblong, oblong to elliptic, acute, subacute or obtuse at the apex; ♂ inflorescence not very ample; flowers with petals 2·7–3·7 mm. long; small trees - - - - - - - - - - - var. *foveolata*
- Leaves narrower, up to 27 × 4·7 cm., lanceolate, acute at the apex and base; ♂ inflorescence very ample of axillary and terminal panicles; flowers smaller with petals 2–3 mm. long; shrubs (or sometimes suffrutices?) - - var. *mossambicensis*

Indumentum of the under surface consisting only of appressed short whitish hairs (sometimes a few stiff hairs present); small trees - - - - var. *cinerea*

Var. **foveolata.** TAB. **126.**
Type as above.

Heeria pulcherrima sensu Burtt Davy & Hoyle, N.C.L.: 29 (1936) pro parte.
Heeria reticulata sensu Meikle in Mem. N.Y. Bot. Gard. **8,** 3: 244 (1953).

N. Rhodesia. N: 29 km. S. of Old Fife on Isoka road, fr. 21.v.1958, *Angus* 2006 (FHO; K; SRGH). W: Ndola, Mpongwe, st. 27.ix.1949, *Hoyle* 1231 (FHO). C: Fiwila, Mkushi, ♀ fl. 8.i.1958, *Robinson* 2685 (K; SRGH). E: Petauke, 62·5 km. from Luangwa R., st. 18.iv.1952, *White* 2410B (FHO; K). S: Mazabuka, fr. 1931, *Stevenson* 220/31 (BM; K). **Nyasaland.** N: Rumpi, fr. 1.v.1952, *White* 2544 (FHO; K). C: Kasungu Hill, fr. 28.viii.1946, *Brass* 17453 (BM; K; SRGH); Kasungu, Chipala Hill, ♀ fl. 14.i.1959, *Robson* 1177 (K; LISC). **Mozambique.** N: Vila Cabral Massangulo, fr. iii.1933, *Gomes e Sousa* 1301 (COI; K); Amaramba, Mandimba Valley, 33·5 km. E. of Nyasaland border, ♂ fl., *Hornby* 2393 (PRE).

In woodlands of several types and *Uapaca* bush.

Var. **cinerea** R. & A. Fernandes in Bol. Soc. Brot., Sér. 2, **38**: 177, t. 35 (1965). Type: Nyasaland, Vintukutu Forest Reserve, near Deep Bay, *Chapman* 106 (BR; FHO, holotype; K).

N. Rhodesia. N: Kalambo Falls, ♀ fl. 12.i.1958, *Lawton* 329 (FHO). **Nyasaland.** N: Vintukutu Forest Reserve, fr. vi.1953, *Chapman* 106 (BR; FHO; K). C: near Salima, ♀ fl. & fr. 14.ii.1959, *Robson* 1603 (K; LISC).

Also in Tanganyika. In *Burkea-Brachystegia* woodland on sandy soils.

Var. **mossambicensis** R. & A. Fernandes in Bol. Soc. Brot., Sér. 2, **38**: 177, t. 36–38 (1965). Type: Mozambique, Namaita, between Nampula and Murrupula, *Torre & Paiva* 11395 (COI; K; LISC, holotype; SRGH).

Mozambique. N: Marrupa, R. Lugenda, fr. 12.vi.1948, *Pedro & Pedrógão* 4277

Tab. 126. OZOROA RETICULATA subsp. FOVEOLATA var. FOVEOLATA. 1, branch with female inflorescence ($\times \frac{2}{3}$) *Robinson* 2685; 2, part of under surface of lamina showing reticulate venation ($\times 3$) *Robinson* 2685; 3, vertical tangential section of male flower ($\times 6$) *Richards* 12007; 4, female flower ($\times 6$) *Robinson* 2685; 5, vertical tangential section of female flower ($\times 6$) *Robinson* 2685; 6, part of an infructescence ($\times 1$) *Richards* 1418; 7, fruit ($\times 2$) *Richards* 1418.

(EA); Nampula, ♀ fl. 2.xi.1937, *Torre* 1224 (COI; LISC); between Nampula and Mecuburi, ♂ fl. 20.i.1937, *Torre* 1244 (COI; LISC).
In dense xerophytic forests.

2. **Ozoroa sphaerocarpa** R. & A. Fernandes in Bol. Soc. Brot., Sér. 2, **38**: 178, t. 39 (1965). Type: Mozambique, Lourenço Marques, Goba, *Torre* 2016 (BM; COI; K; LISC, holotype).

A small tree up to 7·5 m. tall. Young branches greyish-brown to cinnamon, cylindric, striate, densely lenticellate, ± pilose, the oldest glabrescent to glabrous. Leaves alternate, upright; petiole 0·3–0·6(1) cm. long, dorsally convex, flattened at the base, grooved above, ± densely pilose, especially when young, the hairs whitish, patent to subappressed; lamina (3)4·4–11·2(14) × 1·5–3(4·8) cm., narrowly oblong, oblong to elliptic, rounded at both ends or a little narrowed towards the base, mucronulate at the apex, with a much thickened yellowish rather undulate margin, submembranous and soft when young, very coriaceous and rigid when older, dark brown to brown-ochraceous, ± pilose (mainly when young) and dull or ± shining above, greyish to greyish-brown and covered by a double indumentum (a layer of very short whitish appressed hairs and another of longer ± patent or subappressed ones, mainly on the nerves and veins) below; midrib impressed above, conspicuously raised beneath; lateral nerves and reticulation ± impressed above, mainly near the margin where the lamina is almost bullate, conspicuously raised below, the lateral nerves at an angle of 45°–50° with the midrib, straight or more usually somewhat arcuate, (2·5)4–8 mm. apart in the median part of the lamina, 1–2-forked near the margin. Inflorescences terminal and axillary, shorter than the leaves, in the upper axils, forming a terminal panicle leafy at the base, the ♀ ones laxer than the ♂ ones; branches slender, shortly villous; pedicels of the ♀ flowers 2·5–4 mm. long, those of the ♂ ones shorter. Calyx-segments c. 2 × 1·25 mm., lanceolate-triangular, acute, externally densely villous. Petals 3·5 × 2 mm., oblong-rectangular, obtuse and flat at the apex. Stamens with filaments c. 1·5 mm. long; anthers c. 0·75 mm. long. Drupe black, somewhat shining, 7–9 × 10 mm., globose or subglobose.

Mozambique. LM: Goba-Namaacha road, ♂ fl. 3.x.1961, *Balsinhas* 517 (BM); Sábiè, between Boane and Moamba, fr. 28.iv.1944, *Torre* 6516 (BR; LISC; LMJ; PRE; SRGH).
Southern Mozambique, Swaziland and the Transvaal. In bush and woodlands.

3. **Ozoroa gomesiana** R. & A. Fernandes in Bol. Soc. Brot., Sér. 2, **38**: 155, t. 11 (1965). Type: Mozambique, Sul do Save, between Mavume and Mapinhane, *Gomes e Sousa* 2223 (COI, holotype; K).

Shrub; branches cylindric, densely patent-hairy in the upper part, glabrescent to glabrous below. Leaves alternate or opposite to 3-verticillate; petiole 6–7 mm. long, broadly canaliculate above, convex on the back, densely patent-hairy; lamina dark green above, whitish-sericeous to brownish beneath, 2·3–6 × 1·2–2·5 cm., obovate, roundish, obtuse or subacute at the apex, margin very conspicuously ciliate and revolute, cuneate at the base; indumentum of both surfaces double, consisting of a lower layer of short appressed hairs and another of long rather sparse patent ones; midrib impressed above, raised and patent-hairy beneath; lateral nerves somewhat arcuate, usually forked not far from the margin, impressed above, raised and spreading-hairy below, 1·5–3(4·5) mm. apart; reticulation not visible. Flowers white (immature).

Mozambique. SS: between Mavume and Mapinhane, c. 40 km. from Mapinhane, fl. immat. ii.1939, *Gomes e Sousa* 2223 (COI; K).

4. **Ozoroa longipes** (Engl. & Gilg) R. & A. Fernandes in Bol. Soc. Brot., Sér. 2, **38**: 159 (1965). Type from Angola.
Heeria longipes Engl. & Gilg in Warb., Kunene-Samb.-Exped. Baum: 287 (1903).—Meikle in Exell & Mendonça, C.F.A. **2**: 117 (1954).—White, F.F.N.R.: 209 (1962). Type as above.

Shrub or small tree up to 3 m. tall; branches dark brown, terete, striate, glabrous. Leaves from alternate (the lowest) to approximate or verticillate (the uppermost); petiole 1·5–4 cm. long, slender, usually c. 1 mm. in diam., glabrous or sparsely and shortly appressed-pubescent; lamina green and shining above (olive-green to brownish-green when dried), glabrous above, silvery-sericeous below, broadly elliptic (2·7–9 × 1·5–5·5 cm.) or subcircular (3–6·5 cm. in diam.), rounded and emarginate or shortly apiculate or mucronate (mucro 1–2 mm. long) at the apex, rounded or somewhat acute at the base, chartaceous; midrib slightly raised above, very prominent beneath; lateral nerves very slender, in 20–30 pairs 1–3 mm. apart, straight, at an angle of 70°–80° with the midrib, slightly prominent and lighter in colour than the lamina above, a little more raised and yellowish beneath; reticulation not visible. Inflorescences terminal and axillary (1–2 per axil), many-flowered, with the axis and branches minutely and sparsely appressed-pilose; pedicels c. 2·5 mm. long. Calyx-segments 1·2–1·3 mm. long, ovate-triangular. Petals whitish, 2·8–3 mm. long, oblong, obtuse and flat at the apex. Drupe black, shining, 7–9 × 10–12 mm., reniform.

Caprivi Strip. 112 km. from Katima on the road to Singalamwe, ♂ fl. 3.xii.1958; *Killick & Leistner* 3203 (M; SRGH). **Bechuanaland Prot.** N: Khardoum Valley, 29 km. E. of SW. African border, fr. 14.iii.1965, *Wild & Drummond* 1052 (LISC; SRGH). **N. Rhodesia.** B: Sesheke, 22·5 km. W. of Katima Mulilo, fr. 1.ii.1952, *White* 2005 (FHO; K).

Also in Angola and SW. Africa. In *Pterocarpus angolensis*-woodland on Kalahari Sand,

5. **Ozoroa longepetiolata** R. & A. Fernandes in Bol. Soc. Brot., Sér. 2, **38**: 158, t. 15 (1965). Type: S. Rhodesia, Sipolilo, Mpingi Pass, Great Dyke, *Wild* 5774 (COI; SRGH, holotype).

Tree up to 6·5 m. tall, with cinnamon-coloured terete striate branches, the youngest ones shortly appressed-cinereous-pubescent, the oldest ones glabrous. Leaves alternate or opposite or sub-3-verticillate; petiole 2·5–6·5 cm. long, dorsally convex, furrowed at the top, compressed and widened towards the base, glabrous and cinnamon-coloured or shortly appressed-pubescent and ± cinereous; lamina very discolorous when dried (olive-green, light green or yellow-green, shining and glabrous above, silvery-greenish-sericeous due to short dense appressed whitish hairs beneath), 6·5–17 × 1·9–4 cm., lanceolate-attenuate, very acute and usually falcate at the cuspidate apex, with the margin slightly thickened, usually rounded at the base, papyraceous or coriaceous; midrib impressed or a little raised above, very prominent beneath, lateral nerves slender, usually straight, at right angles to the midrib towards the leaf base, 2–6 mm. apart, forked just near the margin, ± raised on both surfaces; reticulation not or slightly impressed above and slightly raised beneath. Panicles up to 15 × 10 cm., terminal, shorter than the leaves, with the axis and branches appressed-pubescent [mature flowers not seen]. Drupe dark-brown to black, shining, up to 7 × 8·5 mm., subcircular in outline, compressed.

S. Rhodesia. N: Lomagundi, Umvukwes, fr. 19.iv.1929, *Eyles* 6339 (K; SRGH); Darwin, Umvukwes, Mvurkwe, ♂ fl. 21.xii.1952, *Wild* 3977 (K; LISC; SRGH; UPS).

Known only from our area. On chrome hills.

6. **Ozoroa obovata** (Oliv.) R. & A. Fernandes in Bol. Soc. Brot., Sér. 2, **38**: 161, t. 16–17 (1965). Type: Mozambique, between Tete and the sea coast, *Kirk* (K, lectotype).
Rhus insignis var. *obovata* Oliv., F.T.A. **1**: 437 (1868). Syntypes: Mozambique, *Kirk* (K) and *Forbes* (K).
Anaphrenium abyssinicum var. *obovatum* (Oliv.) Engl. in A. & C. DC., Mon. Phan. **4**: 358 (1883). Syntypes as above.
Anaphrenium abyssinicum var. *mucronatum* (Bernh. ex Krauss) Engl., loc. cit. (" mucronifolium ") pro parte quoad specim. mossamb.
Heeria mucronata var. *obovata* (Oliv.) Engl., Pflanzenw. Afr. **3**, 2: 198 (1921). —Brenan, T.T.C.L.: 33 (1949). Syntypes as for *Rhus insignis* var. *obovata*.

Much-branched shrub 1·5–5 m. tall or tree 6–8(10–12) m. high; branchlets ferruginous-ochraceous or greyish, cylindric, ± densely lenticellate, glabrescent; terminal and flowering branchlets subterete on the lower part, striate or grooved and angular, closely and densely leafy, ± densely pubescent on the upper part. Leaves alternate to subverticillate; petiole 0·4–1·7(2) cm. long, slender or broad and flattened towards the base, dorsally convex and ± appressed-pilose, flat or grooved and covered by ± patent hairs on the upper surface; lamina subconcolorous (yellow-green and covered with ± patent hairs mainly on the midrib and lateral nerves on both faces) or discolorous (olive-green to nearly black above, pale green to pale- or dark-grey-sericeous beneath, covered by very short and appressed whitish hairs), 2·5–12(17) × 1·5–4·5 cm., obovate or obovate-oblong, elliptic to oblong-elliptic or broadly elliptic, rounded or retuse rarely acute at the apex, with the margin not or only a little thickened, generally somewhat revolute, mainly towards the base, usually not undulate, cuneate or rounded at the base, papyraceous to subcoriaceous; midrib impressed or slightly raised above, very prominentb eneath; lateral nerves very slender (0·1–0·2 mm. thick), 1–2·5(4) mm. apart in the median part of the lamina, straight, at an angle of (55°)70°–75°(80°) to the midrib, 1–2 forked at 1–3 mm. from the margin, impressed or very slightly raised and puberulous above, not at all prominent or slightly so beneath, covered, as also the midrib, by short appressed hairs like those of the lamina, denser on the young leaves, ± sparse and sometimes patent on the oldest ones. Panicle terminal, pyramidal, leafy at the base; axis and branches angular, greyish-puberulous as also the articulated pedicels. Calyx-segments 1·5 × 1–1·25 mm., ovate-triangular, silky outside. Petals 2–2·5 × 0·75–1·25 mm., oblong, obtuse, densely silky outside. Ovary compressed. Drupes at first red, finally black, shining, 6–7 × 8–9 mm., transversally reniform, compressed, wrinkled.

Leaf-lamina obovate to obovate-oblong, cuneate at the base; floriferous branches stout, up to c. 7 mm. thick at the base - - - - - - var. *obovata*
Leaf-lamina elliptic to oblong-elliptic, usually rounded at both ends; panicle-axis slender, c. 3 mm. thick at the base - - - - - - - var. *elliptica*

Var. **obovata.** TAB. **127.**

Mozambique. N: Palma (Tungue), near Nangade, fr. 17.ix.1948, *Barbosa* 2200 (LISC; LMJ); Ibó I., ♂ fl. 1884–85, *Carvalho* (COI); between Mocímboa da Praia and Diaca, ♂ fl. 25.iii.1961, *Gomes e Sousa* 4663 (K; M); Nampula, Serra da Mesa, fr. 3.iv.1964, *Torre & Paiva* 11597 (COI; K; LISC; SRGH). Z: Ile, fr. 31.v.1949, *Andrada* 1553 (COI; LISC); 34·2 km. from Mocuba, ♀ fl. 27.v.1949, *Barbosa & Carvalho* in *Barbosa* 2890 (K; LISC; LMJ); Maganja da Costa, ♂ fl. 20.iv.1943, *Torre* 5207 (LISC). MS: Cheringoma, Chiniziúa, fr. 1.v.1957, *Gomes e Sousa* 4379 (COI; FHO; K; PRE); Búzi 19·2 km. on Nova Lusitânia road, ♂ fl. 25.iii.1960, *Wild & Leach* 5210 (K; SRGH). SS: between Chicumbane and the Limpopo mouth, fr. 17.vi.1960, *Lemos & Balsinhas* 129 (BM; COI; K; LMJ; SRGH). LM: Maputo, Bela Vista, Tinonganine, ♂ fl. 28.iii.1957, *Barbosa & Lemos* in *Barbosa* 7576 (COI; LISC; LMJ); Inhaca I., fr. 31.viii.1959, *Watmough* 372 (COI; SRGH).

Also in Kenya, Tanganyika, Zanzibar and Natal. In coastal sandy bush and forests of several types usually not far from the sea.

Var. **elliptica** R. & A. Fernandes in Bol. Soc. Brot., Sér. 2, **38**: 165, t. 18–21 (1965). Type: Mozambique, Tete, between Mutarara Velha and Sinjal, *Barbosa & Carvalho* 3133 (K; LISC, holotype; LMJ).

S. Rhodesia. S: Nuanetsi, fr. 1.vii.1960, *Farrell* 226 (LISC; SRGH). **Mozambique** Z: Mocuba, Namagoa, fr. vi–vii.1949, *Faulkner* 232 (K; PRE; SRGH). T: Mutarara, Ancuaze Road, fr. 18.iv.1949, *Andrada* 1602 (COI; LISC). MS: Dondo, R. Muda, near Lamego, ♀ fl. & fr. 18.iv.1948, *Garcia* 932 (LISC); Dondo, Serra do Chiburo, ♂ fl. 16.iv.1948, *Mendonça* 3965 (LISC); Báruè, Macossa, fr. 1.vii.1941, *Torre* 2971 (LISC). SS: Guijá, Caniçado, fr. 4.vii.1947, *Pedro & Pedrógão* 1288 (COI; K; LISJC; LMJ; SRGH). LM: near Magude on Mahel road, ♂ fl. 7.i.1948, *Torre* 7069 (LISC).

Known only from our area. In bush, savannas and woodland and by streams.

7. **Ozoroa engleri** R. & A. Fernandes in Bol. Soc. Brot., Sér. 2, **38**: 153, t. 10 (1965). Type: Mozambique, Lourenço Marques, Sábiè, between Moamba and Ressano Garcia, *Torre* 7416 (LISC, holotype).

Small tree 6–8 m. tall or a robust shrub; branches with many densely leafy branchlets, brownish- or reddish- or greyish-appressed-puberulous to glabrescent, cylindric, slightly striate on the lower parts, a little more striate or sulcate on the upper parts. Leaves alternate or the uppermost ones subverticillate or verticillate; petiole 1–3 cm. long, slender, dorsally convex, flattened near the base and narrowly furrowed to the apex on the upper surface, from cinnamon-coloured to dark brown, very sparsely and appressed-pilose, rarely with somewhat longer patent hairs; lamina very discolorous (dark green to nearly black and glabrous on the upper surface, distinctly cinereous when young, sometimes greyish-green when older on the under surface, i.e. densely covered by minute very appressed hairs),

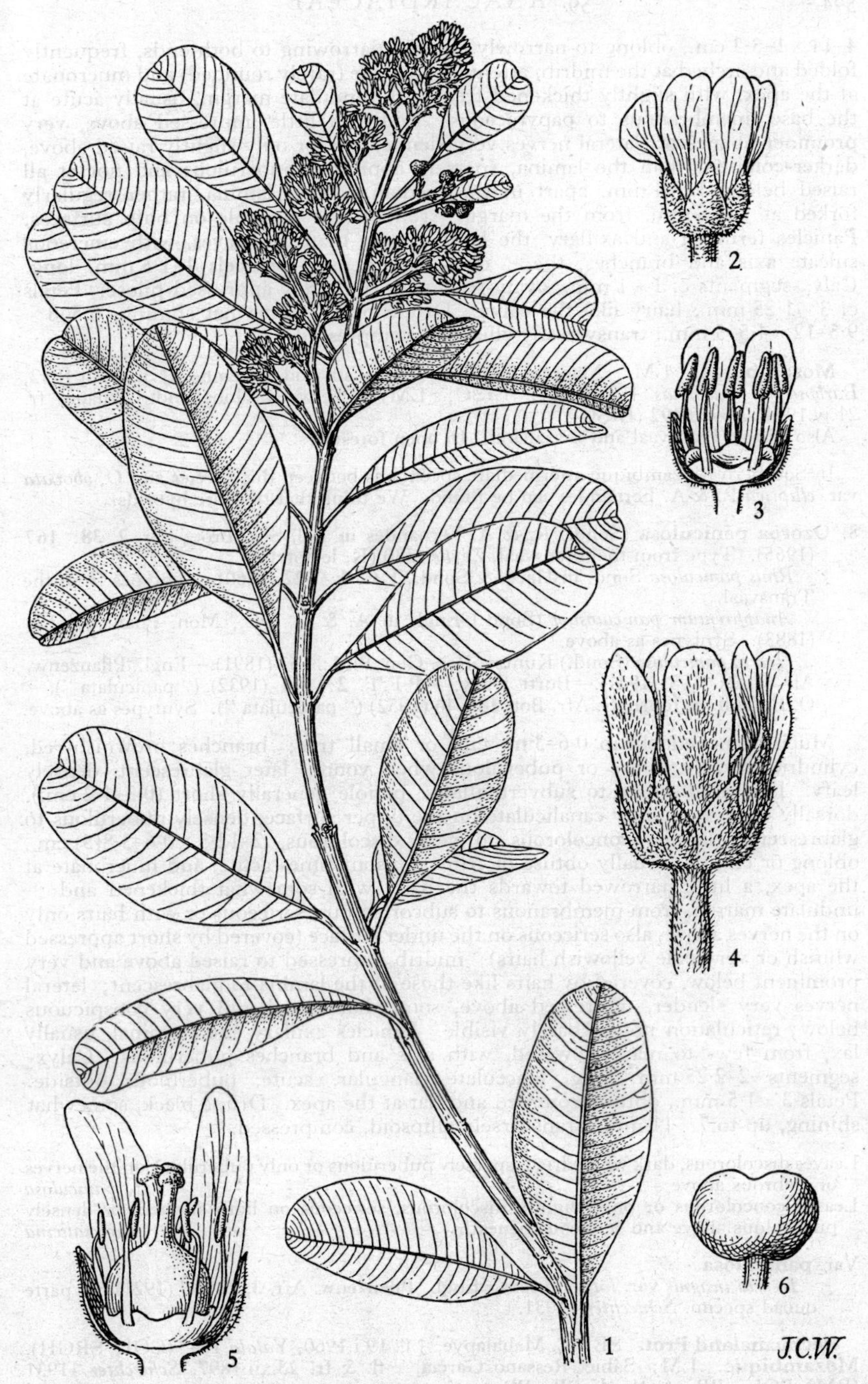

Tab. 127. OZOROA OBOVATA var. OBOVATA. 1, female flowering branch (× $\frac{2}{3}$) *Torre* 3987; 2, male flower (× 8) *Gomes e Sousa* 4663; 3, same, front sepals and petals removed (× 8) *Gomes e Sousa* 4663; 4, female flower (× 8) *Torre* 3987; 5, same, front sepals and petals removed (× 8) *Torre* 3987; 6, fruit (× 2) *Lemos & Balsinhas* 129.

4–14 × 1–3·3 cm., oblong to narrowly elliptic, narrowing to both ends, frequently folded and arched at the midrib, attenuate to acute (rarely rounded) and mucronate at the apex, with slightly thickened frequently undulate margin, usually acute at the base, membranous to papyraceous; midrib a little impressed above, very prominent beneath; lateral nerves very slender, not or only slightly raised above, darker-coloured than the lamina, sparsely appressed-puberulous and not at all raised below, 1·5–4 mm. apart in the median part of lamina, rather regularly forked at 2–2·5 mm. from the margin; reticulation invisible on both surfaces. Panicles terminal and axillary, the later shorter than the leaves, with cinereous sulcate axis and branches, the ♂ ones very dense; pedicels 1–1·5 mm. long. Calyx-segments c. 1 × 1 mm., ovate-triangular, minutely appressed-pilose. Petals c. 3 × 1·25 mm., hairy like the calyx. Drupe black, somewhat shining, 6·5–8 × 9·5–12 × 4·5–5 mm., transversally ellipsoid, compressed.

Mozambique. LM: Maputo, between Bela Vista and Catembe, ♂ fl. 17.ii.1952, *Barbosa & Balsinhas* 4733 (BM; LISC; LMJ); between Goba and Catuane, fr. 21.iv.1944, *Torre* 6492 (LISC).
Also in the Transvaal and Swaziland. In open forests.

In Southern Mozambique intermediate specimens between this species and *O. obovata* var. *elliptica* R. & A. Fernandes can be found. We think that they are hybrids.

8. **Ozoroa paniculosa** (Sond.) R. & A. Fernandes in Bol. Soc. Brot., Sér. 2, **38**: 167 (1965). Type from the Transvaal, *Zeyher* 330 (K, lectotype).
Rhus paniculosa Sond. in Harv. & Sond., F.C. **1**: 522 (1860). Syntypes from the Transvaal.
Anaphrenium paniculosum (Sond.) Engl. in A. & C. DC., Mon. Phan. **4**: 358 (1883). Syntypes as above.
Heeria paniculosa (Sond.) Kuntze, Rev. Gen. Pl. **1**: 152 (1891).—Engl., Pflanzenw. Afr. **3,** 2: 195 (1921).—Burtt Davy, F.P.F.T. **2**: 511 (1932) (" paniculata ").—O. B. Mill. in Journ. S. Afr. Bot. **18**: 46 (1952) (" paniculata "). Syntypes as above.

Much-branched shrub 0·6–3 m. tall, or small tree; branches brownish-red, cylindric, shortly pilose or puberulous when young, later glabrescent, densely leafy. Leaves alternate to subverticillate; petiole generally short (0·3–1·2 cm.), dorsally convex, slightly canaliculate on the upper surface, densely puberulous to glabrescent; lamina concolorous to very discolorous, 2–12·5 × 0·6–3·8(5) cm., oblong or elliptic, usually obtuse or truncate, sometimes acute, and mucronate at the apex, a little narrowed towards the base, with somewhat thickened and ± undulate margin, from membranous to subcoriaceous, sericeous or with hairs only on the nerves above, also sericeous on the under surface (covered by short appressed whitish or very pale yellowish hairs); midrib impressed to raised above and very prominent below, covered by hairs like those of the lamina to glabrescent; lateral nerves very slender, impressed above, somewhat raised and very conspicuous below; reticulation not or barely visible. Panicles axillary and terminal, usually lax, from few- to many-flowered, with axis and branches puberulous. Calyx-segments 2–2·25 mm. long, lanceolate-triangular, acute, puberulous outside. Petals 3 × 1·5 mm., obtuse, rounded and flat at the apex. Drupe black, somewhat shining, up to 7 × 11 mm., transversely ellipsoid, compressed.

Leaves discolorous, dark when dried, sparsely puberulous or only puberulous on the nerves or glabrous above - - - - - - - - - - - - var. *paniculosa*
Leaves concolorous or only slightly discolorous, sericeous on both surfaces or densely puberulous above and sericeous beneath - - - - - - - var. *salicina*

Var. **paniculosa**
Heeria insignis var. *latifolia* sensu Engl., Pflanzenw. Afr. **3,** 2: 197 (1921) pro parte quoad specim. *Schlechter* 11931.

Bechuanaland Prot. SE: E. Mahalapye, ♂ fl. 19.i.1960, *Yalala* 104* (COI; SRGH). **Mozambique.** LM: Sábiè, Ressano Garcia, ♀ fl. & fr. 25.xii.1897, *Schlechter* 11931 (BM; BOL; BR; COI; K; P; W).
Also in the Transvaal, Swaziland, Natal and Cape Prov. Savanna woodlands, in rocky places.

The specimen " Bechuanaland Prot., Mahalapye, Morale Pas. Station, fr. 16.iii.1961,

* Transitional to *O. hereroensis* (Schinz) R. & A. Fernandes.

Yalala 132" (COI; SRGH) is intermediate between *O. paniculosa* var. *paniculosa* and *O. reticulata* var. *reticulata*. We therefore presume that it is a hybrid.

Var. **salicina** (Sond.) R. & A. Fernandes in Bol. Soc. Brot., Sér. 2, **38**: 167, t. 22 (1965). Type from the Transvaal, Magalisberg, *Burke* (K).

Rhus salicina Sond. in Harv. & Sond., F.C. **1**: 522 (1860). Type as above.

Anaphrenium abyssinicum var. *salicinum* (Sond.) Engl. in A. & C. DC., Mon. Phan. **4**: 358 (1883). Type as above.

Heeria salicina (Sond.) Burtt Davy, F.P.F.T. **2**: 511 (1932).—O. B. Mill. in Journ. S. Afr. Bot. **18**: 46 (1952). Type as above.

Heeria insignis sensu O. B. Mill., loc. cit.

Bechuanaland Prot. SW: Tshabong, ♀ fl. 25.ii.1960, *Wild* 5149 (K; SRGH); Tshabong on Cape border, fr. 22.ii.1960, *de Winter* 7484 (K; M; SRGH). SE: 61 km. N. of Gaberones, ♂ fl. 19.i.1960, *Leach & Noel* 234 (COI; SRGH); near Kanye, ♀ fl. xi.1940, *Miller* B/235 (FHO).

Also in the Transvaal and SW. Africa. Open savanna woodlands in rocky and sandy places.

9. **Ozoroa nigricans** (Van der Veken) R. & A. Fernandes in Bol. Soc. Brot., Sér. 2, **38**: 160 (1965). Type from Katanga.

Heeria nigricans Van der Veken in Bull. Jard. Bot. Brux. **29**: 232 (1959); F.C.B. **9**: 18 (1960). Type as above.

Var. **nigricans**

Heeria sp. 2.—White, F.F.N.R.: 209 (1962).

A shrub (?) or a many-stemmed suffrutex; stems from a woody rootstock, virgate, unbranched or nearly so, up to 60 cm. long, subterete and glabrescent below, ± angular and pale-yellowish-pubescent to densely fulvous-lanate in the upper part. Leaves alternate, opposite or sub-3-verticillate; petiole 0·3–1·7 cm. long, robust, grooved, densely fulvous-villous-lanate to tomentose; lamina very discolorous, dark brown to blackish, dull or somewhat shining and glabrous above, whitish or fulvous-tomentose (densely covered by ± long ± crisped or straight hairs) below, (3·5)6–14 × (1·7)3–7·5 cm., broadly elliptic to oblong-elliptic or ovate, rounded or obtuse or subacute and mucronate at the apex, not thickened at the margin, rounded or a little cuneate at the base, membranous to papyraceous; midrib impressed in the upper surface, very prominent below; lateral nerves 2–7 mm. apart, straight or somewhat arcuate near the margin, at an angle of 65°–75° to the midrib, impressed above, raised below; tertiary nerves sometimes a little impressed above. Inflorescences terminal and axillary, the latter nearly sessile and glomerate or up to 3 cm. long and shortly pedunculate; peduncle, axis, branches and pedicels (0·3–1·5 mm. long) densely fulvous-tomentose. Calyx-segments 2–2·5 × 1·5–1·7 mm., ovate-triangular, acute, tomentose outside. Petals 3–4 mm. long, oblong, obtuse, dorsally appressed-pilose, inflexed at the apex. Drupes black, shining, 4–5 × 6–7 mm., transversely ellipsoid.

N. Rhodesia. S: Mumbwa, Nambala Mission, ♂ fl. 16.ix.1947, *Brenan* 7867 (EA; FHO; K); 48 km. W. of Kafue Hook pontoon on road to Mankoya, ♀ fl. & fr. 7.xi.1959, *Drummond & Cookson* 6207 (SRGH); near Mumbwa, ♂ fl. & fr. 1911, *Macaulay* 769 (K).

Also in Katanga. In swamps and *Brachystegia* woodlands.

Var. *elongata* (Van der Veken) R. & A. Fernandes is confined to Katanga and is distinguished by longer petioles and more numerous lateral nerves.

10. **Ozoroa bredoi** R. & A. Fernandes in Bol. Soc. Brot., Sér. 2, **38**: 151, t. 9 (1965). Type: N. Rhodesia, between Kunfundu and Chienge, *Bredo* 3941 (BR, holotype).

Suffrutex; branches cylindric in the lower part, ± striate in the upper portion, with a double indumentum consisting of rather short whitish appressed hairs and stiff reddish patent ± curled long ones. Leaves alternate or sub-3-verticillate; petiole 1–3 mm. long, stout, densely hairy; lamina dull green above, ochraceous beneath, 6–11 × 1–1·6 cm., linear, acute and mucronate at the apex, mucro up to 2 mm. long, with a ± thickened undulate margin, roundish to slightly cuneate at the base; upper surface with an indumentum of ± dense patent hairs, the under surface with a double indumentum of minute scales (always?) and long stiff hairs on the nerves and veins; midrib slightly impressed above, much raised below; lateral nerves 2·5–3·5 mm. apart, slightly raised above, very prominent below as also the reticulation. ♂ panicles long (up to 25 cm.) and narrow, with the axis with an indumentum like that of the stems; pedicels short; bracts up to 5 mm. long, linear. Calyx-segments c. 2·5 × 1·5 mm., hairy outside. Petals with an inflexed triangular apex. Stamens with anthers c. 1 mm. long. Disk cup-shaped. ♀ flowers and drupe unknown.

N. Rhodesia. N: between Kunfundu and Chienge, ♂ fl. 21.ii.1940, *Bredo* 3941 (BR). Known only from N. Rhodesia. Ecology unknown.

11. **Ozoroa kassneri** (Engl. & v. Brehm.) R. & A. Fernandes in Bol. Soc. Brot., Sér. 2, **38**: 157 (1965). Type from Katanga.
Heeria kassneri Engl. & v. Brehm. in Engl., Bot. Jahrb. **54**: 327 (1917).—Engl., Pflanzenw. Afr. **3**, 2: 195 (1921).—Van der Veken in F.C.B. **9**: 14, t. 2 (1960). Type as above.

A suffrutex up to 1·8 m. high (or shrub up to 3 m. tall?), with simple or ± branched stems, greyish to cinnamon-coloured, cylindric in the lower portion, ± angular in the upper part, puberulous to ± densely villous, usually densely leafy, growing from a rootstock. Leaves alternate to opposite or the uppermost ones subverticillate; petiole 0·2–0·8(1·2) cm. long, densely villous or puberulous to glabrescent; lamina usually not very discolorous, 5·5–19 × 2·5–9·5 cm., subcoriaceous to coriaceous, with the margin slightly thickened and revolute, frequently a little bullate on the upper face, dull and ± pubescent above, with a double indumentum below consisting of a layer of very short ± dense whitish appressed hairs on the lamina and another of longer stronger yellowish or whitish hairs mainly on the nerves and reticulation, patent to subappressed, either sparse and not concealing the short ones or denser and ± concealing them; midrib slightly prominent above, much raised below; lateral nerves 2–8 mm. apart, arcuate, forked near the margin, impressed to somewhat raised above, rather to very prominent below as is also the ± close venation. Inflorescences in the upper axils, shorter than the leaves, very compact, subsessile to ± pedunculate, or terminal and arranged in verticils, rarely all the lateral and terminal ones together forming a rather large leafy panicle; axis and branches angular, ± villous to glabrescent; bracts linear, villous; pedicels 0·3–1 mm. long. Calyx-segments 2–2·5 mm. long, ovate-triangular, puberulous. Petals creamy-white, 2·5–3·5 × 1–1·5 mm., oblong, with an inflexed triangular apex, externally appressed-hairy. Drupe black, shining, 5–7 × 6–8 × 3–4 mm., transversely ellipsoid, compressed.

Leaves broadly elliptic to obovate, not more than twice as long as broad, rounded or truncate at the apex, rounded or cuneate at the base - - - var. *kassneri*
Leaves elliptic to narrowly elliptic, more than twice as long as broad, narrowing to both ends, obtuse to acute - - - - - - - - - - var. *rhodesica*

Var. **kassneri**
Heeria argyrochrysea sensu White, F.F.N.R.: 208 (1962).
Heeria sp. 1.—White, op. cit.: 209 (1962) pro parte quoad specim. *White* 3085.

N. Rhodesia. N: near Kalambo R. above Kalambo Falls, ♂ fl. 29.iii.1955, *E. M. & W.* 1300 (LISC; SRGH); Samfya Mission, shore of Lake Bangweulu, fr. 19.viii.1952, *White* 3085 (BM; COI; FHO; K). C: mile 116 on the Lusaka-Fort Jameson road, ♂ fl. 16.iv.1952, *White* 2694 (BM; FHO; K).
Also in Congo. Woodlands on fixed sand-dunes and in rocky places.

Var. **rhodesica** R. & A. Fernandes in Bol. Soc. Brot., Sér. 2, **38**: 157, t. 13 (1965). Type: N. Rhodesia, Abercorn, Ndundu, *Richards* 15224 (K, holotype).
Heeria sp. 1.—White, F.F.N.R.: 209 (1962) pro parte quoad specim. *White* 3184.

N. Rhodesia. N: Fort Rosebery, near Samfya Mission, Lake Bangweulu, ♂ fl. 1.ix.1952, *White* 3184 (BM; FHO; K). C: Rufunsa, fr. 6.vi.1958, *Fanshawe* (FHO; K).
Known only from N. Rhodesia. In scrub or open woodlands, edge of riverine forests and escarpments on sandy rocky ground and on fixed sand-dunes.

12. **Ozoroa nitida** (Engl. & v. Brehm.) R. & A. Fernandes in Bol. Soc. Brot., Sér. 2, **38**: 160 (1965). Type: N. Rhodesia, Kankaso Stream, *Kassner* 2102 (K, lectotype; P).

Heeria nitida Engl. & v. Brehm. in Engl., Bot. Jahrb. **54**: 328 (1917).—Engl., Pflanzenw. Afr. **3**, 2: 196 (1921).—White, F.F.N.R.: 209 (1962). Type as above.

A many-stemmed shrub up to 1·20 m. high; stems woody, cylindric, thick, densely spreading, yellowish-hairy. Leaves alternate or rarely subopposite; petiole (0·6)1–2 cm. long, subterete, thick, densely yellowish-patent-hairy; lamina light green and usually shining above, paler and dull below, 7·5–20 × 4–12·3 cm., usually broadly elliptic or obovate or rarely oblong, obtuse, rounded, truncate or emarginate and mucronate at the apex, bullate towards the margin and with a much thickened, bone-coloured and undulate or crisped margin, rounded or somewhat acute at the base, very coriaceous, glabrous or with patent, weak, sparse hairs above, and with a double indumentum (very short appressed whitish hairs on the lamina and long patent yellowish hairs mainly on the nerves and reticulation, not concealing the short ones) below; midrib impressed above, very prominent beneath; lateral nerves (0·3)0·5–1·2 cm. apart, at an angle of 53°–75° with the midrib, forked near the margin, pale and prominulous above, rather prominent below, tertiary nerves very oblique; reticulation impressed above, rather prominent beneath. Panicles terminal and axillary, congested, shorter than the leaves, with the axis and branches densely villous as also the subulate bracts. Calyx-segments 1·5–2 mm. long, lanceolate-ovate, acute, villous outside. Petals cream, oblong, appressed-hairy. Drupe black, shining, 5·5–7 × 7–9 mm., compressed-depressed, not or only a little emarginate.

N. Rhodesia. B: Kabompo, ♂ fl. 26.xii.1952, *Holmes* 1033 (FHO; K). W: Luanshya, ♂ fl. 4.xi.1953, *Fanshawe* 478 (EA; K; SRGH). S: Choma, Siamambo Forest Reserve, ♀ fl. 13.xii.1952, *Angus* 927 A (FHO; K); Livingstone, Dambwa Forest Reserve, fr. 10.i.1952, *White* 1892 (FHO; K).
Known only from N. Rhodesia. Mixed woodlands and edges of dambos (swamps).

13. **Ozoroa marginata** (Van der Veken) R. & A. Fernandes in Bol. Soc. Brot., Sér. 2, **38**: 159 (1965). Type from Katanga.
Heeria marginata Van der Veken in Bull. Jard. Bot. Brux. **29**: 231 (1959); F.C.B. **9**: 22 (1960). Type as above.
Heeria verticillata sensu White, F.F.N.R.: 209 (1962).

Suffrutex with stems arising from a fleshy to woody rootstock; stems brown or reddish-brown, up to 0·3–0·9 m. tall, cylindric in the lower part, subangulate-furrowed in the upper part, striate, sparsely to ± densely hairy, the hairs appressed to ± patent, short, weak, whitish to yellowish. Leaves alternate or opposite to verticillate, upright; petiole yellowish, pubescent, 1–8 mm. long, furrowed; lamina discolorous, dull or somewhat shining, olive-green or reddish, (4·2)7–14 × 2–6 cm., oblong-elliptic or oblong-obovate, rounded or retuse and mucronate at the apex, with ± thickened and ± undulate margin, narrowed at the base, subcoriaceous to coriaceous, glabrous or minutely puberulous above, ± whitish-sericeous beneath (densely covered on the lamina by white minute appressed hairs either single or mixed, mainly on the lateral nerves, with sparser longer yellowish ones); midrib not or only slightly prominent above, much raised and minutely pubescent or with rather long hairs beneath; lateral nerves 3–5(7) mm. apart, prominulous and lighter than the lamina above, very prominent below, somewhat oblique to the midrib and arcuate; reticulation not visible. Inflorescences terminal and axillary, 1–2 cm. long, congested, with appressed-minutely-pilose axis and branches. ♂ flowers: calyx-segments c. 2 mm. long, ovate-lanceolate, acute, appressed-pilose outside; petals whitish, c. 3 mm. long, oblong, obtuse, densely appressed-pilose outside. ♀ flowers: pedicels up to 1 mm. long; calyx-segments whitish, 2–2·5 mm. long, ovate-lanceolate, acute, pilose; petals oblong, obtuse, externally appressed-pilose, with inflexed apex. Drupe black, shining, 7–10 × 9–12 mm., transversely ellipsoid.

N. Rhodesia. N: Abercorn, fr. iii.1930, *Miller* 60/30 (FHO). W: Solwezi Distr., Mutanda Bridge, fr. 10.vii.1930, *Milne-Redhead* 695 (K). C: 105 km. NE. of Serenje Corner, fr. 16.vii.1950, *Hutchinson & Gillett* 3753 (K). S: Chilila, ♂ fl. 11.ii.1955, *Richards* 4452 (K).
Also in Katanga. Woodlands.

14*. **Ozoroa viridis** R. & A. Fernandes in Bol. Soc. Brot., Sér. 2, **38**: 181, t. 40 (1965). Type: N. Rhodesia, Mkushi Boma, *Angus* 2532 (FHO, holotype).

Suffrutex up to 2·1 m. tall; stems simple, brownish, cylindric in the lower, striate in the upper portion, sparsely lenticellate, densely and shortly appressed-hairy. Leaves alternate or subopposite; petiole 10–12 mm. long, canaliculate above, densely and shortly appressed-hairy; lamina discolorous (green and shortly appressed-hairy above, whitish-sericeous with short appressed hairs beneath), 4·5–8 × 2·5–4·5 cm., elliptic or obovate, roundish or emarginate, sometimes mucronate at the apex, thickened at the yellowish almost flat margin, cuneate at the base; midrib yellowish, raised above nearly to the apex, very prominent below; lateral nerves raised above, very prominent beneath, shortly appressed-hairy, somewhat arcuate, 2–3·5 mm. apart; reticulation raised above and ± concealed by the indumentum beneath. ♂ panicle 20 × 7 cm., leafy at the base, with the axis sulcate and shortly appressed-hairy; bracts up to 2 mm. long, linear, dorsally hairy; pedicels up to 2 mm. long. Calyx-segments 2–2·75 × 1–1·75 mm., ovate to broadly ovate, dorsally hairy. Petals c. 3 × 1·5 mm., broadly elliptic, with a triangular inflexed apex, dorsally appressed-hairy. Filaments 1·5 mm. long; anthers c. 1 mm. long. Disk c. 1·5 mm. in diam., crenate, with a vestigial 3-lobulate gynoecium.

N. Rhodesia. C: Mkushi Boma, ♂ fl. 27.iii.1961, *Angus* 2532 (FHO).

Fanshawe 1369 (BR; K), from Ndola, differs from this species mainly by the longer and ferruginous hairs of the stem, panicle-axis and petiole; petiole shorter (up to 4 mm.) and stouter; lamina not cuneate at the base; longer hairs in the under surface and reticulation of veins more conspicuous below. We think, therefore, that it may be a hybrid between *O. kassneri* and this species.

15. **Ozoroa pwetoensis** (Van der Veken) R. & A. Fernandes in Bol. Soc. Brot., Sér. 2, **38**: 168, t. 3, 4 (1965). Type from Katanga.

Heeria pwetoensis Van der Veken in Bull. Jard. Bot. Brux. **29**: 235 (1959); F.C.B. **9**: 26 (1960). Type as above.

Suffrutex up to 1·5 m. tall with woody rootstock and creeping roots; stems sparsely branched, ± dark brown, terete, sometimes angulate at the upper part, striate, ± pilose to glabrescent. Leaves alternate or opposite or the uppermost ones subverticillate; petiole 0·1–1·2 cm. long, grooved above, ± densely and subappressed-fulvous-hairy; lamina discolorous, green or reddish or pale to dark brown and glabrous above, silvery- or pale-fulvous-sericeous beneath (densely covered with appressed whitish or yellowish usually somewhat long hairs but sometimes short, not so dense on the midrib), (5·5)8–18 × 1·5–5·5 cm., elliptic or oblong to oblong-lanceolate or lanceolate, acute or subobtuse and mucronate at the apex, acute or subobtuse at the base, subcoriaceous to coriaceous; midrib impressed above, much raised and yellowish beneath; lateral nerves slender, straight, almost at right angles to the midrib or rather oblique, forked near the margin, 2–3 mm. apart, slightly raised on the upper surface, ± prominent beneath; tertiary nerves and veins ± raised below. Inflorescences congested, axillary in the uppermost leaf-axils, shorter than the leaves, and also terminal, arranged in whorls, with the axis and the very short branches subappressed-fulvous-hairy. Calyx-segments 1·8–2·5 mm. long, ovate-triangular, externally puberulous. Petals whitish, 3–3·5 mm. long, oblong, inflexed at the apex, dorsally appressed-hairy. Drupe, black, shining, 5–6 × 8–9 mm., transversely ellipsoid.

Lamina not shining above:
- Lateral nerves at an acute angle to the midrib; lamina (5·5)8–14 × (1)1·5–5·5 cm., cuneate to subobtuse at the base - - - - - - - var. *pwetoensis*
- Lateral nerves almost at right angles to the midrib; lamina narrower, 6–18 × 0·5–1(2·2) cm., obtuse at the base - - - - - - - var. *angustifolia*

Lamina somewhat shining above: - - - - - - - - var. *nitidula*

* 14a. **Ozoroa macrophylla** R. & A. Fernandes, in Bol. Soc. Brot., Sér. 2, **39**: 248 (1965).

Ad *O. viridem* indumento, inflorescentia et floribus accedit sed ab ea foliis subsessilibus, lamina ovata usque ad 16·5 × 9 cm., apice obtusa vel acutiuscula, basi rotundata, nervis lateralibus inter sese plus distantibus et reticulo paginae superioris minus conspicuo differt.

N. Rhodesia. N: Chishinga Ranch, near Luwingu, fl. ♂ et fr. immat. 15.v.1961, *Astle* 630 (K, holotype; SRGH).

Var. **pwetoensis**

N. Rhodesia. N: 49·5 km. N. of Isoka, ♂ fl. 31.iii.1961, *Angus* 2613 (FHO; K; SRGH). **Nyasaland.** N: Fort Hill, ♀ fl. & fr. vii.1896, *Whyte* (K).
Also in Katanga. In *Brachystegia* woodlands and among rocks.

Var. **angustifolia** R. & A. Fernandes in Bol. Soc. Brot., Sér. 2, **38**: 168, t. 23–24 (1965). Type: N. Rhodesia, Katete, St. Francis' Hospital, *Wright* 164 (K, holotype).
Heeria homblei sensu White, F.F.N.R.: 208 (1962) pro parte quoad specim. *White* 2442.

N. Rhodesia. E: 11 km. E. of Katete, fr. 8.x.1958, *Robson* 18 (K; LISC); Fort Jameson, ♂ fl. 21.iv.1952, *White* 2442 (FHO; K). **Mozambique.** T: Macanga, near Furancungo, fr. 13.vii.1949, *Barbosa & Carvalho* in *Barbosa* 3591 (K; LISC; LMJ).
In *Julbernardia paniculata* and *Brachystegia* woodlands.

Var. **nitidula** R. & A. Fernandes in Bol. Soc. Brot., Sér. 2, **38**: 169, t. 25 (1965). Type: N. Rhodesia, Chati Forest Reserve, *Fanshawe* 3360 (BR; K, holotype; SRGH).
Heeria homblei sensu White, F.F.N.R.: 208 (1962) pro parte quoad specim. *Fanshawe* 3360 et *Holmes* 717.

N. Rhodesia. N: Chati Forest Reserve, ♂ fl. & fr. 17.vii.1957, *Fanshawe* 3360 (BR; K; SRGH). W: Ndola, Sosu Hills, ♀ fl. & fr. 1.v.1952, *Holmes* 717 (FHO; K), ♂ fl., 718 (FHO; K).
Known only from N. Rhodesia. Woodlands and in crevices of rocky outcrops.

17. **Ozoroa homblei** (De Wild) R. & A. Fernandes in Bol. Soc. Brot., Sér. 2, **38**: 156 (1965). Type from Katanga.
Heeria homblei De Wild. in Fedde Repert. **13**: 107 (1914); in Ann. Soc. Sci. Brux. **38,** 2: 385 (1914).—Meikle in Exell & Mendonça, C.F.A. **2,** 1: 117 (1954).—Van der Veken, F.C.B. **9**: 20 (1960). Type as above.

Suffrutex with annual shoots up to 2 m. tall arising from a woody rootstock; stems and branches red-brown to dark-brown, cylindric and glabrous on the lower part, subangular and ± shortly puberulous or glabrescent towards the apex, striate, rather densely leafy. Leaves mostly subverticillate; petiole 0·2–0·5 cm., grooved on the upper face, convex beneath, puberulous or minutely appressed-hairy to glabrescent; lamina very discolorous, cinnamon-coloured or greenish-brown to dark brown, shining and glabrous above, ± whitish beneath (densely covered with minute and appressed hairs), (3)5–16 × (0·3)0·8–2 cm., linear to oblong-lanceolate, frequently infolded at the midrib, attenuate, very acute and mucronate at the apex, with thickened pale undulate or flat margin, ± acute at the base, subcoriaceous to coriaceous, very rigid and brittle when dry; midrib impressed above, very prominent and ± sparsely appressed-hairy beneath; lateral nerves 1–1·5 mm. apart, slightly prominent above, a little more so below, usually not forked until the margin; tertiary nerves and veins not visible. Inflorescences terminal and subverticillate and axillary, the latter generally shorter than the leaves, up to 3 cm. long, on peduncles 0·5–9 cm. long, the axis and branches striate, minutely appressed-hairy. Flowers shortly pedicellate. Calyx-segments 1·6–2·2 × 0·7–1·3 mm., oblong-ovate to lanceolate, acute, externally appressed-hairy. Petals whitish, c. 3 × 1 mm., oblong-lanceolate, externally appressed-hairy, with a triangular inflexed apex. Drupe black, shining, 6 × 7 mm., broadly obovate, compressed, a little asymmetric, retuse or emarginate, wrinkled.

N. Rhodesia. W: Namulilo, Kamwedzi R., fr. 20.vi.1953, *Fanshawe* 102 (K).
Also in Angola and Katanga. In savannas, woodland and riverine forests (Van der Veken).

18. **Ozoroa stenophylla** (Engl. & Gilg) R. & A. Fernandes in Bol. Soc. Brot., Sér. 2, **38**: 180 (1965). Type from Angola.
Heeria stenophylla Engl. & Gilg in Warb., Kunene-Samb.-Exped. Baum: 287 (1903).—Meikle in Exell & Mendonça, C.F.A. **2,** 1: 117 (1954).—White, F.F.N.R.: 208 (1962). Type as above.
Anaphrenium kienerae Sacleux in Bull. Mus. Hist. Nat. Par., Sér. 2, **6**: 454 (1934). Type: " Haut-Zambèse ", *Kiener* (P).

Many-stemmed suffrutex; stems unbranched, brownish, tufted from a woody underground rootstock, up to 1 m. tall, cylindric, striate, puberulous. Leaves in whorls of 5–10 or subverticillate, the basal ones alternate; petiole 1–6 mm. long;

lamina discolorous, glabrous, almost black (when dried) and shining above and silvery-sericeous below (covered by short appressed whitish hairs), 2·5–16 × 0·1–0·2 cm., acicular, upright, the floral ones somewhat falcate, very rigid, acute and mucronate, with the margins revolute; midrib impressed on the upper surface yellowish-appressed-pubescent, and very prominent beneath; lateral nerves and reticulation not visible on either surface. Inflorescences up to 3 cm. long, in the upper leaf-axils, the uppermost ones verticillate and leafless, with puberulous axis and branches. Calyx-segments triangular, slightly puberulous. Petals c. 2·5 × 1 mm., oblong, inflexed at the apex, sericeous outside. Filaments 1·5 mm. long, anthers c. 0·75 mm. long. Drupe black, shining, 7·5 × 8 mm., broadly obovate, truncate, much compressed.

N. Rhodesia. B: Kalabo, Sikongo Forest Reserve, ♂ fl. 14.ii.1952, *White* 2076 (BR; FHO; K), ♂ fl. & fr. 15.ii.1952, *White* 2078 (FHO; K).

Also in Angola. In open *Burkea africana* or *Baikiaea plurijuga-Burkea africana* woodlands on Kalahari Sand.

MATERIAL INSUFFICIENT

Ozoroa sp. A. aff. kassneri

Shrublet 60 cm. high; branches erect, simple, densely fulvous-hairy with a double indumentum consisting of a layer of short whitish appressed hairs and another of fulvous ± crisped and ± patent long ones. Leaves opposite or 3-verticillate; petioles up to 5 mm. long, stout, densely fulvous-hairy like the stems; lamina discolorous (dark when young and brownish slightly shining later above, greenish beneath), up to 10 × 3 cm., narrowly ovate, acute at the apex, with a thickened hairy recurved margin, cuneate at the base, upper surface with hairs only on the midrib and the lateral nerves, the under surface with a double indumentum of short whitish appressed hairs and long ± patent fulvous ones distributed mainly along the midrib, lateral nerves and veins; midrib impressed above, very prominent below and densely fulvous-tomentose; lateral nerves impressed above, raised beneath; reticulation not visible above, prominulous beneath. Flowers not seen. Drupe dark, shining, 6 × 9 × 4 mm.

N. Rhodesia. N: Mporokoso, 8 km. on the road from Chienge to Kafulwe, fr. 5.xii.1952, *Angus* 719 B (K).

In *Brachystegia* woodland and rocky hills of Luapula porphyry.

Ozoroa sp. B.

Small tree 3 m. tall; branches greyish, cylindric, glabrous, sparsely lenticellate; branchlets dark brown, puberulous. Leaves alternate or opposite or sub-3-verticillate; petiole 7–11 mm. long, slightly grooved above, convex below, densely puberulous; lamina discolorous when dried (dark brownish above, yellowish-green below), 3·5–11 × 1·5–5 cm., elliptic to oblong-elliptic, obtuse or roundish at the apex, with the margin somewhat thickened and slightly undulate, cuneate at the base, membranous; upper surface sparsely hairy, under surface with a double indumentum of short appressed hairs (usually disappearing in the old leaves) and of stiff patent long ones mainly on the nerves and veins; midrib impressed above, very prominent beneath; lateral nerves ascendent, very slightly raised above, prominent below, (1·5)2–4 mm. apart; reticulation not or barely visible above, visible beneath. Flowers and fruits unknown.

N. Rhodesia. N: Abercorn, Ulungu, near Mbwilo village, st. 22–xi–1948, *Glover* in *Bredo* 6344 (BR).

Ecology unknown.

10. RHUS L.

Rhus L., Sp. Pl. **1**: 265 (1753); Gen. Pl. ed. 5: 129 (1754).

Shrublets, suffrutices, shrubs or trees. Leaves alternate, simple, 3-foliolate or imparipinnate, rarely digitately 5-foliolate. Panicles terminal, axillary or both, ± branched. Flowers unisexual, exceptionally bisexual, very small. ♂ flowers: calyx (4)5(6)-partite, the segments imbricate; petals (4)5(6), longer than the calyx, imbricate; stamens 5; filaments subulate, inserted below the disk; anthers ovate, dorsifixed, introrse; disk patelliform or cupuliform; pistillode usually absent. ♀ flowers: perianth similar to that of the ♂; staminodes frequently present; ovary ovoid to subglobose, usually 1-locular, with the ovule pendent from an ascendent funicle inserted at the base of the loculus; styles 3, apical, free or connate at the base; stigmas ± capitate. Drupe globose or ovoid and compressed, frequently asymmetrical, glabrous or hairy; mesocarp fleshy, ± resinous, sometimes ± dry; endocarp bony or crustaceous. Seed ovoid or reniform, compressed, with a thin testa; cotyledons very compressed.

All the African species (c. 130), belong to Sect. *Gerontogeae* Engl., characterized by 3-foliolate leaves (5-digitate in *R. montana* Diels) and, usually, unisexual flowers with a 5-merous perianth.

Leaflets lanate or tomentose below:
- Leaflets densely felted-tomentose below, glabrous or glabrescent above, the median one petiolulate; lateral nerves and reticulation barely visible on the upper surface; flowers not glomerate; drupe tomentose; shrub or tree - - - 1. *tomentosa*
- Leaflets ± densely lanate below, glabrescent above, the median one not petiolulate; lateral nerves and reticulation much raised and mostly visible; flowers in glomerules; drupe glabrous; suffrutex or shrub - - - - - - 2. *kirkii*

Leaflets neither lanate nor tomentose below:
- Plants with petioles ± winged at least towards the apex, usually short (rarely above 2 cm. long); leaflets broadly obovate, obcordate or spathulate, cuneate at the base, subcoriaceous to coriaceous, frequently covered with a shining layer of resin; reticulation inconspicuous or barely visible; branchlets generally angular; panicles small; unarmed shrubs:
  - Leaflets shallowly crenate in the upper part, the median one up to 3 × 1·5 cm. broadly obovate, roundish or truncate at the apex, pale green (when dried) branchlets puberulous; shrublet c. 20 cm. tall - - - - 3. *wildii*
  - Leaflets not crenate:
    - Leaflets ± broadly obcordate or obovate and emarginate at the apex, glaucous or greyish at least above, somewhat undulate between the lateral nerves; branchlets and panicles glabrous - - - - - - - - 4. *glauca*
    - Leaflets spathulate to obovate-cuneate, obtuse or emarginate at the apex, frequently reddish-brown (when dried), flat between the lateral nerves; branchlets and panicles shortly puberulous - - - - - - - - 5. *lucida*
- Plants not uniting the above characters:
  - Leaflets broadly obovate to obovate-cuneate, generally with (1)2–4(5) mucronate teeth on each side of the upper part, ± membranous, hirsute mainly on the nerves (rarely glabrous), with 6–7 slender nerves on each side; reticulation inconspicuous or barely visible; median leaflet 1·7–8(9) × 0·9–5 cm., the lateral ones much smaller and making right or obtuse angles with it; branches and petioles hirsute to glabrous - - - - - - - 6. *dentata*
  - Leaflets not uniting the above characters:
    - Young branches greyish or whitish; leaflets glabrous or with ± sparse hairs, obtuse, subobtuse or emarginate at the apex, with inconspicuous reticulation; shrubs frequently thorny:
      - Leaflets up to 1·5 cm. in breadth (generally less), oblong or oblong-elliptic, the median one c. 3·5–4 times longer than broad; shrub usually with thorny lateral branches - - - - - - 7. *gueinzii* var. *spinescens*
      - Leaflets relatively shorter and broader, the median one up to c. 2·5 times longer than broad:
        - Leaflets glabrous, (1)1·8–2·3(3·5) cm. broad, entire or crenate at the margin, the median one elliptic to obovate, rounded, obtuse or emarginate at the apex - - - - - - - - - 8. *natalensis*
        - Leaflets hairy:
          - Leaflets elliptic or ovate-elliptic, narrowed at both ends, obtuse or acute at the apex, the median one 6–11·5 × 3·2–4·3 cm.; petiole 1·3–3·7 cm. long - - - - - - 11. *tenuinervis* var. *meikleana*
          - Leaflets smaller, the median one rather cuneate at the base and rounded or truncate at the apex:
            - Leaflets 2–4·5 × 1–1·8 cm., entire or crenulate around the upper part, subcoriaceous, nearly concolorous - - - - 9. *marlothii*
            - Leaflets 1·8–3·3 × 0·4–1·4 cm., entire or 3-crenate at the apex, membranous to papyraceous, somewhat discolorous - 10. *pentheri*
    - Young branches reddish to ± dark-brown or chocolate-coloured, if whitish, then leaflets not as above:
      - Leaflets (at least some of them) more than 3 times longer than broad:
        - Margin of the leaflets erose-dentate or serrate-dentate:

Leaflets coriaceous, usually linear, with the margin erose-dentate, the median one 6–12 × 0·2–0·9 cm.; petiole rather thick, up to 3 cm. long - 13. *erosa*
Leaflets membranous to papyraceous, with the margin serrate-dentate, the median one up to 12 × 1·5(2) cm.; petiole very slender, up to 5 cm. long - - - - - - - - 14. *tenuipes*
Margin of the leaflets entire or (rarely) shallowly dentate:
Median leaflets lanceolate to linear-lanceolate, up to 0·8 cm. broad, entire:
Leaflets up to 13 × 0·6 cm., whitish-furfuraceous on both surfaces, very coriaceous, not hairy - - - - - 15. *trifoliolata*
Leaflets up to 5 × 0·8 cm., concolorous, usually acute and mucronulate, papyraceous, sometimes covered with resin, hispid on the lamina and ciliate along the margin - - - - - - - 16. *ciliata*
Median leaflets usually broader than 0·8 cm.:
Leaflets linear-lanceolate, lanceolate or oblong-lanceolate:
Petiole up to 10·5 cm. long, rather more than half the length of the median leaflet; leaflets lanceolate, acuminate-cuspidate, with subfiliform apex and entire whitish ciliate margin, subconcolorous (brownish-red when dried) - - - 17. *acuminatissima*
Petiole shorter, usually less than half the length of the median leaflet; leaflets not so acute, discolorous:
Leaflets lanceolate to linear-lanceolate, (3)6–15(20) times longer than broad, the median one 7–24·5 × 0·7–2·8 cm., entire, dark green above and pale or yellowish-green beneath; reticulation easily visible above - - - - - - - 18. *lancea*
Leaflets lanceolate to oblong-lanceolate, 3–9 times longer than broad, the median one 4–13·5 × 0·6–4·3 cm., entire or shallowly dentate, dark-brownish above (when dried) and paler below; reticulation not or slightly visible on both surfaces - 19. *leptodictya*
Leaflets elliptic or oblanceolate to spathulate, the median one 3–7(9) × 1·5–2·5 cm., acute or obtuse - - - - 20. *oblanceolata*
Leaflets relatively shorter and broader than as above:
Flowers, at least the ♂ ones, glomerate:
Petiole somewhat winged at the apex; branches and petioles glabrous; leaflets obovate to obovate-lanceolate, subacute to subobtuse, entire, glabrous or nearly so; lateral nerves and reticulation very prominent below; suffrutex or shrub - - - - - 2. *kirkii*
Petiole not winged; branches and petioles ± pubescent; leaflets entire or with a few teeth, usually pubescent; reticulation of the adult leaves visible, either raised or not; shrubs or small trees:
Leaflets from lanceolate and ± acute to rhombic and acute or subobtuse, sericeous-pubescent or subappressed-hairy 21. *pyroides* var. *gracilis*
Leaflets ovate to obovate, rounded to subobtuse at the apex, cuneate to subpetiolulate at the base; lateral and tertiary nerves impressed above and very prominent below, sometimes concealed by the indumentum; hairs of the leaves and petioles ± patent - - - - 22. *vulgaris*
Flowers not glomerate:
Leaflets generally crenate or sublobulate at least towards the apex, subcoriaceous to coriaceous; drupe compressed, subquadrate in outline:
Leaflets 2–3 times longer than broad, ovate to elliptic, dentate-crenate, the median one up to 11·5 × 4·3 cm. 11. *tenuinervis* var. *meikleana*
Leaflets relatively shorter and broader, obtuse, rounded or truncate, sometimes subacute at the apex:
Leaflets crenate, duplicate-crenate or sublobulate in the upper 2/3, broadly obovate to oblong-obovate, rarely ovate or broadly elliptic, ± pubescent on both surfaces, the median one 1·5–6(10) × 1–4(7) cm.; branches and petioles ± pubescent
11. *tenuinervis* var. *tenuinervis*
Leaflets very shallowly crenate or entire, subcircular to oblong-obovate, glabrous or nearly so, the median one (2·5)4·5–8·5 × 1·5–5 cm.; branches and petioles glabrous or with sparse hairs
12. *lucens*
Leaflets entire or with only a few crenations or teeth; drupe subglobose or, if compressed, circular in outline:
Reticulation of the adult leaves not or only slightly visible:
Leaflets acute, broader at or below the middle:
Leaflets mostly large, the median one up to 16 × 7 cm., elliptic or ovate or ovate-elliptic, dark green to reddish-brown (when dried), not furfuraceous below - - - - - - 23. *chirindensis*

Leaflets smaller, the median one up to 8·5 × 4 cm., narrowly elliptic to subrhombic, dark green above and yellowish- or orange-furfuraceous below mainly when young - - 25. *quartiniana*

Leaflets obtuse to truncate, broader above the middle:

Leaflets sparsely pubescent, entire or 3-crenate at the apex, membranous, the median one up to 3·3 × 1·4 cm., obovate-cuneate; lateral nerves indistinct or barely visible; drupe compressed 10. *pentheri*

Leaflets ± pubescent mainly beneath; lateral nerves distinctly raised beneath; drupe globose:

Leaflets subconcolorous, the median one obovate-cuneate, much narrowed to the base, truncate and with a few teeth at the apex; lateral nerves about 6–9 pairs, at an angle of c. 45° to the midrib - - - 26. *rehmanniana* var. *longecuneata*

Leaflets discolorous, obovate-cuneate, not so narrowed to the base, rounded at the apex, entire, the median one up to 3·3 × 1·5 cm.; lateral nerves 4–6 pairs, slightly impressed above, at a less acute angle to the midrib - - - - - 27. *refracta*

Reticulation of the adult leaves distinctly visible, either raised or not:

Reticulation close:

Leaflets small, obovate, the median one usually less than 4·5 × 1·5 cm.:

Leaflets not hairy, densely whitish-furfuraceous on both surfaces, very coriaceous; compact shrublet - 28. *magalismontana*

Leaflets sparsely ciliate along the nerves and margin, furfuraceous-ferruginous-glandular on both surfaces; laxly branched shrub 29. *milleri*

Leaflets larger:

Leaflets ± furfuraceous-glandular or lepidote below even when adult:

Leaflets glabrous or nearly so even when young, reddish- or whitish-lepidote below; petioles narrowly winged at the apex - - - - 30. *fanshawei*

Leaflets ± densely pubescent, the median one ovate or obovate, up to 11 × 5 cm., ochraceous-lepidote beneath, coriaceous; petiole not winged - - - 31. *ochracea*

Leaflets (adult) not furfuraceous-glandular or lepidote below:

Leaflets contracted into a very acute acumen, somewhat undulate between the lateral nerves, always entire, with an often folded and falcate apex and revolute margin; petiole more than half as long as the median leaflet; branches, petioles and leaflets glabrous or with sparse hairs 24. *monticola*

Leaflets not contracted into an acumen, acute or obtuse; panicles ample, the terminal one much longer than the leaves:

Petioles up to 1·5 cm. long; leaflets ovate or obovate-cuneate, obtuse to subobtuse, the median one up to 4·5 × 2·2 cm. 32. *macowanii*

Petioles longer than 1·5 cm.; leaflets generally larger than above:

Leaflets papyraceous to subcoriaceous, the median one rhombic, acute to subobtuse, not or indistinctly petiolulate; calyx-segments ± hispidulous - 33. *anchietae*

Leaflets coriaceous, glabrous, the median one up to 10 × 5·5 cm., broadly elliptic or obovate to subcircular, rounded or obtuse at the apex, cuneate and narrowed into a distinct petiolule at the base; calyx-segments glabrous or ciliate on the margin - - - 34. *culminum*

Reticulation lax:

Panicles small, as long or only slightly longer than the leaves; undershrub with rhizomatose base and virgate branches; leaflets entire or with some cuspidate teeth at the apex, cuneate at the base, ± hispidulous as well as the branches and petioles; petioles rather shorter than a half of the median leaflet 35. *rogersii*

Panicles very ample, much longer than the leaves; shrubs or small trees:

Leaflets contracted into a long and acute acumen, the median one up to 16 × 7 cm., elliptic, ovate or ovate-elliptic, dark green to reddish-brown (when dried) - - 23. *chirindensis*

Leaflets not contracted into a long acute acumen:

Petals c. 0·75 mm. long; leaflets oblong to oblong-lanceolate, acute and mucronulate, papyraceous, densely to ± sparsely appressed-hairy, the median one 4–9(11) × 1·3–4 cm.; petiole 1·5–4 cm. long - - - - - - 36. *microcarpa*

Petals 1–1·5 mm. long; leaflets elliptic to obovate (not or only a little cuneate at the base), generally obtuse (rarely subacute or acute), papyraceous to subcoriaceous, the median one up to 12·7 × 7·5 cm; petiole 1·5–7·3 cm. long - 37. *longipes*

1. **Rhus tomentosa** L., Sp. Pl. **1**: 266 (1753).—Engl. in A. & C. DC., Mon.Phan. **4**: 407 (1883); Pflanzenw. Afr. **3,** 2: 202, fig. 97 A-D (1921).—Diels in Engl., Bot. Jahrb. **24**: 572, fig. 3 (1898).—Eyles in Trans. Roy. Soc. S. Afr. **5**: 403 (1916).—Schonl. in Bothalia, **3,** 1: 97, fig. pp. 98 et 99 (1930).—Burtt Davy, F.P.F.T. **2**: 510 (1932). Type from S. Africa (Cape Prov.).

Shrub or small tree up to 4·5 m. tall; branches cylindric or ± angular, glabrous to ± densely rufo-tomentose, the oldest dull-green, lenticellate. Petiole often reddish, 1·5–4 cm. long, slender, subterete, slightly canaliculate above, densely pilose. Leaflets very discolorous, greyish-green, glaucous or dull-green, sparsely pubescent with ± appressed hairs or, with age, glabrescent and ± shining above, densely and shortly whitish- or fulvous-tomentose below, with the margin somewhat revolute or flat, entire or with 1–2(3) teeth above the middle, rigidly coriaceous; midrib impressed or slightly raised towards the base on the upper surface, very prominent on the under surface; lateral nerves not or slightly visible on the upper surface, prominent beneath; reticulation not or scarcely visible above, concealed below; median leaflet (2·5)5–8·5 × (0·9)2–3·8 cm., lanceolate-elliptic to obovate-lanceolate, narrowing to both ends, acute and mucronate or sometimes obtuse at the apex, cuneate and ± shortly petiolulate. Panicles terminal, longer than the leaves, ± dense, much branched, with the axis, branches, pedicels and bracts densely pilose; pedicels c. 1–2 mm. long. Calyx-segments c. 0·75 mm. long, ovate, obtuse, dorsally pilose. Drupe 5–6 × 4 × 3 mm., subglobose, somewhat asymmetrical, densely greyish-tomentose.

S. Rhodesia. E: Inyanga, Pungwe Source, ♂ fl. 19.x.1946, *Wild* 1411 (K; SRGH); Inyanga, Troutbeck, fr. xi.1951, *Miller* B/1218 (K; PRE).

Also in Cape Prov., Transvaal and Natal. On rocky slopes of high mountains (1800–2490 m.) and in forests.

2. **Rhus kirkii** Oliv., F.T.A. **1**: 439 (1868).—Engl. in A. & C. DC., Mon. Phan. **4**: 426 (1893); Pflanzenw. Afr. **3,** 2: 211 (1921).—Burkill in Johnston, Brit. Centr. Afr., Suppl.: 241 (1897).—Meikle in Exell & Mendonça, C.F.A. **2,** 1: 108 (1954).—Van der Veken, F.C.B. **9**: 30 (1960).—White, F.F.N.R.: 213 (1962).—R. & A. Fernandes in Webbia, **19,** 2: 700 (1964). Type: S. Rhodesia, near Victoria Falls, *Kirk* (K, holotype).

Rhus welwitschii Engl. in A. & C. DC., Mon. Phan. **4**: 428 (1883).—Monro in Proc. & Trans. Rhod. Sci. Ass. **8**: 71 (1908). Type from Angola.

Rhus welwitschii var. *angustifoliolata* (" angustifoliola ") Bak. f. in Journ. of Bot. **37**: 429 (1899).—Eyles in Trans. Roy. Soc. S. Afr. **5**: 403 (1916). Type: S. Rhodesia, Bulawayo, *Rand* 97 (BM, holotype).

Rhus polyneura Engl. & Gilg in Warb., Kunene-Samb.-Exped. Baum: 288 (1903).—Meikle in Exell & Mendonça, C.F.A. **2,** 1: 102 (1954).—Van der Veken, F.C.B. **9**: 30 (1960).—White, F.F.N.R.: 213 (1962). Type from Angola.

Rhus polyneura var. *hylophila* Engl. & Gilg. in Warb., Kunene-Samb.-Exped. Baum: 289 (1903).—R. E. Fr. in Wiss. Ergebn. Schwed. Rhod.-Kongo-Exped. **1**: 127 (1914).—Engl., Pflanzenw. Afr. **3,** 2: 213 (1921). Type from Angola.

Rhus eylesii Hutch., Botanist in S. Afr.: 495 (1946). Type: S. Rhodesia, Salisbury Distr., *Eyles* 2042 (K, holotype).

Rhus discolor sensu Suesseng. in Proc. & Trans. Rhod. Sci. Ass. **43**: 35 (1951).

Suffrutex or shrub 0·3–1(3) m. high, with trailing woody roots and cylindric ± lanate or pubescent or glabrous stem and branches. Petiole (0·5)1–4 cm. long, semi-terete, broadly canaliculate on the upper side, slightly marginate or somewhat winged, ± densely covered with a yellowish- or ferruginous-lanate tomentum, or glabrescent or glabrous. Leaflets discolorous, green or brownish, coriaceous, entire, arachnoid-hairy, floccose-tomentose to glabrous above, ± densely yellowish- or ferruginous-lanate-tomentose with crisped hairs to glabrescent or glabrous below; midrib prominulous above, strongly prominent below; lateral nerves very slender and prominulous above, very prominent beneath; reticulation visible on both surfaces; median leaflet 5–12 × 1·6–4(5·5) cm., oblanceolate, oblong, elliptic or subrhombic, acuminate and very acute to subobtuse and apiculate at the apex, sessile, cuneate or much narrowed at the base; lateral leaflets 3–9·5 × 1·3–4 cm., oblanceolate to ovate, asymmetric, acute to obtuse and apiculate at the apex, cuneate or somewhat rounded at the base. Panicles terminal and axillary, generally with a long axis and short lateral branches with the flowers in glomerules; pedicels very short. ♂ flowers: calyx-segments c. 0·5 mm. long, lanate to glabrous, ovate, obtuse; petals whitish, or greenish or brownish, c. 1·3–1·8 mm. long, oblong, obtuse. ♀ flowers: ovary globose; styles very short. Drupe cinnamon-brownish, shining, 5–7 mm. in diam., subglobose.

N. Rhodesia. B: Balovale, fr. 18.vii.1954, *Gilges* 339 (K; PRE; SRGH); Mankoya 16 km. W. of Luampa Mission, fl. 22.ii.1952, *White* 2115 (FHO; K). W: Ndola, fr 3.vi.1953, *Fanshawe* 59 (EA; K; LISC); Mwinilunga, Matonchi Farm, ♂ fl. 5.x.1937, *Milne-Redhead* 2575 (K). C: 44·8 km. SW. of Broken Hill, fr. 13.vii.1930, *Hutchinson & Gillett* 3617 (BM; K). S: Choma, Siamambo Forest Reserve, fr. 10.iv.1952, *White* 2636 (FHO; K). **S. Rhodesia.** N: Miami, fr. vi.1926, *Rand* 189 (BM). W: Bulawayo, ♂ fl. 7.i.1898, *Rand* 97 (BM). C: Beatrice, ♂ fl. 15.xii.1947, *Rattray* 1504 (K; SRGH). S: Buhera, fr. 26.ii.1954, *Masterson* 46 (SRGH).

Also in Angola and the Congo. In open mainly *Brachystegia* woodlands, grasslands and banks of rivers.

3. **Rhus wildii** R. & A. Fernandes in Bol. Soc. Brot., Sér. 2, **38**: 192, t. 58 (1965). Type: S. Rhodesia, Sipolilo, Mpingi Pass, *Wild* 5776 (COI; PRE; SRGH, holotype).

Shrublet c. 20 cm. high, with dark grey fissured glabrescent spreading branches and somewhat angular puberulous branchlets. Petiole 5–8 mm. long, dorsally convex, canaliculate and winged at the apex above, puberulous only in the groove. Leaflets concolorous, pale green (when dried), small, the median one 1·5–3 × 0·8–1·5 cm., obovate, rounded or truncate at the top, shallowly crenate in the upper half, entire in the lower half, contracted to a cuneate base, subcoriaceous, glabrous; midrib prominulous above, rather prominent below, lateral nerves very slender, prominulous on both surfaces; venation inconspicuous. Panicles up to 5 cm. long, terminal and axillary, lax, with slender shortly and sparsely puberulous or glabrous reddish axis and branches; pedicels 1–1·5 mm. long, glabrous. ♂ flowers: calyx-segments c. 0·5 mm. long, roundish, glabrous; petals yellowish-green, c. 1·25 × 0·75 mm., oblong, obtuse. ♀ flowers and drupes not seen.

S. Rhodesia. N: Sipolilo, Mpingi Pass, ♂ fl. 17.v.1962, *Wild* 5776 (COI; PRE; SRGH).

Known only from S. Rhodesia. On exposed chrome ridge.

4. **Rhus glauca** Desf.*, Hist. Arb. **2**: 326 (1809).—Engl. in A. & C. DC., Mon. Phan. **4**: 411 (1883); Pflanzenw. Afr. **3**, 2: 203, fig. 99 C, F (1921).—Diels in Engl., Bot. Jahrb. **24**: 573, fig. 7 C, D (1898).—Schonl. in Bothalia, **3**, 1: 61, fig. p. 61 (1930). Type from S. Africa (Cape Prov.).
Rhus (? *sp. nov.*) *near R. undulata* Jacq.—O. B. Mill. in Journ. S. Afr. Bot. **18**: 47 (1952) pro parte quoad specim. *Miller* B/908 (specim. *Miller* B/881 n.v.).
Rhus sp.—O. B. Mill., loc. cit.

Shrub or (rarely) a small tree up to 7·5 m. tall, unarmed, glabrous in all parts; branchlets greyish or brown, angular. Petiole 0·5–1·2(3) cm. long, canaliculate and winged above. Leaflets glaucous or greyish (in dry state somewhat brownish on the under surface), papyraceous or subcoriaceous, the median one 1·5–2(5) × 1–1·2(3) cm., obcordate-cuneate to broadly obovate-cuneate, emarginate at the apex, the lateral ones obovate or obcordate, c. 2/3 as long as the median one or sometimes smaller, all somewhat undulate between the lateral nerves; midrib and the slender lateral nerves prominulous on both surfaces; venation not visible. Panicles terminal and axillary, many-flowered, the terminal ones longer than the leaves; pedicels 1–2 mm. long; bracts ovate or lanceolate. ♀ flowers: calyx-segments 0·5 mm. long, ovate; petals 1–1·5 mm. long, oblong. Drupe reddish, shining, 5–6 mm. in diam., globose.

* Substitute the earlier **R. glauca** Thunb. apud F. Hoffm. in Phyt. Blätt. **1**: 27 (1803).

Bechuanaland Prot. SE: Gaberones, Mokhoro valley, Baratani Hill, st. xii.1947, *Miller* B/561 (PRE); Kanye, Pharing, ♀ fl. vii.1949, *Miller* B/908 (K; PRE).

Also in S. Africa (Cape Prov.). Slopes of hills, sometimes near streams.

If the information on the label be correct, it is doubtful whether *Miller* B/908 should be referred to *R. glauca*, because, according the monographs, this species is a shrub, not a tree. The leaves, however, in form, colour and venation, the petioles in length, glabrousness and possession of wings and the angular branchlets are characteristic of *R. glauca*.

5. **Rhus lucida** L., Sp. Pl. **1**: 267 (1753).—Engl. in A. & C. DC., Mon. Phan. **4**: 413 (1883); Pflanzenw. Afr. **3**, 2: 203, fig. 99 A, B (1921).—Diels in Engl., Bot. Jahrb. **24**: 574, fig. 7 A, B (1898).—Eyles in Trans. Roy. Soc. S. Afr. **5**: 402 (1916).—Schonl. in Bothalia, **3**, 1: 54, fig. p. 55, 56 (1930).—Burtt Davy, F.P.F.T. **2**: 503 (1932).—Goodier & Phipps in Kirkia, **1**: 58 (1961). Type from S. Africa (Cape Prov.).

Shrub up to 3 m. tall, or a small tree, with the young parts frequently covered with shining resin; old branches brown or greyish, cylindric, glabrous, irregularly fissured; branchlets ± angular, brownish-red or greyish, shortly puberulous with white hairs. Petiole 0·5–1·7 cm. long, canaliculate above, winged in the upper part, glabrous. Leaflets nearly concolorous when dried (dull brownish-red) or somewhat glaucescent or greyish above, spathulate, narrowly to broadly obovate-cuneate or subrhombic, blunt or emarginate at the apex, entire, sessile, subcoriaceous to coriaceous, glabrous on both surfaces, the median one (1·5)2–7 × (0·5)0·7–2·7 cm., the lateral ones c. 2/3 as long as the median one; midrib raised on both surfaces; lateral nerves very slender, subprominent; venation usually not visible. Panicles axillary and terminal, shorter to slightly longer than the leaves, puberulous, sometimes resinous; pedicels 1–2 mm. long. Calyx-segments c. 0·5 mm. long, triangular-ovate, frequently puberulous. Petals c. 1·5 mm. long, oblong. Drupe reddish, shining, 3·5 mm. in diam., subglobose.

S. Rhodesia. E: Chimanimani Mts., ♀ fl. 26.ix.1906, *Swynnerton* 635 (BM; K). S: Belingwe, Emberengwe Mt., fr. 20.x.1959, *Wild* 4842 (LISC; SRGH).

Also in coastal districts from Cape Town to Natal and in the Transvaal. In scrub and forests, sometimes at high altitudes (1590–2100 m.).

The plants of our area correspond with the type of the species. Some varieties occur in S. Africa (v. Schonland, loc. cit.).

6. **Rhus dentata** Thunb., Prodr. Pl. Cap. **1**: 52 (1794).—Engl. in A. & C. DC., Mon. Phan. **4**: 435 (1893).—Schonl. in Bothalia, **3**, 1: 37, fig. p. 37, 39 (1930).—Burtt Davy, F.P.F.T. **2**: 499 (1932). Type from S. Africa (Cape Prov.).

Shrub 1–2·5 m. high or small tree up to 6 m. tall; old branches dull greyish-brown, smooth or slightly striated, lenticellate, glabrous, sometimes thorny; young branches and petioles reddish or fulvous, glabrous to ± densely pilose or hispid-villous, all the hairs very short and crisped or mixed with long patent ones. Petiole 0·6–4 cm. long, slender, dorsally convex, slightly canaliculate above. Median leaflet 1·7–8(9) × 0·9–5 cm., lanceolate-elliptic, elliptic, or more frequently obovate, narrowing to a triangular-cuneate base, entire or with 1–4(5) obtuse or acute and mucronate teeth in the upper part; lateral leaflets 1–3 × 0·6–2·3 cm., rarely subequalling the median leaflet or 1/3–2/3 shorter, at a right or obtuse angle to the median one, narrowly to broadly elliptic, acute or obtuse at the apex, dull green on the upper surface, paler below, entire or with 1–3(4) teeth in the upper part, somewhat cuneate or rounded at the base, all membranous to ± rigid, glabrous or more frequently ± pilose or hispid mainly on the midrib and nerves; midrib impressed or prominulous above, prominent beneath; lateral nerves visible above but not raised, prominent beneath; reticulation inconspicuous above, immersed below. Panicles terminal and axillary, shorter to longer than the leaves, dense, with the axis and branches glabrous or sparsely pilose to ± densely hispid, the patent hairs usually mixed with short crisped ones; pedicels 1·5–2 mm. long. ♂ flowers: calyx-segments 0·75 mm. long, triangular, hispid on the back; petals greenish-yellow, 1·5–2 mm. long, ovate, obtuse; filaments c. 1 mm. long; disk 5-lobulate, lobules crenulate. ♀ flowers: ovary subglobose; styles c. 0·5 mm. long, slightly connate at the base; stigmas capitate; disk cupuliform, 5-lobulate. Drupe reddish, shining, 4 mm. in diam., subglobose.

S. Rhodesia. C: 36·8 km. E. of Rusape, ♂ fl. xi.1957, *Miller* 4749 (SRGH). E: Umtali, Penhalonga, fr. 11.xi.1956, *Chase* 6239 (K; LISC; LMJ; SRGH). **Mozambique.** MS: Manica, Mavita, Moçambize Valley, ♂ fl. 25.x.1944, *Mendonça* 2606 (COI; K; LISC; LMJ; SRGH).

Also in Cape Prov., Natal, Swaziland, Basutoland and the Transvaal. In *Acacia*-bush, secondary thickets, forest edges and termite mounds, at 1000–2100 m.

Separation into varieties based on indumentum, size of leaves, etc. would be artificial as there is continuous variation between extreme forms.

7. **Rhus gueinzii** Sond. in Harv. & Sond., F.C. **1**: 515 (1860).—Schonl. in Bothalia, **3**, 1: 79 (1930) pro parte.—Burtt Davy, F.P.F.T. **2**: 507 (1932) pro parte. Type from Natal.

Var. **spinescens** (Diels) R. & A. Fernandes in Bol. Soc. Brot., Sér. 2, **38**: 186 (1965). Type from the Transvaal (Komati Poort).
Rhus spinescens Diels in Engl., Bot. Jahrb. **40**: 87 (1907).—Schonl. in Bothalia, **3,** 1: 70, fig. p. 70 (1930).—Burtt Davy, F.P.F.T. **2**: 505 (1932). Type as above.
Rhus simii Schonl., tom. cit.: 69, cum fig. (1930). Type from the Transvaal.
Rhus simii var. *lydenburgensis* Schonl., tom. cit.: 70 (1930). Type from the Transvaal.

Shrub or small tree up to 8 m. tall, often with spinescent branches (spine, 3·5–7·5 cm. long, usually stout, cylindric-conic), leafless or leafy; bark of the stems branches and spines pale grey with numerous and very prominent lenticels; branchlets whitish, shortly pubescent. Petiole 1–4 cm. long, slender, glabrous, dorsally convex, canaliculate above. Leaflets pale glaucous or greyish-green above, paler below, lanceolate or oblong, cuneate at the base or the lateral ones ovate to ovate-lanceolate and not so cuneate as the median, all blunt and emarginate at the apex, frequently with the margin entire or shallowly crenate-dentate, membranous or ± rigid, rarely subcoriaceous, the median one 3·5–7 × 0·6–1·3 cm., the lateral ones smaller; midrib slender, prominent on both surfaces; lateral nerves very delicate, 5–7 mm. apart, slightly conspicuous; reticulation barely or not at all visible. Panicles terminal and axillary, shorter than the leaves or the terminal ones somewhat longer, ± dense, with the branches slightly scrofulous and pilose; pedicels 1–1·5 mm. long. ♂ flower: calyx-segments c. 0·5 mm. long, ovate; petals yellowish, c. 1·5 mm. long, oblong; filaments c. 1 mm. long; disk 5-lobulate, lobules emarginate. Drupe shining, 3·5 mm. in diam., globose, glabrous.

S. Rhodesia. S: Gwanda, Tuli Circle, st. v.1959, *Thompson* T/40/59 (K; LISC; PRE; SRGH). **Mozambique.** SS: Limpopo, 10 km. from Aldeia da Barragem, near R. Limpopo, fr. 20.xi.1957, *Barbosa & Lemos* in *Barbosa* 8216 (COI; K; LISC; LMJ); Gaza, suburbs of Vila João Belo, ♂ fl. 10.iii.1942, *Torre* 3992 (LISC). LM: Maputo, Santaca, ♂ fl. 29.x.1948, *Gomes e Sousa* 3871 (COI; K; LISC; PRE; SRGH).

Also in the Transvaal. In deciduous semi-evergreen and evergreen thickets, termite mounds, *Acacia*-savanna, open forests, etc.

Distinguished from var. *gueinzii* by the spinescent branches and the leaflets usually entire and not so discolorous.

8. **Rhus natalensis** Bernh. ex Krauss in Flora, **27**: 349 (1844).—Engl. in A. & C. DC., Mon. Phan. **4**: 421 (1883); Pflanzenw. Afr. **3,** 2: 206 (1921).—Schonl. in Bothalia, **3,** 1: 68, fig. p. 68 (1930).—Brenan, T.T.C.L.: 37 (1949).—Wild, Guide Fl. Vict. Falls: 152 (1953).—Meikle in Exell & Mendonça, C.F.A. **2,** 1: 100 (1954).—Williamson, Useful Pl. Nyasal.: 105 (1955).—Mogg in Macnae & Kalk, Nat. Hist. Inhaca I. Moçamb.: 148 (1958).—Van der Veken, F.C.B. **9**: 40 (1960).—White, F.F.N.R.: 212 (1962) pro parte quoad specim. *Michelmore* 366.—Goodier & Phipps in Kirkia, **1**: 58 (1961). Type from Natal.
Rhus glaucescens A. Rich., Tent. Fl. Abyss. **1**: 143 (1847).—Oliv., F.T.A. **1**: 437 (1868) pro parte excl. syn. *R. quartiniana* A. Rich. et *R. gueinzii* Sond.—Burtt Davy & Hoyle, N.C.L.: 29 (1936).—Schinz & Junod in Mém. Herb. Boiss. **7,** 10: 48 (1900). Type from Ethiopia.
Rhus glaucescens var. *natalensis* (Bernh. ex Krauss) Engl., Pflanzenw. Ost-Afr. **C**: 245 (1895).—R. E. Fr. in Wiss. Ergebn. Schwed. Rhod.-Kongo-Exped. **1**: 127 (1914). Type as for *R. natalensis*.

Shrub 2·5–3 m. high or small tree up to 8 m. tall with the bark of the branchlets greyish or whitish and those of the old branches dull-grey, lenticellate and rough. Petiole 1·5–3·5(4) cm. long, convex below, canaliculate above. Leaflets slightly or conspicuously discolorous (greenish-grey on both sides or dull green or brownish above and greenish-grey below), entire or undulate-crenate at the margin, papyraceous or subcoriaceous, with the midrib slightly raised on the upper surface, prominent below, the lateral nerves slightly raised on both surfaces, more so on the upper one, reticulation scarcely or not at all visible; median leaflet 2·5–9 × 1–3·5 cm., obovate, oblong or elliptic, obtuse or rounded and sometimes emarginate at the apex (rarely acute), ± cuneate at the base; lateral leaflets (1·8)2·5–7 × 1–3·5 cm., sessile, ovate, obovate or oblong-obovate, rounded and sometimes emarginate at the apex, somewhat cuneate at the base. Panicles lax, up to 12 cm. long, generally shorter than the leaves, with the axis and branches ± pilose; pedicels 1–2 mm. long. ♀ flowers: calyx-segments 0·3–0·5 mm. long, ovate; petals whitish or greenish, 1–1·5 mm. long, oblong; disk shallowly 5-lobulate; ovary subglobose; styles 0·5–0·7 mm. long, reflexed; stigmas subcapitate; staminodes present. Drupe shining, 5–6 mm. in diam., glabrous.

N. Rhodesia. N: Chipili stream, *Miller* D.125 (FHO). E: Fort Jameson, fr. 1.vi.1958, *Fanshawe* 4487 (FHO; K). S: 6·4 km. up from Kafue Road Bridge, fr. 24.v.1957, *Angus* 1605 (FHO; K). **S. Rhodesia.** W: Bulawayo, Khamsi Ruins, st. 11.viii.1946, *Gouveia & Pedro* in *Pedro* 1628 (LMJ); Victoria Falls, st. 24.vii.1950, *Robertson & Elffers* 36 (K; PRE; SRGH). E: Umtali, ♂ fl. v.1947, *Chase* 355 (BM; COI; K; SRGH). S: Zimbabwe, ♂ fl. 23.v.1951, *Seymour-Hall* 14/51 (K; SRGH). **Nyasaland.** C: Dedza, Chongoni Mt., ♂ fl. 29.v.1960, *Chapman* 720 (COI; FHO; SRGH). S: Mlanje Mt., ♂ fl. 13.v.1958, *Chapman* 568 (FHO; K). **Mozambique.** N: Ibo I., st. 5.ix.1948, *Pedro & Pedrógão* 5049 (EA); between the lighthouse of Cabo Delgado and Palma, ♂ fl. 17.iv.1964, *Torre & Paiva* 12102 (K; LISC; LMJ; SRGH). Z: Mopeia, Campo, ♂ fl. 23.xii.1901, *Le Testu* 931 (P); Quelimane, Madal Estates, Ibangulue, ♂ fl. 9.iv.1958, *Pedro* 68 (PRE). T: Angónia, fr. 17.vii.1949, *Andrada* 1778 (COI; LISC); Zóbuè, ♂ fl. 17.vi.1947, *Hornby* 2757 (K; PRE; SRGH). MS: Chimoio, near Chibata, ♀ fl. & fr. 27.iv.1948, *Andrada* 1198 (COI; K; LISC; LMJ; SRGH). SS: between road-crossing Vilanculos-Membone-Maboti and Maboti, fr. 28.iii.1952, *Barbosa & Balsinhas* 5045 (BM; LISC; LMJ). LM: Maputo, Ponta do Ouro, ♂ fl. 27.xii.1948, *Gomes e Sousa* 3914 (COI; K; PRE; SRGH); Inhaca I., ♂ fl. & fr. 22.xii.1956, *Mogg* 4594 (J; K).

Widespread in tropical and S. Africa. In coastal scrub of sea and lakes, *Julbernardia-Brachystegia* woodlands, evergreen forests on the top and slope of mountains, riverine thickets and forests, termite mounds, granitic rocky places, etc.; from sea level to 2100 m.

9. **Rhus marlothii** Engl., Bot. Jahrb. **10**: 37 (1889); Pflanzenw. Afr. **3,** 2: 208, fig. 102 (1921).—Schonl. in Bothalia, **3,** 1: 71, fig. p. 71 (1930).—Burtt Davy, F.P.F.T. **2**: 505 (1932). Type from SW. Africa (Hereroland).

Shrub with cylindrical greyish branches; branchlets, petioles and panicles puberulous or ± hispidulous. Petiole 0·5–1·2 cm. long, slightly canaliculate or marginate above. Leaflets concolorous to subconcolorous, membranous to subcoriaceous, and ± sparsely hairy on both surfaces, with the margin slightly crenate in the upper part, the median one 2–4·5 × 1–1·8 cm., ovate or oblong, rounded at the apex, ± cuneate at the base, the lateral ones c. 2/3 the size of the median one and a little asymmetric at the base; midrib raised on both surfaces; lateral nerves slender, sometimes prominulous, venation not visible. Axillary panicles shorter than the leaves, the terminal ones a little longer. Calyx-segments c. 0·5 mm. long, ovate, puberulous. Petals c. 1·25 mm. long, oblong, obtuse. Drupe cinnamon-coloured, c. 4 mm. in diam., compressed-subquadrate, glabrous.

Bechuanaland Prot. SE: Mochudi, fr. i.iv.1914, *Harbor* in *Rogers* 6628 (BM; PRE).
Also in SW. Africa and the Transvaal. In dry places among rocks and in the bushveld.

10. **Rhus pentheri** Zahlbr. in Ann. Hofmus. Wien, **15**: 52 (1900).—Schonl. in Bothalia, **2,** 1: 48, fig. p. 48 (1930).—Burtt Davy, F.P.F.T. **2**: 503 (1932). Type from Natal.

Shrub up to 3 m. tall; branches greyish to brownish, cylindric, glabrous or puberulous; branchlets sometimes densely cinereous- or whitish-villous and somewhat spinescent. Petiole 0·5–1·5 cm. long, subterete, canaliculate above, slender, sparsely hairy. Leaflets usually discolorous (dark green above, paler below), membranous, at first ± sparsely and subappressed-pubescent and scrofulous on both surfaces, later glabrescent or glabrous, margin entire or with 3 blunt teeth at the apex; midrib prominent on both surfaces, lateral nerves not or very slightly raised on both surfaces, tertiary nerves and reticulation not visible; median leaflet

1·8–3·3 × 0·4–1·4 cm., elongate-cuneate, the lateral ones c. half as long as the median one, obovate and not so cuneate at the base. Panicles terminal and axillary, usually shorter than the leaves, pubescent. Calyx-segments c. 0·5 mm. long, triangular, obtuse, pubescent. Petals 0·75–1 mm. long, ovate, obtuse. Drupe cinnamon-brown to dark-brown, shining, c. 3–3·5 mm. in diam., compressed, subcircular in outline.

Mozambique. LM: Maputo, R. Changalane, ♂ fl. 23.iii.1947, *Hornby* 2659 (SRGH); Namaacha, Goba, st. 8.i.1947, *Pedro & Pedrógão* 407 (LMJ).

Also in Cape Prov., Orange Free State, Natal, Swaziland and the Transvaal. On stony ground near seasonally-flowing streams.

11. **Rhus tenuinervis** Engl. in A. & C. DC., Mon. Phan. **4**: 423 (1883); Pflanzenw. Afr. **3,** 2: 208 (1921).—Monro in Proc. & Trans. Rhod. Sci. Assoc. **8**: 71 (1908).—Eyles in Trans. Roy. Soc. S. Afr. **5**: 403 (1916).—Meikle in Exell & Mendonça, C.F.A. **2,** 1: 100 (1954).—Van der Veken, F.C.B. **9**: 35 (1960).—White, F.F.N.R.: 212 (1962). Type from Angola (between Benguela and R. Catumbela).

A much-branched sometimes thorny shrub or small tree up to 8 m. tall; branches with dull-greyish rough lenticellate bark, glabrous or glabrescent; branchlets ± densely patent-pilose, the hairs slender, yellowish. Petiole 1–3 cm. long, dorsally convex, slightly canaliculate, ± pilose. Leaflets discolorous (glaucous or dull green above, paler and generally yellowish below), papyraceous to subcoriaceous, ± patent-pubescent on both surfaces, rarely glabrescent or almost glabrous, with the margin of the upper 2/3 coarsely crenate or bicrenate or almost lobed with crenate lobes, the lower portion entire or (rarely) the leaflets quite entire; midrib and lateral nerves prominulous above, prominent below, venation scarcely visible; median leaflet 2·5–11·5 × 1·5–5(7) cm., ovate, obovate, subcircular, ovate-oblong or elliptic, obtuse, rounded or truncate at the apex, sometimes subobtuse, subacute or acute, cuneate at the base, often much contracted and subpetiolulate; lateral leaflets 1·5–6 × 1–4 cm., oblong-ovate to subcircular, cuneate at the base. Panicles up to 15 cm. long, axillary and terminal, lax, the axis and branches pilose; pedicels 1–2 mm. long, articulated below the calyx. Calyx-segments c. 0·5 mm. long, ovate, obtuse. Petals 1·25–1·5 mm. long, obtuse. Disk patelliform, 5-lobulate. Ovary globose; styles c. 0·4 mm. long. Drupe brownish, up to 5 × 7(8·5) mm., compressed, subquadrangular or subcircular in outline, pruinose.

Var. **tenuinervis**

Rhus commiphoroides Engl. & Gilg in Warb., Kunene-Samb.-Exped. Baum: 289 (1903).—Engl., Pflanzenw. Afr. **3,** 2: 208, fig. 103 A–D (1921).—Schonl. in Bothalia, **3,** 1: 71, fig. p. 72 (1930).—Burtt Davy, F.P.F.T. **2**: 505 (1932).—O. B. Mill., B.C.L.: 34 (1948); in Journ. S. Afr. Bot. **18**: 46 (1952).—Pole Evans in Bot. Surv. C. S. Afr. Mem. **21**: 31 (1948). Type from Angola (Bié).

Rhus kwebensis N. E. Br. in Kew Bull. **1909**: 100 (1909).—Burtt Davy & Hoyle, N.C.L.: 29 (1936).—O. B. Mill., B.C.L.: 34 (1948). Type: Bechuanaland Prot., Kwebe Hills, *Lugard* 200 (K, holotype).

Caprivi Strip. 13 km. NW. of Ngoma Ferry, st. 16.vii.1952, *Codd* 7078 (BM; K; PRE; SRGH). **Bechuanaland Prot.** N: Kazungula Thorn Forest, fr. iv.1936, *Miller* B 133 (BM). SW: 64 km. N. of Kang on road to Ghanzi, ♂ fl. 19.ii.1960, *de Winter* 7343 (M; PRE; SRGH). SE: Lower Ngwezumbo, ♀ fl. iii.1938, *Miller* B 186 (PRE); 194 km. NW. of Molepolole, fr. 15.vi.1955, *Story* 4889 (COI; PRE; SRGH). **N. Rhodesia.** B: Senanga, Lilengo Forest Reserve, ♂ fl. 6.ii.1952, *White* 2022 (FHO; K). C: 17·6 km. S. of Lusaka, near Mt. Makulu Research Station, ♂ fl. 6.v.1956, *Angus* 1277 (FHO; K). S: Mazabuka, Choma, ♂ fl. 28.iii.1956, *Robinson* 1381 (K; SRGH). **S. Rhodesia.** N: E. side of Umvukwe Mts., near Dawsons, fr. 29.iv.1948, *Rodin* 4467 (K). W: Nyamandlhovu, Pasture Research Station, fr. iv.1956, *Plowes* 1937 (K; SRGH). C: 21 km. E. of Gwelo, fr. 3.vii.1930, *Hutchinson & Gillett* 3399 (BM; K). E: Alma Farm, Odzi R., fr. 12.iv.1948, *Chase* 720 (BM; SRGH). S: Ndanga, fr. 24.v.1959, *Noel* 1965 (SRGH). **Nyasaland.** C: Lake Nyasa, coast near Salima, fr. 12.vii.1936, *Burtt* 6073 (BM; K). **Mozambique.** N: Amaramba, Mandimba, ♂ fl. 14.v.1948, *Pedro & Pedrógão* 3412 (EA); Nampula, ♂ fl. 20.vi.1937, *Torre* 1558 (COI; LISC). MS: Chimoio, fr. 6.v.1948, *Andrada* 1224 (COI; K; LISC; SRGH).

Also in Angola, Katanga, SW. Africa, Transvaal, Cape Prov., Tanganyika and Kenya. Widespread on white sand soils, termite mounds, rocky slopes and sometimes on banks of rivers in various woodland types.

Var. **meikleana** R. & A. Fernandes in Bol. Soc. Brot., Sér. 2, **38**: 190, t. 52–54 (1965).

Type: Nyasaland, Likabula Valley, Palombe Mission, *Chapman* 548 (COI; EA; FHO, holotype; K).

Rhus commiphoroides var.—Burtt Davy & Hoyle, N.C.L.: 29 (1936).

Nyasaland. S: Zomba, ♂ fl. iii.1932, *Clements* 265 (FHO). **Mozambique.** N: Maniamba, Metangula, Lake Nyasa shore, ♂ fl. 22.v.1948, *Pedro & Pedrógão* 3843 (EA); 26 km. from Malema, ♂ fl. 24.v.1961, *Leach & Rutherford-Smith* 10993 (COI; K; SRGH). T: Moatize, Zóbuè, fr. 15.vii.1942, *Torre* 4405 (BM; COI; K; LISC; SRGH).

Known only from our area and from Tanganyika. In *Brachystegia* woodland, *Brachystegia-Oxytenanthera* woodland and open savannas with *Albizia* and *Acacia*.

This taxon is in several characters intermediate between *R. tenuinervis* and *R. natalensis*. We therefore think that it may be a hybrid between these two species. The pollen, however, is fertile and the drupes are apparently well developed.

12. **Rhus lucens** Hutch., Botanist in S. Afr.: 480 in adnot. (1946).—O. B. Mill. in Journ. S. Afr. Bot. **19**, 4: 179 (1953).—Wild, Guide Fl. Vict. Falls: 152 (1953).—White, F.F.N.R.: 212 (1962). Type: S. Rhodesia, near Victoria Falls, *Hutchinson & Gillett* 3473 (BM; K, holotype; SRGH).

A shrub or small tree up to 4·5 m. tall; old branches dull-greyish, lenticellate, glabrous, the younger ones brownish-grey, glabrous or sparsely patent-pilose. Petiole 1·5–2·5 cm. long, flat and slightly marginate above, glabrous or pilose. Leaflets glaucous or pale grey-green on the upper surface, paler below, coriaceous, glabrous or very sparsely pilose, rounded or retuse, or emarginate at the apex with the margin shallowly crenate, median leaflet (2·5)4·5–8·5 × 1·5–5 cm., obovate to subcircular, narrowed and cuneate at the base; lateral leaflets obovate to subcircular, rounded and frequently emarginate at the apex, almost cuneate to rounded at the base; midrib and lateral nerves raised mainly on the under surface, reticulation conspicuous above. Panicles shorter than the leaves, lax, with the axis and branches pilose. ♂ flowers: calyx-segments 0·5 mm. long, ovate, obtuse; petals c. 1·25 mm. long, ovate-oblong, obtuse; filaments c. 0·5 mm. long. Drupe yellowish-brown, shining or somewhat pruinose, 4 × 5 mm., compressed, subquadrate, asymmetrical.

Bechuanaland Prot. N: Chobe, S. of Deka, fr. vi. 1952, *Miller* B/1328 (K; PRE). **N. Rhodesia.** C: Bell Point, Mkushi-Lunsemfwa River Junction, ♂ fl. 4.v.1957, *Fanshawe* 3266 (EA; K). S: Livingstone, Katombora, fr. 7.vii.1956, *Gilges* 642 (K; PRE; SRGH). **S. Rhodesia.** N: Urungwe, Kariba Gorge, ♂ fl. 17.iv.1955, *Lovemore* 431 (COI; K; LISC; LMJ; SRGH). W: between Wankie and the Victoria Falls, fr. 6.xii.1930, *Pole Evans* 2737 (COI; K; PRE; SRGH).

Known only from our area. In open spots in dry forest and in escarpment woodlands.

13. **Rhus erosa** Thunb., Fl. Cap. **2**: 212 (1818).—Engl. in A. & C. DC., Mon. Phan. **4**: 439 (1883); Pflanzenw. Afr. **3**, 2: 215, fig. 107 A–C (1921).—Diels in Engl., Bot. Jahrb. **24**: 587, fig. 6 A–C (1898).—Eyles in Trans. Roy. Soc. S. Afr. **5**: 402 (1916).—Schonl. in Bothalia, **3**, 1: 84, fig. p. 84 (1930). Type from S. Africa (Cape Prov.).

A much-branched spreading shrub up to 3 m. tall, glabrous; young branchlets cinnamon-brown, subflexuous, subangular, the oldest ones greyish, rugose, lenticellate. Petiole 2–3 cm. long, stout, rigid, dorsally convex and slightly canaliculate and marginate above towards the apex. Leaflets pale greyish-green (when dried), linear or linear-lanceolate, attenuate and very acute at the apex, margin generally ± erose-dentate or pinnatifid, narrowed towards the base, the teeth somewhat recurved and acute, rarely entire, coriaceous, usually covered with a layer of resin; median leaflet 6·2–12 × 0·2–0·9 cm. (with the teeth), lateral ones smaller and falcate; midrib raised on both sides, lateral nerves very slightly prominent on both surfaces, reticulation usually slightly prominent. Panicles as long as or shorter than the leaves, very lax, branches slender; pedicels 1·5–3 mm., glabrous. Calyx-segments c. 0·5 mm. long, ovate, glabrous. Petals c. 1 mm. long. Filaments c. 1 mm. long; anthers c. 0·5 mm. long, ovate-subcircular in outline. Disk 10-lobulate. Drupe yellowish, shining, c. 4 mm. in diam., subglobose.

Bechuanaland Prot. N: near Bushman's River, st. vii.1899, *Rogers* 4535 (SRGH).

Also in Cape Prov., Basutoland, Orange Free State and the Transvaal. On stony kopjes.

14. **Rhus tenuipes** R. & A. Fernandes in Bol. Soc. Brot., Sér. 2, **38**: 191, t. 55–57 (1965); Type: S. Rhodesia, Charter, Mhlaba Hills, Great Dyke, *Wild* 5607 (COI; PRE. SRGH, holotype). Type: S. Rhodesia.

Shrub or small tree up to 4·2 m. tall, much branched, glabrous; young branches reddish, slender, slightly striate, flexuous, pendulous, sparsely lenticellate, the oldest ones greyish and also lenticellate. Leaves pendent, 3-foliolate; petiole reddish, very slender, up to 5 cm. long, canaliculate above; leaflets subconcolorous or slightly discolorous (when dried), membranous to papyraceous, linear-lanceolate or lanceolate, narrowing to an acute and mucronate apex, with the margin ± irregularly dentate-serrate, the teeth subfalcate and mucronulate, sometimes patent, cuneate at the base; median leaflet 4·5–12 × (0·3)1·5(2) cm., the lateral ones a little smaller; midrib narrow, prominulous on both surfaces; lateral nerves prominulous above, slightly impressed below, sometimes inconspicuous on both surfaces; reticulation not visible. Panicles terminal and axillary, glabrous, lax, as long as or longer than the leaves, with the axis and branches slender; pedicels 1·5–3 mm. long. ♂ flowers: calyx-segments c. 0·5 mm. long, ovate, glabrous; petals c. 1·25 mm. long, ovate-oblong; filaments c. 0·5 mm. long; anthers c. 0·25 mm. long; disk 10-lobulate, c. 1 mm. in diam. Drupe dark-brownish, shining, 3 × 4·5 mm., subquadrate, compressed.

S. Rhodesia. W: Bulawayo, Matobo Distr., st. 22.iii.1950, *Orpen* 3/50 (SRGH). C: Charter, Mhlaba Hills, ♂ fl. 16.i.1962, *Wild* 5607 (COI; PRE; SRGH). E: Insiza, between Filabusi and Shabani, st. 17.ix.1947, *West* 2417 (K; SRGH). S: Victoria, fr., *Monro* 1767 (BM).

Known only from S. Rhodesia. In bush on stony hills and roadsides usually in soil with serpentine and chromium.

15. **Rhus trifoliolata** Bak. f. in Journ. of Bot. **37**: 429 (1899).—Monro in Proc. & Trans. Rhod. Sci. Ass. **8**: 71 (1908).—Eyles in Trans. Roy. Soc. S. Afr. **5**: 403 (1916). Type: S. Rhodesia, Bulawayo, *Rand* 66 (BM, holotype).

Shrublet up to 90 cm. tall, with the young parts scrofulous, the oldest glabrous and greyish; branchlets brownish-red. Petiole 1–4 cm. long, dorsally convex, flat or slightly canaliculate above. Leaflets concolorous, sessile, linear or linear-lanceolate, with the margin entire or slightly undulate, acute and mucronate at the apex, sometimes falcate, narrowed towards the base, very coriaceous, the median one 5–13 × 0·3–0·6 cm., the lateral ones as long as the median or shorter; midrib slender, raised on both surfaces, lateral nerves very branched, fairly prominent on both sides. Panicle very lax, with divaricate scrofulous branches; pedicels 2·5–4 mm. long. ♀ flowers: calyx segments c. 0·75 mm. long, ovate, obtuse, scrofulous; petals c. 1·5 mm., oblong; styles 0·5 mm. long, deflexed; stigmas subcapitate; disk patelliform, shallowly 5-lobulate. Drupe shining, 3–4 mm. in diam., subglobose, glabrous.

S. Rhodesia. W: Bulawayo, Hope Fountain, fr. 18.vi.1947, *Keay* FHI 21328 (FHO; SRGH). C: Gwelo, Lalapanzi, ♀ fl. 31.i.1951, *Greenhow* 8/51 (FHO; K; SRGH).

Known only from S. Rhodesia. On Kalahari Sand in scrub.

16. **Rhus ciliata** Licht. ex Roem & Schult. in L., Syst. Veg. ed. nov. **6**: 661 (1820).—Engl. in A. & C. DC., Mon. Phan. **4**: 418 (1883); Pflanzenw. Afr. **3**, 2: 205 (1921).—Schonl. in Bothalia, **3**, 1: 82 (1930) pro parte excl. syn. *R. tridactyla* Burch.—Burtt Davy, F.P.F.T. **2**: 506 (1932). Type from the Transvaal.

Shrub up to 3 m. tall; branches brownish, spreading, stiff, frequently spinescent or armed with ± long thorns, cylindric, striate, hispidulous. Petiole 0·75–1·5 cm. long, slender, dorsally convex, canaliculate above and somewhat winged towards the apex. Leaflets concolorous (pale-brownish when dried), lanceolate, entire, acute and usually mucronulate at the apex, narrowed at the base, subcoriaceous, hispidulous in both surfaces, ciliate on the margin, frequently covered with a shining layer of resin, the median one up to 5 × 0·8 cm.; midrib impressed above, slightly raised below; lateral nerves barely or not visible. Panicles axillary and terminal, of variable length, very lax, with the axis, branches and the 1·5–4 mm. long pedicels hispidulous. Calyx-segments 0·5–0·75 mm., ovate or rounded, dorsally sparsely ciliate. Petals c. 1·25 mm. long, oblong, obtuse. Drupe pale-brownish, 4–5 mm. in diam., subglobose.

Bechuanaland Prot.: near Mafeking,* ♂ fl. 1.ii.1899, *Bolus* 6404 (PRE).

Also in Cape Prov., Orange Free State, western Transvaal and SW. Africa. Mainly along dry river-beds.

17. **Rhus acuminatissima** R. & A. Fernandes in Bol. Soc. Brot., Sér. 2, **38**: 183, t. 41–43 (1965). Type: Nyasaland, Mlanje Mt., near Likabula, *Clements* 121 (FHO, holotype; K).
Rhus retinorrhoea sensu Burtt Davy & Hoyle, N.C.L.: 29 (1936).

Shrub or small tree up to 5 m. tall; branches brownish-red, glabrescent, cylindric, slightly striate, lenticellate, the lenticels very small, roundish; branchlets hispidulous. Petiole (3·5)5–10·5 cm. long, slender, subcylindric, a little grooved on the upper side, reddish-brown, ± densely hispidulous to glabrous. Leaflets nearly concolorous (brownish on both sides in dry state, sometimes a little darker above), lanceolate-acuminate, narrowed to a very acute subfiliform apex, with entire subtranslucent whitish-edged margin, cuneate at the base, membranous to papyraceous, the youngest ones ± hispid on the midrib and margin and sparsely so on the lamina, the oldest ones glabrous; median leaflet 5–12 × 1·5–3·3 cm., the lateral ones a little smaller or sometimes subequal; midrib slender, prominent on both surfaces, lateral nerves not or only slightly raised on both surfaces, reticulation not visible above, sometimes conspicuous but not raised below. Panicles up to 13 cm. long, terminal and axillary, multiflorous, with the axis, branches and pedicels (1–2 mm. long) hispidulous. Calyx-segments 0·5 mm. long, lanceolate, acute, hispidulous. Petals 1·25 × 0·5 mm., ovate-oblong, subacute; filaments 0·75 mm. long. Drupe immature black, c. 2·5 mm. in diam., circular in outline.

Nyasaland. S: Mwanza Area, W. of Shire R., st., *Topham* 297 (FHO); Mt. Mlanje, Luchenya Plateau, ♂ fl., *Topham* 939 (FHO; K). **Mozambique.** N: Mt. Ribáuè, fr. 25.i.1964, *Torre & Paiva* 10245 (BM; COI; K; LISC; SRGH).

Known only from the S. of Nyasaland and Mt. Ribáuè in Mozambique. In hygrophilous montane forests.

18. **Rhus lancea** L. f., Suppl. Pl.: 184 (1781).—Gibbs in Journ. Linn. Soc., Bot. **37**: 437 (1906).—Monro in Proc. & Trans. Rhod. Sci. Ass. **8**: 71 (1908).—Eyles in Trans. Roy. Soc. S. Afr. **5**: 402 (1916).—Schonl. in Bothalia, **3**, 1: 72, fig. p. 73 (1930).—Burtt Davy, F.P.F.T. **2**: 506 (1932).—Steedman, Trees etc. S. Rhod.: 39 (1933).—O. B. Mill., B.C.L.: 34 (1948); in Journ. S. Afr. Bot. **18**: 47 (1952).—Pardy in Rhod. Agr. Journ. **50,** 5: 366, cum photo. (1953).—Palmer & Pitman, Trees S. Afr.: 290, cum fig. (1961).—White, F.F.N.R.: 212 (1962). Type from S. Africa (Cape Prov.).

Shrub or small tree up to 6 m. high, glabrous; branches and branchlets straight, angular or striate; bark ± dark-reddish-brown, lenticellate. Petiole 2–7 cm. long, canaliculate and marginate above mainly at the apex. Leaflets dark green above and light green or green-yellowish below, lanceolate, narrowing to both ends, acute and mucronate at the apex, cuneate at the base, entire, rigid to subcoriaceous, sessile, straight or falcate; median leaflet 7–24·5 × 0·7–2·8 cm., the lateral ones a little smaller, sometimes arched at the base; midrib prominent on both sides (more raised on the under surface), lateral nerves numerous, 1–3·5(6) mm. apart, prominent above as also the reticulation, less prominent below. Panicles terminal and axillary, ± equalling the leaves, ± lax, with the axis sometimes furfuraceous; pedicels 2–3 mm. long, slender, articulated near the middle. ♂ flower: calyx-segments c. 0·5 mm. long, semicircular, obtuse; petals greenish-yellow, c. 1·5 mm. long, oblong, obtuse; filaments 0·5–0·75 mm. long; anthers 0·3 mm. long, elliptic. ♀ flower: ovary globose; styles thick, recurved; stigmas 2-lobed; staminodes present. Drupe dull greyish to shining brown, 4–6 mm. in diam., subglobose, slightly depressed, often slightly asymmetrical, glabrous.

Bechuanaland Prot. SE: Lobatsi, Government Farm, fr. viii.1940, *Miller* B/203 (BM; PRE; UPS). **N. Rhodesia.** W: Ndola, ♂ fl. 6.viii.1954, *Fanshawe* 1444 (EA; K; LISC; SRGH). S: Choma and Kalomo, ♂ fl. 16.viii.1961, *Bainbridge* 549 (FHO); near Victoria Falls, ♂ fl. & fr. vii.–x.1860, *Kirk* (K). **S. Rhodesia.** N: near Mazoe Dam on Salisbury road, ♀ fl. & fr. 2.ix.1960, *Rutherford-Smith* 40, ♂ fl. 2.ix.1960, 41 (SRGH). W: Matobo, Besner Kobila, ♀ fl. vii.1953, *Miller* 1841 (K; LISC; LMJ; P; SRGH);

* This locality may not be in Bechuanaland Prot.; but as Mafeking is near the frontier, we have decided to include the species as its occurrence in Bechuanaland Prot. is very probable.

Bulawayo, fr. ix.1958, *Miller* 5413 (SRGH). C: Salisbury, ♂ fl. vii.1919, *Eyles* 1759 (BM; COI; PRE; SRGH). E: Umtali, N. side of Christmas Pass, ♂ fl. 31.vii.1955, *Chase* 5695 (BM; COI; LISC; SRGH). S: Fort Victoria, near Baden's Farm, fl. immat. 5.vi.1932, *Cuthbertson* 5867 (BM; K; SRGH).

Also in SW. Africa, the Transvaal, Orange Free State and Cape Prov. In savannas, open woodlands and riverine forests.

19. **Rhus leptodictya** Diels in Engl., Bot. Jahrb. **40**: 86 (1907).—Monro in Proc. & Trans. Rhod. Sci. Ass. **8**: 71 (1908).—Eyles in Trans. Roy. Soc. S. Afr. **5**: 402 (1916).—Engl., Pflanzenw. Afr. **3,** 2: 215 (1921).—R. & A. Fernandes in Webbia, **19,** 2: 697 (1965). Type: S. Rhodesia, Bulawayo, *Engler* 2915 (B†). Neotype from the Transvaal, Pretoria, *Reck* 13 (GRA).

Rhus amerina Meikle in Mem. N.Y. Bot. Gard. **8,** 3: 243 (1953).—Palmer & Pitman, Trees S. Afr.: 290, cum fig. (1961). Type: S. Rhodesia, Matopos, *Hutchinson* 4140 (K, holotype).

Spreading shrub or small tree up to 8·5 m. tall, with rough greyish bark; young branchlets reddish-brown, shining, ± angular, glabrous or sometimes pilose, the old ones subterete, greyish or brownish, lenticellate. Petiole 1·5–5 cm. long, slender, glabrous or sometimes pilose, slightly canaliculate and margined on the upper surface. Leaflets sessile, dull green above, lighter below (sometimes, when dried, brownish or blackish above and greyish or brownish-grey below), linear-lanceolate or oblong-lanceolate, membranous or ± rigid to subcoriaceous, glabrous or sometimes pilose mainly on the midrib; median leaflet 4–10·5(13·5) × (0·6)1–2(4·3) cm., abruptly cuneate at the base, the lateral ones smaller (about 1/3–2/3) and not so cuneate at the base, all acute or subacute at the apex, with the margin slightly revolute, entire or shallowly crenate-dentate; midrib slender, raised on both surfaces; lateral nerves (2–5 mm. apart) and reticulation scarcely visible. Panicles axillary and terminal, nearly as long as the leaves or the terminal ones slightly longer, lax with the axis and branches very sparsely furfuraceous, glabrous or pilose; pedicels 1–2 mm. long. ♂ flowers: calyx-segments white or pale yellow, c. 0·5 mm. long, ovate, obtuse; petals c. 1 mm. long, oblong; filaments c. 0·5 mm. long; anthers c. 0·2 mm. long; disk patelliform, 5-lobulate. ♀ flowers: ovary obliquely ovoid; styles deflexed; staminodes present. Drupe yellow-brown to cinnamon-brown, shining, 4–5 mm. in diam., depressed-globose or subquadrate, slightly asymmetrical, smooth, glabrous.

Bechuanaland Prot. SE: Lobatsi, Mogobane, fr. 3.iv.1957, *de Beer* 11 (K; SRGH); Dikomo Di Ki, ♂ fl. 26.ii.1960, *Wild* 5181 (SRGH). **S. Rhodesia.** N: Banket, st. vii.1922, *Eyles* 6233 (SRGH). W: Ndumba Hill, Inyati, c. 56 km. from Bulawayo, fr. 4.vi.1947, *Keay* FHI 21318 (FHO); Matopos, Mtsheleli Valley, fr. 1.v.1952, *Plowes* 1448 (K; LISC; SRGH). C: Salisbury, ♀ fl. & fr. 15.iv.1922, *Eyles* 3406 (BOL; K; SRGH); Chilimanzi Reserve, ♂ fl. 7.iii.1951, *Wormald* 28/51 (K; LISC; SRGH). E: Umtali, Zimunya's Reserve, fr. 3.vi.1956, *Chase* 6141 (K; LISC; SRGH). S: Ndanga, fr. v.1955, *Armitage* 104/55 (SRGH); Beitbridge, Jopempi Mt., ♂ fl. 26.ii.1961, *Wild* 5440 (COI; LISC; SRGH). **Nyasaland.** C: Kasungu Hill, fr. 28.viii.1946, *Brass* 17451 (BM; K; SRGH). **Mozambique.** SS: Limpopo, Chicualacuala, ♂ fl. vii.1928, *Smuts* (PRE). LM: Namaacha, Goba-Fronteira, st. 14.xii.1947, *Barbosa* 733 (LISC).

Also in the Transvaal and Cape Prov. Among rocks in granite and quartzite kopjes, sandy soils and reddish sandy loams in open savanna-woodland, *Acacia nigrescens-Colophospermum-Commiphora* associations, fringing forests, along streams, etc.

The specimens " N. Rhodesia, 30 miles North of Fort Jameson, fl. ♂ 26.iv.1962, *White* 2472 (FHO; K) ", identified by White (F.F.N.R.: 212, 1962) as *Rhus natalensis* Bernh. ex Krauss, may be a hybrid between this species and *R. leptodictya*.

20. **Rhus oblanceolata** Schinz in Bull. Herb. Boiss., Sér. 2, **8**: 638 (1908).—Burtt Davy, F.P.F.T. **2**: 509 (1932). Type from the Transvaal (Olifants R.).

Shrublet with puberulous to densely hirsute cylindric or slightly angular reddish-brown branchlets. Petiole 1–3 cm. long, stout, dorsally convex, canaliculate above, puberulous to hirsute. Leaflets subconcolorous, narrowly elliptic, oblanceolate or spathulate, subcoriaceous, frequently brown-ferruginous when dried, puberulous to ± densely hirsute and reddish- or whitish-furfuraceous-glandular on both surfaces, the median one 3–7(9) × 1–1·5(2·5) cm., the lateral ones a little smaller, all rounded, obtuse or acute and mucronulate at the apex, margin sometimes ciliate, rather narrowed towards the base, entire; adult leaflets with the midrib and the rather ascendent lateral nerves raised on both sides but mainly below and the reticulation ± prominent above. Panicles axillary, narrow, ± equalling the leaves, with the axis and branches ± pilose and ferruginous-furfuraceous; pedicels 1–2 mm. long. Calyx-segments c. 1 mm. long, lanceolate, subobtuse, hairy outside. Petals 1·25–1·75 × 0·5–1 mm., ovate, obtuse.

S. Rhodesia. W: Matabeleland, ♂ fl. 29.xii.1949, *Hunt* in GHS 26487 (SRGH).
Also in the Transvaal. In rocky places?

21. **Rhus pyroides** Burch., Trav. Int. S. Afr. **1**: 340 (1822).—Engl. in A. & C. DC., Mon. Phan. **4**: 430 (1883); Pflanzenw. Afr. **3,** 2: 211 (1921).—Schonl. in Bothalia, **3,** 1: 29 (1930).—Burtt Davy, F.P.F.T. **2**: 497 (1932). Type from S. Africa (Cape Prov.).
Rhus villosa sensu Oliv., F.T.A. **1**: 439 (1868) pro parte.

Var. **gracilis** (Engl.) Burtt Davy, F.P.F.T. **2**: 497 (1932).—O. B. Mill. in Journ. S. Afr. Bot. **18**: 47 (1952). Type from the Transvaal.
Rhus villosa var. *gracilis* Engl. in A. & C. DC., Mon. Phan. **4**: 425 (1883). Type as above.
Rhus pyroides var. *transvaalensis* Schonl. in Bothalia, **3,** 1: 30, fig. p. 31 (1930). Syntypes from the Transvaal.
Rhus pyroides sensu O. B. Mill. in Journ. S. Afr. Bot. **18**: 47 (1952) pro parte excl. specim. *Miller* B/392.
Rhus sp. near *R. pyroides*.—O. B. Mill., loc. cit.

Shrub or small tree up to 7 m. tall, often thorny, with the young branches brownish and patent-hairy, the oldest ones greyish. Petiole c. half as long as the median leaflet or shorter, slender, terete, pubescent, slightly furrowed above. Leaflets glaucous (when dried pale greenish, slightly darker above), submembranous, entire or with some teeth, appressed-pilose or sometimes densely sericeous, at times nearly glabrous; median leaflet 3–8 × 1–2·3 cm., oblanceolate to elliptic, narrowing at both ends, acute at the apex, cuneate at the base, the lateral ones elliptic, narrower and shorter than the median or ± equal to it; midrib and lateral nerves slender, raised on both surfaces, more so below, reticulation very close, visible on both surfaces. Panicles terminal and axillary, large, pubescent; flowers glomerulate. Calyx-segments 0·5–0·75 mm. long. Petals oblong, 1·25–1·5 mm. long. Drupe c. 3 mm. in diam., subglobose, slightly compressed.

Bechuanaland Prot. N: Ngamiland, ♂ fl. xii.1930, *Curson* 357 (PRE). SE: Palapye, Moeng College, ♂ fl. 29.xi.1957, *de Beer* 516 (K; SRGH); Pharing, ♀ fl. & fr. 14.xi.1948, *Hillary & Robertson* 510 (PRE); 4·8 km. N. of Lobatsi, Farm Knock Duff, ♂ fl. 16.i.1960, *Leach & Noel* 124 (COI; SRGH). **S. Rhodesia.** W: Matopos, Mtsheleli Valley, ♀ fl. & fr. 22.xi.1951, *Plowes* 1323 (LISC; SRGH). C: Chilimanzi, Umvuma, fr. iv.1951, *Gibson* 3/51 (K; SRGH); Salisbury, Lochinvar, ♂ fl. 17.xi.1945, *Wild* 377 (SRGH). **Mozambique.** LM: Libombos, near Namaacha, ♂ fl. 22.ii.1955, *E.M. & W.* 533 (BM; LISC; SRGH).
Also in the Transvaal, Orange Free State, Natal, Swaziland, E. Cape and SW. Africa. In woodlands, forests of various types and thickets, and on termite mounds.

In spite of the fact that some specimens are without flowers, we have referred all the material of our area to var. *gracilis*. It may be, however, that some of these specimens belong to var. *pyroides*, which has flowers not glomerulate.

22. **Rhus vulgaris** Meikle in Kew Bull. **6**: 290 (1951).—Van der Veken, F.C.B. **9**: 32, t. 4 (1960).—White, F.F.N.R.: 213 (1962). Type from Kenya.
Rhus villosa sensu Oliv. in F.T.A. **1**: 439 (1868) pro parte.—Engl. in A. & C. DC., Mon. Phan. **4**: 424 (1883) pro parte; Pflanzenw. Afr. **3,** 2: 209 (1921) pro parte.—Burkill in Johnston, Brit. Centr. Afr.: 241 (1897).—Gibbs in Journ. Linn. Soc., Bot. **37**: 437 (1906).—Eyles in Trans. Roy. Soc. S. Afr. **5**: 403 (1916).—Suesseng. in Proc. & Trans. Rhod. Sci. Ass. **43**: 35 (1951).
Rhus incana sensu Brenan, T.T.C.L.: 37 (1949) pro parte.

Shrub or tree up to 9 m. tall; branches ochraceous or brownish-reddish, ± densely pubescent with somewhat long hairs or puberulous. Petiole 1·5–6 cm. long, subcylindric or nearly so, ± densely hairy, rarely more than half as long as the median leaflet. Leaflets ± discolorous, green or dark-green above and paler or ochraceous below, or concolorous (light green), papyraceous to subcoriaceous, with ± appressed hairs on the upper surface and patent ones on the under surface, entire or with some teeth in the upper part; median leaflet 4–10·5 × 2·4–6 cm., elliptic or ovate to obovate or subcircular, rounded, truncate or emarginate (rarely acute) at the apex, cuneate and sessile at the base, or contracted into a very short petiolule; lateral leaflets c. 1/3–1/2 as long as the median, with ± asymmetric base and obtuse or rarely acute apex; midrib a little raised above, more so below, lateral nerves slender, arcuate, not prominent or sunk on the upper surface, raised beneath, as also the reticulation (sometimes concealed by the indumentum) which is visible on the upper surface in the oldest leaves. Panicles terminal and axillary in the upper leaf-axils, forming a large inflorescence (c. 30 cm. long), ± densely hairy, the terminal ones frequently with some bract-like leaves at the base. Flowers shortly pedicellate or subsessile, glomerulate. Calyx-segments c. 0·5 mm. long, ovate, hairy. Petals yellow or whitish, c. 1 mm. long, oblong-ovate, obtuse. Drupe red or brownish-red, c. 5 mm. in diam., globose.

N. Rhodesia. N: Abercorn, Kali Dambo, fr. 20.i.1952, *Richards* 785 (K). **S. Rhodesia.** W: Matobo, Farm Besner Kobila, ♂ fl. x.1956, *Miller* 1926 (K; LISC; SRGH). C: Chilimanzi, Shasha Kopjes, ♂ fl. 10.xi.1951, *Mylne* 52/51 (K; SRGH). E: Umtali, Zimunya, ♂ fl. 8.i.1956, *Chase* 5941 (BM; LISC; LMJ; SRGH). **Nyasaland.** N: Mpanda Summit, ♂ fl., *Lewis* 104 (FHO). C: Dedza, ♂ fl. 12.xi.1960, *Chapman* 1039 (COI; SRGH). S: Mlanje Mt., ♂ fl. 8.x.1957, *Chapman* 455 (BM; FHO). **Mozambique.** Z: Morrumbala, Muse Valley, ♂ fl. 2.x.1904, *Vasse* 101 (P).

Also in the Cameroun Republic, Ubangi, the Congo, the Sudan, Ethiopia, Uganda, Kenya and Tanganyika. On termite mounds and rocky kopjes, along stream banks and in forests.

23. **Rhus chirindensis** Bak. f. in Journ. Linn. Soc., Bot. **40**: 49 (1911).—Eyles in Trans. Roy. Soc. S. Afr. **5**: 402 (1916).—R. & A. Fernandes in Webbia **19,** 2: 700 (1965). Type: S. Rhodesia, Gazaland, near Chirinda, *Swynnerton* 168 (BM, holotype). TAB. **128.**

Rhus legatii Schonl. in Bothalia, **3,** 1: 51, fig. p. 52 (1930).—Burtt Davy, F.P.F.T. **2**: 502 (1932).—Palmer & Pitman, Trees S. Afr.: 291, cum fig. (1961).—Goodier & Phipps in Kirkia, **1**: 58 (1961). Type from S. Africa (Cape Prov.).

Shrub or small tree up to 5 m. tall; branches dull brown or blackish (when dried), cylindric, pubescent or glabrous. Petiole 1·5–6·5 cm. long, almost cylindric, narrowly canaliculate and marginate above, pubescent or glabrous. Leaflets ± dull red-brown, ovate or ovate-lanceolate, acuminate (acumen flat or sometimes falcate, very acute, apiculate), entire and ± undulate at the margin, membranous to ± rigid or subcoriaceous, glabrous or ± pubescent on the margin, midrib and nerves; median leaflet (3)6–13(16) × (1·2)2·5–4(7) cm., cuneate and frequently petiolulate at the base, the lateral ones (2)2·5–7(12) × (0·8)1·3–3·5(5·5) cm., asymmetric and slightly cuneate or somewhat rounded at the base, very shortly petiolulate to sessile; midrib slightly raised in the upper surface, very prominent below; lateral nerves arcuate, slender, raised on both sides, reticulation lax, almost invisible or sometimes conspicuous. Panicles terminal and axillary, ample, pyramidal, much branched, multiflorous, the terminal ones longer than the leaves, the axillary ones as long as the latter or somewhat longer; pedicels 1–2·5 mm. long. ♂ flowers: calyx-segments 0·5 mm. long, ovate, obtuse, glabrous; petals c. 1·5 mm. long, elliptic, obtuse; filaments c. 1 mm. long. ♀ flowers: ovary ovoid; styles reflexed; disk cupuliform, 5-lobulate; staminodes present. Drupe pinkish-yellow to reddish-brown, shining, (4)5(6) mm. in diam., globose, glabrous.

S. Rhodesia. W: Gwaai, ♀ fl., *Eyles* 6445 (SRGH). C: Selukwe, Ferny Creek, ♂ fl. 8.xii.1953, *Wild* 4300 (K; LISC; SRGH). E: Umtali, E. Vumba Mts., ♂ fl. 4.xii.1961, *Chase & Wild* 7567 (COI; SRGH); Chipinga, fr. ii.1960, *Farrell* 179 (LISC; SRGH). S: Nuanetsi, Lundi R., ♂ fl. xii.1955, *Davies* 1674 (K; SRGH). **Mozambique.** MS: Manica, Mavita, Moçambize Valley, ♂ fl. 26.x.1944, *Mendonça* 2642 (BR; COI; LISC; LMJ); Chimoio, Amatongas, fr. 22.i.1948, *Mendonça* 3670 (COI; K; LISC; LMJ; SRGH); Chimoio, Vila Pery, ♀ fl. & fr. 10.xii.1943, *Torre* 6275 (K; LISC; SRGH). SS: near Inhacondo, fr. 16.iii.1952, *Barbosa & Balsinhas* 4924 (BM; LISC; LMJ). LM: near Lourenço Marques, fr. v.1946, *Pimenta* (LISC).

Also in the Transvaal, Swaziland, Natal and Cape Prov. On termite mounds and in thickets, open woodlands and forests of several types and in exposed places among rocks at the top and on slopes of mountains.

24. **Rhus monticola** Meikle in Mem. N.Y. Bot. Gard. **8,** 3: 242 (1953). Type: Nyasaland, Mlanje Mt., Luchenya Plateau, *Brass* 16656 (K, holotype).

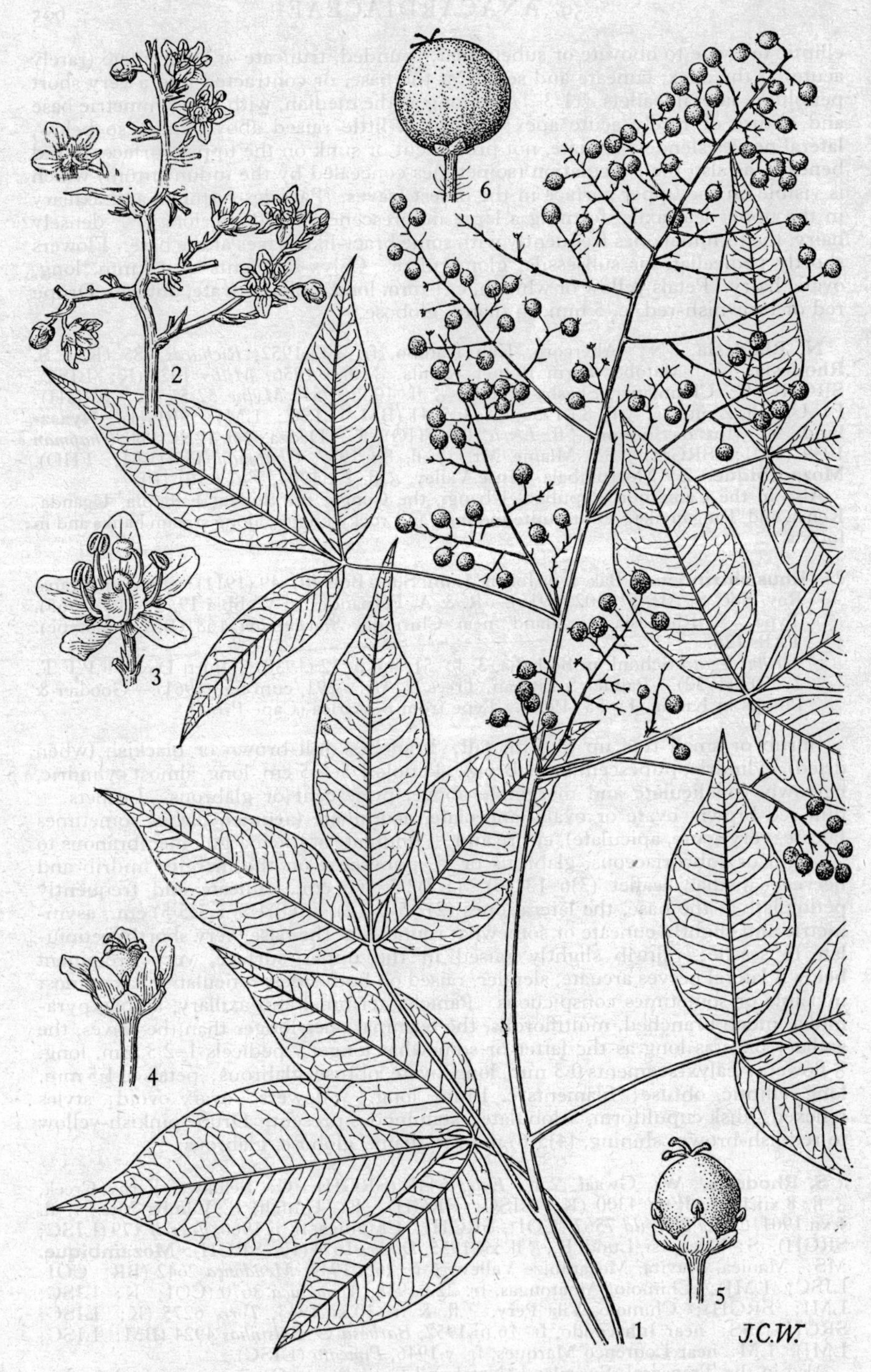

Tab. 128. RHUS CHIRINDENSIS. 1, branch with infructescence (× ⅔) *Andrada* 1007; 2, terminal part of male inflorescence (× 3) *Torre* 6275; 3, male flower (× 10) *Goodier & Phipps* 345; 4, female flower (× 10) *Torre* 6275; 5, immature fruit (× 10) *Torre* 6275; 6, fruit (× 4) *Andrada* 1007.

Shrublet up to 1 m. tall with the branchlets, petioles, axis and branches of the inflorescences ± densely pilose, the hairs slender, whitish, patent or crisped; bark of the stem and older branches reddish-brown. Petiole up to 6 cm. long, subcylindric, flat or slightly canaliculate above, glabrous with age. Leaflets discolorous (brownish-green above, lighter below), obovate to elliptic, narrowing at both ends, acuminate at the apex, the acumen up to 0·8 cm. long, cuspidate, plicate and falcate, with entire revolute margin; median leaflet 2·3–8 × 1–3·5 cm., cuneate at the base and petiolulate, the petiolule up to 0·8 cm. long; lateral leaflets 1·3–7 × 0·6–2·8 cm., not so cuneate or sometimes rounded at the base; midrib and lateral nerves slightly raised above, very prominent below, reticulation fairly visible on both sides. Panicles terminal, up to 10 cm. long, pyramidal; pedicels c. 1·5 mm. long, glabrous. ♂ flowers: calyx-segments c. 0·75 mm. long, ovate, obtuse, glabrous; petals c. 1·5 mm. long, oblong-ovate; filaments c. 1 mm. long. ♀ flowers not seen. Drupe brownish, c. 5 mm. in diam., subglobose.

Nyasaland. S: Mt. Mlanje, ♂ fl. 6.iii.1897, *Adamson* 351 (K); Mt. Mlanje, fr. 26.vi.1958, *Chapman* 606 (FHO; K; SRGH).

Known only from Mlanje Mt. In evergreen forest and rocky situations in grassland.

25. **Rhus quartiniana** A. Rich., Tent. Fl. Abyss. **1**: 141 (1847).—Van der Veken, F.C.B. **9**: 36 (1960).—White, F.F.N.R.: 213 (1962). Type from Ethiopia.

Rhus quartiniana var. *acutifoliolata* (Engl.) Meikle in Exell & Mendonça, C.F.A. **2**, 1: 106 (1954). Type from Angola.

Shrub or small tree up to 7 m. tall, with the old branches brownish, striate, lenticellate, pubescent or glabrous; branchlets cylindric, striate, ± densely and shortly whitish- or yellowish-patent-pilose. Petiole 0·8–2·5 cm. long, semicylindric, canaliculate above, ± densely pilose or tomentose. Leaflets discolorous (dark green or olive-green and dull or shining above, yellowish-green to orange coloured below), lanceolate, oblong-elliptic to subrhombic or ovate, obtuse or acute at the apex, patent-hairy mainly on nerves and margin, covered with resin and in the juvenile state also with ± dense very small scales on the upper surface, sparse or absent in the adult leaves and with dense generally persistent scurf on the under surface; median leaflet 3–7(8·5) × 1–2·7(3·3) cm., cuneate at the base and often petiolulate, the petiolule up to 0·6 cm., the lateral ones 1·5–5·5 × 0·7–2·5 cm., sessile, sometimes asymmetric at the base; midrib raised on both sides, lateral nerves slender, not or slightly prominent above, moderately raised below, reticulation almost invisible. Panicles up to 10 × 7 cm., with the axis and branches densely hispidulous and lepidote; pedicels c. 1 mm. long. ♂ flowers: calyx-segments c. 0·5 mm. long, shortly ovate, hispidulous on the back; petals yellowish-green, c. 1 mm. long, ovate. ♀ flowers: ovary globose. Drupe reddish- or yellowish-brown, c. 3 mm. in diam., subglobose.

Unarmed shrub or small tree; leaflets generally entire - - var. *quartiniana*
Thorny shrub; leaflets thinner and not so furfuraceous, generally irregularly dentate var. *zambesiensis*

Var. **quartiniana.** TAB. **129.**

Caprivi Strip. Lisikili, 24 km. E. of Katima Mulilo, st. 17.vii.1952, *Codd* 7092 (EA; K; PRE; SRGH). **Bechuanaland Prot.** N: 59 km. N. of Kachikan on road to Kazane, st. 11.vii.1937, *Erens* 391 (K; PRE); Kabulabula, Chobe R., fr. vii.1930, *van Son* in Herb. Transv. Mus. 28762 (BM). **N. Rhodesia.** B: Balovale, ♂ fl. iv.1954, *Gilges* 354 (K; PRE; SRGH). N: Fort Rosebery, fr. 9.v.1958, *Fanshawe* 4426 (FHO; K); Abercorn-Mbosi road, ♀ fl. 30.iii.1932, *Main-Thompson* 1108 (K). W: Kitwe, ♀ fl. 2.vi.1957, *Fanshawe* 3305 (K; LISC; SRGH). E: Nyamadzi R., ♂ fl. 25.iii.1955, *E. M. & W.* 1175 (BM; LISC; SRGH). S: Kasungula, ♀ fl. & fr. v., *Gairdner* 554 (K); Livingstone, ♂ fl. iv.1909, *Rogers* 7091 (BOL; K). **S. Rhodesia.** N: Gokwe Pool, Umniati R., st. viii.1947, *Whellan* 75 (K; SRGH). W: Victoria Falls, ♂ fl. 25.iv.1932, *Main-Thompson* 1342 (K). C: Hartley, Poole Farm, ♂ fl. 24.ii.1951, *Hornby* 3223 (K; SRGH); Salisbury, Hunyani R., st. 20.iv.1947, *Wild* 2006 (COI; K; SRGH). E: Ndanga, Sabi R., fr. 10.vi.1950, *Chase* 2282 (BM; SRGH); Ndanga, Chiredzi R., ♂ fl. 30.iii.1959, *Farrell* 78 (SRGH). **Mozambique.** T: Angónia, R. Mauè, fr. 13.v.1948, *Mendonça* 4193 (LISC).

Also in Ethiopia, Kenya, Tanganyika, Congo, Angola and SW. Africa. In riverine bush or sometimes ± inundated forests.

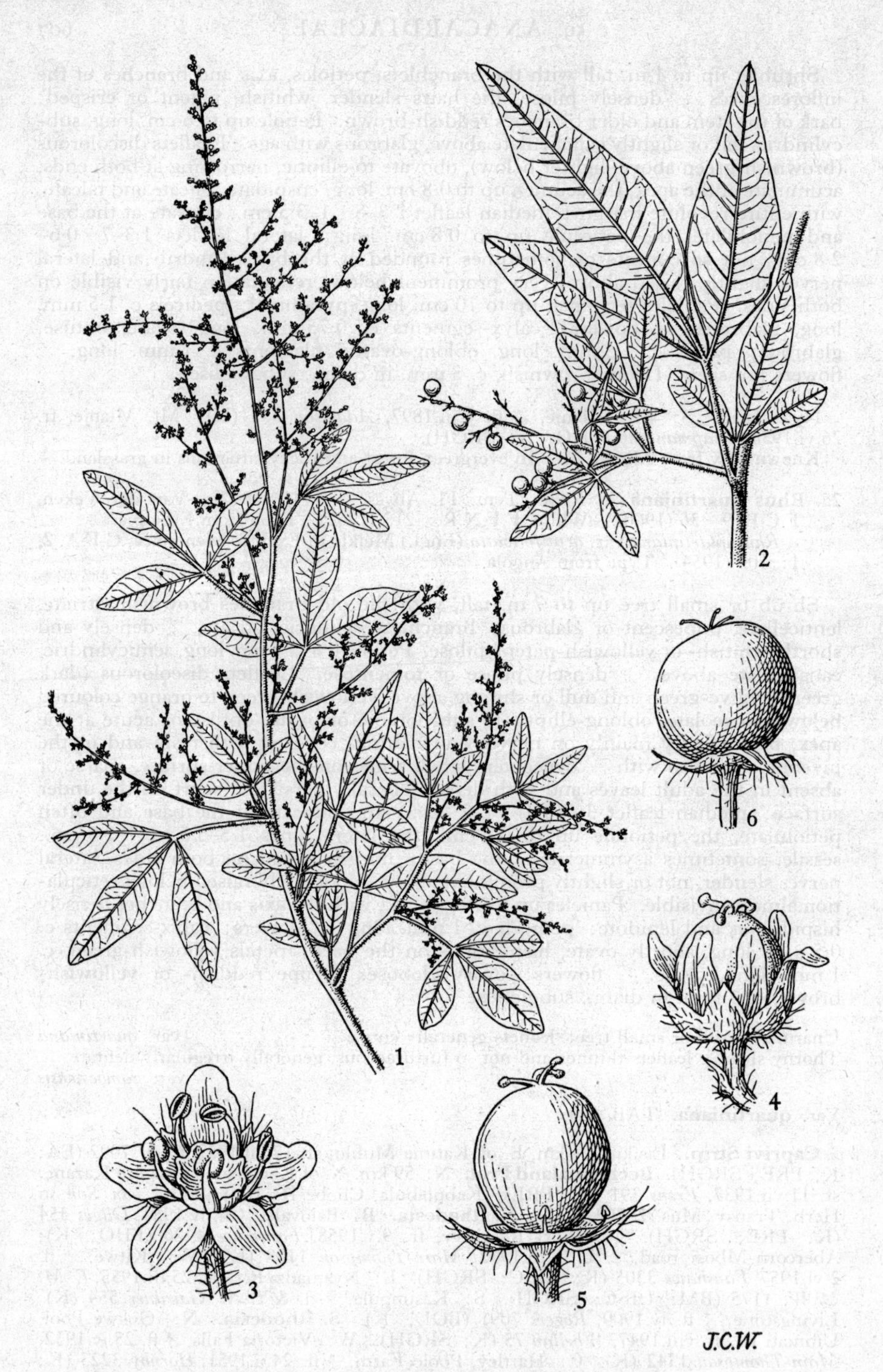

Tab. 129. RHUS QUARTINIANA var. QUARTINIANA. 1, branch with male inflorescences (× $\frac{2}{3}$); 2, branchlet with fruits (× $\frac{2}{3}$); 3, male flower (× 14); 4, female flower with young fruit (× 14); 5, female flower after petal-fall (× 14); 6, fruit (× 6); 1 and 3–6 from *Gilges* 354; 2 from *Flanagan* 3211.

Var. **zambesiensis** R. & A. Fernandes in Bol. Soc. Brot., Sér. 2, **38**: 189, t. 50 (1965). Type: N. Rhodesia, Barotseland, Nangweshi, Zambezi R. bank, *Codd* 7200 (BM, holotype; EA; K; PRE; SRGH).

Caprivi Strip. Mpilila I., st. 12.i.1951, *Killick & Leistner* 3331 (K; M). **N. Rhodesia.** B: near Senanga, st. 2.viii.1952, *Codd* 7351 (BM). S: Livingstone, ♀ fl. 6.iii.1956, *Gilges* 615 (SRGH); Katambora, Zambezi, ♂ fl. 14.iv.1949, *West* 2914 (SRGH). **S. Rhodesia.** W: Victoria Falls, st. 24.vii.1950, *Robertson & Elffers* 38 (K; PRE).

In riverine bush and woodlands.

26. **Rhus rehmanniana** Engl. in A. & C. DC., Mon. Phan. **4**: 422 (1883).—Burtt Davy, F.P.F.T. **2**: 496 (1932). Syntypes from Natal and the Transvaal.
Rhus macowanii forma *rehmanniana* (Engl.) Schonl. in Bothalia, **3**, 1: 26, fig. p. 26 (1930). Syntypes as above.

Var. **longecuneata** R. & A. Fernandes in Bol. Soc. Brot., Sér. 2, **38**: 190 (1965). Type: Mozambique, Lourenço Marques, Maputo, *Hornby* 2597 (K, holotype; LMJ; PRE; SRGH).

Small shrub; branches greyish-brown, sparsely lenticellate, at first villous later glabrescent; branchlets terete, densely and softly villous. Petiole 1–3·5 cm., subterete, slightly furrowed above, villous. Leaflets yellowish-green above, paler below, subcoriaceous, pubescent or villous; median leaflet 2·5–6·5 × 0·5–4·5 cm., obovate-cuneate, truncate or subtruncate and crenulate at the apex, entire and narrowing from the upper part to the cuneate base; lateral leaflets 1·5–4 × 0·6–2·5 cm., obovate, sometimes emarginate at the apex, entire or ± crenate in the upper part, cuneate at the base; midrib and the branched lateral veins slightly prominent on the upper surface very prominent on the lower one, reticulation inconspicuous. Panicles axillary, ± subequalling the leaves, and terminal ones much longer; axis pubescent. Calyx-segments c. 0·5 mm. long. Petals c. 1·5 mm. long.

Mozambique. LM: Maputo, ♂ fl. 13.iii.1947, *Hornby* 2597 (K; LMJ; PRE; SRGH).

Known as yet only from the S. of Mozambique. In bush.

This variety differs from var. *rehmanniana* by the smaller and long-cuneate leaflets.

27. **Rhus refracta** Eckl. & Zeyh., Enum. Pl. Afr. Austr. Extratrop. **2**: 145 (1836).—Sond. in Harv. & Sond., F.C. **1**: 510 (1860).—Engl. in A. & C. DC., Mon. Phan. **4**: 427, 538 (1883).—Schonl. in Bothalia, **3**, 1: 45, fig. p. 45 (1930). Type from S. Africa (Cape Prov.).

Shrub or small tree with cylindric greyish- or pale rufous-villous stems and spinescent often divaricate branches. Petiole 0·7–1·8 cm. long, densely hairy, slender, slightly sulcate above. Leaflets subconcolorous to ± discolorous (the upper surface dark green the under one lighter), rigidly membranous, ± villous (mainly on the nerves below), rounded or emarginate at the apex, with entire or sparsely crenate somewhat revolute margin; median-leaflet 1·2–3·3 × 1–1·5 cm., obovate, cuneate at the base, the lateral ones obovate-oblong, not cuneate and nearly symmetrical; midrib slightly impressed, lateral nerves usually deeply so above, raised beneath, reticulation neither prominent nor visible. Panicles axillary and terminal, shorter to slightly longer than the leaves, hairy. Calyx-segments c. 0·5 mm. long, ± ovate, pilose. Petals 1·25 mm. long, oblong. Drupe blackish at maturity, 4 mm. in diam., subglobose.

Mozambique. MS: Cheringoma, 10 km. from Inhaminga, fr. 11.v.1942, *Torre* 4136 (COI; K; LISC; SRGH).

Also in Natal and the Cape Prov. In woodland.

28. **Rhus magalismontana** Sond. in Harv. & Sond., F.C. **1**: 510 (1860).—Schonl. in Bothalia, **3**, 1: 90 (1930) pro parte excl. syn. *R. burkeana* et *R. oblanceolata*.—Burtt Davy, F.P.F.T. **2**: 508 (1932).—O. B. Mill., B.C.L.: 34 (1948); in Journ. S. Afr. Bot. **18**: 47 (1952). Type from the Transvaal.

A much-branched shrublet up to 0·6 m. tall; branches greyish, cylindric, rugose, hispidulous to glabrescent; branchlets brownish, densely hispidulous and furfuraceous. Petiole 0·1–1·5 cm. long, canaliculate above, slightly marginate and sometimes a little winged towards the apex, ± hispidulous and furfuraceous.

Leaflets concolorous (greenish or pinkish-ochraceous to pale green when dried), rigidly coriaceous, glabrous when old, densely covered on both surfaces with an appressed scrofulous layer of thin whitish or nearly translucent scales; median leaflet (0·6)1·8–3·5(6) × (0·3)0·7–1·4(2·5) cm., obovate-cuneate or oblanceolate-cuneate, obtuse or sometimes acute and mucronulate at the apex, the lateral ones ± equalling the median one and somewhat asymmetric at the base, all entire; midrib and lateral nerves prominent, a little more so on the upper surface; reticulation slightly prominent in both surfaces. Panicles with terminal ones up to 4·5 cm. long, and axillary ones ± equalling the leaves, with the axis, branches and bracts hispidulous and reddish-furfuraceous-glandular; pedicels c. 1 mm. long. ♂ flowers: calyx-segments c. 1 mm. long, ovate, obtuse, reddish-glandular outside. Petals c. 2 mm. long, oblong. Drupe shining, 4 mm. in diam., subglobose.

Bechuanaland Prot. SE: Kanye, Pharing, fr. 13.xi.1948, *Hillary & Robertson* 492 (PRE); 3·2 km. S. of Lobatsi, ♂ fl. 16.i.1960, *Leach & Noel* 113 (SRGH).
Also in the Transvaal. Rocky places on hills.

29. **Rhus milleri** R. & A. Fernandes in Bol. Soc. Brot., Sér. 2, **38**: 188, t. 48 (1964).
Type: Bechuanaland Prot., Kanye, Pharing, *Miller* B/948 (K, holotype; PRE).
Rhus pyroides sensu O. B. Mill. in Journ. S. Afr. Bot. **18**: 47 (1952) pro parte quoad specim. *Miller* B/392.
Rhus pyroides var. *puberula* sensu O. B. Mill., loc. cit. pro parte quoad specim. *Hillary & Robertson* 585.
Rhus (*? sp. nov.*).—O. B. Mill., loc. cit.

Shrub up to 2·4 m. tall; branches cylindric, striate, covered with dark-greyish fissured bark, hispidulous to glabrescent; branchlets reddish, hispidulous and ferrugineous scrofulous-glandular, as also the petioles, panicles and calyx-segments. Petiole 0·5–2 cm. long, canaliculate and marginate or somewhat winged above towards the apex. Leaflets membranous becoming slightly rigid with age, the youngest pale-reddish, the others light-green, hispidulous mainly on the midrib and margins, ± densely-ferrugineous-scrofulous-glandular on both surfaces, entire; median leaflet 1·5–3·5 × 0·6–1·5 cm., obovate or oblanceolate, obtuse, cuneate and somewhat asymmetric at the base; midrib and the delicate lateral nerves prominent on both surfaces; venation raised on both surfaces in the oldest leaves. Panicles terminal and subterminal, up to 9 cm. long, few-branched; pedicels 0·5–1·25 mm. long; bracts 1·5 mm. long. ♂ flowers: calyx-segments c. 0·5 mm. long, ovate, obtuse; petals 1 × 0·5 mm., oblong-obtuse.

Bechuanaland Prot. SE: Kanye, Pharing, ♂ fl. 15.xi.1948, *Hillary & Robertson* 585 (K; PRE).
Known at present only from Bechuanaland Prot. On termite mounds and in savanna-woodlands.

30. **Rhus fanshawei** R. & A. Fernandes in Bol. Soc. Brot., Sér. 2, **38**: 185, t. 46 (1965).
Type: N. Rhodesia, Nkolemfumu, *Fanshawe* 4767 (EA; K, holotype; SRGH).

Suffrutex with upright stems up to 0·9 m. tall arising from a long horizontal woody rhizome; stems unbranched or ± branched, cinnamon-brownish, cylindric, striate, very shortly puberulous, more densely so in the upper parts. Petiole 1·5–5·5 cm. long, canaliculate and inconspicuously to distinctly winged above, puberulous to glabrescent. Leaflets subconcolorous to discolorous, thinly membranous when young, the oldest ones papyraceous to subcoriaceous (sometimes coriaceous), glabrous or very sparsely hispid and densely lepidote mainly in the juvenile state, the scales roundish, appressed, whitish to ferruginous; median leaflet 7–10 × 2–3·5 cm., lanceolate, rhombic to obovate, obtuse or acute and mucronate at the apex, cuneate at the base, the lateral ones oblong to ovate, obtuse, a little to very asymmetric at the base; lateral nerves and venation very conspicuous, prominent on the under surface. Panicle terminal and axillary, lax, pauciflorous, hispidulous; flowers subsessile. Calyx-segments 0·5 mm. long, ovate-lanceolate, subobtuse, lepidote. Petals 1 mm. long, ovate-oblong, obtuse. Drupe brownish-red, shining, c. 4 mm. in diam., globose, glabrous.

N. Rhodesia. B: Masese, fr. 22.vii.1961, *Fanshawe* 6666 (FHO). N: Shiwa Ngandu, fr. 5.ii.1955, *Fanshawe* 1989 (K; SRGH); Chambezi R.-Kasama road, c. 41·5 km. from Kasama, fl. immat. 23.x.1949, *Hoyle* 1315 (FHO).
Known at present only from N. Rhodesia. On rocky hills and in woodlands.

31. **Rhus ochracea** Meikle in Kew Bull. **8**: 107 (1953).—White, F.F.N.R.: 213 (1962). Type: Nyasaland, Fort Hill, *Whyte* (K, holotype).

Suffrutex or small branched shrub 0·6–1·2(3·6) m. tall. Branches reddish-brown, cylindric, ± densely puberulous with somewhat crisped hairs mixed with longer patent ones. Petiole 1–4 cm. long, robust, canaliculate above, ± densely hispidous or puberulous and scrofulous to glabrescent. Juvenile leaves papyraceous with faint venation, the adult ones subcoriaceous or coriaceous, discolorous when dried (dull green or brownish above, yellowish-green, ochraceous or ferrugineous beneath), sparsely to very densely hispid mainly on the midrib and lateral nerves on both surfaces, ± densely and softly scrofulous-glandular (glandular hairs small, many-branched, yellowish, shining) below and not so densely so above, entire, rarely crenate in the upper part; median-leaflet 3·8–11 × 1·7–5 cm., elliptic, obovate to broadly obovate, sometimes oblanceolate, obtuse or subobtuse or rounded or acute at the apex and cuneate at the base and sometimes petiolulate, the lateral ones 1·8–7·5 × 0·9–4 cm., sessile, obovate, obtuse at the apex and asymmetric at the base; midrib, lateral nerves and reticulation raised on both surfaces, more strongly so below. Panicles terminal and axillary (in the upper leaf-axils) usually longer than the leaves, making in all a large terminal leafy inflorescence up to 30 × 25 cm.; axis and pedicels hispid and lepidote-glandular. ♂ flowers: calyx-segments 0·75–1 mm. long, ovate, hispid; petals 1·5 mm. long, oblong. Drupe brown, shining, up to 4 mm. in diam., globose, glabrous.

Shrub; median leaflet up to 11 × 5 cm.; panicles ample, longer than the leaves var. *ochracea*
Suffrutex; median leaflet up to 5 × 2 cm.; panicles generally shorter than the leaves var. *saxicola*

Var. **ochracea**

Rhus villosa var. *tomentosa* sensu R. E. Fr. in Wiss. Ergebn. Schwed. Rhod.-Kongo-Exped. **1**: 127 (1914).
Rhus squalida sensu White, F.F.N.R.: 213 (1962).

N. Rhodesia. N: Mbesuma Ranch, fr. 25.vii.1961, *Astle* 822 (SRGH); Abercorn, ♂ fl. 23.x.1952, *Robertson* 139 (EA; K; LMJ; PRE; SRGH). **Nyasaland.** N: Fort Hill, ♂ fl. vii.1896, *Whyte* (K).

Also in Tanganyika. Mainly in *Brachystegia-Julbernardia* woodlands and among rocks and in grasslands, on termite mounds, etc.

Var. **saxicola** R. & A. Fernandes in Bol. Soc. Brot., Sér. 2, **38**: 189, t. 49 (1965). Type: N. Rhodesia, Mpika, Muchinga Escarpment, *Angus* 865 (FHO; K, holotype).

N. Rhodesia. N: Mpika, Muchinga Escarpment, 48 km. S. of Shiwa Ngandu, ♀ fl. 28.xi.1952, *Angus* 865 (FHO; K).

Known at present only from N. Rhodesia. On tops of rocky hills.

Angus 865 is a doubtful specimen, intermediate between *R. ochracea* and *R. fanshawei*, which it approaches by its scales. In some aspects it is also somewhat similar to *R. quartiniana*, differing in habit and by its broader leaflets. Material is, nevertheless, insufficient to judge whether it should be considered as an independent species or as a hybrid. We therefore consider it to be a variety of *Rhus ochracea*, the nearest species.

32. **Rhus macowanii** Schonl. in Bothalia, **3**, 1: 24, fig. p. 25 (1930) pro parte excl. syn. *R. rehmanniana* Engl.—Burtt Davy, F.P.F.T. **2**: 496 (1932). Type from the Transvaal.

Shrub or small tree up to 9 m. tall. Branches cylindric, fulvous-ochraceous, densely villous. Leaves 3-foliolate (occasionally 4- or 5-foliolate); petiole 1–1·5 cm. long, subterete; leaflets somewhat discolorous (when dried dull-brown on the upper surface, reddish-brown or yellowish-brown below), subcoriaceous, villous with ± patent hairs, entire or sometimes crenate or dentate in the upper part; median leaflet 2·5–4·5 × 1·4–2·2 cm., ovate or obovate-cuneate, rounded, obtuse or subacute at the apex; lateral leaflets c. 2/3 as long as the median one, symmetric or slightly asymmetric at the base; midrib and the arcuate lateral nerves visible, but slightly or not at all raised on the upper surface, rather prominent on the lower; tertiary nerves and reticulation distinct. Panicles many-branched and multiflorous, pubescent, the axillary ones as long as the leaves or longer, the terminal ones much longer; pedicels 0·5–0·75 mm. long. Calyx-segments c. 0·5 mm. long, ovate, obtuse, hairy. Petals 1–1·25 mm. long, oblong. Drupe cream-coloured or reddish, c. 3 mm. in diam., subglobose.

Mozambique. LM: Delagoa Bay, ♂ fl. 1908, *Junod* 360 (G).
In the coastal districts of eastern S. Africa. In open scrub and in riverine bush.

33. **Rhus anchietae** Ficalho & Hiern ex Hiern, Cat. Afr. Pl. Welw. **1**: 184 (1896).—Meikle in Exell & Mendonça, C.F.A. **2,** 1: 106 (1954).—Van der Veken, F.C.B. **9**: 37 (1960).—White, F.F.N.R.: 213 (1962).—R. & A. Fernandes in Webbia, **19,** 2: 701 (1965). Type from Angola (Huila).
Rhus suffruticosa Meikle in Bol. Soc. Brot., Sér. 2, **26**: 287, t. 6 (1952); in Exell & Mendonça, tom. cit.: 103, t. 19 (1954).—Van der Veken, F.C.B. **9**: 31 (1960).—White, F.F.N.R.: 212 (1962). Type: N. Rhodesia, Mwinilunga District, near Dobeka Bridge, *Milne-Redhead* 3609 (K, holotype).

Dense shrub or small tree up to 8 m. tall; branches dark-brown to reddish-brown, cylindric, striate, glabrous or glabrescent; branchlets glabrous to ± densely and shortly fulvous-tomentose. Petiole 2–6 cm. long, canaliculate above, ± fulvous-tomentose to glabrous. Leaflets discolorous (darker above than below), subcoriaceous, generally acute, sometimes obtuse, entire, sparsely pilose to glabrous on the lamina, ± densely pilose on the midrib and nerves, sometimes (when young) covered with resin and shining; median leaflet 6–15 × 3–7·3 cm., rhombic to elliptic, ± narrowed towards the apex, narrowing to the base and sometimes petiolulate, the petiolule up to 1 cm. long; lateral leaflets somewhat smaller than the median one, broadly elliptic to rhombic, asymmetric and scarcely cuneate at the base; midrib and lateral nerves prominulous above, rather prominent below, reticulation close, ± visible on both surfaces. Panicles axillary and terminal up to 30 × 25 cm., lax, with the axis and branches ± pubescent; pedicels 0·5–1·5 mm. long. ♂ flowers: calyx-segments c. 0·5 mm. long, ovate, pubescent; petals greenish to yellowish, c. 1·5 mm. long, oblong-ovate. Drupe brown, 3–5 mm. in diam., glabrous and shining.

N. Rhodesia. N: road from Abercorn to Chila, Lucheche Stream, ♂ fl. 1954, *Richards* 2327 (EA; K); Abercorn, Kambole road, ♀ fl. & fr. 1.i.1955, *Richards* 3822 (K). W: Ndola, Chichele Dambo, fr. 10.iv.1951, *Holmes* 445 (FHO; K); Ndola, Mendolo Forest Reserve, ♀ fl. 20.xii.1951, *Holmes* 447 (FHO; K). C: Broken Hill, fr. 20.vi.1920, *Rogers* 24429 (FHO; PRE). **Nyasaland.** N: Rumpi, Nyika, ♂ fl. 20.vi.1960, *Chapman* 771 (COI; SRGH).
Also in Angola, Congo, Uganda and Tanganyika. In riverine forest, savannas and dambos (swamps).

Rhus suffruticosa Meikle is so closely related to *R. anchietae* Ficalho & Hiern that it is almost impossible to say whether herbarium material belongs to one taxon or the other. In both the branches and leaves may be glabrous or ± pubescent, the form of the leaflets is very similar, the inflorescences, flowers and drupes are identical. We therefore think that *R. suffruticosa* is a savanna and swamp form of *R. anchietae.*

34. **Rhus culminum** R. & A. Fernandes in Bol. Soc. Brot., Sér. 2, **38**: 184, t. 44–45 (1965). Type: S. Rhodesia, Chimanimani Mts., *Chase* 2958 (BM, holotype; COI; SRGH).

Straggling shrub up to 3 m. tall or small tree 3–4·5 m. high; branches reddish-brown, straight or tortuous, cylindric, slightly striate and sparsely lenticellate, glabrous to ± densely puberulous. Petiole reddish, 1·5–5 cm. long, stout, slightly canaliculate above, glabrous to ± puberulous. Leaflets light- to dull-green or reddish-brown above, ochraceous to ferruginous below, rigidly coriaceous, glabrous on both surfaces except for a few hairs at the base, entire; median leaflet 5–10 × 2·8–5·5 cm., elliptic or broadly elliptic or nearly subcircular, roundish or obtuse to subacute and mucronate at the apex, cuneate and narrowed into a petiolule at the base; lateral leaflets a little smaller than the median one, oblong (up to 6 × 2·3 cm.) or elliptic (up to 8·7 × 5·3 cm.), obtuse at the apex, asymmetrical and oblique at the base; midrib slender, with both the lateral nerves and the very close reticulation prominent on both surfaces but more distinctly so below. Panicles terminal, very ample, up to 24 cm. long, the axis and the nearly divaricate branches reddish, rather thick, cylindric, glabrous to hispidulous; bracts 1·5–3 mm. long, lanceolate, acute; pedicels 0·5–1·5 mm. long. Flowers pale yellow. Calyx-segments ovate-lanceolate, subacute, glabrous or ciliate on the margin.

Petals 1·5–1·75 mm. long, ovate, obtuse. Stamens with filaments 0·75 mm. long and anthers c. 0·5 mm. long. Drupe immature, black, subglobose.

S. Rhodesia. E: Inyanga, Pungwe R., ♂ fl. x.1956, *Miller* 3787 (K; LISC; SRGH); Inyanga National Park, fr. xi.1958, *West* 3790 (K; SRGH). **Mozambique.** MS: Gorongosa, near Gogôgo Hill, ♂ fl. 29.ix.1943, *Torre* 5986 (BM; K; LISC).

Known only from the mountainous regions of eastern S. Rhodesia and the Manica e Sofala Distr. of Mozambique. In forests and woodlands and on the tops of mountains among rocks (1350–2100 m.).

35. **Rhus rogersii** Schonl. in Bothalia, **3,** 1: 42, fig. p. 43 (1930).—R. & A. Fernandes in Bol. Soc. Brot., Sér. 2, **38**: 190, t. 51 (1965). Syntypes from the Transvaal and Swaziland.

Suffrutex up to 30 cm. tall with a woody rootstock; stems upright, branched, cinnamon-coloured, cylindric, slightly striate, glabrescent at the base, ± villous at the apex. Petiole 0·3–1·8 cm. long, slightly canaliculate or flat above, ± villous. Leaflets sessile, discolorous (ferruginous above, pale cinnamon-coloured below), not shining, papyraceous, with very slightly revolute margin, sparsely villous on both surfaces, a little more so on the nerves and margin; median leaflet 3·5–8 × 1·7–3·3 cm., broadly to narrowly elliptic, somewhat narrowed at the apex into a short cuspidate acumen, entire or with 1–3 mucronate-cuspidate teeth in the upper part, symmetric and cuneate at the base; lateral leaflets narrower and c. 1/3 shorter than the median one, usually entire or sometimes also with few teeth, asymmetric at the base; midrib, lateral nerves and reticulation paler coloured than the lamina, prominulous on both surfaces. Panicles narrow, the axillary ones shorter, the terminal ones nearly as long as the leaves; axis, branches, bracts and pedicels (1 mm. long) villous. Calyx-segments c. 1 mm. long, ovate-lanceolate, sparsely villous to glabrous. Petals c. 1·5–2 mm. long, oblong, obtuse. ♀ flowers and fruits not seen.

Mozambique. LM: Namaacha, ♂ fl. 22.xii.1944, *Torre* 6943 (K; LISC; SRGH).
Also in the Transvaal and Swaziland. In scrub.

36. **Rhus microcarpa** Schonl. in Bothalia, **3,** 1: 80, fig. p. 79 (1930). Syntypes from Natal.

Shrub 2–6 m. high, sometimes thorny, the thorns c. 1 cm. long and uncinate; branches and branchlets pale yellowish-grey, cylindric, the latter ± densely pubescent and finally glabrescent or always glabrous and brownish. Petiole 1·5–4 cm. long, slender, canaliculate, pubescent to glabrescent or glabrous. Leaflets concolorous to slightly discolorous, papyraceous to coriaceous (in oldest leaves), entire or rarely with 1–2 teeth, the margin revolute, glabrous or ± sparsely pubescent to glabrescent, the hairs appressed, denser on the midrib and nerves; median leaflet (2·2)4–9(11) × 1·3–4 cm., usually 4·5–6 × 1·7–2·7 cm., obovate, rhombic to oblanceolate, rounded or generally acute to very shortly acuminate at the apex, the acumen folded and mucronate, cuneate to subpetiolulate at the base; lateral leaflets usually 3–3·5 × 1·3–2 cm., obovate or elliptic, obtuse or subacute, mucronate, not so cuneate as the median one; midrib, lateral and tertiary nerves raised on both surfaces. ♂ panicles axillary and terminal, the former equal to or longer than the leaves, the latter longer, pyramidal, multiflorous, with the axis and branches sparsely to densely pubescent; ♀ panicles shorter and laxer than the ♂ ones. ♂ flowers: calyx-segments 0·5 mm. long, triangular; petals cream or yellowish-green, c. 0·75 mm. long, ovate. Drupe red, turning blackish when mature, c. 2·5 mm. in diam., subglobose, edible.

Mozambique. SS: Gaza, João Belo, Chipenhe, Chirindzeni Forest, ♂ 1.iv.1959, *Barbosa & Lemos* 8432 (COI; K; LMJ; SRGH). LM: Marracuene, Bobole, ♀ fl. 7.iv.1947, *Barbosa* 137 (COI); Mamaacha, ♂ fl. 20.i.1958, *Barbosa & Lemos* 8247 (COI; K; LISC; LMJ); Inhaca I., fr. 10.vii.1957, *Barbosa* 7633 (LMJ; PRE; SRGH).
Also in Natal. In coastal bush and woodlands, secondary thickets and riverine forests.

37. **Rhus longipes** Engl. in A. & C. DC., Mon. Phan. **4**: 431 (1883); Pflanzenw. Afr. **3,** 2: 212 (1921).—Diels in Engl., Bot. Jahrb. **24**: 583 (1898).—Wild in Rhod. Agr. Journ. **49,** 5: 286 (1952).—Meikle in Mem. N.Y. Bot. Gard. **8,** 3: 241 (1953); in Exell & Mendonça, C.F.A. **2,** 1: 107 (1954).—Van der Veken, F.C.B. **9**: 37 t. 5 (1960).—

White, F.F.N.R.: 213 (1962).—R. Fernandes in Mem. Junta Invest. Ultr., Sér. 2, **38**: 38 (1962). Type from Angola.

Rhus inamoena Standley ex Bullock in Kew Bull. **1933**: 77 (1933) *nom. nud.*—Burtt Davy & Hoyle, N.C.L.: 29 (1936).—Brenan, T.T.C.L.: 37 (1949).

Rhus villosa sensu Oliv., F.T.A. **1**: 439 (1868) pro parte.—Burtt Davy & Hoyle, N.C.L.: 29 (1936), et auct. mult., non L. f.

Rhus villosa var. *grandifolia* Oliv., F.T.A. **1**: 439 (1868).—R. E. Fr. in Wiss. Ergebn. Schwed. Rhod-Kongo-Exped. **1**: 127 (1914).—Burtt Davy & Hoyle, N.C.L.: 29 (1936). Type from Zanzibar.

Rhus longipes var. *grandifolia* (Oliv.) Meikle in Mem. N.Y. Bot. Gard. **8**, 3: 241 (1953). Type as above.

Shrub or small tree up to 9 m. tall; branches brownish-grey, cylindric, slightly furrowed, lenticellate, the youngest ones somewhat angular, dark-brown and sparsely pilose to ± densely fulvous-pilose. Petiole 1·5–6 cm. long, semiterete, canaliculate on the upper surface, ± pilose. Leaflets elliptic, obovate-elliptic to broadly elliptic, obtuse and sometimes emarginate or rarely acute at the apex, entire, the young ones very dark (nearly black when dried), membranous, the older ones brownish-green or ± dark-brownish on the upper surface and paler below, somewhat rigid, with slender hairs on the nerves and margin and sometimes also on the lamina persistent or ± caducous with age (the oldest leaves very frequently glabrous); median leaflet 4·2–12·7 × 1·7–7·5 cm., cuneate and petiolulate at the base; lateral leaflets 2·3–9 × 0·9–5·8 cm., sessile, ± asymmetric at the base; midrib and lateral nerves in the young leaves scarcely visible on the upper surface and prominulous below, the oldest ones with the midrib and nerves raised on both sides, mainly below, the reticulation lax and very conspicuous. Panicles pyramidal lax, the terminal ones generally longer than the leaves, the axillary ones smaller; axis and branches ± densely fulvous-pubescent, with patent and crisped hairs; pedicels c. 1 mm. long, pubescent. ♂ flowers: calyx-segments 0·5 mm. long, ovate, ± pilose outside; petals 1–1·5 mm. long, oblong-ovate, obtuse. ♀ flowers: ovary subglobose, 1-locular with (1–2)3(4) styles or more frequently 2-lobed and 2-locular with (2–3)4–5 styles. Drupe reddish, shining, 3–7 mm. in diam., glabrous or very sparsely pilose, globose, 1-locular and 1-seeded or frequently reniform, 2-locular, with 1 seed in each loculus, or 1 loculus sterile.

Branches, petioles and young leaflets ± pubescent with ± patent hairs; ovaries 2-lobed with 4–5 styles; 2-seeded drupes present to a variable extent - - var. *longipes*
Branches, petioles and leaflets glabrous or nearly so; 2-lobed ovaries and 2-seeded drupes absent - - - - - - - - var. *schinoides*

Var. **longipes**

Rhus villosa sensu R. E. Fr. in Wiss. Ergebn. Schwed. Rhod-Kongo-Exped. **1**: 127 (1914).

Rhus villosa var. *grandifolia* sensu R. E. Fr., loc. cit.

Rhus glaucescens var. *natalensis* sensu Suesseng. in Proc. & Trans. Rhod. Sci. Ass. **43**: 35 (1951).

N. Rhodesia. N: Fort Rosebery, near Samfya Mission, fr. 21.viii.1952, *Angus* 274 (BM; FHO; K). W: Luanshya, ♀ fl. & fr. 29.vii.1954, *Fanshawe* 1407 (EA; K). C: Broken Hill, ♂ fl. 10.ix.1947, *Brenan* 7850 (EA; FHO; K); Chilanga, ♀ fl. & fr. 10.ix.1909, *Rogers* 8460 (BOL; K). E: Nyika Plateau, on path to N. Rukuru Waterfall, ♂ fl. & fr. 27.x.1958, *Robson* 393, fr. 393 A (K; LISC). S: Mazabuka, Siamambo Forest, ♂ fl. 24.vii.1952, *Angus* 29 (BM; FHO; K). **S. Rhodesia.** N: Mazoe, ♂ fl. viii.1906, *Eyles* 398 (BM; BOL; SRGH); Lomagundi, ♀ fl. & fr. x.1920, *Eyles* 2673 (K; SRGH). C: road from Chishawasha Seminary to Gletwyk, fr. 23.viii.1956, *Cowan* 15/56 (K; LISC; SRGH). E: Melsetter, Alfa Farm, Inyambewa Valley, ♂ fl. 6.vii.1955, *Chase* 5640 (BM; K; LISC; LMJ; SRGH). **Nyasaland.** N: 19·2 km. N. of Mzimba, ♂ 30.vii.1960, *Leach & Brunton* 10366 (LISC; SRGH). C: Dedza, Chenkula Village, fr. 12.xi.1958, *Jackson* 2264 (FHO; SRGH). S: Mlanje Mt. near Tuchila R., ♀ fl. & fr. 18.vii.1957, *Chapman* 388 (BM; EA; FHO; K). **Mozambique.** N: Vila Cabral, Massangulo, R. Zuculumezi, fr. 8.x.1942, *Mendonça* 679 (K; LISC; LMJ; SRGH); Macondes, between Mueda and Chomba, fr. 25.ix.1948, *Barbosa & Lemos* in *Barbosa* 2240 (LISC; LM; LMJ). Z: between Gúruè and Ile, fr. 19.x.1949, *Barbosa & Carvalho* 4535 (K; LM; LMJ). T: Marávia, Fíngoè, fr. 24.ix.1942, *Mendonça* 394 (BR; COI; K; LISC; PRE; SRGH). MS: Manica, Vumba Mt., ♂ fl. & fr.. 24.viii.1962, *Gomes e Sousa* 4777 (COI).

Widespread in tropical Africa (West tropical Africa, Angola, Congo, Uganda, Kenya, Tanganyika and Zanzibar). In savannas, thickets, woodlands of various types, forests, etc.

Var. **schinoides** R. Fernandes in Mem. Junta Invest. Ultr., Sér. 2, **38**: 39 (1962). Type: N. Rhodesia, 13 km. NW. of Abercorn, *Hutchinson & Gillett* 4010 (K, holotype).
Rhus schinoides Hutch., Botanist S. Afr.: 524 in adnot. (1946) non Willd. ex Schult. (1820). Type as above.

N. Rhodesia. N: 57·5 km. from Abercorn on the road to Tunduma, fr. 23.x.1947 *Brenan* 8183 (EA; FHO; K).
Known only from N. Rhodesia. In *Brachystegia* woodland and by streams.

SPECIES INSUFFICIENTLY KNOWN

Rhus buluwayensis Diels in Engl., Bot. Jahrb. **40**: 87 (1907).—Engl., Pflanzenw. Afr. **3,** 2: 211 (1921). Type: S. Rhodesia, Bulawayo, ix.1905, *Engler* 2923a (B†).

From the description this species may be *R. tenuinervis* Engl. or *R. vulgaris* Meikle or another species of the same group. We cannot trace any isotype and so cannot decide.

Rhus transvaalensis Engl. in A. & C. DC., Mon. Phan. **4**: 440 (1883); Pflanzenw. Afr. **3,** 2: 215 (1921).—Schonl. in Bothalia, **3,** 1: 53, cum fig. (1930). Type from the Transvaal.

Engler (Pflanzenw. Afr. **3,** 2: 215 (1921)) states that this species occurs in Chirinda Forest, S. Rhodesia, at an altitude of 1200–1300 m. We have not seen any specimen from our area belonging to this species.

60. CONNARACEAE

By E. J. Mendes

Erect or scandent shrubs, small trees or lianes, rarely rhizomatous shrublets. Leaves alternate, imparipinnate or 1–3-foliolate, leaflets entire; stipules absent. Flowers usually bisexual, actinomorphic or subzygomorphic; androecium and gynoecium usually dimorphic. Sepals 5, imbricate or valvate, usually free. Petals 5, free or slightly connate near the base, imbricate or rarely valvate. Stamens 10 or 5, hypogynous to perigynous; filaments often united at the base; anthers 2-thecous, dorsifixed, with longitudinal introrse dehiscence. Disk absent or thin. Carpels 5 or solitary, rarely 3, free, 1-locular; apparently all fertile but only 1–3 ripening into fruit; ovules 2, collateral, erect, attached at the base, near the base or near the middle of the ventral suture. Fruit usually follicular, dehiscing by the ventral suture, sometimes indehiscent or dehiscing irregularly. Seed usually 1 in each follicle, with or without endosperm, often arillate or pseudarillate.
A family easily recognised by the imparipinnate or 1–3-foliolate exstipulate leaves and apocarpous ovaries; mainly tropical, represented in Africa by 17 genera.

Leaves 3-foliolate or imparipinnate; carpels 5:
- Leaves 3-foliolate; fruit 1–3-follicular, follicles fulvo-tomentose - **2. Agelaea**
- Leaves imparipinnate with 2–16 (19) pairs of leaflets:
  - Follicle velutinous, ± rostrate - - - - - - - **1. Cnestis**
  - Follicle glabrous or glabrescent, acute or rounded:
    - Flowering precocious or with young leaves: inflorescence racemose, few-flowered, axillary - - - - - - - - - - - **5. Byrsocarpus**
    - Flowering not precocious; inflorescence paniculate, many-flowered, terminal, subterminal or axillary:
      - Follicle smooth, dehiscing by the ventral suture - - - - **6. Jaundea**
      - Follicle with fine longitudinal striation, dehiscing by irregular basal slits **4. Santaloides**

Leaves 1-foliolate; carpel 1 - - - - - - - - - **3. Burttia**

1. CNESTIS Juss.

Cnestis Juss., Gen. Pl.: 374 (1789).—Schellenb. in Engl., Pflanzenr. IV, 127: 29 (1938).

Shrubs, scandent or ± erect, small trees or woody lianes. Leaves imparipinnate; leaflets opposite or subopposite. Inflorescence terminal or axillary, racemose; racemes often clustered in old leaf-bases on older branches. Flowers 5-merous; androecium and gynoecium dimorphic. Sepals free, valvate or slightly imbricate. Petals shorter than, as long as, or a little longer than the sepals. Stamens 10; filaments free or connate at the base into a short tube. Carpels 5, densely hairy; stigma usually rounded or capitate; ovules basal or on the ventral suture near the base. Fruit a follicle, usually with a curved or twisted rostrum; dehiscence by the ventral suture; pericarp densely velutinous and pilose externally, pilose internally, the hairs often urticating. Seed with a shining testa and a basal aril; cotyledons large, flattened; endosperm present, fleshy.

An Old World genus, mainly tropical, the majority of species from tropical and subtropical Africa.

Cnestis natalensis (Hochst. ex Krauss) Planch. & Sond. ex Sond., F.C. **1**: 528 (1860).—Schellenb. in Mitt. Bot. Mus. Univ. Zür. **50**: 14 (1910); in Engl., Pflanzenw. Afr. **3**, 1: 318 (1915); in Engl., Pflanzenr. IV, 127: 40 (1938).—Bak. f. in Journ. Linn. Soc., Bot., **40**: 50 (1911).—Eyles in Trans. Roy. Soc. S. Afr. **5**, 4: 360 (1916).—Burtt Davy, F.P.F.T. **2**: 511 (1932).—Henkel, Woody Pl. Natal and Zululand: 203 (1934).—Wild, South. Rhod. Bot. Dict.: 65 (1953). TAB. **130**. Syntypes from Natal.

Zanthoxylum natalense Hochst. ex Krauss in Flora, **27**, 1: 304 (1844); Beitr. Fl. Cap- und Natall.: 41 (1846).—Schellenb. in Mitt. Bot. Mus. Univ. Zür. **50**: 14 (1910) (" Xanthoxylum ") in synon.; in Engl., Pflanzenr. IV, 127: 40 (1938) (" Xanthophyllum ") in synon. Syntypes as above.

Large woody climber, scandent shrub or small tree with trailing branches. Leaf-rhachis 6–18 cm. long, with a dense cinnamon indumentum; terminal leaflet 2–5 × 1–2 cm., obovate, equal-sided, apex ± acuminate, base obliquely rounded or very shallowly cuneate, subcoriaceous, reticulate on both surfaces, glabrescent above, whitish-grey-tomentose underneath; lower surface with a prominent rufous midrib. Inflorescences 4–8 cm. long, racemose, produced along leafless usually terminal long shoots, or clustered on short lateral shoots; shoots, inflorescence-rhachis, bracts, bracteoles, pedicels and outer surface of sepals densely cinnamon-tomentose. Sepals 4·5–5 × 1·1–1·6 mm., narrowly triangular. Petals 3·5–4·5 × 0·6–1 mm., narrowly elliptic to narrowly obovate. Long-staminate flowers with 5 antisepalous stamens up to 4·5 mm. long and 5 antipetalous ones up to 4 mm. long; short-staminate flowers with 5 antisepalous stamens up to 3·5 mm. long and 5 antipetalous ones up to 2·6 mm. long; filaments ± flattened, free. Carpels of long-staminate flowers up to 1·8 mm. long, those of short-staminate flowers up to 4·5 mm. long. Follicle up to 2·5 × 1 cm., subrecurved, ± rostrate, sparsely rufo-velutinous. Seed black, shining, up to 1·8 × 0·8 cm.

S. Rhodesia. E: Chirinda, Chipete, immat. fr. 22.x.1947, *Wild* 2126 (K; SRGH). **Mozambique.** MS: Espungabera, fl. 28.viii.1947, *Pimenta* 13 (K; LISC; LM; SRGH).

Also in Natal and S. Africa (Pondoland). Forest margins, riverine fringes and forest patches, 1100–1200 m.

2. AGELAEA Soland. emend. Planch.

Agelaea Soland. emend. Planch. in Linnaea, **23**: 437 (1850).—Schellenb. in Engl., Pflanzenr. IV, 127: 65 (1938).

Woody lianes, scandent shrubs or small trees; usually with fascicled hairs. Leaves 3-foliolate. Inflorescence a large many-flowered panicle. Flowers 5-merous; androecium and gynoecium heteromorphic. Sepals imbricate, with a dense brown indumentum, usually fringed with glandular hairs. Petals longer than the sepals, often slightly connate near the base, glabrous. Filaments connate at the base, terete, glabrous. Carpels pilose; stigma expanded and sometimes flattened; ovules basally inserted. Fruit 1–3-follicular; dehiscence by the ventral suture; pericarp tomentose. Seed ± ovoid, testa dark and shining; aril basal, obliquely cupuliform; cotyledons plano-convex, radicle apical; endosperm absent.

A genus mainly of tropical Africa, extending southwards to Angola, the Rhodesias and Mozambique and northwards to the Sudan; also recorded from the Madagascan region.

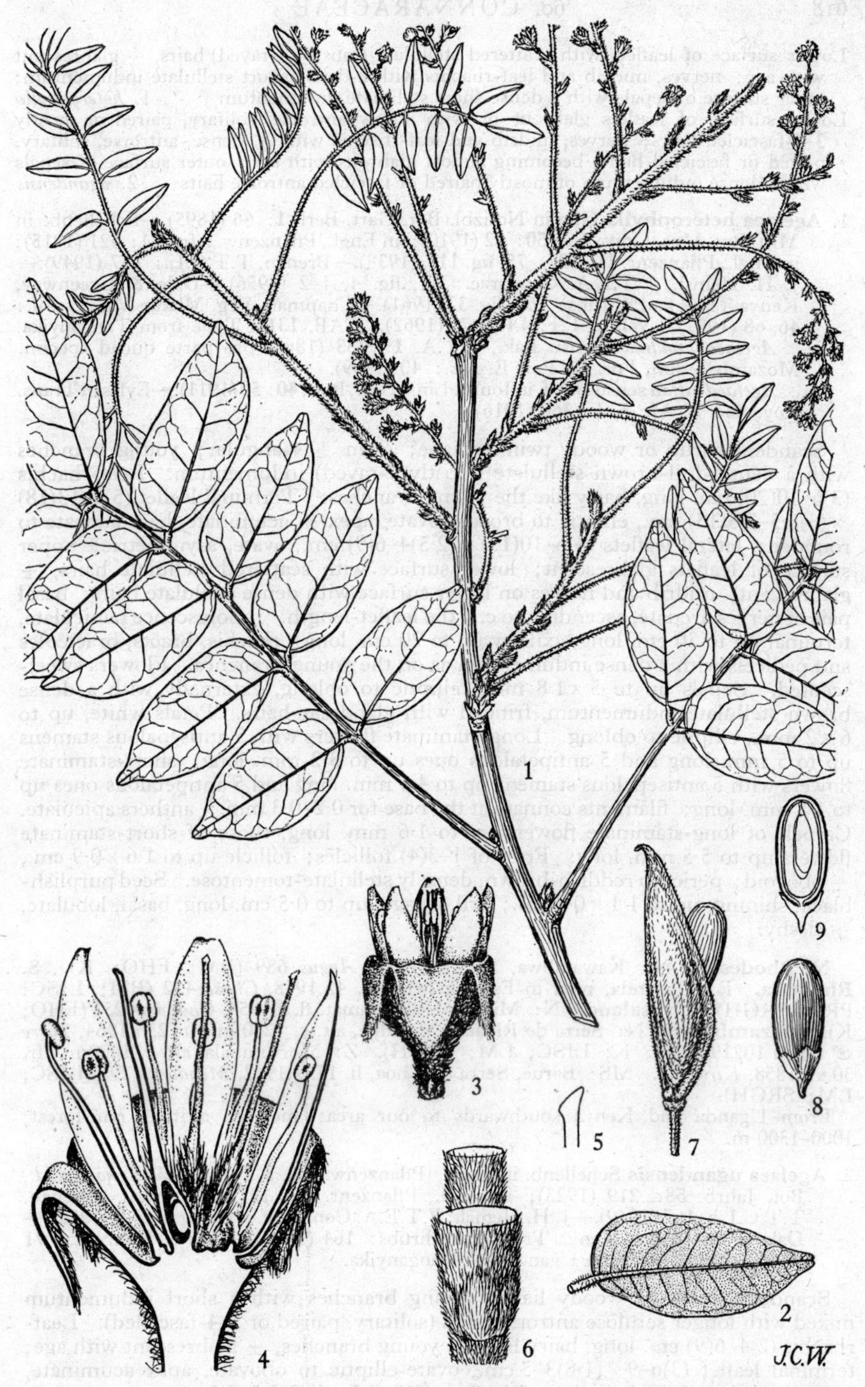

Tab. 130. CNESTIS NATALENSIS. 1, flowering branch ($\times \frac{2}{3}$); 2, lateral leaflet, under surface ($\times \frac{2}{3}$), 1–2 from *Swynnerton* 166; 3, short-staminate flower ($\times 4$); 4, same flower in section ($\times 8$); 5, stigma ($\times 16$); 6, bracteoles ($\times 10$), 3–6 from *Pimenta* 13; 7, mature follicle showing extruding seed ($\times 1$); 8, seed, showing pseudo-aril ($\times 1$); 9, vertical section of seed showing embryo ($\times 1$) 7–9 from *Armitage* 176/55.

Lower surface of leaflets with scattered stellulate (mostly 4-rayed) hairs, ± glabrescent with age; nerves, midrib and leaf-rhachis with a dense short stellulate indumentum; outer surface of sepals with a dense short stellulate indumentum - 1. *heterophylla*
Lower surface of leaflets glabrous or with a few antrorse, solitary, paired or mostly 3–4-fascicled hairs; nerves, midrib and leaf-rhachis with ± dense, antrorse, solitary, paired or fascicled hairs, becoming almost glabrous with age; outer surface of sepals with a dense indumentum of mostly paired or fascicled antrorse hairs 2. *ugandensis*

1. **Agelaea heterophylla** Gilg in Notizbl. Bot. Gart. Berl. **1**: 66 (1895).—Schellenb. in Mitt. Bot. Mus. Univ. Zür. **50**: 62 (1910); in Engl., Pflanzenw. Afr. **3,** 1: 321 (1915); in Engl., Pflanzenr. IV, 127: 75, fig. 11. (1938).—Brenan, T.T.C.L.: 167 (1949).—J. H. Hemsl., F.T.E.A. Connarac.: 11, fig. 4, 1–2 (1956).—Dale & Greenway, Kenya Trees and Shrubs: 164, fig. 32 (1961).—Chapman, Veg. Mlanje Mt. Nyasal.: 46, 68 (1962).—White, F.F.N.R.: 71 (1962). TAB. **131**. Type from Tanganyika.
Agelaea lamarckii sensu Bak., F.T.A. **1**: 453 (1868) pro parte quoad specim. Mozamb.—Sim, For. Fl. Port E. Afr.: 40 (1909).
Agelaea nitida sensu Bak.f. in Journ. Linn. Soc., Bot., **40**: 50 (1911).—Eyles in Trans. Roy. Soc. S. Afr. **5,** 4: 360 (1916).

Scandent shrub or woody twining liane; stem ± 4-angular; young branches with a dense dull-brown stellulate (mostly 4-rayed) indumentum. Leaf-rhachis (3)6–10(20) cm. long, hairy like the young branches. Terminal leaflet (5)9–13(18) × (3)5·5–7·5(13) cm., elliptic to broadly ovate, apex ± acuminate, base cuneate to rounded; lateral leaflets (4)6–10(13) × (2·5)4–6(9) cm., ovate, asymmetric; upper surface of leaflets glabrescent; lower surface with scattered stellulate hairs, ± glabrescent; midrib and nerves on lower surface with dense stellulate hairs; basal pair of nerves arcuate, ascending to c. ½ the leaflet-length. Inflorescence paniculate, terminal up to 30 cm. long, axillary up to 20 cm. long; rhachis, bracts, bracteoles and pedicels with a dense indumentum as on the young branches. Flowers sweet-scented. Sepals up to 5 × 1·8 mm., elliptic to oblong, externally with a dense brown stellulate indumentum, fringed with glandular hairs. Petals white, up to 6 × 2 mm., elliptic to oblong. Long-staminate flowers with 5 antisepalous stamens up to 5 mm. long and 5 antipetalous ones up to 3·2 mm. long; short-staminate flowers with 5 antisepalous stamens up to 4·4 mm. long and 5 antipetalous ones up to 1·8 mm. long; filaments connate at the base for 0·2–0·3 mm.; anthers apiculate. Carpels of long-staminate flowers up to 1·6 mm. long, those of short-staminate flowers up to 5·5 mm. long. Fruit of 1–3(4) follicles; follicle up to 1·6 × 0·9 cm., ± obovoid; pericarp reddish-brown, densely stellulate-tomentose. Seed purplish-black, shining, up to 1·1 × 0·7 cm.; aril orange, up to 0·5 cm. long, basal, lobulate, ± fleshy.

N. Rhodesia. N: Kawambwa, fr. 21.x.1952, *Angus* 689 (BM; FHO; K). **S. Rhodesia.** E: Chirinda, road to Espungabera, fr. 4.i.1948, *Chase* 432 (BM; LISC; PRE; SRGH). **Nyasaland.** N: Misuku Hills, immat. fl. x.1954, *Chapman* 251 (FHO; K). **Mozambique.** N: Serra de Ribáuè, Mepáluè, alt. c. 1250 m., fr. 25.i.1964, *Torre & Paiva* 10219 (COI; K; LISC; LM; SRGH). Z: Morrumbala, alt. c. 1000 m., fr. 30.xii.1858, *Kirk* (K). MS: Báruè, Serra de Choa, fl. 17.ix.1942, *Mendonça* 291 (LISC; LM; SRGH).
From Uganda and Kenya southwards to our area; medium altitude rain-forest; 1000–1300 m.

2. **Agelaea ugandensis** Schellenb. in Engl., [Pflanzenw. Afr. **3,** 1: 322 (1915) *nom. nud.*, Bot. Jahrb. **58**: 219 (1923); in Engl., Pflanzenr. IV, 127: 82 (1938).—Brenan. T.T.C.L.: 167 (1949).—J. H. Hemsl., F.T.E.A. Connarac.: 12, fig. 4, 5–9 (1956).—Dale & Greenway, Kenya Trees and Shrubs: 164 (1961).—White, F.F.N.R.: 71 (1962). Syntypes from Uganda and Tanganyika.

Scandent shrub or woody liane; young branches with a short indumentum mixed with longer setulose antrorse hairs (solitary, paired or 3–4-fascicled). Leaf-rhachis (2)4–6(9) cm. long, hairy like the young branches, ± glabrescent with age; terminal leaflet (3)6–9 × (1·8)3–5 cm., ovate-elliptic to obovate, apex acuminate, base cuneate to rounded; lateral leaflets (2)5–7·5 × (1·2)2·5–4·5 cm. ovate, asymmetric, upper surface glabrous or with a few hairs on the midrib; lower surface almost glabrous or with a few setulose antrorse hairs (solitary, paired or 3–4-fascicled) on the midrib and main nerves; basal pair of lateral nerves very prominent on the lower surface, slightly arcuate, usually ascending to beyond ⅔ of the leaflet-length. Inflorescence paniculate, terminal up to 25 cm. long, axillary up to

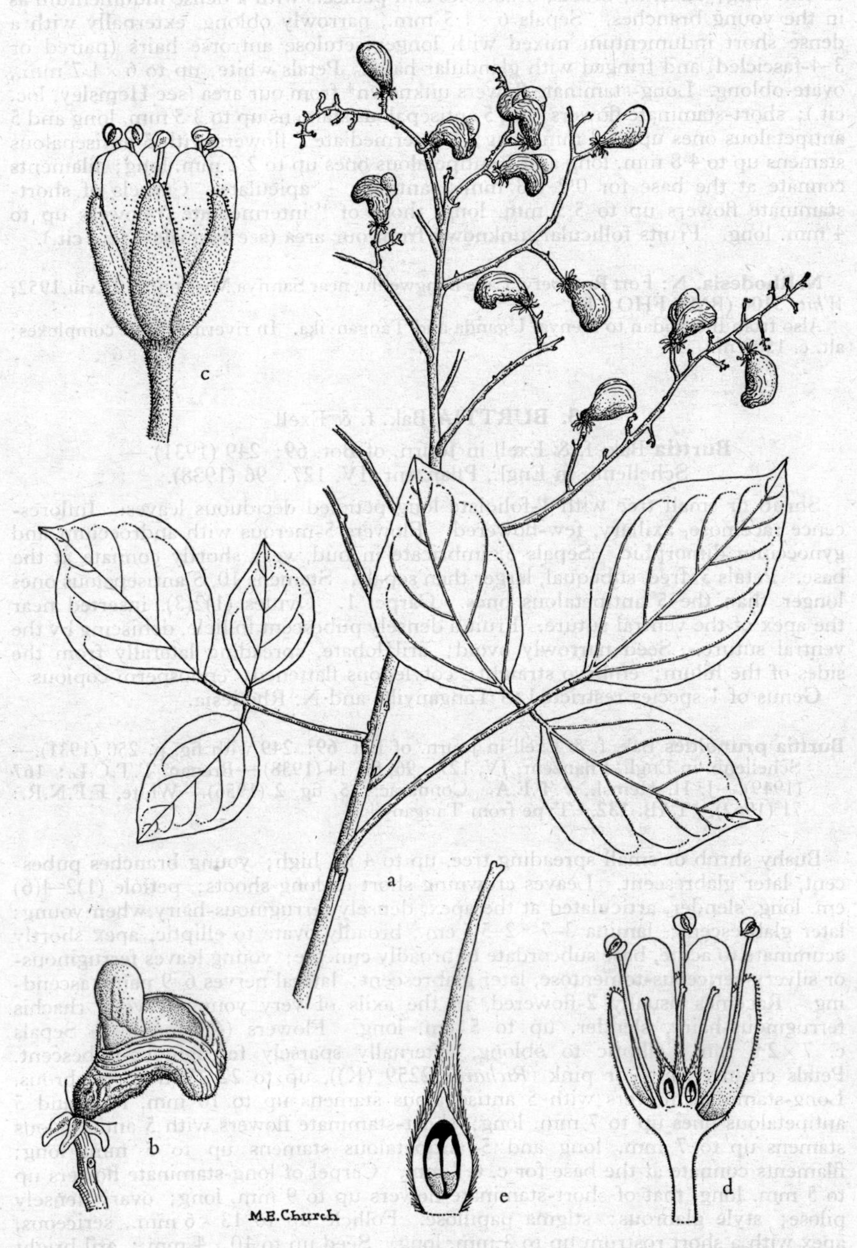

Tab. 131. AGELAEA HETEROPHYLLA. a, fruiting branchlet with leaves (× ½); b, dehiscing fruit showing seed with basal aril (× 2); c, long-staminate flower (× 5); d, same flower in vertical section, with antipetalous stamens removed (× 5); e, carpel in vertical section showing basally inserted ovules (× 15); a–b from *Bally* 8783; c–e from *Eggeling* 6240. From Dale & Greenway, Kenya Trees and Shrubs.

15 cm. long; rhachis, bracts, bracteoles and pedicels with a dense indumentum as in the young branches. Sepals 6 × 1·5 mm., narrowly oblong, externally with a dense short indumentum mixed with longer setulose antrorse hairs (paired or 3–4-fascicled) and fringed with glandular hairs. Petals white, up to 6 × 1·7 mm., ovate-oblong. Long-staminate flowers unknown* from our area (see Hemsley, loc. cit.); short-staminate flowers with 5 antisepalous stamens up to 3·5 mm. long and 5 antipetalous ones up to 2 mm. long; " intermediate " flowers with 5 antisepalous stamens up to 4·8 mm. long and 5 antipetalous ones up to 2·2 mm. long; filaments connate at the base for 0·2–0·3 mm.; anthers ± apiculate. Carpels of short-staminate flowers up to 5·5 mm. long, those of " intermediate " flowers up to 4 mm. long. Fruits follicular, unknown from our area (see Hemsley, loc. cit.).

N. Rhodesia. N: Fort Rosebery, Lake Bangweulu, near Samfya Mission, fl. 21.viii.1952, *White* 3106 (BM; FHO; K).

Also from the Sudan to Kenya, Uganda and Tanganyika. In riverine forest complexes; alt. c. 1500 m.

3. BURTTIA Bak. f. & Exell

Burttia Bak. f. & Exell in Journ. of Bot. **69**: 249 (1931).—Schellenb. in Engl., Pflanzenr. IV, 127: 96 (1938).

Shrub or small tree with 1-foliolate long-petioled deciduous leaves. Inflorescence racemose, axillary, few-flowered. Flowers 5-merous with androecium and gynoecium dimorphic. Sepals 5, imbricate in bud, very shortly connate at the base. Petals 5, free, subequal, larger than sepals. Stamens 10, 5 antisepalous ones longer than the 5 antipetalous ones. Carpel 1. Ovules (1)2(3), inserted near the apex of the ventral suture. Fruit a densely pubescent follicle, dehiscing by the ventral suture. Seed narrowly ovoid; aril lobate, spreading laterally from the sides of the hilum; embryo straight; cotyledons flattened; endosperm copious.

Genus of 1 species restricted to Tanganyika and N. Rhodesia.

Burttia prunoides Bak. f. & Exell in Journ. of Bot. **69**: 249 with fig. p. 250 (1931).—Schellenb. in Engl., Pflanzenr. IV, 127: 96, fig. 14 (1938).—Brenan, T.T.C.L.: 167 (1949).—J. H. Hemsl., F.T.E.A. Connarac.: 5, fig. 2 (1956).—White, F.F.N.R.: 71 (1962). TAB. **132**. Type from Tanganyika.

Bushy shrub or small spreading tree, up to 4 m. high; young branches pubescent, later glabrescent. Leaves crowning short or long shoots; petiole (1)2–4(6) cm. long, slender, articulated at the apex, densely ferruginous-hairy when young: later glabrescent; lamina 3–7 × 2–5·5 cm., broadly ovate to elliptic, apex shortly acuminate to acute, base subcordate to broadly cuneate; young leaves ferruginous- or silvery-sericeous-tomentose, later glabrescent; lateral nerves 6–9 pairs, ascending. Racemes usually 2-flowered, in the axils of very young leaves; rhachis ferruginous-hairy, slender, up to 5 cm. long. Flowers (4)5-merous. Sepals c. 7 × 2·5 mm., elliptic to oblong, externally sparsely ferruginous-pubescent. Petals creamy-white or pink (*Richards* 2259 (K)), up to 22 × 9 mm., glabrous. Long-staminate flowers with 5 antisepalous stamens up to 10 mm. long and 5 antipetalous ones up to 7 mm. long; short-staminate flowers with 5 antisepalous stamens up to 7 mm. long and 5 antipetalous stamens up to 5 mm. long; filaments connate at the base for c. 0·7 mm. Carpel of long-staminate flowers up to 5 mm. long, that of short-staminate flowers up to 9 mm. long; ovary densely pilose; style glabrous; stigma papillose. Follicle up to 13 × 6 mm., sericeous, apex with a short rostrum up to 3 mm. long. Seed up to 10 × 4 mm.; aril bright red at maturity.

N. Rhodesia. N: Bulaya-Mwewe, fl. 22.x.1949, *Bullock* 1340 (K; SRGH); Lake Tanganyika, st. 18.v.1936, *Burtt* 5997 (BM; K).

Also from Tanganyika. Deciduous thickets and woodlands; 1000–1500 m.

* We examined the only 6 flowering gatherings known from our area and noticed that 3 of them showed short-staminate flowers, the other 3 belonging to the third, " intermediate ", stamen-style relationship (see Hemsley, loc. cit.).

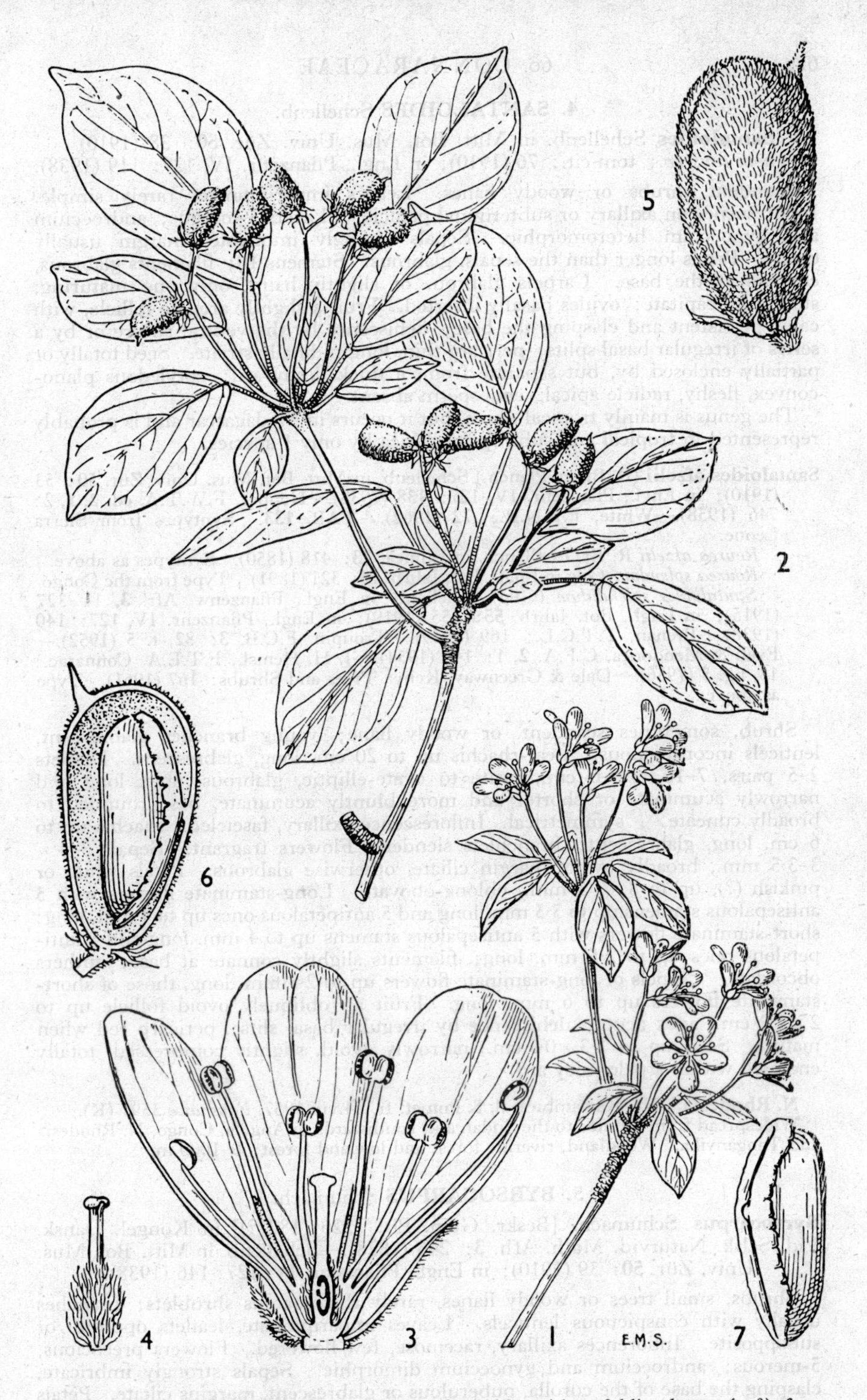

Tab. 132. BURTTIA PRUNOIDES. 1, flowering branchlet with unfolding leaves (×⅔), from *Burtt* 3521; 2, branch with mature leaves and fruits (×⅔), from *Burtt* 5148; 3, long-staminate flower in vertical section (×5), from *Burtt* 3521; 4, carpel from long-staminate flower (×5), from *Burtt* 3521; 5, fruit (×2½); 6, fruit with half of pericarp removed to show insertion of seed (×2½); 7, seed (×2½), 5–7 from *Burtt* 5148. Adapted from F.T.E.A.

4. **SANTALOIDES** Schellenb.

Santaloides Schellenb. in Mitt. Bot. Mus. Univ. Zür. **50**: 38 (1910) *nom conserv.*; tom cit.: 76 (1910); in Engl., Pflanzenr. IV, 127: 119 (1938).

Scandent shrubs or woody lianes. Leaves imparipinnate, rarely simple. Inflorescence an axillary or subterminal panicle. Flowers 5-merous; androecium and gynoecium heteromorphic. Sepals strongly imbricate, margin usually ciliate. Petals longer than the sepals, glabrous. Stamens 10; filaments glabrous, connate at the base. Carpels glabrous or slightly hairy, only one maturing; stigma subcapitate; ovules basally inserted. Fruit a slightly arcuate follicle, with calyx persistent and clasping the base; dehiscence by the ventral suture or by a series of irregular basal splits; pericarp thin, longitudinally striate. Seed totally or partially enclosed by, but separate from, a thick fleshy aril; cotyledons plano-convex, fleshy, radicle apical; endosperm absent.

The genus is mainly tropical Asian, but it occurs in Madagascar and is probably represented in tropical and subtropical Africa by only 1 species.

Santaloides afzelii (R. Br. ex Planch.) Schellenb. in Mitt. Bot. Mus. Univ. Zür. 50: 53 (1910); in Engl., Pflanzenr. IV, 127: 138 (1938).—Hepper, F.W.T.A. ed. 2, **1**, 2: 746 (1958).—White, F.F.N.R.: 72 (1962). TAB. **133**. Syntypes from Sierra Leone.

Rourea afzelii R. Br. ex Planch. in Linnea, **23**: 418 (1850). Syntypes as above.

Rourea splendida Gilg in Engl., Bot. Jahrb. **14**: 321 (1891). Type from the Congo.

Santaloides splendidum (Gilg) Schellenb. in Engl., Pflanzenw. Afr. **3**, 1: 327 (1915); in Engl., Bot. Jahrb. **55**: 455 (1919); in Engl., Pflanzenr. IV, 127: 140 (1938).—Brenan, T.T.C.L.: 169 (1949).—Troupin, F.C.B. **3**: 82, t. 5 (1952).—Exell & Mendonça, C.F.A. **2**, 1: 148 (1954).—J. H. Hemsl., F.T.E.A. Connarac.: 13, fig. 5 (1956).—Dale & Greenway, Kenya Trees and Shrubs: 167 (1961). Type as above.

Shrub, sometimes scandent, or woody liane; young branches glabrescent, lenticels inconspicuous. Leaf-rhachis up to 20 cm. long, glabrescent. Leaflets 2–5 pairs, 7–10 × 3·5–6 cm., ovate to ovate-elliptic, glabrous, apex long and narrowly acuminate or shorter and more bluntly acuminate, base rounded to broadly cuneate, ± symmetrical. Inflorescence axillary, fascicled; rhachis up to 6 cm. long, glabrescent; branchlets slender. Flowers fragrant. Sepals 3–4 × 3–3·5 mm., broadly ovate, margin ciliate, otherwise glabrous. Petals white or pinkish (?), up to 8 × 3·5 mm., oblong-obovate. Long-staminate flowers with 5 antisepalous stamens up to 5·5 mm. long and 5 antipetalous ones up to 4 mm. long; short-staminate flowers with 5 antisepalous stamens up to 4 mm. long and 5 antipetalous ones up to 2·5 mm. long; filaments slightly connate at base; anthers obcordate. Carpels of long-staminate flowers up to 2·5 mm. long, those of short-staminate flowers up to 6 mm. long. Fruit an obliquely ovoid follicle up to 2 × 1·2 cm., apex acute; dehiscence by irregular basal slits; pericarp red when mature. Seed up to 1·3 × 0·7 cm., narrowly ovoid, slightly compressed, totally enclosed within a pale juicy aril.

N. Rhodesia. N: Kawambwa, fl. & immat. fr. 24.viii.1957, *Fanshawe* 3591 (K).
Widespread from Guinée to the Sudan and southwards to Angola, Congo, N. Rhodesia and Tanganyika. Woodland, riverine forest and lowland forest; 0–1000 m.

5. **BYRSOCARPUS** Schumach.

Byrsocarpus Schumach. [Beskr. Guin. Pl.: 226 (1827?)] in Kongel. Dansk. Vid. Selsk. Naturvid. Math. Afh. **3**: 246 (1828).—Schellenb. in Mitt. Bot. Mus. Univ. Zür. **50**: 39 (1910); in Engl., Pflanzenr. IV, 127: 146 (1938).

Shrubs, small trees or woody lianes, rarely rhizomatous shrublets; branches usually with conspicuous lenticels. Leaves imparipinnate, leaflets opposite or subopposite. Inflorences axillary, racemose, few-flowered. Flowers precocious, 5-merous; androecium and gynoecium dimorphic. Sepals strongly imbricate, clasping the base of the corolla, puberulous or glabrescent, margins ciliate. Petals longer than the sepals. Stamens 10, filaments connate into a short basal tube, glabrescent. Ovary hairy; styles terete; stigma capitate or subcapitate; ovules 2, inserted basally. Fruit a ± ovoid follicle, ± curved, with calyx persistent; dehiscence by the ventral suture, with seed usually extruded and held by the peri-

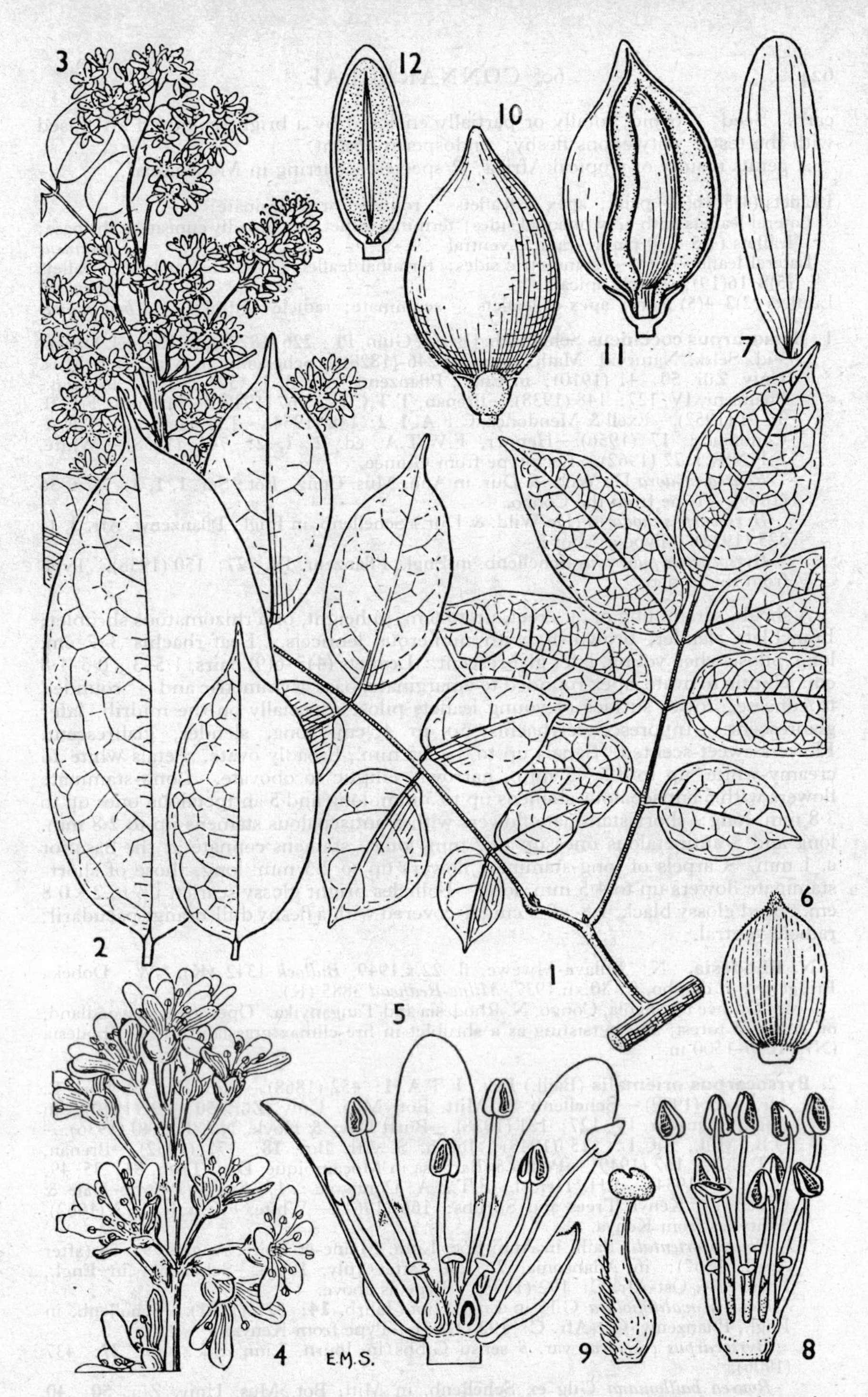

Tab. 133. SANTALOIDES AFZELII. 1, leafy branchlet ($\times \frac{2}{3}$); 2, leaflets to show size variation ($\times \frac{2}{3}$), 1–2 from *Drummond & Hemsley* 3427; 3, inflorescence ($\times \frac{2}{3}$); 4, part of inflorescence to show flower arrangement ($\times 2$); 5, long-staminate flower in vertical section ($\times 8$); 6, sepal ($\times 8$); 7, petal ($\times 8$); 8, long-staminate flower with sepals and petals removed ($\times 8$); 9, single carpel of long-staminate flower ($\times 16$), 3–9 from *Eggeling* 5548; 10, fruit ($\times 2$); 11, fruit with part of pericarp and aril cut away to show seed ($\times 2$); 12, seed in vertical section ($\times 2$), 10–12 from *Drummond & Hemsley* 3427. Adapted from F.T.E.A.

carp. Seed ± ovoid, totally or partially enclosed by a bright-coloured aril fused with the testa; cotyledons fleshy; endosperm absent.

A genus mainly of tropical Africa; 2 species occurring in Madagascar.

Leaflets (4)5–16(19) pairs; apex of leaflets ± rounded or emarginate:
Lateral leaflets with very unequal sides; terminal leaflet very broadly cuneate at the base; leaflets (4)5–6(9) pairs; radicle ventral - - - - - - 1. *coccineus*
Lateral leaflets with ± symmetric sides; terminal leaflet ± rounded at base; leaflets (5)8–16(19); radicle apical - - - - - - - - 2. *orientalis*
Leaflets (2)3–4(5) pairs; apex of leaflets ± acuminate; radicle ventral 3. *boivinianus*

1. **Byrsocarpus coccineus** Schumach. [Beskr. Guin. Pl.: 226 (1827?)] in Kongel. Dansk. Vid. Selsk. Naturvid. Math. Afh. **3**: 246 (1828).—Schellenb. in Mitt. Bot. Mus. Univ. Zür. **50**: 41 (1910); in Engl., Pflanzenw. Afr. **3,** 1: 324 (1915); in Engl., Pflanzenr. IV, 127: 148 (1938).—Brenan, T.T.C.L.: 167 (1949).—Troupin, F.C.B. **3**: 91 (1952).—Exell & Mendonça, C.F.A. **1,** 2: 148 (1954).—J.H. Hemsl., F.T.E.A. Connarac.: 17 (1956).—Hepper, F.W.T.A. ed. 2, **1,** 2: 741 (1958).—White, F.F.N.R.: 72 (1962). Lectotype from Guinée.

Rourea inodora De Wild. & Dur. in Ann. Mus. Cong., Bot., Sér. 1, **1,** 3: 71, t. 36 (1899). Type from the Congo.

Byrsocarpus inodorus (De Wild. & Dur.) Schellenb. in Engl., Pflanzenw. Afr. **3,** 1: 325 (1915). Type as above.

Byrsocarpus puberulus Schellenb. in Engl., Pflanzenr. IV, 127: 150 (1938). Type from the Congo.

A shrub, scrambling or ± erect, up to 6 m. in height, or a rhizomatous shrublet; branchlets slender, brownish, with numerous lenticels. Leaf-rhachis 3–7 cm. long, pilose when young, later glabrescent. Leaflets (4)5–6(9) pairs, 1·5–3 × 0·5–1·8 cm., elliptic to ovate, apex rounded to emarginate, base asymmetric and ± rounded to cuneate; lower surface of young leaflets pilose especially on the midrib, later glabrescent. Inflorescence-rhachis up to 4 cm. long, slender, glabrescent. Flowers sweet-scented. Sepals up to 3 × 2·5 mm., broadly ovate. Petals white to creamy-white, up to 8·5 × 3 mm., narrowly elliptic to obovate. Long-staminate flowers with 5 antisepalous stamens up to 5 mm. long and 5 antipetalous ones up to 3·8 mm. long; short-staminate flowers with 5 antisepalous stamens up to 2·8 mm. long and 5 antipetalous ones up to 2 mm. long; stamens connate at the base for c. 1 mm. Carpels of long-staminate flowers up to 2·3 mm. long, those of short-staminate flowers up to 4·5 mm. long. Follicles bright glossy scarlet, up to 2 × 0·8 cm. Seed glossy black, 1·8 × 0·6 cm., $\frac{2}{3}$-covered with a fleshy dull-orange pseudaril; radicle ventral.

N. Rhodesia. N: Bulaya-Mwewe, fl. 22.x.1949, *Bullock* 1342 (K). W: Dobeka Bridge, N. of dambo, fr. 30.xii.1937, *Milne-Redhead* 3885 (K).

From Guinée to Angola, Congo, N. Rhodesia and Tanganyika. Open areas in woodland, on edges of forest, and persisting as a shrublet in fire-climax grassland in N. Rhodesia (N); 1000–1500 m.

2. **Byrsocarpus orientalis** (Baill.) Bak., F.T.A. **1**: 452 (1868).—Sim, For. Fl. Port. E. Afr.: 40 (1909).—Schellenb. in Mitt. Bot. Mus. Univ. Zür. **50**: 42 (1910); in Engl., Pflanzenr. IV, 127: 151 (1938).—Burtt Davy & Hoyle, N.C.L.: 40 (1936).—O.B. Mill., B.C.L.: 15 (1948); Journ. S. Afr. Bot. **18**: 171 (1952).—Brenan, T.T.C.L.: 167 (1949).—Pedro & Barbosa in Moçambique, Doc. Trim. **81**: 35, 49, 54 et 57 (1954).—J.H. Hemsl., F.T.E.A. Connarac.: 17, fig. 6 (1956).—Dale & Greenway, Kenya Trees and Shrubs: 166 (1961).—White, F.F.N.R.: 72 (1962). Syntypes from Kenya.

Rourea orientalis Baill. in Ann. Soc. Linn. Maine-et-Loire, Angers, **9**: 54 (after April, 1867); in Adansonia, **7**: 230 (after July, 1867).—Schellenb. in Engl., Pflanzenw. Ost-Afr. **C**: 192 (1895). Type as above.

Rourea ovalifoliolata Gilg in Engl., Bot. Jahrb. **14**: 327 (1891).—Schellenb. in Engl., Pflanzenw. Ost-Afr. **C**: 192 (1895). Type from Kenya.

Byrsocarpus coccineus var. *β* sensu Gibbs in Journ. Linn. Soc., Bot. **37**: 437 (1906).

Rourea bailloniana Gilg ex Schellenb. in Mitt. Bot. Mus. Univ. Zür. **50**: 40 (1910) *in synon.*

Byrsocarpus baillonianus Schellenb., loc. cit., *nom. nud.*

Byrsocarpus ovalifoliolatus (Gilg) Schellenb., tom. cit.: 42 (1910). Type from Kenya.

Byrsocarpus coccineus sensu Eyles in Trans. Roy. Soc. S. Afr. **5,** 4: 360 (1916) (" Brysocarpus ").

Byrsocarpus coccineus var. *parviflorus* sensu Eyles, loc. cit. (" Brysocarpus ").
Rourea sp.—Eyles, loc. cit.
Byrsocarpus tomentosus Schellenb. in Engl., Bot. Jahrb. **55**: 452 (1919); in Engl., Pflanzenr. IV, 127: 151, fig. 261 (1938).—R.E. Fr., Wiss. Ergebn. Schwed. Rhod.-Kongo-Exped. **1,** Ergänzungsheft: 11 (1921).—Burtt Davy & Hoyle, N.C.L.: 40 (1936).—Brenan, T.T.C.L.: 168 (1949).—O.B. Mill., Journ. S. Afr. Bot. **18**: 17 (1952).—Troupin, F.C.B. **3**: 92 (1952).—Wild, Guide Fl. Vict. Falls: 152 (1953).—Exell & Mendonça, C.F.A. **1,** 2: 149 (1954). Type from Tanganyika.

Small tree up to 6 m. high, or a shrub, often scandent, or a rhizomatous shrublet; branches brownish-purplish, conspicuously and densely lenticellate; young twigs sparsely hairy to densely pubescent. Leaf-rhachis (4)9–20 cm. long, almost glabrous or pubescent. Leaflets (5)8–16(19) pairs, (0·5)1·5–4 × (0·4)0·8–2 cm., narrowly to broadly elliptic or oblong, rarely ovate or obovate, pubescent or almost glabrous, rounded and usually mucronate or acute at the apex; subcordate, rounded or broadly cuneate at the base; midrib sometimes densely hairy especially on the lower surface. Inflorescence-rhachis up to 6 cm. long, pubescent or glabrous. Flowers sweet-lemon-scented. Sepals pale green to pinkish-brown, up to 4·3 × 3 mm., ovate. Petals white to yellow, rarely pinkish-tinged, up to 12 × 3·5 mm., ligulate or narrowly elliptic, spreading to form a flat star. Long-staminate flowers with 5 antisepalous stamens up to 7 mm. long and 5 antipetalous ones up to 4 mm. long; short-staminate flowers with 5 antisepalous stamens up to 3 mm. long and 5 antipetalous ones up to 2 mm. long; stamens connate at the base for 1–2 mm. Carpels of long-staminate flowers up to 1·5 mm. long, those of short-staminate flowers up to 4·5 mm. long; ovary densely pilose. Mature follicle reddish-brown, up to 2·3 × 1·2 cm., persistent calyx often spreading at maturity. Seed up to 1·6 × 0·8 cm., totally enclosed within a bright scarlet pseudaril; radicle apical.

Caprivi Strip: Mpilila I., Chobe R., alt. c. 900 m., fr. 15.i.1959, *Killick* 3389 (PRE; SRGH). **Bechuanaland Prot.** N: Kazungula, fl. xi.1935, *Miller* B79 (FHO; K). **N. Rhodesia.** B: Balovale, alt. c. 1000 m., fr. xii.1953, *Gilges* 314 (K; PRE; SRGH). N: Mwenzo, alt. c. 1500 m., fl. 27.ix.1938, *Greenway* 5788 (FHO; K). W: Dobeka Bridge, fr. 14.xii.1937, *Milne-Redhead* 3660 (K). C: Mount Makulu Research Station, near Chilanga, Makulu Stream, fr. 20.xi.1957, *Angus* 1790 (FHO; K; PRE). E: Lundazi-Mzimba road, 4·8 km., fr. 13.ii.1957, *Angus* 1522 (FHO). S: Kandahar I., fl. 11.x.1911, *Rogers* 5469 (BM; SRGH). **S. Rhodesia.** N: Urungwe, Msukwe R., fl. 26.ix.1952, *Phelps* 29 (SRGH). W: Victoria Falls, Cataract I., fl. 23.xi.1949, *Wild* 3169 (LISC; SRGH). E: southern Melsetter, fl. ix.1912, *Swynnerton* (BM). **Nyasaland.** N: Rumpi, Deep Bay, Lake Nyasa, st. 11.v.1952, *White* 2821c (FHO; K). C: Lake Nyasa Hotel, near Salima, alt. 470 m., fr. 12.ii.1959, *Robson & Steele* 1614 (K; LISC). S: Fort Johnston, fl. 23.xi.1954, *Jackson* 1397 (FHO; K). **Mozambique.** N: Nacala, between Fernão Veloso and Quissangulo, fr. 15.x.1948, *Barbosa* 2422 (FHO; K; LISC; LM; LMJ; SRGH). Z: between Pebane and Mocubela, fl. 25.x.1942, *Torre* 4679 (COI; FHO; K; LISC; LM; SRGH). T: Moatize, Zóbuè, fl. 21.x.1941, *Torre* 3696 (COI; FHO; LISC; LM; SRGH). MS: Gorongosa, fl. 7.x.1944, *Mendonça* 2381 (COI; FHO; LISC; PRE).

From the Congo and Angola to Kenya, Tanganyika and Mozambique. Open woodland and bushland complexes, forest margins, thickets fringing water-courses and persisting in fire-climax grassland as a shrublet in N. Rhodesia (N, W and B) and in Nyasaland (N); 200–1600 m.

3. **Byrsocarpus boivinianus** (Baill.) Schellenb. in Mitt. Bot. Mus. Univ. Zür. **50**: 40 (1910); in Engl., Pflanzenr. IV, 127: 155 (1938).—Brenan, T.T.C.L.: 167 (1949).—J.H. Hemsl., F.T.E.A. Connarac.: 16 (1956).—Dale & Greenway, Kenya Trees and Shrubs: 166 (1961). TAB. **134**. Type from Kenya.

Rourea boiviniana Baill. in Ann. Soc. Linn. Maine-et-Loire, Angers, **9**: 55 (after April 1867); Adansonia, **7**: 231 (after July 1867). Type as above.
Byrsocarpus ovatifolius Bak., F.T.A. **1**: 452 (1868).—Sim, For. Fl. Port. E. Afr.: 40 (1909). Type from the Mozambique-Tanganyika border, R. Rovuma, *Meller* (K, holotype).
Byrsocarpus maximus Bak., tom. cit.: 453 (1868).—Sim, loc. cit.—Schellenb. in Engl., Pflanzenw. Afr. **3,** 1: 324, fig. 211 (1915); in Engl., Bot. Jahrb. **55**: 454 (1919); in Engl., Pflanzenr. IV, 127: 155 (1938).—Brenan, T.T.C.L.: 167 (1949). Type from the Mozambique-Tanganyika border, R. Rovuma, *Kirk* (K, holotype).
Rourea maxima (Bak.) Gilg in Engl., Pflanzenw. Ost-Afr. **C**: 192 (1895). Type as above.
Rourea ovatifolia (Bak.) Gilg, loc. cit.—Burkill, List Pl. Brit. Centr. Afr.: 241 (1897). Type as for *B. ovatifolius*.

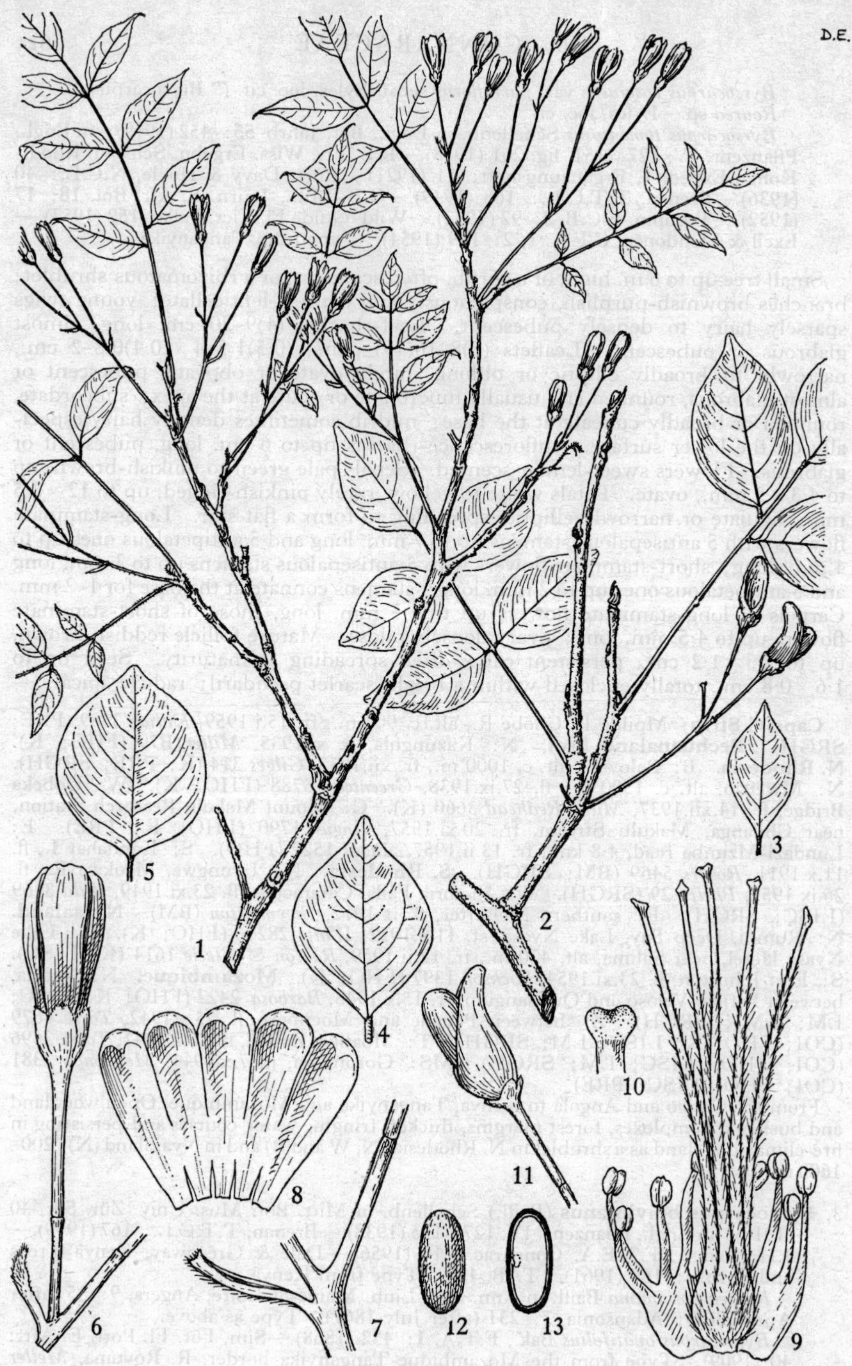

Tab. 134. BYRSOCARPUS BOIVINIANUS. 1, flowering branch bearing very young leaves ($\times \frac{2}{3}$), from *Mendonça* 998; 2, fruiting branch ($\times \frac{2}{3}$), from *Kirk* s.n.; 3, 4 & 5, mature terminal leaflets ($\times \frac{2}{3}$), respectively from *Kirk* s.n., *Kirk* s.n. and *Gomes e Sousa* 4638; 6, part of inflorescence showing one flower, articulations of pedicels, and the bract ($\times 3$); 7, bract and articulated pedicel ($\times 4$); 8, displayed corolla showing connate petals ($\times 3$); 9, displayed androecium and gynoecium of a short-staminate flower ($\times 6$); 10, stigma (c. $\times 16$), 6–10 from *Mendonça* 998; 11, mature follicle with extruding seed ($\times 1$); 12 & 13, seed and its vertical section showing ventral radicle ($\times 1$), 11–13 from *Kirk* s.n.

Shrub, scandent or ± erect, up to 4 m. high; young stems reddish-brown with numerous small lenticels, older stems grey. Leaf-rhachis 6–11 cm. long, glabrescent. Leaflets (2)4(5) pairs, (1·5)2·5–5·5 × (0·8)1·8–3 cm., elliptic to ovate or broadly ovate, apex acuminate to subacuminate, base rounded or subcordate or broadly cuneate; lower surface of young leaflets sparsely hairy especially on midrib, later glabrous. Sepals 3–3·5 × 2·2–3 mm., ovate to very broadly ovate. Petals c. 11 × 2 mm., narrowly elliptic to narrowly obovate. Long-staminate flowers with 5 antisepalous stamens up to 8·2 mm. long and 5 antipetalous ones up to 7 mm. long; short-staminate flowers with 5 antisepalous stamens up to 3·5 mm. long and 5 antipetalous ones up to 2·5 mm. long; stamens connate at the base for c. 1 mm. Carpels of long-staminate flowers up to 3 mm. long, those of short-staminate flowers up to 9·5 mm. long. Follicle yellow to red when mature, up to 2 × 1 cm. Seed up to 16 × 8 mm., almost enclosed within a bright-scarlet pseudaril.

Mozambique. N: between Palma and Nangade, fl. 20.x.1942, *Mendonça* 998 (COI; K; LISC; LM; SRGH); Eráti, between Namapa and Nacarôa, Muchamapa, alt. c. 400 m., fr. immat. 14.xii.1963, *Torre & Paiva* 9583 (COI; K; LISC; LM; SRGH).

Also in Kenya and Tanganyika. Scattered in coastal forest and bushland but up to 200 km. inland along the main river systems, in forest and woodland complexes; 0–750 m.

6. JAUNDEA Gilg

Jaundea Gilg in Engl. & Prantl, Nat. Pflanzenfam. **3,** 3: 388 (1894); in Notizbl. Bot. Gart. Berl. **1**: 66 (1895).—Schellenb. in Engl., Pflanzenr. IV, 127: 161 (1938).

Shrubs, sometimes scandent, small trees or lianes. Leaves imparipinnate; leaflets opposite or subopposite. Inflorescence paniculate, axillary or terminal. Flowers not precocious, 5-merous; androecium and gynoecium dimorphic. Sepals imbricate, puberulous externally. Petals longer than the sepals, glabrous. Stamens 10; filaments glabrous, shortly connate at the base. Carpels pilose; stigma capitate; ovules inserted basally. Fruit a follicle, rounded at the apex, glabrous, with calyx persisting and clasping the base; dehiscing by the ventral suture. Seed ovoid; testa and aril fused to form a fleshy pseudaril enclosing the seed; cotyledons fleshy; radicle ventral; endosperm absent.

A tropical African genus of 3 species.

Jaundea pinnata (Beauv.) Schellenb. in Candollea, **2**: 92 (1925); in Engl., Pflanzenr. IV, 127: 164, fig. 29 (1938).—Troupin, F.C.B. **3**: 87, t. 6 (1952).—Exell & Mendonça, C.F.A. **1,** 2: 151 (1954).—J.H. Hemsl., F.T.E.A. Connarac.: 21, fig. 7 (1956).—Hepper, F.W.T.A. ed. 2, **1,** 2: 742 (1958).—Dale & Greenway, Kenya Trees and Shrubs: 167 (1961).—White, F.F.N.R.: 72, fig. 14 (1962). TAB. **135.** Type from Nigeria.

Cnestis pinnata Beauv., Fl. Owar. & Benin, **1**: 98, t. 60 (1804). Type as above.

Rourea monticola Gilg in Notizbl. Bot. Gart. Berl. **1**: 68 (1895). Syntypes from Tanganyika.

Byrsocarpus monticola (Gilg) Schellenb. in Mitt. Bot. Mus. Univ. Zür. **50**: 44 (1910); in Engl., Pflanzenw. Afr. **3,** 1: 323 (1915). Syntypes as above.

Jaundea monticola (Gilg) Schellenb. in Engl., Bot. Jahrb. **55**: 461 (1919); in Engl., Pflanzenr. IV, 127: 166 (1938).—Robyns, Fl. Parc Nat. Albert, **1**: 258, t. 24 (1948).—Brenan, T.T.C.L.: 168 (1949).—Troupin, F.C.B. **3**: 88 (1952). Syntypes as above.

Scandent shrub, small tree with drooping branches or twining woody liane; young branches puberulous, soon glabrescent. Leaf-rhachis 7–18 cm. long, glabrous. Leaflets (2)3(4) pairs, (3)5–8(13) × (2)3–4(6) cm., ovate, elliptic or oblong-elliptic, apex ± acuminate or apiculate, base rounded to cuneate; lateral nerves 4–6 pairs, prominent on lower surface, arcuate, ascending. Inflorescences axillary, solitary or fascicled; rhachis to 10 cm. long, sparsely pilose to densely pubescent. Sepals up to 3 × 2 mm., ovate, margin ciliate. Petals white to creamy-yellow, up to 10 × 2·5 mm., narrowly elliptic. Long-staminate flowers unknown from our area (see Hemsley, loc. cit.); short-staminate flowers with 5 antisepalous stamens up to 1·5 mm. long and 5 antipetalous ones up to 1 mm. long; filaments flattened near the base, shortly connate at base. Carpels of short-staminate flowers up to 3·5 mm. long. Follicle up to 2 × 1·2 cm.; pericarp greenish-yellow

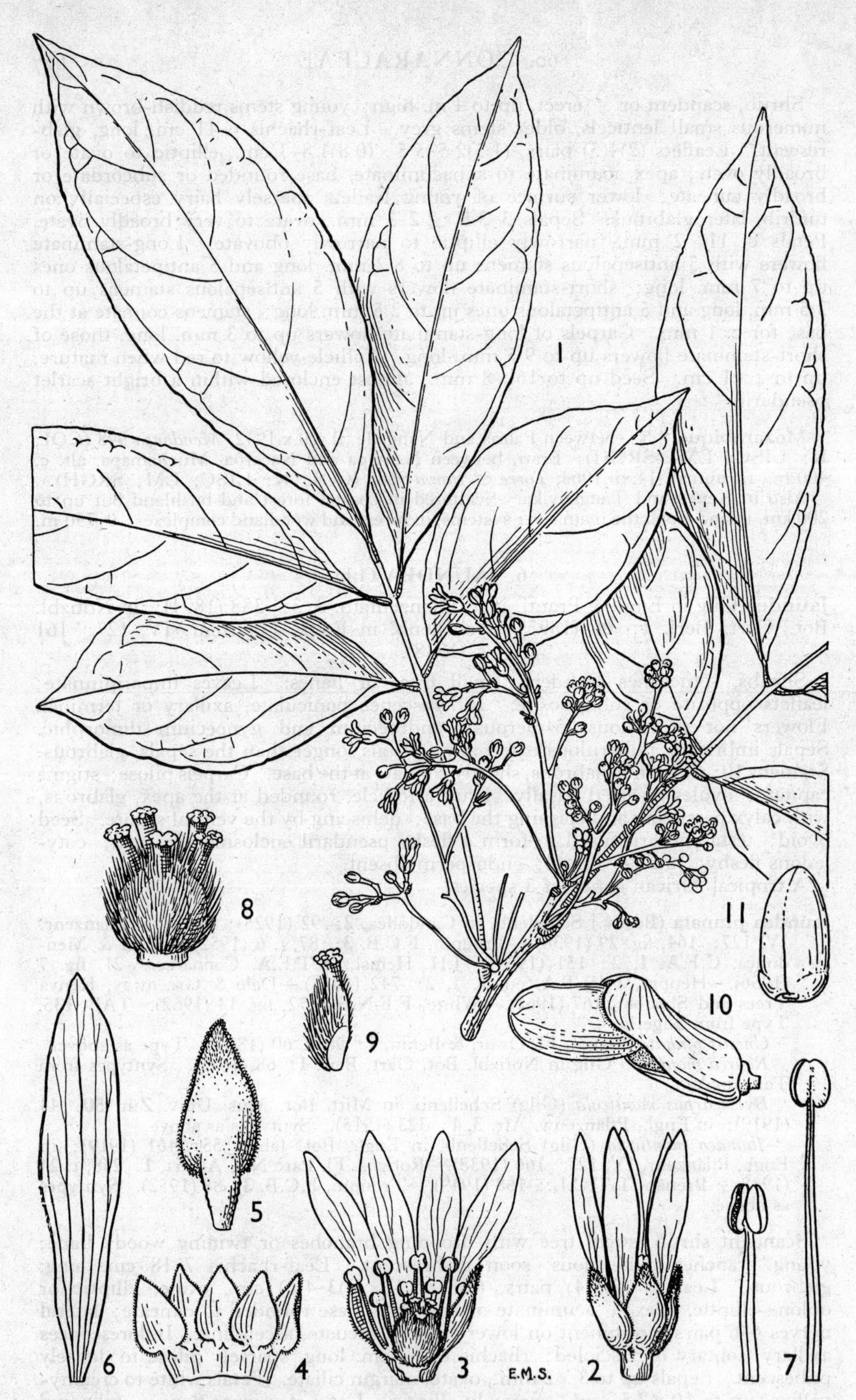

Tab. 135. JAUNDEA PINNATA. 1, flowering branch with leaves (×⅔); 2, flower (×4); 3, long-staminate flower with some sepals and petals removed (×4); 4, calyx opened out to show inner surface of sepals (×4); 5, sepal to show exterior surface (×8); 6, petal (×8); 7, stamens (×16); 8, gynoecium (×8); 9, single carpel (×8), 1–9 from *Wallace* 452; 10, mature fruit with extruding seed (×1); 11, seed (×1), 10–11 from *Drummond & Hemsley* 4375. From F.T.E.A.

or reddish (?). Seed up to 1·8 × 0·9 cm.; pseudaril orange, almost enclosing the seed.

N. Rhodesia. E: Nyika Plateau, near the Rest House, alt. 2150 m., fl. 15.xi.1958, *Robson & Fanshawe* 615A (K; LISC). **Nyasaland.** N: Misuku Hills, Walindi Forest, alt. 1650 m., galled carpels 12.i.1959, *Richards* 10629 (K)

From Guinée to Angola and eastwards to the Sudan, Kenya, Tanganyika and Nyasaland. In lowland and upland rain-forest as a liane, and in forest remnants as a shrub or small tree; up to 2200 m.

GENUS UNKNOWN

Scandent shrub. Leaf-rhachis up to 8 cm. long, glabrescent. Leaflets 1–2 pairs, 4–10 × 2–4·5 cm., ovate to elliptic, asymmetric, apex bluntly acuminate, base rounded to cuneate, glabrescent. Inflorescences axillary, fascicled; rhachis to 4 cm. long, brown hairy. Sepals 2·5–3·0 × 1·5–2·0 mm., ovate, apex acute and densely ciliate, otherwise sparsely ciliate. Petals cream, 7·0 × 1·8 mm., narrowly lanceolate, corrugate towards the acute apex. Pedicel c. 3 mm. long, articulated above $\frac{2}{3}$ of the length. Short-staminate flowers with 5 antisepalous stamens 1·6 mm. long and 5 antipetalous ones 1·1 mm. long; stamens free; filaments terete, glabrous. Disk of 10 small teeth alternating with the stamens. Carpels of short-staminate flowers 3·5 mm. long; ovaries densely hairy; styles glabrous; stigmata grooved, oblique. Long-staminate flowers unknown. Fruit unknown.

N. Rhodesia. N: Abercorn, Ndundu, 1740 m., fl. 27.v.1961, *Richards* 15140 (K).

In dense riverine forest, in damp loam soil.

The only specimen known is a very poor one and it cannot be assigned to any of the Connaraceae known to occur in our area.

ADDITIONS AND CORRECTIONS

VOLUME 1, PART 1

p. 116, line 10. For " *Barbosa* 1148 " read " *Mendonça* 1148 ".

p. 151, line 4. For " Stamens 6–9 " read " Stamens 3–9 ".

p. 163. Before "**9. COCCULUS** DC." insert:

4. **Tinospora fragosa** (Verdoorn) Verdoorn & Troupin, Acad. Sci. Outre-Mer, Cl. Sci. Nat. Méd. **13,** 2: 196 (1962). Type from the Transvaal.
Desmonema fragosum Verdoorn in Journ. S. Afr. Bot. **7**: 209 (1941). Type as above.

Climber with thick stems and with lateral branches much abbreviated and swollen. Leaves produced after the flowers. Stamens 6, connate at the base.

S. Rhodesia. E: Chipinga Distr., road to Sabi Valley Development Scheme, 26 km. S. of junction of Chipinga road, fl. 17.viii.1961, *Chase* 7525 (K; SRGH). S: Beitbridge, Chikwarawara, fr. xi. 1956, *Davies* 2186 (SRGH).
Also in the Transvaal.

p. 176, line 49. For " *Gomes e Sousa* 4329 " read " *Gomes e Sousa* 3429 ".

p. 179, line 22. For " *Torre* 181 " read " *Torre* 183 ".

p. 203, line 59. For " mature capsule 7–10 mm. long " read " mature capsule 7–10 cm. long ".

p. 205, line 26. Transfer " *Mendonça* 3187 " to var. *macrophylla* (above).

p. 207, line 28. For " *Sousa* 45 " read " *Sousa* 47 ".

p. 219, line 4. Transfer " *Barbosa* 2350 " to *Maerua cerasicarpa* where it is already cited on p. 218.

p. 234, line 5. Insert:

Mozambique. T: Maravia, fl. & fr. 25.ix.1942, *Mendonça* 401 (LISC).

p. 282. Before " 2. **Dovyalis lucida** Sim " insert:

1a. **Dovyalis longispina** (Harv.) Warb. in Engl. & Prantl, Nat. Pflanzenfam. **3,** 6a: 44 (1895) (" Doryalis ").—Paiva in Mem. Junta Invest. Ultram., Sér. 2, No. 28: 42 (1961). Type from S. Africa.
Aberia? longispina Harv. in Harv. & Sond., F.C. **2**: 585 (1862). Type as above.

Mozambique. LM: *Barbosa* 443 (COI; LM); *Mendonça* 3551 (LISC); *Mendonça* 4488 (LISC); *Pimenta* s.n. (LISC; LM).
For description and further details see Paiva (loc. cit.).

p. 307. Under " **Muraltia flanaganii** " add:

Mozambique. MS: Mavita, fl. & fr. 7.vii.1949, *Pedro & Pedrógão* 7333 (LMJ) (see Paiva in Mem. Junta Invest. Ultram., Sér. 2, No. 28: 60 (1961)).

p. 310, line 14. For " 31. *seminuda* " read " 31. *leptophylla* ".

p. 324. Under " **Polygala usafuensis** " add:

Mozambique. N: various specimens identified by Paiva (in Mem. Junta Invest. Ultram., Sér. 2, No. 28: 69 (1961)).

p. 326, line 21. Replace " **Polygala seminuda** Harv." by " 31. **Polygala leptophylla** Burch., Trav. Int. S. Afr. **1**: 400 (1822) ".

p. 327. **Polygala filicaulis** Baill. Paiva (in Mem. Junta Invest. Ultram., Sér. 2, No. 28: 70 (1961)) considers that the specimen cited (*Mendonça* 2064) belongs to *P. capillaris* E. Mey.

VOLUME 1, PART 2

p. 362. After line 40 add:

Corrigiola paniculata Peter in Fedde, Repert. Beih. **40,** 2: 31 (1932). Type from Tanganyika.
Corrigiola barotsensis Wild in Kirkia, **4**: 160 (1964). Type: N. Rhodesia, Barotseland, Machili, *Fanshawe* 6147 (K; SRGH, holotype).

An annual herb resembling the perennial *C. drymarioides* Bak. f. but with distinctly petiolate leaves and smaller flowers in very lax inflorescences.

N. Rhodesia. B: Machili, 17.ii.1961, *Fanshawe* 6275 (SRGH).
Also in Tanganyika.

p. 369. After " **Anacampseros rhodesica** " insert:

2. **Anacampseros subnuda** Poellnitz in Engl., Bot. Jahrb. **65**: 429 (1933).—Pole Evans, Fl. Pl. S. Afr. **15**: t. 576 (1935).
Syntypes from Cape Prov. and the Transvaal.

Leaves 1·5–2 × 0·5 cm., thick and fleshy, covered towards the apex with long greyish hair and minute tubercles terminated by a cluster of short hairs; stipules filamentous, exceeding the leaves. Inflorescence a 2–6-flowered racemoid cyme c. 8 cm. long.

S. Rhodesia. W: Matobo, Terrington Farm, fl. xi.1961, *Miller* 8041 (SRGH).
Also in the Transvaal, Cape Prov. and the Orange Free State.
An attractive pink-flowered species growing in the shallow soil over rocky outcrops

p. 382. Under **Hypericum conjungens** add:

Nyasaland N: Nyika Plateau, near Nganda road, fl. 8.vi.1960, *Chapman* 748 (BM; SRGH).

p. 432, line 28. Delete " *Barbosa in Mendonça* 1772 (LISC) ". This is a specimen of a cultivated species of *Gossypium* not *Gossypioides kirkii*.

p. 444, line 50. Transfer the MS reference (*Junod* 366) to SS.

p. 448, line 60. Transfer " *Drummond & Seagrief* 5215 (CAH; K; PRE; SRGH) " from *Hibiscus pusillus* to *Cienfugosia digitata*.

p. 453. After " **Hibiscus sabiensis** " insert:

26a. **Hibiscus coddii** Exell in Bol. Soc. Brot., Sér. 2, **33**: 177 (1959). Type from the Transvaal.
S. Rhodesia. S: Beitbridge Distr., Umzingwane, Fulton's Drift, fl. & fr. 26.iii.1961, *Wild* 5433 (BM; SRGH).
Also in the Transvaal.

p. 454. Footnote. Under " **Hibiscus meyeri** subsp. **transvaalensis** " add:

S. Rhodesia. S: Beitbridge Distr., Chiturapazi, fl. & fr. 22.ii.1961, *Wild* 5332 (BM; SRGH). Also in the Transvaal.

p. 457, line 41. For " *Mendonça* 165 " read " *Mendonça* 516 ".

p. 461, line 27. For " *Torre* 4098 " read " *Torre* 4089 ".

p. 462, line 45. For " bracts of the calyx " read " bracts of the epicalyx ".

p. 471, last line. Transfer *Swynnerton* 2054 from SS to MS.

p. 481, line 26. Transfer " *Mendonça* 3891 (LISC; SRGH) " to *Sida rhombifolia* (see Gonçalves in Mem. Junta Invest. Ultram., Sér. 2, **41**: 105 (1963)).

p. 489, line 21, *Robson* 308 (BM; K; LISC; SRGH) should be transferred to *Abutilon longicuspe* fide H. Wild.

p. 495, line 35. Transfer the T reference to MS.

p. 499, line 55. Transfer the LM reference to SS.

p. 523. After line 24 add:

7a. **Dombeya leachii** Wild in Kirkia, **4**: 160 (1964). Type: Mozambique, Niassa, Mt. Ribaué, *Leach & Schelpe* 11403 (K; SRGH, holotype).

Slender shrub 3 m. tall with pale-rose-coloured flowers and with chocolate-coloured young stems, stipules and bracteoles. Distinguished from *D. nyasica* by its broader glabrous stipules and persistent glabrous characteristically coloured bracteoles.

Mozambique. N: Mt. Ribaué, 19.vii.1962, *Leach & Schelpe* 11403 (K; SRGH).

Known only from the locality cited.

p. 531. Under " **Melhania griquensis** " add:

Mozambique. SS: Guijá, fl. & fr. vii.1915, *Gazaland Expedition* 337 (LMM) (see Gonçalves in Mem. Junta Invest. Ultram., Sér. 2, **41**: 136 (1963)).

p. 548. Under **Hermannia kirkii** add:

Nyasaland. S: Port Herald, S. of Lilange R., fl. & fr., 25.iii.1960, *Phipps* 2711 (BM; SRGH).

p. 562. Delete " 2. **?Pterygota alata** (Roxb.) R. Br.". The specimen cited (*Pedro & Pedrógão* 4611 (EA; LMJ)) is *Caloncoba welwitschii*.

VOLUME 2, PART 1

p. 5. Add to the " List of new names ": *Fagaropsis angolensis* var. *mollis* (Suesseng.) Mendonça, comb. nov. . . . *page* 191.

p. 14. line 13. For " Folhas imparipinadas " read " Folhas paripinadas ".

p. 92. **Hugonia grandiflora** N. Robson. Substitute *Barbosa & Lemos* in *Barbosa* 2247 (LISC) as the holotype.

p. 153, line 2. Delete " TAB. **22** fig. A ".

p. 170. Replace " **Impatiens brachycentra** Schulze & Launert " by:

Impatiens richardsiae Launert, nom. nov. Type: N. Rhodesia, Nyika Plateau, *Richards* 10423 (K, holotype).

Impatiens brachycentra Schulze & Launert in Bol. Soc. Brot., Sér. 2, **36**: 62 (1962) non Kar. & Kir. (1842). Type as above.

The blame for this mistake is entirely mine [E. L.].

p. 176, line 29. Paragraph starting " **S. Rhodesia.** C: Salisbury " refers to subsp. **cecilii** Group B and should be inserted after line 18.

p. 223. After line 56 add:

Also in Egypt, Israel, Arabia, Kenya, Sudan Republic, Tanganyika, the Congo and Angola. Dry woodland with bushes, open woodland or grassland; 0–1500 m.

p. 233. Under **Ochna mossambicensis** add:

S. Rhodesia. E: Chipinga Distr., Msilizwe R. bank, 750 m., fl. xi.1962, *Goldsmith* 216/62 (BM; SRGH).

p. 237, line 50. For " *Kirkman* " read " *Kirkham* ".

p. 286, line 24. For " *Cedrela ordorata* " read " *Cedrela odorata* ".

p. 292. Replace " 3. **Entandrophragma stolzii** " by:

3. **Entandrophragma excelsum** (Dawe & Sprague) Sprague in Kew Bull. **1910**: 180 (1910). Type from Uganda.

Pseudocedrela excelsa Dawe & Sprague in Journ. Linn. Soc., Bot. **37**: 511 (1906). Type as above.

Also in Uganda, the Congo and Tanganyika.

It now appears certain that *E. stolzii* is a synonym of the earlier *E. excelsum*.

p. 293, line 4. Read " **Lovoa** Harms in Engl. & Prantl, Nat. Pflanzenfam. **3**, 4: 307 (June, 1896)". A slightly earlier citation than the one originally given.

p. 297, line 18. For " 1936 " read " 1934 ".

p. 299, line 45. For " **7** " read " **8** " and make the same correction on p. 310, line 30, p. 312, line 27, p. 313, line 24, p. 316, line 29 and p. 318, line 36.

p. 300, line 33. Delete the *Bellingham* citation. Lukoma is in Tanganyika.

p. 308, line 62. For " 320 " read " 321 ".

p. 316, line 30. For " **1**: 487 (1926) " read " **2**: 487 (1932) ".

INDEX TO BOTANICAL NAMES